AF412844

**Current Topics in
Developmental Biology**

Volume 69

Current Topics in Developmental Biology

Volume 69

Edited by

Gerald P. Schatten
Director, PITTSBURGH DEVELOPMENTAL CENTER
Deputy Director, Magee-Women's Research Institute
Professor and Vice-Chair of Ob-Gyn-Reproductive
  Sci. & Cell Biol.-Physiology
University of Pittsburgh School of Medicine
Pittsburgh, Pennsylvania 15213

ELSEVIER
ACADEMIC
PRESS

AMSTERDAM • BOSTON • HEIDELBERG • LONDON
NEW YORK • OXFORD • PARIS • SAN DIEGO
SAN FRANCISCO • SINGAPORE • SYDNEY • TOKYO

Elsevier Academic Press
525 B Street, Suite 1900, San Diego, California 92101-4495, USA
84 Theobald's Road, London WC1X 8RR, UK

This book is printed on acid-free paper. 

For all information on all Elsevier Academic Press publications visit our Web site at www.books.elsevier.com

ISBN-13: 978-0-12-153169-0
ISBN-10: 0-12-153169-4

PRINTED IN THE UNITED STATES OF AMERICA
05 06 07 08 09 9 8 7 6 5 4 3 2 1

Contents

3

Glia–Neuron Interactions in Nervous System Function and Development

Shai Shaham

4

The Novel Roles of Glial Cells Revisited: The Contribution of Radial Glia and Astrocytes to Neurogenesis

Tetsuji Mori, Annalisa Buffo, and Magdalena Götz

5

Classical Embryological Studies and Modern Genetic Analysis of Midbrain and Cerebellum Development

Mark Zervas, Sandra Blaess, and Alexandra L. Joyner

6

Brain Development and Susceptibility to Damage; Ion Levels and Movements

Maria Erecinska, Shobha Cherian, and Ian A. Silver

7

Thinking about Visual Behavior; Learning about Photoreceptor Function

Kwang-Min Choe and Thomas R. Clandinin

8

Critical Period Mechanisms in Developing Visual Cortex

Takao K. Hensch

9

Brawn for Brains: The Role of MEF2 Proteins in the Developing Nervous System
Aryaman K. Shalizi and Azad Bonni

10

Mechanisms of Axon Guidance in the Developing Nervous System
Céline Plachez and Linda J. Richards

Contributors

Numbers in parentheses indicate the pages on which the authors' contributions begin.

Sandra Blaess (101), Howard Hughes Medical Institute, Developmental Genetics Program, Skirball Institute of Biomolecular Medicine, Department of Cell Biology, New York University School of Medicine, New York, New York 10016

Azad Bonni (239), Department of Pathology, Harvard Medical School, Boston, Massachusetts 02115

Annalisa Buffo (67), Institute for Stem Cell Research, GSF-National Research Center for Environment and Health, D-85764 Neuherberg, Munich, Germany

Shobha Cherian (139), Department of Neonatal Medicine, University Hospital of Wales, Cardiff CF14 4XW, United Kingdom

Kwang-Min Choe (187), Department of Neurobiology, Stanford University, Stanford, California 94305

Thomas R. Clandinin (187), Department of Neurobiology, Stanford University, Stanford, California 94305

Claude Desplan (1), Center for Developmental Genetics, Department of Biology, New York University, New York, New York 10003

Maria Erecinska (139), Department of Anatomy, School of Veterinary Science, Bristol BS2 8EJ, United Kingdom

Magdalena Götz (67), Institute for Stem Cell Research, GSF-National Research Center for Environment and Health, D-85764 Neuherberg, Munich, Germany

Jean M. Hébert (17), Departments of Neuroscience and Molecular Genetics, Albert Einstein College of Medicine, Bronx, New York 10461

Takao K. Hensch (215), Laboratory for Neuronal Circuit Development, RIKEN Brain Science Institute, Saitama 351-0198, Japan

Alexandra L. Joyner (101), Howard Hughes Medical Institute, Developmental Genetics Program, Skirball Institute of Biomolecular Medicine, Departments of Cell Biology and Physiology and Neuroscience, New York University School of Medicine, New York, New York 10016

Tamara Mikeladze-Dvali (1), Center for Developmental Genetics, Department of Biology, New York University, New York, New York 10003

Tetsuji Mori (67), Institute for Stem Cell Research, GSF-National Research Center for Environment and Health, D-85764 Neuherberg/Munich, Germany

Daniela Pistillo (1), Center for Developmental Genetics, Department of Biology, New York University, New York, New York 10003

Céline Plachez (267), Department of Anatomy and Neurobiology, The University of Maryland, School of Medicine, Baltimore, Maryland 21201

Linda J. Richards (267), University of Queensland, School of Biomedical Sciences and The Queensland Brain Institute, St. Lucia, Queensland 4072, Australia

Shai Shaham (39), Laboratory of Developmental Genetics, The Rockefeller University, New York, New York, 10021

Aryaman K. Shalizi (239), Biological and Biomedical Sciences Program, Harvard Medical School, Boston, Massachusetts 02115

Ian A. Silver (139), Department of Anatomy, School of Veterinary Science, Bristol BS2 8EJ, United Kingdom

Mark Zervas (101), Howard Hughes Medical Institute, Developmental Genetics Program, Skirball Institute of Biomolecular Medicine, Department of Cell Biology, New York University School of Medicine, New York, New York 10016

1

Flipping Coins in the Fly Retina

Tamara Mikeladze-Dvali, Claude Desplan, and Daniela Pistillo
Center for Developmental Genetics, Department of Biology
New York University, New York, New York 10003

Color vision in *Drosophila melanogaster* relies on the presence of two different subtypes of ommatidia: the "green" and "blue." These two classes are distributed randomly throughout the retina. The decision of a given ommatidium to take on the "green" or "blue" fate seems to be based on a stochastic mechanism. Here we compare the stochastic choice of photoreceptors in the fly retina with other known examples of random choices in both sensory and other systems. © 2005, Elsevier Inc.

I. Introduction

Development of a multicellular organism depends on the proper generation of different cell types. During their life span, cells constantly have to make decisions. These decisions affect cell survival, the commitment to a specific cell fate, and subsequent differentiation, and are made both non-cell autonomously and cell autonomously. In the first case, extrinsic factors, including instructive signals from other cells or tissues and environmental cues, determine cell fate. In the second case, cells make a decision independently of the environment. These intrinsic decisions can be lineage dependent, implying the retention of a molecular memory, or can rely on a stochastic event. In the latter case, the choice can occur between two or more states, and can be preferentially biased toward one of them.

In the case of the *Drosophila melanogaster* color vision system, each ommatidium has to make a stochastic, biased choice between the "blue"

Current Topics in Developmental Biology, Vol. 69   0070-2153/05 $35.00
    1   DOI: 10.1016/S0070-2153(05)69001-1

or "green" subtype. This choice does not affect the neighboring ommatidia, which each make their own intrinsic decision. Here we discuss this stochastic choice and compare it to examples of stochastic choice in other systems.

II. "Green" or "Blue": A Stochastic Choice in the Fly Retina

Drosophila acquires visual information through an array of about 800 ommatidia. Each ommatidium is a single eye unit that has eight photoreceptor cells, a lens, four lens-secreting cone cells, and eight other accessory cells. The eight photoreceptors (R1–8) have widely expanded membranes forming the rhabdomere that harbors the photosensitive G-protein-coupled, seven-transmembrane domain receptor rhodopsins (Rh). Six of the eight photoreceptors (R1 to R6) are involved in motion detection and image formation. The other two photoreceptors, R7 and R8, are involved in color vision and polarized light detection. R1–R6 are called "outer" photoreceptors due to their position within the ommatidium. Their rhabdomeres span the entire thickness of the retina and project their axons to the lamina part of the optic lobe. R1–R6 all express *rh1*, one of the five rhodopsins expressed in the fly eye (Fig. 1a and b) (Hardie, 1985; O'Tousa *et al.*, 1985; Zucker *et al.*, 1985). The morphology and the type of opsin expressed in R1–R6 is invariant in all 800 ommatidia.

R7 and R8 are located in the center of the ommatidium and are therefore called "inner" photoreceptors (Fig. 1a and b). The rhabdomeres of R7 and R8 are much shorter than those of R1–R6, with the photoreceptors projecting to a deeper part of the optic lobe, the medulla. The rhabdomeres of R7 and R8 are positioned on top of each other, R7 being more distal and R8 more proximal (Fig. 1a). An important property of the two inner photoreceptors is that they share a common optic path. When a light beam hits an ommatidium, it first passes through R7 and then R8. It is believed that the fly is able to distinguish colors by comparing the inputs of R7 and R8 coming from one ommatidium (Strausfeld, 1989). R7 and R8 each express only one of four color-sensitive opsins (*rh3, rh4, rh5, rh6*) in a highly regulated manner (Chou *et al.*, 1996; Franceschini *et al.*, 1981; Hardie, 1979, 1985; Papatsenko *et al.*, 1997).

At first glance, the fly retina appears to be a homogeneous structure. However, a close examination reveals that there are three different subtypes of ommatidia (Fig. 1c). The differences are due to rhodopsin expression in the inner photoreceptors (R7 and R8) and their physiological function. Two of the three subtypes, the "green" and the "blue," are involved in color vision, and their cell fate is chosen by a stochastic event (see below) (Franceschini *et al.*, 1981; Kirschfeld *et al.*, 1978). The third subtype, known as the dorsal rim area (DRA), contributes to the compass of the fly (Labhart

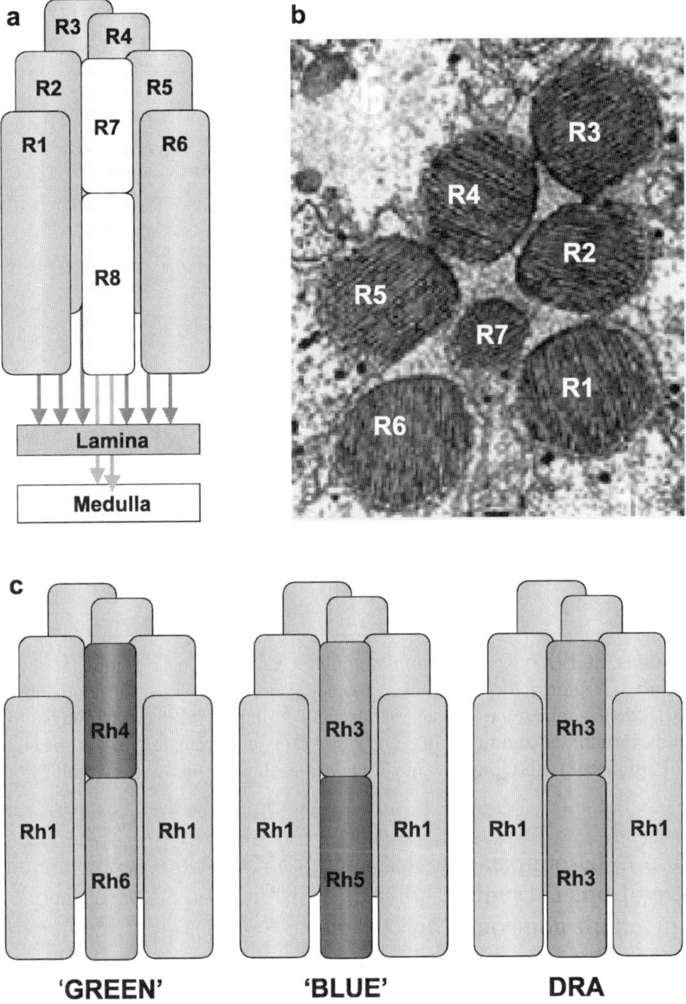

Figure 1 The three subtypes of ommatidia present in the fly retina. (a): schematic representation of the position, morphology and axonal projection of the outer photoreceptors (R1 to R6) and of the inner photoreceptors (R7 and R8). (b): electron micrograph of a cross-section through an ommatidium. (c): schematic representation of the three ommatidial subtypes present in the retina. In the 'green' subtype (left) R7 expresses UV-*rh4* and R8 Green-*rh6*; in the 'blue' subtype (center) R7 expresses UV-*rh3* and R8 Blue-*rh5*, in the Dorsal Rim Area ommatidia (DRA, right) both R7 and R8 express *rh3*. (See Color Insert.)

and Meyer, 1999). DRA ommatidia differentiation is defined by positional cues rather than by a stochastic event (Wernet *et al.*, 2003). The DRA ommatidia form one or two rows at the most dorsal part of the eye. Inner photoreceptors (R7 and R8) of all DRA ommatidia express *rh3* and have a distinct morphology, allowing them to detect the polarization vector of reflected sunlight.

The other two subclasses of inner photoreceptors are involved in color vision. We will refer to them as the blue and the green subtypes; the colors reflect the sensitivity of their respective R8 rhodopsins. Morphologically, the blue and the green subtypes are very similar, the main difference lying in the rhodopsin expression of the inner photoreceptors. In the blue subtype, R7 expresses the UV-sensitive rhodopsin *rh3* and R8 expresses the blue-sensitive *rh5*. In the green subtype, the R7 expresses the UV-sensitive *rh4* (which has a slight shift in the absorbance maximum from *rh3*) and R8 expresses the green-sensitive *rh6*. As in most other sensory systems, each photoreceptor expresses only a single rhodopsin. However, rhodopsin expression within the green and blue subtypes is highly stereotyped. The R7 and R8 rhodopsins are always coupled within one subtype, so that *rh3* is always associated with *rh5* in the blue subtype and *rh4* with *rh6* in the green subtype; however, for example, the combination of *rh4* and *rh5* is never observed in wild-type eyes (Chou *et al.*, 1996, 1999; Papatsenko *et al.*, 1997). Thus, the association of a given R8 rhodopsin with its R7 partner must have a physiological relevance for the fly color vision system. Interestingly, the two subtypes are not represented equally in the retina: 70% of the ommatidia are of the green subtype and 30% are of the blue.

Work over the past few years has elucidated a stepwise genetic model for photoreceptor terminal differentiation. In the first step, the transcription factor *spalt* induces inner photoreceptor (R7 and R8) fate. In the absence of *spalt,* photoreceptors develop into outer photoreceptors (R1–R6) (Mollereau *et al.*, 2001). Then, the transcription factor *prospero* defines the R7 fate by preventing R8 opsins from being expressed in R7 (Cook *et al.*, 2003). After these two steps of cell fate decisions, a photoreceptor knows that it is an inner photoreceptor and that it has become R7 or R8 (Fig. 2a). The ommatidium then has to make one final decision and commit either to the green or to the blue subtype.

Because the two inner photoreceptors of a given ommatidium share one optic path and have to express rhodopsins of the same subtype, the decision must affect both R7 and R8 and must be coordinated between them. Two models can be envisioned: in one, the decision can be made by both cells individually and then coordinated; in the other, the choice is made by one of the cells and is then imposed upon the other one.

The latter appears to be the case. In a *sevenless* mutant in which no R7 cell is present, all R8 cells express *rh6*. In the opposite situation, when R8 is genetically ablated and only R7 develops, both green and blue rhodopsins are expressed in R8. Based on these experiments, the following model was proposed: at the beginning, a stochastic choice between the green and

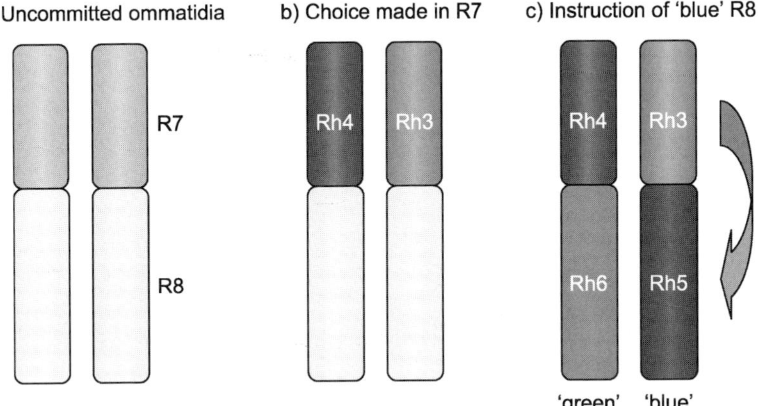

Figure 2 A two-step model for the cell fate decision 'green' *versus* 'blue' in inner photoreceptors. (a): in ommatidia that have not yet committed to the 'green' or 'blue' fate, R7 and R8 do not express any rhodopsin. (b): a stochastic event in R7 induces expression of either *rh4* or *rh3*. (c): an *rh3* expressing R7 cell sends a signal to the underlying R8 cell inducing *rh5* expression. In the absence of signal (*i.e.* when R7 expresses *rh4*), the R8 cell expresses *rh6*. (See Color Insert.)

blue fate is made by R7 (Chou *et al.*, 1996, 1999; Papatsenko *et al.*, 1997) (Fig. 2b). Once an R7 chooses the blue fate (30% of the cases), it sends an instructive signal to R8. Upon receiving the signal, R8 commits to the same blue fate and expresses *rh5*. In the absence of the R7 signal (i.e., when R7 expresses *rh4*), R8 becomes green (Fig. 2c). This mechanism ensures the correct coupling of rhodopsins between R7 and R8 and does not allow ambiguity. It should be stressed that the stochastic choice is made by each R7 independent of its neighbor, with a bias toward the green subtype, causing it to be chosen twice as frequently as the blue one.

III. Is the R7 Decision Purely Stochastic?

"Stochastic variation implies randomness as opposed to a fixed rule or relation" (Webster's Encyclopedic Unabridged Dictionary, 1989, pg. 1398). Is the choice really stochastic? So far, the molecular mechanism of the green/ blue choice in R7 has not been elucidated. The distribution of the blue and green ommatidia within the retina allows us to speculate about the nature of the event. The overall distribution of the two subtypes is homogenous over the retina and does not follow any obvious pattern or rule (Fig. 3). No mathematical model has been developed that would predict the fate of a green or blue ommatidium in a specific retinal position, and we can assume that there is no (or only minimal) positional information that influences the R7 decision.

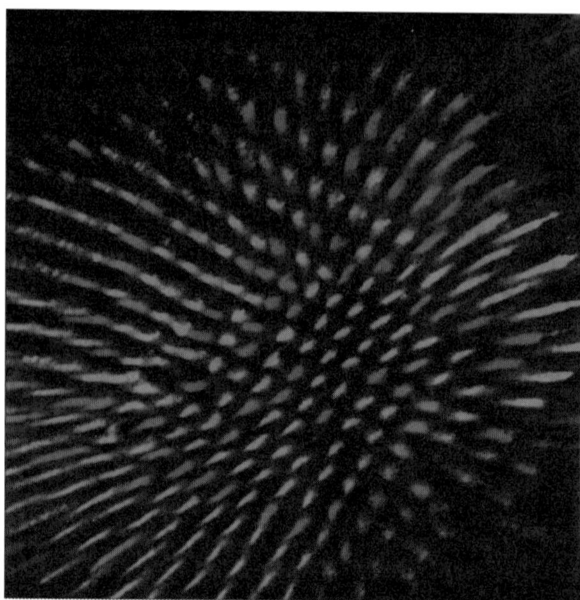

Figure 3 Stochastic distribution of the 'green' and 'blue' subtypes in the retina. Confocal image of a wild type whole mount retina stained with anti-Rh5 in blue ('blue' subtype) and anti-Rh6 in red ('green' subtype). No pattern or rule can be found in the distribution of the two subtypes. (See Color Insert.)

Stochastic choices occur in other circumstances. For instance, in the *Drosophila* nervous system, a single cell is selected randomly from an equivalence group to undergo a specific cell fate. This cell, in turn, prevents the neighboring cells from adopting the same fate through a mechanism known as lateral inhibition (reviewed in Simpson, 1997). In the fly retina, however, ommatidia of the same subtype can easily be found adjacent to each other, just as ommatidia of one subtype can be completely surrounded by ommatidia of the other. In other words, the fate chosen by a given ommatidium does not prevent adjacent ommatidia from making the same decision, and the ommatidium does not induce its neighbors to make the same choice. This indicates that a mechanism of cell selection followed by lateral inhibition can not apply to the R7 decision: the decision made by a given R7 appears intrinsic, and one can look at each ommatidium as an independent unit.

We assume that the green versus blue choice is based on a stochastic event in R7, with the green subtype accounting for 70% of the ommatidia and the blue for 30%. Therefore, the distribution of ommatidia is stochastic, but is biased toward the green outcome. It is important to stress here that despite the fact that the outcome is binary, the molecular mechanism underlying the choice need not be binary, as more complex scenarios could also lead to a two-state outcome. A hypothetical example is a stochastic expression of one out of ten transcriptional activators, seven of which would lead to the green fate and three to the blue one. In this hypothetical situation, the event leading to the choice is stochastic and unbiased (0.1 probability for each activator); however, the outcome of the choice is biased toward the green fate.

We can therefore argue that each ommatidium makes an independent decision to become green or blue. The choice seems to rely on a stochastic (random) event. The probability that a given ommatidium becomes green is 0.7 and blue is 0.3. There are other examples in biology where a stochastic choice is made, and knowledge about the underlying biological mechanisms in those examples is useful in helping us understand the development of the fly retina.

IV. How to Choose One out of Two: A Binary Choice in the Primate Retina

Trichromatic color vision is a recently evolved trait in mammals. In primates, red-green color vision has evolved in two different ways.

New World monkeys possess a single X chromosome-linked green-encoding opsin gene. Within these species, multiple alleles encode different spectral variations of the green opsin. Whereas males possess only one X chromosome and are dichromates, females with a heterozygous set of alleles become trichromates, as different cones express different alleles of the green opsin gene depending on which X chromosome is inactivated (Jacobs *et al.*, 1996; McMahon *et al.*, 2004; Smallwood *et al.*, 2003; Wang *et al.*, 1999).

A different mechanism has evolved in Old World monkeys, as well as in humans. In Old World primates, trichromacy relies on the acquisition of a red type (L) of cones in addition to the blue (S) and green (M) cones found in many diurnal mammals.[1] An unequal crossover of two X-linked polymorphic alleles resulted in a head-to-tail arrangement of an M (green) and L (red) pigment gene (Wang *et al.*, 1999). Having the M and L genes on one chromosome requires a mechanism to ensure the expression of one gene in each cone in addition to X-inactivation. The current model for the mutually exclusive expression of the M and L genes involves a shared upstream enhancer termed locus control region (LCR) that escaped duplication (Nathans *et al.*, 1989; Smallwood *et al.*, 2002; Wang *et al.*, 1999). The LCR regulates the expression of the tandem genes but is able to contact only one of the two promoters through a looping mechanism. The LCR can

---

[1] In the primate retina, each cone has to make two binary choices: first S versus M/L, then M versus L. The mechanism underlying the first choice is poorly understood and therefore is not discussed here (for further reading see Bumsted and Hendrickson, 1999).

a   How to choose one from many: the Locus Control Region

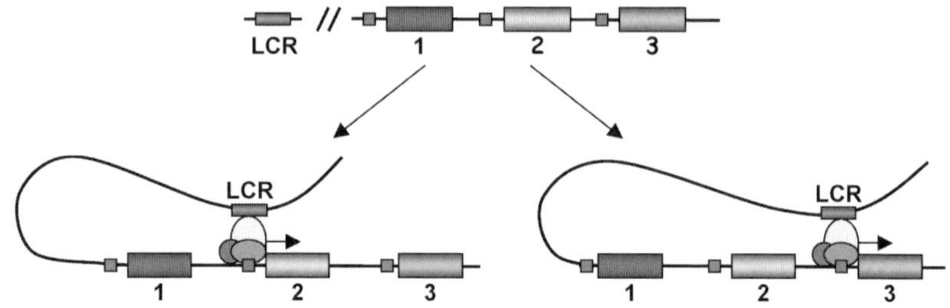

b   How to make many from one: DNA Recombination

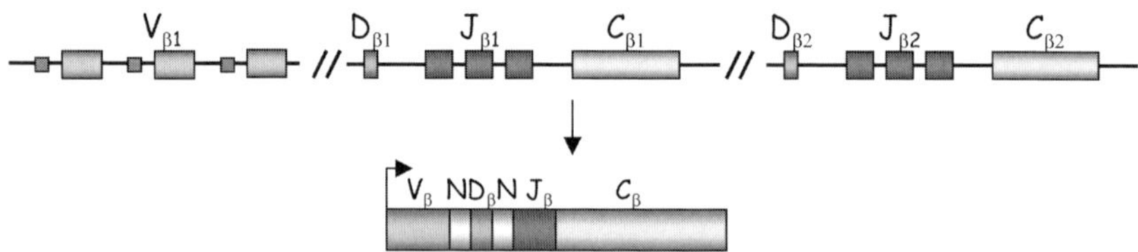

Figure 4 How to choose one gene from many and how to make many variants from one gene. (a): schematic drawing of a choice involving a Locus Control Region model. Though a looping mechanism, the LCR can contact only one promoter at a time, inducing the expression of only one gene in a cluster. (b): Schematic drawing of the recombination model in the immune system. The V (Variable), D (Diversity), J (Junction) and C (constant) segments that compose the β chain of the T cell antigen receptor are brought together by DNA rearrangements. Nucleotide addiction and deletion (N) in the joining region further increases variability. (See Color Insert.)

function as a stochastic selector for the expression of a single pigment gene from each X chromosome by contacting either the M or the L gene promoter. This allows males to be trichromatic, which is essential for fruit gatherers. Females' X-inactivation is also required so that only one gene is expressed per photoreceptor (Fig. 4a).

The ratio of L and M cones in the human retina is highly variable (Roorda and Williams, 1999). The percentage of L cones is most frequently 65–70% but can range from 50 to 92% (McMahon *et al.*, 2004). How is the different ratio of the two populations generated? McMahon *et al.* (2004) tested the hypothesis that the promoters of the M and L genes carry sequence differences that would allow differential binding to the LCR. Upon closer examination of the 236-bp-long L and M gene promoters from 73 humans, they concluded that sequence polymorphisms could not account for the variability of M and L gene expression. Another hypothesis is that preferential expression of the L gene simply relies on the proximity to the LCR. Although there is huge diversity in the number of L or M genes in the human X-chromosome locus, this is not sufficient to account for the large variability in the ratio (Smallwood *et al.*, 2002), so other factors outside the LCR–L/M region must also contribute to generate these fluctuations. Interestingly, the extreme differences in the red:green distribution do not affect the color discrimination ability of humans (McMahon *et al.*, 2004; Neitz *et al.*, 2002).

V. How to Choose One out of Many: Receptor Selection in the Olfactory System

Olfactory receptors (ORs) are, like the opsin proteins, seven transmembrane G-protein-coupled receptors (Buck and Axel, 1991). In flies, about 60 OR genes have been identified, while in vertebrates, the number of identified genes ranges from about 100 in fish to about 1000 in mice and humans. In both cases, the OR genes are distributed throughout the genome, although they are often organized into clusters (Clyne *et al.*, 1999; Gao and Chess, 1999; Glusman *et al.*, 2000; Rouquier *et al.*, 1998; Sullivan *et al.*, 1996).

It is thought that each olfactory neuron expresses only a single OR. Moreover, in mice, only one of the two alleles of each OR gene is expressed in each neuron, a phenomenon known as allelic exclusion, and the choice of which of the two alleles is expressed appears to be random (Chess *et al.*, 1994).

In rodents, the olfactory epithelium can be divided into four zones on the basis of the expression profile of the different ORs: each OR is expressed in only one zone. OR genes can be subdivided into approximately 100 subfamilies, with genes belonging to the same subfamily tending to be clustered together in the genome and expressed in the same zone (Ressler *et al.*, 1993). Within a zone, each gene is then expressed in a certain number of neurons in a stochastic way. When a given neuron expresses one OR gene, it excludes all others, including the other allele of the gene. Therefore, there must be two mechanisms of gene expression regulation: one that ensures that a given OR is expressed in the appropriate zone, and another that is responsible for the stochastic expression of ORs within a zone and for the exclusion of all others.

The presence of *cis*-regulatory elements able to drive expression of a reporter gene in a tissue-specific, zonal, and punctuate fashion similar to the expression pattern of an endogenous OR has been reported for several OR genes (Qasba and Reed, 1998; Serizawa *et al.*, 2003).

Several models that explain expression of a single OR per olfactory neuron have recently been reviewed by Serizawa *et al.* (2004).

The current model for OR selection also involves the presence of an LCR. A 2 kb sequence located 75 kb upstream of the *mOR28* gene (H region) is necessary to induce expression of *mOR28* and of other genes present in the same cluster. This region can activate a single OR gene in the cluster by making contacts with only one promoter at a time, probably through a looping mechanism similar to the one described for the primate retina

(Fig. 4a). The long distance between the LCR and the OR cluster is required to ensure a random selection of all the genes in the cluster instead of a bias toward the most proximal gene. Experimental reduction of the distance between the H region and the OR cluster leads to preferential activation of the most proximal gene in the cluster (Serizawa *et al.*, 2003).

In the mouse and human genomes, there are several OR pseudogenes, some of which can be transcribed. One could imagine that the promoter of a pseudogene instead of a functional OR might trap the LCR, suggesting that until the expression of a functional OR is achieved, the activation process remains active. Upon activation of a functional OR, the process halts. Moreover, expression of a functional receptor initiates a negative feedback loop generating a signal that inhibits expression of the second allele and of OR genes in other clusters (Serizawa *et al.*, 2003).

In other words, in the mouse olfactory epithelium, stochastic expression of an OR gene is achieved through a two-step mechanism: first, a *cis*-regulatory element, the LCR, makes contact and activates only one OR in a cluster; second, the presence of an OR protein somehow inhibits the expression of other OR genes. How this repression is achieved is still under investigation.

VI. How to Make Many from One: Recombination in the Immune System

Stochastic cell fate choices are not restricted to sensory systems. Another example of large receptor diversity is found in the vertebrate immune system. The mechanism underlying the choice of a single antigen receptor in the B and T lymphocytes is very different from OR selection. While in the olfactory system the choice of one receptor is at the level of gene selection, the diversity of antigen receptors in the immune system is generated by random DNA rearrangement of a single variable coding region.

The T-cell antigen receptor (TCR) rearrangement serves as a powerful illustration of the recombination phenomenon. The variety of the TCR heterodimers (composed of α- and β-chains) is assembled by somatic recombination from a pool of discontinuous variable (V), joining (J), and diversity (D) gene segments (Fig. 4b). The V_α-J_α and V_β-D_β-J_β rearrangement is based on a stochastic event. The V, D, and J segments are flanked by recombinatorial signal sequences (RSSs), which are recognized by the recombination activating proteins RAG-1 and RAG-2 (Oettinger *et al.*, 1990; Schatz *et al.*, 1989). Further variation is introduced by imprecision in the joining of the coding segments. This junctional diversity is due to nucleotide addition and deletion at the broken DNA ends during recombination. In the case of TCRβ, allelic exclusion ensures that only a single antigen receptor is expressed in a given cell (reviewed in Khor and Sleckman, 2002; Jung and Alt, 2004; Oettinger, 2004).

The theoretical value of combinatorial diversity is calculated to be 5.2×10^{13} possible $\alpha\beta$TCR variants in humans. Positive and negative intrathymic selection limit the enormous variability of the T-cells (Cohn, 2004; Nikolich-Zugich *et al.*, 2004).

To summarize, a stochastic somatic recombination mechanism in the immune system generates a vast diversity of proteins from one single coding region.

VII. How to Make Many from One: Alternative Splicing of *Dscam*

Another mechanism that produces a large population of different proteins from one coding region is found in the *Drosophila* Down syndrome cell adhesion molecule (DSCAM). Here the variety of proteins is generated from a single coding region by alternative splicing of the mRNA.

*Dscam*s are cell-surface proteins containing ten immunoglobulin domains and six fibronectin domains in the extracellular region (Schmucker *et al.*, 2000). They appear to be involved in axon guidance (Hummel *et al.*, 2003; Schmucker *et al.*, 2000; Wang *et al.*, 2002). Due to alternative splicing of various exons (e.g., exon 6 has 48 alternative variants; exon 9 has 33) *Dscam* is capable of generating 38,016 possible alternative splice forms, and this diversity is supposed to contribute to the specificity of neuronal connectivity (Neves *et al.*, 2004; Schmucker *et al.*, 2000). Neves *et al.* (2004) performed analysis of *Dscam* expression in single cells and homogenous cell populations using quantitative RT-PCR and oligonucleotide microarrays. They found that "a given cell type expresses a broad, yet distinctive, spectrum of splice variants." As an example, a certain photoreceptor cell may express 14–50 distinct mRNAs from a pool of thousands of exon variants characteristic for its cell type. Thus, the process involves stochastic generation of several splicing isoforms; however, it also implies a random expression of more than one alternative *Dscam* protein from a pool that is specific for the given cell type.

VIII. Conclusions

Based on the distribution of the green and blue ommatidia, we assume that the R7 choice is a stochastic event, but the exact molecular mechanism underlying the choice is poorly understood. Comparing different systems that base their intrinsic cell decisions on a stochastic event might help us to understand the processes in the fly retina. The examples listed above elucidate two qualitatively different mechanisms: in primate L/M gene selection and in OR gene selection, the choice is based on selection of one gene among two or among many, respectively. The second case involves the choice from multiple alternatives of a single gene. In the immune system the random selection employs somatic recombination of variable coding segments of a single gene, thus allowing one single choice that is irreversible. In *Dscam* the regulation is posttranscriptional, with alternative exons that are randomly spliced. Moreover, the expression of a subset of *Dscam* splicing variants per cell type, which might also change with developmental timing, adds another level of complexity to the system.

What can we learn from the mechanisms described above? As mentioned before, the blue versus green choice happens in the R7 cell of each ommatidium independently and is then imposed onto the R8, which expresses the matching rhodopsin. In other words, the outcome of the R7 choice is reflected in the expression of *rh3* or *rh4* and leads to the choice of other characters such as the generation of the instructive, blue-specific signal in the *rh3* expressing R7 and the synthesis of a filtering pigment in green R7. One could imagine that the stochastic event selects the blue or green fate at the level of the two rhodopsin genes, as seen in the M/L and OR gene selection. However, the molecular mechanism underlying the phenomenon is clearly distinct: the fly rhodopsin genes do not form clusters and are located on different arms of one chromosome, making the LCR model rather unlikely. On the other hand, no DNA rearrangement nor splicing isoforms have been found in the *rh3* and *rh4* genes. If a similar system were to be used in the fly retina, it would require the regulation of upstream genes rather than that of the rhodopsin genes themselves. In fact, we have recently obtained evidence that a regulator of rhodopsin genes is expressed stochastically in a subset of R7 and precludes the expression of *rh4* (Wernet and Desplan, in preparation). However, the exact biological mechanism for the choice of green versus blue in R7 still remains to be elucidated.

Acknowledgments

The authors thank Arzu Celik, Ben Collins, Esteban Mazzoni, and Satoko Yamaguchi for helpful discussion and comments to the manuscript. This work was supported by NIH grant ROI-EY13012 to C.D. D.P. was supported by a fellowship from EMBO.

References

Buck, L., and Axel, R. (1991). A novel multigene family may encode odorant receptors: A molecular basis for odor recognition. *Cell* **65**(1), 175–187.

Bumsted, K., and Hendrickson, A. (1999). Distribution and development of short-wavelength cones differ between Macaca monkey and human fovea. *J. Comp. Neurol.* **403**(4), 502–516.

Chess, A., Simon, I., Cedar, H., and Axel, R. (1994). Allelic inactivation regulates olfactory receptor gene expression. *Cell* **78**(5), 823–834.

Chou, W., Huber, A., Bentrop, J., Schultz, S., Chadwell, L. V., Paulsen, R., and Britt, S. (1999). Patterning of the R7 and R8 cells of Drosophila: Evidence for induced and default cell-fate specification. *Development* **126**, 606–616.

Chou, W. H., Hall, K. J., Wilson, D. B., Wideman, C. L., Townsons, S. M., and Britt, S. G. (1996). Identification of a novel Drosophila opsin reveals specific patterning of the R7 and R8 photoreceptor cells. *Neuron* **17**(6), 1101–1115.

Clyne, P. J., Warr, C. G., Freeman, M. R., Lessing, D., Kim, J., and Carlson, J. R. (1999). A novel family of divergent seven-transmembrane proteins: Candidate odorant receptors in Drosophila. *Neuron* **22**(2), 327–338.

Cohn, M. (2004). An alternative to current thinking about positive selection, negative selection and activation of T cells. *Immunology* **111**(4), 375–380.

Cook, T., Pichaud, F., Sonneville, R., Papatsenko, D., and Desplan, C. (2003). Distinction between color photoreceptor cell fates is controlled by Prospero in Drosophila. *Dev. Cell* **4**(6), 853–864.

Franceschini, N., Kirschfeld, K., and Minke, B. (1981). Fluorescence of photoreceptor cells observed *in vivo*. *Science* **213**(11), 1264–1267.

Gao, Q., and Chess, A. (1999). Identification of candidate Drosophila olfactory receptors from genomic DNA sequence. *Genomics* **60**(1), 31–39.

Glusman, G., Bahar, A., Sharon, D., Pilpel, Y., White, J., and Lancet, D. (2000). The olfactory receptor gene superfamily: Data mining, classification, and nomenclature. *Mamm. Genome* **11**(11), 1016–1023.

Hardie, R. (1979). Electrophysiological analysis of fly retina. I. Comparative properties of R1–R6 and R7–R8. *J. Comp. Physiol.* **129**, 19–33.

Hardie, R. (1985). Functional organization of the fly retina. *In* "Progress in Sensory Physiology" (D. Ottoson, Ed.), Vol. 5, pp. 1–79. Springer, New York.

Hummel, T., Vasconcelos, M. L., Clemens, J. C., Fishilevich, Y., Vosshall, L. B., and Zipursky, S. L. (2003). Axonal targeting of olfactory receptor neurons in Drosophila is controlled by Dscam. *Neuron* **37**(2), 221–231.

Jacobs, G. H., Neitz, M., Deegan, J. F., and Neitz, J. (1996). Trichromatic colour vision in New World monkeys. *Nature* **382**(6587), 156–158.

Jung, D., and Alt, F. W. (2004). Unraveling V(D)J recombination; insights into gene regulation. *Cell* **116**(2), 299–311.

Khor, B., and Sleckman, B. P. (2002). Allelic exclusion at the TCRbeta locus. *Curr. Opin. Immunol.* **14**(2), 230–234.

Kirschfeld, K., Feiler, R., and Franceschini, N. (1978). A photostable pigment within the rhabdomere of fly photoreceptor NO R7. *J. Comp. Physiol.* **125**, 275–284.

Labhart, T., and Meyer, E. P. (1999). Detectors for polarized skylight in insects: A survey of ommatidial specializations in the dorsal rim area of the compound eye. *Microsc. Res. Tech.* **47**(6), 368–379.

McMahon, C., Neitz, J., and Neitz, M. (2004). Evaluating the human X-chromosome pigment gene promoter sequences as predictors of L:M cone ratio variation. *J. Vis.* **4**(3), 203–208.

Mollereau, B., Dominguez, M., Webel, R., Colley, N. J., Keung, B., de Celis, J. F., and Desplan, C. (2001). Two-step process for photoreceptor formation in Drosophila. *Nature* **412**(6850), 911–913.

Nathans, J., Davenport, C. M., Maumenee, I. H., Lewis, R. A., Hejtmancik, J. F., Litt, M., Lovrien, E., Weleber, R., Bachynscki, B., Zwas, F., Klingaman, R., and Fishman, G. (1989). Molecular genetics of human blue cone monochromacy. *Science* **245**(4920), 831–838.

Neitz, J., Carroll, J., Yamauchi, Y., Neitz, M., and Williams, D. R. (2002). Color perception is mediated by a plastic neural mechanism that is adjustable in adults. *Neuron* **35**(4), 783–792.

Neves, G., Zucker, J., Daly, M., and Chess, A. (2004). Stochastic yet biased expression of multiple Dscam splice variants by individual cells. *Nat. Genet.* **36**(3), 240–246.

Nikolich-Zugich, J., Slifka, M. K., and Messaoudi, I. (2004). The many important facets of T-cell repertoire diversity. *Nat. Rev. Immunol.* **4**(2), 123–132.

Oettinger, M. A. (2004). How to keep V(D)J recombination under control. *Immunol. Rev.* **200**, 165–181.

Oettinger, M. A., Schatz, D. G., Gorka, C., and Baltimore, D. (1990). RAG-1 and RAG-2, adjacent genes that synergistically activate V(D)J recombination. *Science* **248**(4962), 1517–1523.

O'Tousa, J. E., Baehr, W., Martin, R. L., Hrish, J., Pak, W. L., and Applebury, M. L. (1985). The Drosophila ninaE gene encodes an opsin. *Cell* **40**(4), 839–850.

Papatsenko, D., Sheng, G., and Desplan, C. (1997). A new rhodopsin in R8 photoreceptors of Drosophila: Evidence for coordinate expression with Rh3 in R7 cells. *Development* **124**(9), 1665–1673.

Qasba, P., and Reed, R. R. (1998). Tissue and zonal-specific expression of an olfactory receptor transgene. *J. Neurosci.* **18**(1), 227–236.

Ressler, K. J., Sullivan, S. L., and Buck, L. B. (1993). A zonal organization of odorant receptor gene expression in the olfactory epithelium. *Cell* **73**(3), 597–609.

Roorda, A., and Williams, D. (1999). The arrangement of the three cone classes in the living human eyes. *Nature* **397**(6719), 520–522.

Rouquier, S., Taviaux, S., Trask, B. J., Brand-Arpon, V., van den Engh, G., Demaille, J., and Giorgi, D. (1998). Distribution of olfactory receptor genes in the human genome. *Nat. Genet.* **18**(3), 243–250.

Schatz, D. G., Oettinger, M. A., and Baltimore, D. (1989). The V(D)J recombination activating gene, RAG-1. *Cell* **59**(6), 1035–1048.

Schmucker, D., Clemens, J. C., Shu, H., Worby, C. A., Xiao, J., Muda, M., Dixon, J. E., and Zipursky, S. L. (2000). Drosophila Dscam is an axon guidance receptor exhibiting extraordinary molecular diversity. *Cell* **101**(6), 671–684.

Serizawa, S., Miyamichi, K., Nakatamy, H., Suzuki, M., Saito, M., Yoshihara, Y., and Sakano, H. (2003). Negative feedback regulation ensures the one receptor-one olfactory neuron rule in mouse. *Science* **302**(5653), 2088–2094.

Serizawa, S., Miyamichi, K., and Sakano, H. (2004). One neuron-one receptor rule in the mouse olfactory system. *Trends Genet.* **20**(12), 648–653.

Simpson, P. (1997). Notch signaling in development. *Perspect. Dev. Neurobiol.* **4**(4), 297–304.

Smallwood, P. M., Wang, Y., and Nathans, J. (2002). Role of a locus control region in the mutually exclusive expression of human red and green cone pigment genes. *Proc. Natl. Acad. Sci. USA* **99**(2), 1008–1011.

Smallwood, P. M., Olveczky, B. P., Williams, G. L., Jacobs, G. H., Reese, B. E., Meinster, M., and Nathans, J. (2003). Genetically engineered mice with an additional class of cone photoreceptors: Implications for the evolution of color vision. *Proc. Natl. Acad. Sci. USA* **100**(20), 11706–11711.

Strausfeld, N. (1989). Beneath the compound eye: Neuroanatomical analysis and physiological correlates in the study of insect vision. *In* "Facets of Vision" (D. Stavenga and R. Hardie, Eds.), pp. 317–359. Springer, New York.

Sullivan, S. L., Adamson, M. C., Ressler, K. J., Kozak, C. A., and Buck, L. B. (1996). The chromosomal distribution of mouse odorant receptor genes. *Proc. Natl. Acad. Sci. USA* **93**(2), 884–888.

Wang, Y., Smallwood, P., Cowan, M., Blesh, D., Lawler, A., and Nathans, J. (1999). Mutually exclusive expression of human red and green visual pigment-reporter transgenes occurs at high frequency in murine cone photoreceptors. *Proc. Natl. Acad. Sci. USA* **96**(9), 5251–5256.

Wang, J., Zugates, C. T., Liang, I. H., Lee, C. H., and Lee, T. (2002). Drosophila Dscam is required for divergent segregation of sister branches and suppresses ectopic bifurcation of axons. *Neuron* **33**(4), 559–571.

Wernet, M., Labhart, T., Baumann, F., Mazzoni, E. O., Pichaud, F., and Desplan, C. (2003). Homothorax switches function of Drosophila photoreceptors from color to polarized light sensors. *Cell* **115**(3), 267–279.

Zucker, C., Cowman, A., and Rubin, G. M. (1985). Isolation and structure of a rhodopsin gene from D. melanogaster. *Cell* **40**, 851–858.

"Webster's Encyclopedic Unabridged, Dictionary of the English, Language." Gramercy Books, New York.

2

Unraveling the Molecular Pathways That Regulate Early Telencephalon Development

Jean M. Hébert
Departments of Neuroscience and Molecular Genetics
Albert Einstein College of Medicine, Bronx, New York 10461

The telencephalon, at the rostral end of the developing central nervous system, starts off as a sheet of neuroepithelial cells. During development, this sheet of cells becomes patterned and morphologically partitioned into areas that give rise to the adult cerebral hemispheres. How does this happen? How are telencephalic precursor cells instructed to generate myriad neural cell types in different areas and at different times as well as to change their rates of cell proliferation, differentiation, and death? The molecular pathways required for patterning the telencephalic neuroepithelium and forming the cerebral hemispheres are beginning to be unraveled. © 2005, Elsevier Inc.

I. Introduction

It is important to understand how a simple sheet of neuroepithelial precursor cells gives rise to our complex adult cerebral hemispheres, the seat of our highest intellectual functions. The telencephalon, at the rostral end of the developing neural tube, is the embryonic precursor to the cerebral hemispheres. Recent studies are beginning to shed light on several questions regarding the earliest steps of telencephalon development. Early telencephalon development can be covered by three broad questions. First, how is the telencephalon induced? Second, how do telencephalic precursor cells acquire positional identities that define different areas? And third, within each of these areas, how do precursor cells progressively generate specific neural subtypes? As development proceeds, these more differentiated telencephalic areas provide the cues necessary for neurons to generate their complex networks of connections that ultimately form the mature cerebral hemispheres and underlie the vast range of human behaviors. The adult cerebral hemispheres are composed largely of the neocortex, which is the center of higher cognitive and perceptual functions, and the basal ganglia, which have multiple functions such as coordinating motor and emotional outputs. The hemispheres also include other important adult structures, such as the hippocampus, which is essential in memory acquisition. Elucidating the mechanisms that underlie how neural precursor cells generate the mature cerebral hemispheres not only is a challenging intellectual endeavor, but also will be key in designing effective regenerative therapies for a range of forebrain disorders, from developmental to degenerative ones.

The telencephalon first becomes morphologically apparent at the anterior end of the neural tube as an inflated sheet of neuroepithelial cells surrounding bilateral ventricles. Even at this earliest stage of its development, the telencephalon shows evidence of patterning, such as the restricted expression domains of certain genes (Monuki and Walsh, 2000; Ragsdale and Grove, 2001; Rubenstein *et al.*, 1998). For example, the homeobox genes *Gsh2*, *Pax6*, and *Emx2* are each expressed in specific regions of the telencephalon and are essential for its normal patterning (Bishop *et al.*, 2000; Gulisano *et al.*, 1996; Mallamaci *et al.*, 2000; Muzio *et al.*, 2002a,b; Szucsick *et al.*, 1997; Toresson *et al.*, 2000; Walther and Gruss, 1991; Yun *et al.*, 2001). In addition to being defined by specific patterns of gene expression, the various telencephalic regions also exhibit different rates of cell proliferation, differentiation, and programmed death, leading to distinct morphologies. Despite recent advances, the precise mechanisms by which telencephalic areas are specified and patterned remain only superficially understood.

It has been proposed that signaling centers in the midline play an important role in shaping and patterning the telencephalon (e.g., Monuki and

Walsh, 2000; Ohkubo *et al.*, 2002; Ragsdale and Grove, 2001; Storm *et al.*, 2003; Wilson and Rubenstein, 2000). The molecular pathways that form these signaling centers and pattern the telencephalon appear to be largely conserved among vertebrates. The developmental steps from gastrulation leading up to the appearance of the telencephalon have recently been insightfully reviewed (Wilson and Houart, 2004). This chapter focuses on some of the better-understood aspects of telencephalic induction and patterning.

II. Induction of the Telencephalon

The anterior edge of the developing neural plate, where ectoderm abuts neurectoderm, is composed of a group of specialized cells that form the anterior neural ridge (ANR) in mouse or the anterior neural border (ANB) in zebrafish. These cells are necessary to induce telencephalic character in anterior neural tissue. In mice, removal of the ANR results in loss of expression of *Foxg1* (Shimamura and Rubenstein, 1997), a transcription factor gene of the forkhead family whose expression delineates most of the embryonic telencephalon (Tao and Lai, 1992). Likewise, ablation of the ANB in zebrafish leads to a loss of *emx1* expression, which marks the dorsal telencephalon, greatly reduced levels of *dlx2*, which marks the ventral forebrain, and an increase in cell death in anterior neural tissue (Houart *et al.*, 1998). In addition, transplanting the ANB to more posterior neural tissue results in ectopic expression of *emx1*, *dlx2*, and *foxg1*, suggesting that the ANB is sufficient to induce telencephalic character (Houart *et al.*, 1998, 2002).

What are the factors responsible for the inductive properties of the ANR and ANB? In zebrafish, compelling evidence suggests that the Wingless/Int (Wnt) antagonist Tlc, a secreted Frizzled-related protein, is the responsible factor (Houart *et al.*, 2002). First, *tlc* is expressed in the ANB; second, antisense morpholinos against *tlc* lead to a loss of the telencephalon; third, *tlc*-expressing cells can rescue the loss of telencephalon and cell death in ANB ablated embryos; and finally, *tlc*-expressing cells, such as the ANB, can induce ectopic expression of *emx1* and *foxg1* in more posterior neural tissue. Tlc is likely to be acting as a true Wnt antagonist since transplanting cells expressing Wnts into the ANB can inhibit expression of telencephalic genes, including *fgf8* (Houart *et al.*, 2002). Other evidence also supports the notion that inhibiting Wnt activity is necessary to induce the telencephalon. Zebrafish embryos mutant for the *masterblind* gene, which encodes Axin, a negative regulator of Wnt signaling, and mouse embryos mutant for *Six3*, a direct repressor of Wnt gene expression, both lack a telencephalon (Lagutin *et al.*, 2003; Masai *et al.*, 1997). It remains to be determined, however, if Wnt antagonism via secreted Frizzled-related proteins is also required to specify the telencephalon in species other than zebrafish.

Another gene expressed in the ANR and ANB is *Fgf8*. In mice, FGF8-soaked beads placed on the anterior neural plate in cultured explants can induce *Foxg1* expression (Shimamura and Rubenstein, 1997). Hence it is possible that fibroblast growth factor (FGF) signaling, like Wnt antagonism, induces the telencephalon. To date, however, no FGF-signaling mutant in mice or zebrafish lacks a telencephalon, casting doubt on this possibility. On the other hand, in each of the mutants generated to date, telencephalic tissue may still be induced due to functional compensation by related genes. Consistent with this possibility, three FGF receptor genes are expressed in neuroepithelial precursor cells, and at least five FGF ligand genes are expressed at the anterior end of the developing neural tube (Hébert *et al.*, 2003; Maruoka *et al.*, 1998; McWhirter *et al.*, 1997; Orr-Urtreger *et al.*, 1991; Peters *et al.*, 1992; Shinya *et al.*, 2001). Thus, whether FGF signaling is required to induce the telencephalon remains an open question. Nevertheless, it is likely that FGF signaling acts downstream of Wnt antagonism, at least in zebrafish, to promote telencephalon development, since *tlc* is both necessary and sufficient to induce *fgf8* expression in anterior neural tissue (Houart *et al.*, 2002). Interestingly, like *tlc* in zebrafish, *Fgf8* expression in the mouse anterior neural plate has been shown to regulate cell survival, presumably by regulating *Foxg1* expression (Storm *et al.*, 2003).

Little is known about the mechanisms that regulate formation of the ANB or ANR itself. In zebrafish, *tlc* expression is likely to be induced by a threshold level of bone morphogenetic proteins (BMPs) emanating from the lateral ectoderm flanking the neural plate. In *bmp2b* mutant embryos, although the neural plate is expanded at the expense of ectoderm, a

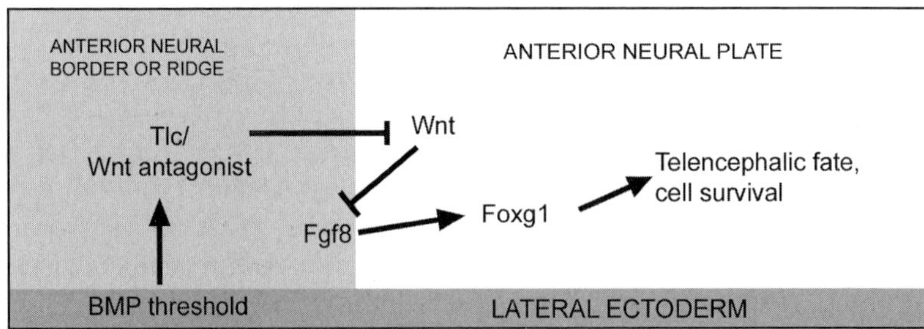

Figure 1 Model of telencephalon induction. Expression of the Wnt antagonist, Tlc, is induced in the anterior neural border by a threshold level of BMP signaling emanating from the lateral ectoderm. Tlc, as well as other forms of Wnt antagonism (see text), inhibits Wnt function and promotes *Fgf8* expression in the anterior neural border and *Foxg1* expression in cells of the anterior neural plate that adopt a telencephalic fate.

telencephalon does not form (Barth *et al.*, 1999). In addition, *tlc* expression in the ANB is lost in the presence of ectopic Noggin activity (Houart *et al.*, 2002). Together, these studies point to a model of genetic interactions leading to telencephalic induction (Fig. 1).

III. Formation of Midline Signaling Centers and Their Interactions

A distinguishing feature of the telencephalon when compared to the rest of the developing brain is its bilateral ventricles. Understanding how the telencephalon becomes split into two hemispheres and how the midline structure between them forms is of particular interest because midline cells are believed to secrete factors that pattern the early telencephalon. The telencephalic midline is first identified as the seam at the rostral end of the closing neural tube. The ventral portion of the midline may also be derived in part from the medial-most portion of the neural plate. Early in the development of the midline, cells undergo increased cell death and reduced proliferation compared to their lateral neighbors, causing the midline to remain thin and constricted while the surrounding neuroepithelium expands to form the bilateral ventricles. The midline also generates unique cell types. Dorsally, the most medial precursor cells form the choroid plexus, which secretes the cerebrospinal fluid into the ventricles, and precursors immediately lateral to these form the cortical hem, whose progeny are likely to form the fimbria. Ventrally, medial precursor cells generate part of the septum and medial ganglionic eminence.

Within the midline reside three signaling centers suspected of shaping and patterning the early telencephalon. The dorsal midline expresses intercellular signaling molecules of the BMP and Wnt families, the rostral midline expresses FGFs, and the prechordal plate ventral to the telencephalon, as well as the ventral telencephalon itself express Sonic Hedgehog (SHH). Are these signaling molecules required to generate and/or maintain the midline with its unique properties?

A. The Dorsal Midline

The dorsal midline expresses several Bmp genes: *Bmp2, Bmp4, Bmp5, Bmp6*, and *Bmp7* (Furuta *et al.*, 1997). Furthermore, BMP4-soaked beads placed on cultured explants of lateral telencephalon can induce dorsal midline features, such as increased cell death, reduced proliferation, expression of the midline marker *Msx1*, and repression of the non-midline marker *Foxg1* (Furuta *et al.*, 1997). However, whether BMPs, which are also expressed in the lateral ectoderm prior to neural tube closure, are required to form or maintain the dorsal midline *in vivo*, remains to be demonstrated. A telencephalon specific knockout of the *Bmpr1a* gene has demonstrated that BMP signaling is required for formation of at least the most medial structure of the dorsal midline, the choroid plexus (Hébert *et al.*, 2002). However, in this mutant, the cortical hem appears normal and hemisphere separation still occurs, perhaps due to functional compensation by *Bmpr1b*, which is also expressed in telencephalic precursor cells.

In addition to Bmp genes, several Wnt genes, including *Wnt2b*, *Wnt3a*, *Wnt5a*, and *Wnt8b*, are also expressed in the lateral anterior neural plate prior to neural tube closure and in the dorsal midline once the anterior neural tube is closed (Grove *et al.*, 1998; Lee *et al.*, 2000). Although their full role in generating or maintaining the early midline remains obscure, Wnt signaling is clearly required for forming at least one dorso-medial structure; mice mutant for components of the Wnt signaling pathway, including *Wnt3a*, lack all or part of the hippocampus (Galceran *et al.*, 2000; Lee *et al.*, 2000). Interestingly, loss of a splice variant of *Rfx4*, which encodes a transcription factor expressed at high levels in the early dorsal midline, leads to loss of both morphological and molecular features of the midline, including loss of *Wnt3a* expression (Blackshear *et al.*, 2003). This implicates *Rfx4* as an important regulator of dorsal midline formation that is likely to act upstream of Wnt signaling (Blackshear *et al.*, 2003).

B. The Rostral Midline

As the neural tube closes, the ANR becomes the rostral midline. Both the ANR and the rostral midline express several Fgf genes: *Fgf3*, *Fgf8*, *Fgf15*, *Fgf17*, and *Fgf18* (Crossley *et al.*, 2001; Maruoka *et al.*, 1998; McWhirter *et al.*, 1997; Shinya *et al.*, 2001). Only *fgf3* in zebrafish and *Fgf8* in both zebrafish and mice have been shown to be required for normal telencephalon development (see Section V), but their role, if any, in specifically generating or maintaining the midline is unclear. FGF8-soaked beads placed in the lateral dorsal prosencephalon of chick embryos have the potential to induce a sulcus with features that resemble a rostral midline (Crossley *et al.*, 2001). In addition, the expression levels of particular *Fgf8* alleles were found to be critical for regulating cell death and midline morphology, suggesting that FGF signaling may be important in shaping the rostro-dorsal midline (Storm *et al.*, 2003). Recent evidence also suggests that FGF8 plays a role in specifying ventro-medial cell types (see Section IV).

C. The Ventral Midline

SHH signaling is required for the formation of the ventral telencephalic midline (Chiang *et al.*, 1996; Ericson *et al.*, 1995). Surprisingly, the disruption of SHH signaling also leads to a loss of the dorsal midline (reviewed in Hayhurst and McConnell, 2003). How *Shh*, which is only expressed ventrally, is required for the dorsal midline to form remains an intriguing question. Nevertheless, this finding suggests that the signaling centers located along the ventral-to-dorsal midline might interact.

D. Interactions between Midline Signaling Centers

Supporting the premise that midline signaling centers interact, mice mutant for *Shh* lose *Fgf8* expression in the rostral midline (Aoto *et al.*, 2002; Ohkubo *et al.*, 2002). In addition, it appears that both these genes act, at least in part, antagonistically with Bmp and Wnt genes expressed in the dorsal midline. Bmp4-soaked beads placed in the embryonic chicken forebrain repress expression of *Fgf8* and *Shh* (Ohkubo *et al.*, 2002), and increased *Bmp4* expression due to loss of megalin results in loss of *Shh* expression (Spoelgen *et al.*, 2005). Conversely, ectopic expression of Noggin, a protein that directly binds and inhibits BMPs, leads to an expansion of the *Fgf8* domain in the anterior forebrain of mouse and chicken embryos (Ohkubo *et al.*, 2002; Shimogori *et al.*, 2004). Consistent with this antagonism, an inverse correlation between *Fgf8* and Bmp/Wnt gene expression is found in mice mutant for *Gli3*. In the *Gli3* mutant, the *Fgf8* expression domain is expanded, whereas Bmp and Wnt gene expression are reduced or lost (Aoto *et al.*, 2002; Grove *et al.*, 1998; Kuschel *et al.*, 2003; Theil *et al.*, 1999).

It should be noted, however, that the antagonism between the expression of *Fgf8* and at least one Bmp gene, *Bmp4*, appears dose dependent. A loss of *Fgf8* expression in mouse telencephalic tissue due to a knockout of this gene leads to an increase of *Bmp4* expression in the domain compared to wild-type controls, whereas a greatly reduced level of *Fgf8* expression due to a hypomorphic allele results in the opposite phenotype, a loss of *Bmp4* expression (Storm *et al.*, 2003). Equally unexpected, overexpression of *Fgf8* had an effect similar to that of complete loss of *Fgf8*. This study underscores the value of using several alleles in genetically assessing the function of a gene. The results obtained with the different *Fgf8* alleles, although at first glance contradictory, can be reconciled by postulating that FGFs act through two intercellular pathways, the first of which is activated by low, but not null, levels of FGF8 and results in repression of *Bmp4*, and the second of which is activated by higher levels of FGF8 and antagonizes the first pathway, leading to derepression of *Bmp4* (Storm *et al.*, 2003). The importance of

                                                                Jean M. Hébert

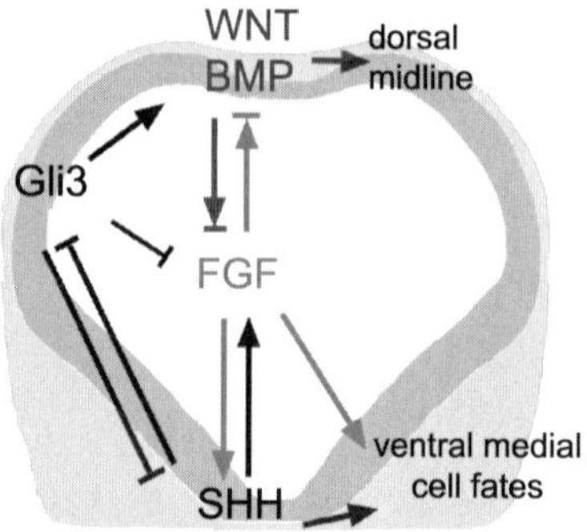

Figure 2 Model of the interactions between SHH, FGF, and BMP/WNT in forming the telencephalic midline. Schematic of a coronal section through the early telencephalon with dorsal up. Blue and red arrows and bars represent hypothesized roles for BMP/Wnt and FGF signaling, respectively. BMPs and FGFs are hypothesized to interact antagonistically or protagonistically depending on the stages of development and their levels of expression (an early threshold level of BMP is required to induce FGF8 in the ANB [Fig. 1], whereas BMPs later appear to repress FGFs; conversely, FGFs regulate BMPs in a dose-dependent manner via *Foxg1*, not shown, with low and high levels inhibiting and promoting BMPs, respectively). SHH is required to maintain expression of at least one Fgf gene, *Fgf8*, and is required to specify ventral medial cell fates. *Gli3* has been implicated in regulating the function of all three signaling centers. (See Color Insert.)

regulating levels of FGF signaling is illustrated by another finding: simply reducing the level of signaling by deleting one of three FGF receptors expressed in the telencephalon results in an increase, rather than a decrease, in cell proliferation at the rostro-medial end of the telencephalon (Hébert *et al.*, 2003).

Together, these findings suggest a model of how BMP/WNT, FGF, and SHH might interact to form the midline (Fig. 2). This model also provides an explanation for how ventral SHH is required to form the dorsal midline and, conversely, how the dorsal midline might influence ventral midline development. In the *Shh* mutant, *Fgf8* expression is initiated but not maintained (Aoto *et al.*, 2002; Ohkubo *et al.*, 2002). In these embryos, the low levels of other FGFs (or the lingering levels of FGF8) could then inhibit expression of BMPs, which would lead to a loss of dorsal midline structures (Storm *et al.*, 2003). Similarly, BMPs could have an effect on not only the dorsal midline, but also the ventral midline via regulation of FGFs and SHH. However, whether BMP signaling is essential for dorsal midline formation remains unclear. In addition, the extent to which BMP signaling is required to regulate expression of Fgf genes and *Shh* remains unknown.

The three groups of cells that express *Shh*, or Fgf genes, or Bmp and Wnt genes are the only putative signaling centers identified to date that are likely to play roles in forming the midline and patterning the early telencephalon.

However, additional signaling centers are likely to appear as the telencephalon grows and becomes morphologically more complex. For example, a discrete population of cells at the lateral border between ventral and dorsal telencephalon express *Fgf7*, the WNT antagonist *Sfrp2*, and several EGF family members (Assimacopoulos *et al.*, 2003). It will be interesting to determine whether these factors and others also play essential roles in patterning the developing telencephalon.

IV. Specification of Dorsal vs. Ventral Telencephalon

Aside from the specialized cell types found all along the midline, the telencephalon is grossly subdivided into dorsal and ventral areas. Dorsal and ventral precursor cells give rise to different cell types and structures. For example, dorsally they give rise to glutamatergic neurons and form the neocortex and hippocampus, whereas ventrally they form the medial and lateral ganglionic eminences (MGE and LGE, respectively), which form the striatum and globus pallidus and generate the GABAergic neurons, which migrate and populate vast areas of the telencephalon (Marin and Rubenstein, 2001). How are telencephalic precursor cells specified as dorsal or ventral? Specific transcription factors, and the extracellular signals that are thought to regulate their expression, have recently been implicated in this process.

A. The Dorsal-Ventral Border

Pax6 encodes a homeobox transcription factor expressed at high levels in the dorsal telencephalon. *Gsh2*, another homeobox gene, is expressed specifically in the ventral telencephalon. Both of these genes are essential for specifying the fates of the cells near the boundary between the dorsal and ventral telencephalon. In the *Pax6* mutant, the ventral-most area of the developing cortex acquires the molecular characteristics of the dorsal LGE, whereas in the *Gsh2* mutant, the reverse occurs: the dorsal LGE is transformed into ventral cortex (Corbin *et al.*, 2000; Stoykova *et al.*, 2000; Toresson *et al.*, 2000; Yun *et al.*, 2001). In embryos mutant for both *Pax6* and *Gsh2*, patterning of the dorsal-ventral border is not as severely disrupted compared to either single mutant, further illustrating the functional antagonism between these two genes (Toresson *et al.*, 2000).

Other factors are likely to participate in setting up the dorsal-ventral boundary, as well. For instance, embryos mutant for the orphan nuclear receptor gene *Tlx* display a slight ventralization and a loss of features characteristic of the dorsal-ventral border, a defect that is worsened when combined with the loss of one allele of *Pax6* (Stenman *et al.*, 2003).

In addition, retinoic acid is required for inducing characteristics of the dorsal-ventral border in chick embryos (Marklund *et al.*, 2004).

B. Specifying Dorsal Telencephalon

Pax6 acts in combination with other genes to specify the dorsal telencephalon. In embryos mutant for *Pax6* and another homeobox gene, *Emx2*, dorsal precursor cells fail to adopt or maintain a cortical fate and instead assume, at least in part, a ventral fate (Muzio *et al.*, 2002b). One copy of either gene is sufficient to maintain a cortical identity. It is important to note that the size of the dorsal telencephalon in the double mutant is drastically smaller, suggesting that *Emx2* and *Pax6* are required to maintain not only the cortical fate of these cells, but also their proliferative state. In addition to genes required to inhibit cortical cells from adopting a ventral fate, at least one other gene prevents cortical cells from adopting midline fates. *Lhx2*, which encodes a LIM homeodomain transcription factor, is required to keep cortical precursor cells from becoming dorsal midline cells. In embryos that lack *Lhx2*, the entire area that normally becomes cortex is lost at the expense of an expanded choroid plexus and cortical hem (Monuki *et al.*, 2001). Hence, *Pax6*, *Emx2*, and *Lhx2* all act to maintain cells as cortical precursors, rather than as ventral or dorso-medial ones.

C. Specifying Ventral Telencephalon

Two major subdivisions of the ventral telencephalon are the MGE and LGE. The delineation between these two areas is likely to also be regulated by homeobox transcription factor genes. For instance, *Nkx2.1* , which is expressed specifically in the MGE, is essential for specifying the fate of MGE precursor cells. In the *Nkx2.1* mutant, the MGE assumes the molecular characteristics of the LGE and generates striatal rather than pallidal neurons (Sussel *et al.*, 1999). Two other transcription factor genes, *Gsh1* and *Gsh2*, are together required to specify the fate of LGE precursors. Although *Gsh1* is mainly expressed in the MGE, in embryos deficient for *Gsh2*, *Gsh1* expression expands into the LGE and rescues the fate of LGE precursors, as demonstrated in a mutant that lacks both *Gsh1* and *Gsh2* (Toresson and Campbell, 2001; Yun *et al.*, 2003).

The previously mentioned transcription factors, as well as others (Zaki *et al.*, 2003), are clearly important in specifying broad areas of the telencephalon. Moreover, based on their combined expression patterns, the telencephalon can be further divided into subdomains (Campbell, 2003; Marin and Rubenstein, 2001; Puelles *et al.*, 2000; Schuurmans and Guillemot, 2002;

Yun *et al.*, 2001). What is not entirely clear is how the restricted expression pattern of these transcription factors is generated. In some cases, their patterns of expression are maintained by cross-repressing each other, as for *Gsh2* and *Pax6* as well as *Pax6* and *Emx2* (Muzio *et al.*, 2002a; Torreson *et al.*, 2000; Yun *et al.*, 2001). In addition, secreted factors emanating from midline signaling centers are likely to act upstream of at least some of these transcription factors to establish and further maintain their patterns of expression.

D. Regulation of Laterally Expressed Transcription Factors by Midline Signals

Midline factors regulate the expression of transcription factor genes in the lateral areas of the telencephalon and, as a result, pattern these areas. For instance, the fate of precursor cells in the dorsal telencephalon is likely to be regulated at least in part by dorsal midline factors. Beads that are soaked with these factors and placed in the ventral forebrain of chicks promote the development of telencephalic precursor cells with inappropriate dorsal idenities. Bmp4- and Bmp5-soaked beads disrupt the fate of ventral cells (Golden *et al.*, 1999), and loss of *megalin*, which results in increased *Bmp4* expression, results in a loss of ventral cell types (Spoelgen *et al.*, 2005). Likewise, Wnt3A beads induce dorsal features, such as *Pax6* expression, in ventral cells (Gunhaga *et al.*, 2003). Furthermore, Wnt followed by FGF signaling appears to be required to promote early expression of *Pax6* in the prospective dorsal telencephalon and later expression of the neocortical marker *Emx1* (Gunhaga *et al.*, 2003). Consistent with this finding, gain- and loss-of-function mutations in β-catenin, a downstream effector of Wnt signaling, lead to gain and loss of dorsal telencephalic cell identities, respectively (Backman *et al.*, 2005). Thus, factors such as BMPs, Wnts, and FGFs emanating from midline signaling centers can affect expression of transcription factors in lateral telencephalic areas. The regulation of dorsally expressed transcription factors by midline signals is discussed further in Section V.

The ventral midline factor SHH is essential for specifying ventral cell fates. Loss of SHH signaling leads to a loss of ventral cells that express *Dlx2*, *Gsh2*, and *Nkx2.1* at the expense of dorsal cells that express *Emx2* and *Pax6* (Chiang *et al.*, 1996; Ericson *et al.*, 1995; Fuccillo *et al.*, 2004; Ohkubo *et al.*, 2002). Likewise, ectopic expression of *Shh* can induce *Dlx2* and *Nkx2.1* in the dorsal telencephalon of zebrafish and mice (Barth and Wilson, 1995; Ericson *et al.*, 1995; Hauptmann and Gerster, 1996; Kohtz *et al.*, 1998; Shimamura and Rubenstein, 1997). Moreover, *Shh* would actually ventralize a much greater area of the telencephalon than it normally does if it were not for the zinc finger transcription factor gene *Gli3*. *Gli3* acts to antagonize *Shh* and dorsalize the telencephalon (Aoto *et al.*, 2002; Grove *et al.*, 1998; Kuschel *et al.*, 2003; Theil *et al.*, 1999). Remarkably, embryos mutant for both *Shh* and *Gli3* show normal dorsal-ventral patterning, suggesting that additional factors can induce dorsal and ventral fates in the telencephalon (Rallu *et al.*, 2002).

A good candidate for a ventralizing signal other than SHH is FGF. In zebrafish, *fgf8* and *fgf3* are required for the formation of the ventral telencephalon (Shanmugalingam *et al.*, 2000; Shinya *et al.*, 2001; Walshe and Mason, 2003). Furthermore, in mice, FGF8-soaked beads can induce ventral markers dorsally even in the absence of SHH signaling (Kuschel *et al.*, 2003). This raises the question of whether SHH itself could be acting in part through FGFs to specify ventral cell fates (Fig. 2). Like *Shh*, *Fgf8* is also antagonized by *Gli3*. In mice, loss of *Gli3* leads to an expansion of *Fgf8* expression (Aoto *et al.*, 2002; Kuschel *et al.*, 2003; Theil *et al.*, 1999), suggesting that *Gli3* inhibits ventralization of the embryo not only by blocking *Shh* function, but also by repressing *Fgf8*.

V. Patterning the Anterior-Posterior Axis of the Dorsal Telencephalon

Studies addressing how the telencephalon is patterned along the anterior-posterior axis have focused on the dorsal telencephalon. Less is known about anterior-posterior patterning of the ventral telencephalon. The homeobox genes *Emx2* and *Pax6* are expressed in counter gradients in the cerebral cortex and confer regional identities to cortical precursor cells. In the *Emx2* mutant, caudo-medial regions are lost at the expense of rostro-lateral regions, and vice versa for the *Pax6* mutant, demonstrating that these genes have essential roles in specifying positional identities within the cortex (Bishop *et al.*, 2000, 2002; Hamasaki *et al.*, 2004; Mallamaci *et al.*, 2000). What regulates the expression of these transcription factors?

Compelling evidence indicates that both BMP and WNT signaling directly promote expression of *Emx2* in the dorsal telencephalon (Theil *et al.*, 2002). An enhancer from the *Emx2* gene, which can drive expression of a reporter specifically in the dorsal telencephalon, contains binding sites for transcriptional factors that mediate BMP and Wnt signaling, SMAD and LEF/TCF, respectively. When these sites are mutated, enhancer activity is lost. In addition, simultaneous and ectopic activation of the BMP and Wnt pathways, via expression of a constitutively active BMP receptor and β-catenin, is sufficient to turn on the *Emx2* enhancer outside of its normal expression domain (Theil *et al.*, 2002). Together, these data show that BMP and Wnt

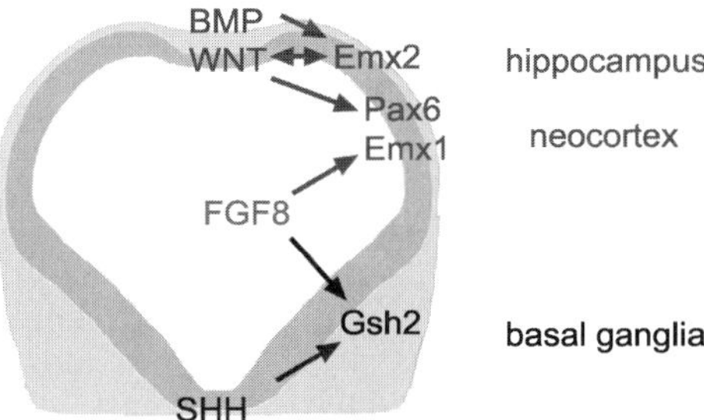

Figure 3 Model of the likely interactions between certain dorsalizing (blue) and ventralizing (black) factors that generate the major subdivisions of the telencephalon. FGFs (red) may affect both dorsal and ventral processes. Interactions between the transcription factors themselves have been omitted, and only the interactions between midline-associated secreted factors and transcription factors are illustrated. (See Color Insert.)

signaling directly promote *Emx2* expression. Interestingly, *Emx2* itself is required to maintain normal levels of midline *Wnt2b*, *Wnt3a*, and *Wnt8b* expression (Muzio *et al.*, 2005; Shimogori *et al.*, 2004), suggesting that there is a positive feedback loop between Wnt expression in the midline and *Emx2* expression in adjacent cells (Fig. 3). In the *Emx2* mutant, Noggin expression is upregulated, suggesting that *Emx2* may also provide feedback to promote BMP activity by inhibiting Noggin expression (Shimogori *et al.*, 2004).

A puzzling question remains regarding the regulation of dorsally expressed genes such as *Emx1*, *Emx2*, and *Lhx2*. Although they are expressed in a gradient in the cerebral cortex with highest levels present caudally and medially, their expression is excluded from the midline itself, where Bmp and Wnt genes are expressed (Shinozaki *et al.*, 2004). What keeps *Emx1*, *Emx2*, and *Lhx2* from being expressed in the dorsal midline? *Fgf8* is required for patterning the cerebral cortex along its anterior-posterior axis (Garel *et al.*, 2003; Shimogori and Grove, 2001). In mouse embryos hypomorphic for *Fgf8*, gradients of transcription factors, such as *Emx2*, shift anteriorly, suggesting that *Fgf8* represses *Emx2* (Garel *et al.*, 2003). In addition, ectopic expression of *Fgf8* inhibits *Emx2* (Crossley *et al.*, 2001; Storm *et al.*, 2003). Since FGF8 concentrations in the dorsal telencephalon are likely to be highest in the rostral midline area, this could explain why *Emx2* expression is excluded from at least the rostro-dorsal midline and is lower in the rostral telencephalon in general.

Lhx2 expression may or may not be excluded from the dorsal midline by a mechanism similar to that used for *Emx2*. BMP-soaked beads can regulate expression of *Lhx2* in a concentration-dependent manner in cultured explants, with high concentrations of BMP repressing and lower concentrations activating *Lhx2* expression, suggesting that *Lhx2* expression may be excluded from the midline due to high levels of BMP signaling and strongly expressed just adjacent to the midline due to lower levels of BMP signaling (Monuki *et al.*, 2001). However, a requirement for BMPs in regulating *Lhx2 in vivo* is less likely because a reduction in the level of BMP signaling in mice mutant for *Bmpr1a* has no effect on *Lhx2* expression (Hébert *et al.*, 2002). It remains possible, though, that like *Emx2*, *Lhx2* is regulated by a combination of BMP and Wnt signaling.

In regulating the expression of patterning genes such as *Emx2*, BMPs and Wnts are likely to also play a key role in promoting development of the hippocampus. In the rostro-dorsal telencephalon, the hippocampus develops immediately adjacent to the cortical hem, a source of Wnts (Grove *et al.*, 1998). In mice lacking *Wnt3a*, the precursor cells of the hippocampus fail to proliferate, leading to a lack of hippocampal structures (Lee *et al.*, 2000). Consistent with this finding, mice that are homozygous for a dominant negative allele of *Lef1*, which blocks the function of β-catenin, a downstream mediator of WNT signaling, also lack a hippocampus (Galceran *et al.*, 2000).

VI. Neuronal Fate Determination

One major challenge for the future is to decipher how these early patterning events translate into the generation of specific types of neurons and glia in each area of the telencephalon. In the mouse telencephalon, neurogenesis begins just after midgestation and is essentially over by birth (Gillies and Price, 1993). Oligodendrogenesis and gliogenesis, excluding the early generation of radial glia, starts several days later than neurogenesis and peaks postnatally. The production of neuronal and glial cell types depends on both the location of the precursor cells in the telencephalon and the age of the embryo. For example, glutamatergic neurons are generated dorsally in the cerebral cortex, whereas GABAergic neurons are primarily generated ventrally in the ganglionic eminence (Marin and Rubenstein, 2001). In addition, the production of particular neuronal subtypes is tightly linked to embryonic age. For example, in the mouse neocortex, glutamatergic layer 6 neurons that project primarily subcortically are produced before glutamatergic layer 2/3 neurons that project primarily to other cortical areas (Gillies and Price, 1993; McConnell, 1995). What factors instruct neural precursor cells to become a given cell type in each area and at each developmental stage?

A first step in bridging the gap between early patterning processes and the generation of specific types of differentiated neural cells is likely to come from an understanding of what regulates the expression of certain basic helix-loop-helix (bHLH) transcription factors (Schuurmans and Guillemot, 2002). Several of these transcription factor genes, including *Ngn1* and *Ngn2* (expressed dorsally) and *Mash1* (expressed primarily ventrally), are essential for specifying subtypes of neurons generated in the telencephalon. In the *Ngn2* mutant, and even more so in the *Ngn1;Ngn2* double mutant, dorsal precursor cells lose their dorsal characteristics and adopt ventral ones, including expression of *Mash1* (Fode *et al.*, 2000). In this context, *Mash1* is necessary for dorsal precursors to adopt ventral fates, since in the *Mash1; Ngn2* double mutant dorsal cells do not adopt ventral fates. *Mash1* is also sufficient to induce ventral fates, as shown in a mutant in which *Mash1* is expressed dorsally, leading to the ectopic production of GABAergic neurons (Fode *et al.*, 2000; Parras *et al.*, 2002).

Little is known about what directly regulates the expression of these bHLH transcription factors. In the ventral telencephalon, the *Mash1* expression domain is reduced without *Gsh2*, and even more so without both *Gsh1* and *Gsh2* (Toresson and Campbell, 2001; Toresson *et al.*, 2000; Yun *et al.*, 2001). Conversely, in the dorsal telencephalon of the *Pax6* mutant, the *Ngn1* and *Ngn2* expression domain is reduced. Even more dramatically, in the *Pax6;Emx2* double mutant, expression of *Ngn1* and *Ngn2* is lost (Muzio *et al.*, 2002b). Together, these results indicate that *Pax6* and *Emx2* act upstream to induce *Ngn1* and *Ngn2* expression dorsally and that *Gsh1* and *Gsh2* (and other yet-to-be-identified factors) act upstream to induce expression of *Mash1* ventrally.

The question of whether these homeobox transcription factors directly regulate expression of the bHLH transcription factors or whether they are acting indirectly through other factors to regulate their expression remains unanswered. For instance, in the *Gsh2* and *Gsh1;Gsh2* double mutants, not only is *Mash1* expression greatly reduced, but also expression of Dlx genes, which are also required for differentiation of ventral neurons (Yun *et al.*, 2002). Hence it is possible that *Gsh1* and *Gsh2* regulate *Mash1* expression indirectly through Dlx genes. In the case of *Pax6* promoting *Ngn2* expression, this interaction is more than likely to be direct (Scardigli *et al.*, 2003).

In addition to promoting ventral vs. dorsal neuronal fates, bHLH genes also play a role in a precursor cell's decision to adopt a neuronal, rather than a glial, cell fate. Forced expression of *Ngn1* not only induces neurogenesis in neural precursor cells, but also inhibits the generation of astrocytes (Sun *et al.*, 2001). In addition, in embryos mutant for both *Ngn2* and *Mash1*, neurogenesis is reduced and there is a precocious accumulation of astrocytic precursors (Nieto *et al.*, 2001). Other key signaling pathways such as BMP and Notch are also involved in the neuronal versus glial lineage decision (reviewed in Schuurmans and Guillemot, 2002). In addition to the neuronal-glial decision, different types of neurons are produced at different times in development. For example, in the neocortex, the different types of neurons that populate the cortical layers are born at different times (reviewed in McConnell, 1995). bHLH and homeobox genes have also recently been implicated in specifying stage-specific neuronal subtypes. For instance, *Foxg1* has been shown to actively repress the production of some of the earliest-born neurons in the telencephalon, the Cajal Retzius neurons (Hanashima *et al.*, 2004). Also, *Ngn1* and *Ngn2* are required for specifying deep, but not superficial, neocortical neurons, whereas *Pax6* and *Tlx* are required for specifying superficial neurons (Schuurmans *et al.*, 2004). In most cases, however, little is known about what regulates the stage-specific generation of neuronal subtypes.

VII. Conclusion

The molecular pathways involved in the early steps of telencephalon development are emerging. In this chapter, we have examined some of the better-understood mechanisms underlying how the telencephalon is induced, the telencephalic midline forms, the signaling centers within the midline interact and pattern lateral areas, and the first steps in regulating area and stage-specific neurogenesis. The telencephalon is induced by repression of Wnt signaling. Formation of the midline involves the interaction of three signaling centers that express the secreted factors *Shh* ventrally, Fgf genes rostrally, and Bmp and Wnt genes laterally and dorsally. These signaling centers regulate the expression of transcription factors such as *Emx2*, *Pax6*, and *Gsh2*, which are required for specifying major areas of the telencephalon and regulating the expression of region-specific neurogenic genes such as *Ngn2* dorsally and *Mash1* ventrally. These recent advances bring us closer to an understanding of how different combinations and sequences of signals drive precursor cells to generate particular neural cell types.

However, several questions regarding how a range of extracellular signals over time are integrated within a precursor cell to yield a reproducible cell fate outcome for a given telencephalic area and developmental stage remain unanswered. For example, what sequence of extracellular and intracellular signaling pathways is required to turn a naïve neural precursor cell into a neocortical layer 4 neuron? To further understand the mechanisms behind telencephalic development, it will be necessary to determine how the known patterning factors regulate each other's function, what other factors are involved, and how they ultimately exert their regulatory effects on the downstream molecules that directly impart the full differentiated features of a given neural cell type.

In humans, disorders affecting either patterning or neurogenesis in the cerebral hemispheres have devastating consequences. For example, holoprosencephaly, in which midline defects lead to incomplete separation of the hemispheres, results in perinatal lethality or a range of disorders, including mental retardation and severe epilepsy. Other disorders, such as micrencephaly, macrencephaly, tuberous sclerosis, and a range of neoplastic growths, are thought to result from abnormal cell proliferation or neurogenesis. In most cases, the molecular mechanisms underlying these disorders remain poorly understood. A deeper understanding of how the fate of neural precursor cells is regulated *in vivo* will undoubtedly lead to new therapeutic strategies for treating certain cerebral disorders.

Acknowledgments

The author thanks Zaven Kaprielian, Gordon Fishell, Susan McConnell, John Rubenstein, and lab members for helpful comments and discussions.

References

Aoto, K., Nishimura, T., Eto, K., and Motoyama, J. (2002). Mouse GLI3 regulates Fgf8 expression and apoptosis in the developing neural tube, face, and limb bud. *Dev. Biol.* **251,** 320–332.

Assimacopoulos, S., Grove, E. A., and Ragsdale, C. W. (2003). Identification of a Pax6-dependent epidermal growth factor family signaling source at the lateral edge of the embryonic cerebral cortex. *J. Neurosci.* **23,** 6399–6403.

Backman, M., Machon, O., Mygland, L., van den Bout, C. J., Zhong, W., Taketo, M. M., and Krauss, S. (2005). Effects of canonical Wnt signaling on dorso-ventral specification of the mouse telencephalon. *Dev. Biol.* **279,** 155–168.

Barth, K. A., and Wilson, S. W. (1995). Expression of zebrafish nk2.2 is influenced by sonic hedgehog/vertebrate hedgehog-1 and demarcates a zone of neuronal differentiation in the embryonic forebrain. *Development* **121,** 1755–1768.

Barth, K. A., Kishimoto, Y., Rohr, K. B., Seydler, C., Schulte-Merker, S., and Wilson, S. W. (1999). Bmp activity establishes a gradient of positional information throughout the entire neural plate. *Development* **126,** 4977–4987.

Bishop, K. M., Goudreau, G., and O' Leary, D. D. M. (2000). Regulation of area identity in the mammalian neocortex by *Emx2* and *Pax6*. *Science* **288,** 344–349.

Bishop, K. M., Rubenstein, J. L., and O'Leary, D. D. (2002). Distinct actions of *Emx1, Emx2*, and *Pax6* in regulating the specification of areas in the developing neocortex. *J. Neurosci.* **22,** 7627–7638.

Blackshear, P. J., Graves, J. P., Stumpo, D. J., Cobos, I., Rubenstein, J. L. R., and Zeldin, D. C. (2003). Graded phenotypic response to partial and complete deficiency of a brain-specific transcript variant of the winged helix transcription factor RFX4. *Development* **130,** 4539–4552.

Campbell, K. (2003). Dorsal-ventral patterning in the mammalian telencephalon. *Curr. Opin. Neurobiol.* **13,** 50–56.

Chiang, C., Litingtung, Y., Lee, E., Young, K. E., Corden, J. L., Westphal, H., and Beachy, P. A. (1996). Cyclopia and defective axial patterning in mice lacking Sonic hedgehog gene function. *Nature* **383,** 407–413.

Corbin, J. G., Gaiano, N., Machold, R. P., Langston, A., and Fishell, G. (2000). The Gsh2 homeodomain gene controls multiple aspects of telencephalic development. *Development* **127,** 5007–5020.

Crossley, P. H., Martinez, S., Ohkubo, Y., and Rubenstein, J. R. (2001). Coordinate expression of Fgf8, Otx2, Bmp4, and Shh in the rostral prosencephalon during development of the telencephalic and optic vesicles. *Neuroscience* **108,** 183–206.

Ericson, J., Muhr, J., Placzek, M., Lints, T., Jessell, T. M., and Edlund, T. (1995). Sonic hedgehog induces the differentiation of ventral forebrain neurons: A common signal for ventral patterning within the neural tube. *Cell* **81,** 747–756.

Fode, C., Ma, Q., Casarosa, S., Ang, S. L., Anderson, D. J., and Guillemot, F. (2000). A role for neural determination genes in specifying the dorsoventral identity of telencephalic neurons. *Genes Dev.* **14,** 67–80.

Fuccillo, M., Rallu, M., McMahon, A. P., and Fishell, G. (2004). Temporal requirement for hedgehog signaling in ventral telencephalic patterning. *Development* **131,** 5031–5040.

Furuta, Y., Piston, D. W., and Hogan, B. L. M. (1997). Bone morphogenetic proteins (BMPs) as regulators of dorsal forebrain development. *Development* **124,** 2203–2212.

Galceran, J., Miyashita-Lin, E. M., Devaney, E., Rubenstein, J. L., and Grosschedl, R. (2000). Hippocampus development and generation of dentate gyrus granule cells is regulated by LEF1. *Development* **127,** 469–482.

Garel, S., Huffman, K. J., and Rubenstein, J. L. (2003). Molecular regionalization of the neocortex is disrupted in Fgf8 hypomorphic mutants. *Development* **130,** 1903–1914.

Gillies, K., and Price, D. J. (1993). The fates of cells in the developing cerebral cortex of normal and methylazoxymethanol acetate-lesioned mice. *Eur. J. Neurosci.* **5,** 73–84.

Golden, J. A., Bracilovic, A., McFadden, K. A., Beesley, J. S., Rubenstein, J. L., and Grinspan, J. B. (1999). Ectopic bone morphogenetic proteins 5 and 4 in the chicken forebrain lead to cyclopia and holoprosencephaly. *Proc. Natl. Acad. Sci. USA* **96,** 2439–2444.

Grove, E. A., Tole, S., Limon, J., Yip, L., and Ragsdale, C. W. (1998). The hem of the embryonic cerebral cortex is defined by the expression of multiple Wnt genes and is compromised in Gli3-deficient mice. *Development* **125,** 2315–2325.

Gulisano, M., Broccoli, V., Pardini, C., and Boncinelli, E. (1996). *Emx1* and *Emx2* show different patterns of expression during proliferation and differentiation of the developing cerebral cortex in the mouse. *Eur. J. Neurosci.* **8,** 1037–1050.

Gunhaga, L., Marklund, M., Sjodal, M., Hsieh, J. C., Jessell, T. M., and Edlund, T. (2003). Specification of dorsal telencephalic character by sequential Wnt and FGF signaling. *Nat. Neurosci.* **6,** 701–707.

Hamasaki, T., Leingartner, A., Ringstedt, T., and O'Leary, D. D. (2004). EMX2 regulates sizes and positioning of the primary sensory and motor areas in neocortex by direct specification of cortical progenitors. *Neuron* **43,** 359–372.

Hanashima, C., Li, S. C., Shen, L., Lai, E., and Fishell, G. (2004). Foxg1 suppresses early cortical cell fate. *Science* **303,** 56–59.

Hauptmann, G., and Gerster, T. (1996). Complex expression of the zp-50 pou gene in the embryonic zebrafish brain is altered by overexpression of sonic hedgehog. *Development* **122,** 1769–1780.

Hayhurst, M., and McConnell, S. K. (2003). Mouse models of holoprosencephaly. *Curr. Opin. Neurol.* **16,** 135–141.

Hébert, J. M., Mishina, Y., and McConnell, S. K. (2002). BMP signaling is required locally to pattern the dorsal telencephalic midline. *Neuron* **35,** 1029–1041.

Hébert, J. M., Lin, M., Partanen, J., Rossant, J., and McConnell, S. K. (2003). FGF signaling through FGFR1 is required for olfactory bulb morphogenesis. *Development* **130,** 1101–1111.

Houart, C., Westerfield, M., and Wilson, S. W. (1998). A small population of anterior cells patterns the forebrain during zebrafish gastrulation. *Nature* **391,** 788–792.

Houart, C., Caneparo, L., Heisenberg, C., Barth, K., Take-Uchi, M., and Wilson, S. (2002). Establishment of the telencephalon during gastrulation by local antagonism of Wnt signaling. *Neuron* **35,** 255–265.

Kohtz, J. D., Baker, D. P., Corte, G., and Fishell, G. (1998). Regionalization within the mammalian telencephalon is mediated by changes in responsiveness to Sonic Hedgehog. *Development* **125,** 5079–5089.

Kuschel, S., Ruther, U., and Theil, T. (2003). A disrupted balance between Bmp/Wnt and Fgf signaling underlies the ventralization of the *Gli3* mutant telencephalon. *Dev. Biol.* **260,** 484–495.

Lagutin, O. V., Zhu, C. C., Kobayashi, D., Topczewski, J., Shimamura, K., Puelles, L., Russell, H. R., McKinnon, P. J., Solnica-Krezel, L., and Oliver, G. (2003). Six3 repression of Wnt signaling in the anterior neuroectoderm is essential for vertebrate forebrain development. *Genes Dev.* **17,** 368–379.

Lee, S. M. K., Tole, S., Grove, E., and McMahon, A. P. (2000). A local Wnt3a signal is required for development of the mammalian hippocampus. *Development* **127,** 457–467.

Mallamaci, A., Muzio, L., Chan, C.-H., Parnavelas, J., and Boncinelli, E. (2000). Area identity shifts in the early cerebral cortex of Emx2-/- mutant mice. *Nat. Neurosci.* **3,** 679–686.

Marin, O., and Rubenstein, J. L. (2001). A long, remarkable journey: Tangential migration in the telencephalon. *Nat. Rev. Neurosci.* **2,** 780–790.

Marklund, M., Sjodal, M., Beehler, B. C., Jessell, T. M., Edlund, T., and Gunhaga, L. (2004). Retinoic acid signalling specifies intermediate character in the developing telencephalon. *Development* **131,** 4323–4332.

Maruoka, Y., Ohbayashi, N., Hoshikawa, M., Itoh, N., Hogan, B. M., and Furuta, Y. (1998). Comparison of the expression of three highly related genes, Fgf8, Fgf17 and Fgf18, in the mouse embryo. *Mech. Dev.* **74,** 175–177.

Masai, I., Heisenberg, C. P., Barth, K. A., Macdonald, R., Adamek, S., and Wilson, S. W. (1997). Floating head and masterblind regulate neuronal patterning in the roof of the forebrain. *Neuron* **18,** 43–57.

McConnell, S. K. (1995). Constructing the cerebral cortex: Neurogenesis and fate determination. *Neuron* **15,** 761–768.

McWhirter, J. R., Goulding, M., Weiner, J., Chun, J., and Murre, C. (1997). A novel fibroblast growth factor gene expressed in the developing nervous system is a downstream target of the chimeric homeodomain oncoprotein E2A-Pbx1. *Development* **124,** 3221–3232.

Monuki, E. S., and Walsh, C. A. (2000). Proto-mapping the areas of cerebral cortex: Transcription factors make the grade. *Nat. Neurosci.* **3,** 640–641.

Monuki, E. S., Porter, F. D., and Walsh, C. A. (2001). Patterning of the dorsal telencephalon and cerebral cortex by a roof plate-Lhx2 pathway. *Neuron* **32,** 591–604.

Muzio, L., DiBenedetto, B., Stoykova, A., Boncinelli, E., Gruss, P., and Mallamaci, A. (2002a). Emx2 and Pax6 control regionalization of the pre-neuronogenic cortical primordium. *Cereb. Cortex* **12,** 129–139.

Muzio, L., DiBenedetto, B., Stoykova, A., Boncinelli, E., Gruss, P., and Mallamaci, A. (2002b). Conversion of cerebral cortex into basal ganglia in Emx2(-/-) Pax6(Sey/Sey) double-mutant mice. *Nat. Neurosci.* **5,** 737–745.

Muzio, L., Soria, J. M., Pannese, M., Piccolo, S., and Mallamaci, A. (2005). A mutually stimulating loop involving Emx2 and canonical Wnt signaling specifically promotes expansion of occipital cortex and hippocampus. *Cereb. Cortex* [Epub ahead of print].

Ohkubo, Y., Chiang, C., and Rubenstein, J. L. (2002). Coordinate regulation and synergistic actions of BMP4, SHH and FGF8 in the rostral prosencephalon regulate morphogenesis of the telencephalic and optic vesicles. *Neuroscience* **111,** 1–17.

Orr-Urtreger, A., Givol, D., Yayon, A., Yarden, Y., and Lonai, P. (1991). Developmental expression of two murine fibroblast growth factor receptors, *flg* and *bek*. *Development* **113,** 1419–1434.

Parras, C. M., Schuurmans, C., Scardigli, R., Kim, J., Anderson, D. J., and Guillemot, F. (2002). Divergent functions of the proneural genes Mash1 and Ngn2 in the specification of neuronal subtype identity. *Genes Dev.* **16,** 324–338.

Peters, K. G., Werner, S., Chen, G., and Williams, L. T. (1992). Two FGF receptor genes are differentially expressed in epithelial and mesenchymal tissues during limb formation and organogenesis in the mouse. *Development* **114,** 233–243.

Puelles, L., Kuwana, E., Bulfone, A., Shimamura, K., Keleher, J., Smiga, S., Puelles, E., and Rubenstein, J. L. R. (2000). Pallial and subpallial derivatives in the embryonic chick and mouse telencephalon, traced by the expression of the Dlx-2, Emx-1, Nkx-2.1, Pax-6 and Tbr-1 genes. *J. Comp. Neurol.* **424,** 409–438.

Ragsdale, C. W., and Grove, E. A. (2001). Patterning the mammalian cerebral cortex. *Curr. Opin. Neurobiol.* **11,** 50–58.

Rallu, M., Machold, R., Gaiano, N., Corbin, J. G., McMahon, A. P., and Fishell, G. (2002). Dorsoventral patterning is established in the telencephalon of mutants lacking both Gli3 and Hedgehog signaling. *Development* **129,** 4963–4974.

Rubenstein, J. L. R., Shimamura, K., Martinez, S., and Puelles, L. (1998). Regionalization of the prosencephalic neural plate. *Ann. Rev. Neurosci.* **21,** 445–478.

Scardigli, R., Baumer, N., Gruss, P., Guillemot, F., and Le Roux, I. (2003). Direct and concentration-dependent regulation of the proneural gene Neurogenin2 by Pax6. *Development* **130,** 3269–3281.

Schuurmans, C., and Guillemot, F. (2002). Molecular mechanisms underlying cell fate specification in the developing telencephalon. *Curr. Opin. Neurobiol.* **12,** 26–34.

Schuurmans, C., Armant, O., Nieto, M., Stenman, J. M., Britz, O., Klenin, N., Brown, C., Langevin, L.M, Seibt, J., Tang, H., Cunningham, J. M., Dyck, R., Walsh, C., Campbell, K., Polleux, F., and Guillemot, F. (2004). Sequential phases of cortical specification involve neurogenin-dependent and -independent pathways. *EMBO J.* **23,** 2892–2902.

Shanmugalingam, S., Houart, C., Picker, A., Reifers, F., Macdonald, R., Barth, A., Griffin, K., Brand, M., and Wilson, S. W. (2000). Ace/Fgf8 is required for forebrain commissure formation and patterning of the telencephalon. *Development* **127,** 2549–2561.

Shimamura, K., and Rubenstein, J. L. R. (1997). Inductive interactions direct early regionalization of the forebrain. *Development* **124,** 2709–2718.

Shimogori, T., and Grove, E. A. (2001). Neocortex patterning by the secreted signaling molecule FGF8. *Science* **294,** 1071–1074.

Shimogori, T., Banuchi, V., Ng, H. Y., Strauss, J. B., and Grove, E. A. (2004). Embryonic signaling centers expressing BMP, WNT and FGF proteins interact to pattern the cerebral cortex. *Development* **131,** 5639–5647.

Shinozaki, K., Yoshida, M., Nakamura, M., Aizawa, S., and Suda, Y. (2004). Emx1 and Emx2 cooperate in initial phase of archipallium development. *Mech. Dev.* **121,** 475–489.

Shinya, M., Koshida, S., Sawada, A., Kuroiwa, A., and Takeda, H. (2001). Fgf signalling through MAPK cascade is required for development of the subpallial telencephalon in zebrafish embryos. *Development* **128,** 4153–4164.

Spoelgen, R., Hammes, A., Anzenberger, U., Zechner, D., Andersen, O. M., Jerchow, B., and Willnow, T. E. (2005). LRP2/megalin is required for patterning of the ventral telencephalon. *Development* **132,** 405–414.

Stenman, J., Yu, R. T., Evans, R. M., and Campbell, K. (2003). Tlx and Pax6 co-operate genetically to establish the pallio-subpallial boundary in the embryonic mouse telencephalon. *Development* **130,** 1113–1122.

Storm, E., Rubenstein, J. L., and Martin, G. R. (2003). Dosage of Fgf8 determines whether cell survival is positively or negatively regulated in the developing forebrain. *Proc. Natl. Acad. Sci. USA* **100,** 1757–1762.

Stoykova, A., Treichel, D., Hallonet, M., and Gruss, P. (2000). Pax6 modulates the dorsoventral patterning of the mammalian telencephalon. *J. Neurosci.* **20,** 8042–8050.

Sun, Y., Nadal-Vicens, M., Misono, S., Lin, M. Z., Zubiaga, A., Hua, X., Fan, G., and Greenberg, M. E. (2001). Neurogenin promotes neurogenesis and inhibits glial differentiation by independent mechanisms. *Cell* **104,** 365–376.

Sussel, L., Marin, O., Kimura, S., and Rubenstein, J. L. (1999). Loss of Nkx2.1 homeobox gene function results in a ventral to dorsal molecular respecification within the basal telencephalon: Evidence for a transformation of the pallidum into the striatum. *Development* **126,** 3359–3370.

Szucsick, J. C., Witte, D. P., Li, H., Pixley, S. K., Small, K. M., and Potter, S. S. (1997). Altered forebrain and hindbrain development in mice mutant for the Gsh-2 homeobox gene. *Dev. Biol.* **191,** 230–242.

Tao, W., and Lai, E. (1992). Telencephalon-restricted expression of BF-1, a new member of the HNF-3/fork head gene family, in the developing rat brain. *Neuron* **8,** 957–966.

Theil, T., Alvarez-Bolado, G., Walter, A., and Ruther, U. (1999). Gli3 is required for Emx gene expression during dorsal telencephalon development. *Development* **126,** 3561–3571.

Theil, T., Aydin, S., Koch, S., Grotewold, L., and Ruther, U. (2002). Wnt and Bmp signalling cooperatively regulate graded *Emx2* expression in the dorsal telencephalon. *Development* **129,** 3045–3054.

Toresson, H., and Campbell, K. (2001). A role for Gsh1 in the developing striatum and olfactory bulb of Gsh2 mutant mice. *Development* **128,** 4769–4780.

Toresson, H., Potter, S., and Campbell, K. (2000). Genetic control of dorsal–ventral identity in the telencephalon: Opposing roles for Pax6 and Gsh2. *Development* **127,** 4361–4371.

Walshe, J., and Mason, I. (2003). Unique and combinatorial functions of Fgf3 and Fgf8 during zebrafish forebrain development. *Development* **130,** 4337–4349.

Walther, C., and Gruss, P. (1991). *Pax-6,* a murine paired box gene, is expressed in the developing CNS. *Development* **113,** 1435–1449.

Wilson, S. W., and Houart, C. (2004). Early steps in the development of the forebrain. *Dev. Cell* **6,** 167–181.

Wilson, S. W., and Rubenstein, J. L. R. (2000). Induction and dorsoventral patterning of the telencephalon. *Neuron* **28,** 641–651.

Yun, K., Potter, S., and Rubenstein, J. L. R. (2001). Gsh2 and Pax6 play complementary roles in dorsoventral patterning of the mammalian telencephalon. *Development* **128,** 193–205.

Yun, K., Fischman, S., Johnson, J., Hrabe de Angelis, M., Weinmaster, G., and Rubenstein, J. L. R. (2002). Modulation of the notch signaling by *Mash1*and *Dlx1/2* regulates sequential specification and differentiation of progenitor cell types in the subcortical telencephalon. *Development* **129,** 5029–5040.

Yun, K., Garel, S., Fischman, S., and Rubenstein, J. L. R. (2003). Patterning of the lateral ganglionic eminence by the *Gsh1* and *Gsh2* homeobox genes is required for histogenesis of the striatum and olfactory bulb and the growth of axons through the basal ganglia. *J. Comp. Neurol.* **461,** 151–165.

Zaki, P.A, Quinn, J.C, and Price, D. J. (2003). Mouse models of telencephalic development. *Curr. Opin. Genet. Dev.* **13,** 423–437.

3

Glia–Neuron Interactions in Nervous System Function and Development

Shai Shaham
The Rockefeller University, New York, New York, 10021

I. Introduction
II. Defining Neurons and Glia
III. Glial Roles in Synaptogenesis
IV. Glial Modulation of Synaptic Activity
V. Glial Effects on Neuronal Conduction
VI. Glial Regulation of Neuronal Migration and Process Outgrowth
VII. Reciprocal Control of Cell Survival between Neurons and Glia
VIII. Genetic and Functional Studies of Glia in the Nematode *Caenorhabditis elegans*
    A. Anatomy
    B. Functional Studies
IX. Summary
    Acknowledgments
    References

Nervous systems are generally composed of two cell types—neurons and glia. Early studies of neurons revealed that these cells can conduct electrical currents, immediately implying that they have roles in the relay of information throughout the nervous system. Roles for glia have, until recently, remained obscure. The importance of glia in regulating neuronal survival had been long recognized. However, this trophic support function has hampered attempts to address additional, more active functions of these cells in the nervous system. In this chapter, recent efforts to reveal some of these additional functions are described. Evidence supporting a role for glia in synaptic development and activity is presented, as well as experiments suggesting glial guidance of neuronal migration and process outgrowth. Roles for glia in influencing the electrical activity of neurons are also discussed. Finally, an exciting system is described for studying glial cells in the nematode *C. elegans,* in which recent studies suggest that glia are not required for neuronal viability. © 2005, Elsevier Inc.

I. Introduction

Glia were described as components of the spinal cord nearly 160 years ago by the German pathologist Rudlof Virchow (1846). Virchow and others (e.g., Cajal, 1913) elaborated on these initial studies to show that glial matter

Current Topics in Developmental Biology, Vol. 69
Copyright 2005, Elsevier Inc. All rights reserved.

0070-2153/05 $35.00
DOI: 10.1016/S0070-2153(05)69003-5

pervade the nervous systems of vertebrates. Although glia were recognized as the predominant cell type in the vertebrate brain, early experimentation aimed at elucidating their functions was unsuccessful. Part of the problem was noted by Santiago Ramón y Cajal in his classic volume *Histology of the Nervous System* (Cajal, 1911). Cajal stated: "What is the function of glial cells in neural centers? The answer is still not known, and the problem is even more serious because it may remain unsolved for many years to come until physiologists find direct methods to attack it. Neuronal function was clarified by the phenomena of conduction ... But how can the physiology of glia be clarified if they cannot be manipulated?" Even today, Cajal's dilemma reflects the central problem in understanding glial function: what is the readout for glial activities?

There are three broad possibilities concerning the roles of glia in the nervous system: (1) they may have no role, (2) they may have a role that is completely independent of the neurons with which they physically associate, or (3) they might function in concert with neurons to perform nervous system tasks. Although recent studies have begun to hint at intimate functional connections between glia and neurons, it is somewhat surprising that a hundred years after Cajal's writings, we still lack clear-cut evidence to distinguish among the possibilities described above. Indeed, given our current state of understanding, it is still very possible that glia perform both neuron-dependent and neuron-independent functions in the nervous system. Nonetheless, because of the spatial proximity of glial cells to neurons, as seen most clearly with the myelin-forming glia, it has been a central assumption in the field that glia must function, at least in part, to regulate neuronal parameters. How can this hypothesis be addressed experimentally? One approach would be to examine neurons *in vitro* or *in vivo* in the presence or absence of glia and compare their properties and development. Although a completely reasonable approach, this simple strategy has, in many cases, failed because neurons usually died when cultured without glia or in mutants lacking glia (e.g., Hosoya *et al.*, 1995; Jones *et al.*, 1995; Ullian *et al.*, 2001). Thus, although it is clear that glia provide survival capacity to neurons, this very property often makes it impossible to study roles for glia in regulating neuronal function.

How to proceed, then? A number of strategies to overcome neuronal death upon glial removal have recently been employed. Substances that promote neuronal survival, some of which are of glial origin, have been added to neuronal preparations, allowing neurons to be cultured without the physical presence of glia (e.g., Meyer-Franke *et al.*, 1995; Ullian *et al.*, 2001). These studies have yielded important information; however, significant caveats remain. For example, might survival factors function in other capacities to alter neuronal physiology? Might dissociation of primary nervous tissue to its cellular components affect critical neuronal properties?

Glial function has also been perturbed more subtly so that neuronal death will not result. For example, chemicals that specifically inhibit glial proteins without affecting neuronal survival have been used to explore glial effects on neuronal parameters (e.g., McBean, 1994; Robitaille, 1998). It should also be possible, in principle, to generate glia harboring mutations that affect neuronal function but not survival.

Although these approaches have proven quite informative, it is yet unclear how relevant these studies are to the functioning of the nervous system *in vivo*. Glial alterations leading to obvious organismal consequences have been described. Demyelinating diseases, such as multiple sclerosis, Dejerine-Sottas syndrome, or Guillain-Barre syndrome, that result from alteration of glial-derived myelin, severely affect organismal motor and sensory behaviors (Franklin, 2002; Newswanger and Warren, 2004; Plante-Bordeneuve and Said, 2002). However, in these diseases, neuronal death often occurs. As in other areas of biological inquiry, theories regarding glial function will ultimately be tested by generating animals harboring specific glial deficits that do not affect neuronal survival and looking for behavioral and/or developmental abnormalities.

A third approach to circumvent the effects of glia on neuronal survival has been to search for a natural setting in which glia are not required for neuronal survival. Recent studies have demonstrated that such a setting exists in the nematode *Caenorhabditis elegans* (T. Bacaj and S. Shaham, unpublished results; Perens and Shaham, 2005). Furthermore, anecdotal reports as well as more comprehensive recent studies (T. Bacaj and S. Shaham, unpublished results) suggest that glial deficits in *C. elegans* have clear behavioral and developmental consequences. These observations, combined with the facility of genetic studies in *C. elegans*, suggest that this organism may provide an exciting new system in which to decipher both the roles of glia in the nervous system and the molecular effectors of these roles.

Insights into the roles of glia–neuron interactions in nervous system function are examined in this chapter. It is not the purpose of this chapter to provide a comprehensive review of all aspects of glia–neuron interactions; rather, the intention is to point out some of the salient and novel functions that have recently been attributed to glia in the control of neuronal function and development. In the following sections, highlights of recent progress are presented, beginning with insights into the roles of glia in neuronal development and regulation of synaptic and conductive activities of neurons. Discussions of more indirect roles for glia as an energy source for neurons and as regulators of neuronal cell survival follow. The chapter concludes with a description of new studies on glial function in *C. elegans*.

II. Defining Neurons and Glia

Before embarking on a discussion of glia–neuron interactions, it is important to define each cell type. This is no small matter, since valid comparisons of glia–neuron interactions across different species rest on the assumption that the cell types under study are fundamentally similar. Neurons are, in some sense, easier to define than glia. Although these cells come in myriad shapes and sizes, they share a number of basic properties. Neurons conduct fast currents and connect to other neurons, or to terminal cells (such as muscles or gland cells), by synapses or gap junctions. They also extend processes. The molecular mechanisms controlling these basic properties are generally conserved in neurons of different organisms; however, some widely used functional and molecular markers are probably not appropriate neuronal identifiers. For example, while the action potential and its associated voltage-gated sodium channel are hallmarks of neurons in vertebrates, neither exists in neurons of the nematode *C. elegans* (Bargmann, 1998; Goodman *et al.*, 1998). However, *C. elegans* clearly possesses cells that elaborate processes, connect by synapses and gap junctions, and conduct fast currents (Goodman *et al.*, 1998; Lockery and Goodman, 1998; White *et al.*, 1986). Cell shape criteria can also lead to confusion. For example, the neuroepithelial cells housed in vertebrate taste buds are not usually classified as neurons, primarily for morphological reasons. However, these cells possess sensory receptors (for detection of taste substances) and synapse onto neurons (Barlow, 2003), suggesting that they must share basic neuronal properties.

Vertebrate glia are generally classified according to morphological and molecular criteria. In vertebrates, glia of the peripheral nervous system (PNS) are termed Schwann cells. These extend processes that ensheath or myelinate axons but can also ensheath synapses between neurons. Glia of the central nervous system (CNS) generally fall into three categories: oligodendrocytes, which myelinate CNS axons; astrocytes, which extend many processes that contact both blood vessels and neurons; and microglia, cells thought to be of mesodermal origin that are hypothesized to function in an immune capacity in the CNS (Peters *et al.*, 1991). Microglia may, thus, not be truly glial cells. A popular marker for vertebrate glia is the glial fibrillary acidic protein (GFAP), an intermediate filament protein found in some but not all glia (Eng *et al.*, 1970, 1971).

The morphological and molecular markers described above do not account for all vertebrate cells that have been termed glia, however. For example, olfactory ensheathing cells are neither oligodendrocytes nor astrocytes by morphology, yet they express GFAP and are intimately associated with olfactory neurons. Similar observations hold for Müller glia in the retina, Bergmann glia in the cerebellum, and support cells of the inner hair cells. Furthermore, GFAP expression fails to mark some cells considered glial in nature, and GFAP is not expressed in astrocytes and radial glia of some vertebrates (Dahl *et al.*, 1985).

All glia, however, meet three criteria, which do not also apply to cells of non-glial nature. First, glia are always physically associated with neurons. Second, glia are not neurons themselves; they generally do not transmit fast currents or form presynaptic structures (although neuronal synapses onto glia have been documented). Third, glia and neurons are lineally related. Recent studies on the nature of stem cells in the vertebrate brain have revealed that glia and neurons often arise from common ectodermally derived precursor cells such as radial glial cells (Alvarez-Buylla *et al.*, 2002; Doetsch, 2003). In the PNS, many glia and neurons are derived from the neural crest—a developmentally discrete ectodermal cell population originating near the neural tube in early vertebrate development (Le Douarin and Dupin, 2003; Le Douarin *et al.*, 1991). In the fruit fly *Drosophila melanogaster*, glia and neurons also arise from common precursor cells (Jones, 2001). Thus, kinship between glia and neurons is an important aspect of glial identity. In the following sections of this chapter, all references to glia and neurons, regardless of organismal origin, conform to the definitions elaborated in this section.

III. Glial Roles in Synaptogenesis

Ultrastructural studies of the vertebrate CNS have shown that glial processes, usually those associated with astrocytes, can be found adjacent to, or ensheathing, synaptic connections between neurons (Peters *et al.*, 1991; Spacek, 1985; Ventura and Harris, 1999; Wolff, 1976). In the periphery, synaptic Schwann cells envelop most neuromuscular junctions (Herrera *et al.*, 2000; Hirata *et al.*, 1997; Kelly and Zacks, 1969). These observations have led to the hypothesis that glia may play important roles in synaptogenesis and synaptic function.

A number of recent observations have provided evidence that glia in the CNS can promote synaptogenesis. Purified cultured postnatal rat retinal ganglion cells (RGCs) that have been separated from their glial components can be kept alive in culture using a number of survival factors, including brain-derived neurotrophic factor (BDNF) and ciliary neurotrophic factor (CNTF) (Meyer-Franke *et al.*, 1995). These cultured RGCs normally form functional synapses only inefficiently, as assessed by electrophysiological criteria and by localization of pre- and postsynaptic proteins. However, when these cells are co-cultured with glia from the RGC target region, synaptic efficacy is dramatically enhanced (Fig. 1) (Pfrieger and Barres, 1997). Specifically, the frequency of spontaneous postsynaptic currents in such cultures is increased 70-fold, and current amplitudes are increased 5-fold. In addition, a larger number of synapses can be visualized in such cultures (Nagler *et al.*, 2001; Ullian *et al.*, 2001). Incubation of RGCs with glia-conditioned medium reproduced the effects seen in the co-culture experiments, suggesting that a soluble factor or factors were required for the increase in synapse number and efficacy. Biochemical purification

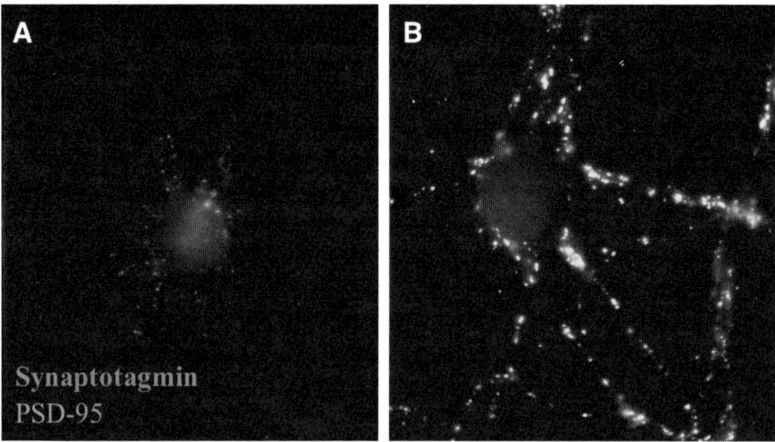

Figure 1 Glia promote synaptogenesis. (A) Retinal ganglion cells (RGCs) cultured in the absence of glia, stained using anti-synaptotagmin (a pre-synaptic marker; red) and anti-PSD-95 (a postsynaptic marker; green). Note few yellow puncta. (B) RGCs cultured in the presence of glia. Note large increase in yellow puncta. (Image courtesy of Erik Ullian and Ben Barres.) (See Color Insert.)

approaches suggested that one relevant component of the glia-conditioned medium was cholesterol bound to the apoE lipid-carrying protein (Mauch *et al.*, 2001). It is unclear whether cholesterol exerts specific roles in this setting, or whether it plays a more general role. For example, cholesterol could be a limiting component of synaptic vesicle membranes. Interestingly, the epsilon4 allele of apoE has been implicated in susceptibility to Alzheimer's disease, in which reduction in synaptic efficacy is observed (Myers and Goate, 2001). Thus, glia may underlie some aspects of this disease.

A second glia-derived component that is sufficient to induce the formation of postsynaptically silent RGC synapses of normal morphology has also been recently identified (Ullian *et al.*, 2004a). The secreted protein, thrombospondin, a large, extracellular matrix component best known for its roles in clotting, may serve to stabilize physical interactions between neurons at the synapse. The fact that thrombospondin-induced synapses are postsynaptically silent suggests that yet another glial component must allow for activation of synapses stabilized by this protein.

A number of *in vitro* studies also suggest that Schwann cells can promote synaptogenesis. For example, Ullian *et al.* (2004b) showed that Schwann cells induce the formation of glutamatergic synapses between cultured spinal motor neurons. Furthermore, Schwann cell-conditioned medium induced synapse formation between cultured *Xenopus laevis* motor neurons and muscle cells (Peng *et al.*, 2003). Selective ablation of perisynaptic Schwann cells *in vivo* using antibody-driven complement-induced lysis revealed that growth and addition of synapses significantly decreased, and existing synapses often retracted (Reddy *et al.*, 2003). Thus, Schwann cells seem to be important for maintaining synapses *in vivo*, although whether the effects observed in this study were secondary to a general deterioration of neuronal health is not clear.

IV. Glial Modulation of Synaptic Activity

The modulation of synaptic activity is hypothesized to be an essential component of nervous system versatility. The reigning hypothesis suggests that alterations in synaptic efficacy, as manifested by the ability of a post-synaptic cell to respond to a presynaptic cell, are essential for complex phenomena such as learning and memory. A number of recent observations suggest that glia may play an important role in regulating synaptic efficacy. Such a function at the *Xenopus* neuromuscular junction has been described. In *Xenopus*, as in other vertebrates, the synapse between a motor neuron and a muscle fiber is associated with a perisynaptic Schwann cell (Couteaux and Pecot-Dechavassine, 1974). High-frequency stimulation of the motor neuron leads to a decrease in muscle fiber activity, as measured by decreases in the end-plate potential following neurotransmitter release (Colomar and Robitaille, 2004; Robitaille, 1998). This phenomenon is often termed long-term depression (LTD), and similar phenomena in CNS neurons have been suggested to play essential roles in memory acquisition (Zucker and Regehr, 2002). Release of Ca^{2+} from intracellular stores within the perisynaptic Schwann cell is also observed during high-frequency presynaptic stimulation (Jahromi *et al.*, 1992; Reist and Smith, 1992; Rochon *et al.*, 2001). This observation shows that the perisynaptic Schwann cell can somehow monitor synaptic activity and suggests that Ca^{2+} release and LTD may be related. Robitaille (Bourque and Robitaille, 1998; Robitaille, 1998) hypothesized that Schwann cells at the neuromuscular junction detect presynaptic activity using G-protein-coupled receptors (GPCR), such as muscarinic acetylcholine receptors, to sense neurotransmitter release. To assess whether the activity of such G-proteins influenced LTD, he injected Schwann cells with GTP-γS, a G-protein activator, to mimic GPCR activation. Following presynaptic stimulation, an excessive decrease in synaptic activity was observed, consistent with a reduction in neurotransmitter release by the presynaptic cell or increased turnover of released transmitter. Thus, a mimic of GPCR activation was sufficient to cause LTD-like synaptic changes. Furthermore, injection of GTP-βS, a G-protein antagonist, into the Schwann cell resulted in increased synaptic activity, consistent with a predicted decrease in LTD. Taken together, these results suggest that specific manipulation of synaptic glia can affect synaptic activity.

The mechanism by which perisynaptic Schwann cells might regulate neurotransmitter dynamics in the *Xenopus* neuromuscular junction is not clear. However, this instructive activity could very well be modulated at the level of synaptic neurotransmitter levels. There are now numerous examples of how glia might modulate the concentration of neurotransmitter at the synaptic cleft. Glia as well as neurons express a variety of neurotransmitter transporters (Bergles and Jahr, 1997; Huang and Bergles, 2004; Rothstein *et al.*, 1994). Perhaps the most-studied glial transporters have been those involved in clearance of glutamate and glycine. However, glial transporters for gamma amino butyric acid (GABA) have been described as well (Minelli *et al.*, 1995, 1996). Glutamate is the major excitatory neurotransmitter in the vertebrate CNS, and its levels at the synaptic cleft are tightly controlled. Studies of glutamate signaling in the magnocellular nuclei of the rat hypothalamus have provided evidence that glial clearance of glutamate is important in synaptic transmission. The magnocellular nuclei undergo a stereotypical retraction of astrocyte processes from synaptic areas in lactating females (Hatton, 2002; Theodosis and Poulain, 1993). Oliet *et al.* described a feedback mechanism for non-lactating animals whereby pharmacological inhibition of glutamate transporters, causing an increase in synaptic glutamate, resulted in decreased transmitter release from presynaptic neurons (Oliet *et al.*, 2001). The same experiment performed in lactating rats yielded little change in transmitter release, suggesting that the glutamate transporters on astrocytes are responsible for clearance and maintenance of presynaptic neurotransmitter release (Oliet *et al.*, 2001). Although other interpretations of this result are possible, these studies were an important attempt to study glial function in a natural, *in vivo* setting.

Evidence that astrocyte glutamate transporters are important for clearance of synaptic glutamate has also come from antisense studies in the rat. Both *in vitro* and *in vivo* administration of antisense oligonucleotides against the GLAST or GLT-1 glial glutamate transporters resulted in elevated glutamate levels. In living rats, such a blockade resulted in neurodegenerative features characteristic of glutamate-induced neurotoxicity and progressive paralysis (Rothstein *et al.*, 1996). Similar results were observed in mice harboring targeted lesions in transporter genes (Tanaka *et al.*, 1997; Watase *et al.*, 1998).

Glial clearance of glycine, a major CNS inhibitory neurotransmitter, from synapses is also important for regulating synaptic activity. Two glycine transporters, GlyT1 and GlyT2, have been identified in mammals (Guastella *et al.*, 1992; Liu *et al.*, 1992, 1993; Smith *et al.*, 1992). Expression studies suggest that the GlyT1 transporter is widely expressed on glia of the CNS, whereas GlyT2 expression is restricted to CNS neurons (Adams *et al.*, 1995; Zafra *et al.*, 1995a,b). Strikingly, targeted disruption of the GlyT1 transporter leads to severe motor and respiratory deficits in newborn homozygous mice (Gomeza *et al.*, 2003), indicating that glycine uptake by glia may be essential for neuronal function. Taken together, the results discussed here suggest that glial uptake of neurotransmitters is essential for proper synaptic activity, raising the possibility that regulated uptake could modulate synaptic function.

In addition to clearance of neurotransmitter using transporters, glia can also release neurotransmitter inhibitors. For example, in the fresh water snail *Lymnaea stagnalis*, a soluble glia-derived protein similar to the acetylcholine receptor (AChR) can bind synaptic acetylcholine. Although mutant animals lacking this AChR mimic have not been described, *in vitro* studies strongly suggest that this protein can modulate synaptic responses (Smit *et al.*, 2001).

Glia may also influence synaptic activity by secretion of neurotransmitters into the synaptic cleft. There is now ample evidence that glutamate is released from astrocytes in CNS slices and *in vitro* (Araque *et al.*, 1998, 2001; Bezzi *et al.*, 1998; Kang *et al.*, 1998; Liu *et al.*, 2004; Parpura *et al.*, 1994). When this release is studied, it is often coupled to the release of Ca^{2+} from intracellular stores within astrocytes (Araque *et al.*, 2001). Recent studies have also begun to elucidate the mechanism by which glutamate is exported out of astrocytes. It seems that a vesicular compartment is involved in release, and that a vesicular glutamate transporter (VGLUT), previously thought to be expressed and functional only in neurons, participates in glutamate release (Bezzi *et al.*, 2004; Montana *et al.*, 2004).

Glia have been observed to synthesize and release other synaptic mediators such as acetylcholine (Heumann *et al.*, 1981; Lan *et al.*, 1996), GABA (Minchin and Iversen, 1974), and ATP (Newman, 2003; Zhang *et al.*, 2003); however, much less is known about the relevance of this release, and whether it also occurs *in vivo*. Additional studies *in vivo*, examining animals deficient in astrocyte-specific neurotransmitter release, should help in assessing the significance of this glial activity.

V. Glial Effects on Neuronal Conduction

In addition to participating in important regulatory events at the synapse, glia also affect the electrical properties of neurons. Perhaps the best-studied example of such regulation is the role of myelin in insulating axons. Many vertebrate CNS and PNS axons are ensheathed by a specialized glial myelin sheath. Ensheathment is punctuated by gaps, called the nodes of Ranvier. In these gaps, the action potential traveling down the neuron is regenerated. The organization of the nodes, as well as specialized paranodal structures, is mediated by specific neuronal–glial interactions. For example, the neuronal proteins contactin and contactin-associated protein interact with the glial membrane protein neurofascin 155 to form the paranodal regions (Charles *et al.*, 2002). Axons that have been demyelinated propagate currents ineffi-ciently, which has been attributed to leakage of current as it proceeds down the axon shaft; thus, glia may play important roles as electrical insulators. However, inefficient conduction could also be a consequence of the dis-organized localization of the voltage-gated sodium channels and other relevant channels that mediate action potential generation (Arroyo *et al.*, 2002; Ulzheimer *et al.*, 2004). Careful measurements of currents along such demyelinated axons could test the validity of this hypothesis.

The extent of myelin ensheathment directly correlates with axonal dia-meter and activity. Signaling between the neuronal factor neuregulin-1 (NRG1), and the ErbB receptor family is important for conveying informa-tion regarding axon thickness to the surrounding myelin (Michailov *et al.*, 2004). Thus, myelinating glia are able to measure axon dimensions and calculate myelin thickness. Myelin thickness, in turn, is a relevant parameter in assessing axonal conduction efficiencies, further supporting the hypothesis that developmental signals between glia and neurons regulate the conductive properties of neurons.

Glia also affect neuronal excitability by regulating the levels of potassium ions that bathe neurons. For example, in Müller glia in the retina, K^+ released by neurons is taken up from the extracellular environment by a host of glial-specific and non-glial-specific K^+ channels. In the eye there is a correlation between the loss of inwardly rectifying K^+ currents in Müller glia and glaucoma (Francke *et al.*, 1997), suggesting that glial regulation of K^+ levels may be a component of the mechanism leading to neuronal loss and dysfunction in this disease.

In addition to indirect regulation of axonal currents by glia, functional coupling between neurons and glia has also been documented. In mammali-an embryonic brain cultures, stimulation of calcium waves in astrocytes can induce current propagation in neurons, suggesting electrical coupling be-tween glia and neurons (Nedergaard, 1994). More direct evidence for such coupling was provided by examination of the rat locus ceruleus (LC) nucle-us. Neurons in this brain region have previously been shown to fire synchro-nously as a result of interneuronal gap junctions. Interestingly, recordings from glia adjacent to synchronously firing LC neurons demonstrated oscil-lating glial membrane potentials that were temporally correlated with neu-ronal firing events (Alvarez-Maubecin *et al.*, 2000). Dye injected into LC glia could be found in LC neurons after sufficiently timed incubations. Further-more, immunoelectron microscopy using antibodies against connexins, the principle components of gap junctions, revealed glial and neuronal connexin immunoreactivity at sites of glia–neuron membrane apposition. Functional studies suggest that glia of the LC reduce neuronal excitability (Alvarez-Maubecin *et al.*, 2000). Thus, it seems that at least in some instances, glia can modulate conductive properties of neurons by direct electrical coupling.

Synchronous firing of neurons can also be achieved by secretion of glutamate from astrocytes. Studies of hippocampal CA1 neurons have shown that glutamate secreted onto these neurons and binding to extrasynaptic NMDA receptors may allow synchronous CA1 firing (Fellin *et al.*, 2004).

The experiments described here suggest that firing properties of neurons can be regulated by glia by both direct electrical and chemical coupling, or by control of myelin development. As with other studies presented here, the consequences of these glial activities in intact animals have not yet been analyzed; however, the identification of specific pathways and molecules involved in these processes should aid in designing the relevant *in vivo* experiments.

VI. Glial Regulation of Neuronal Migration and Process Outgrowth

It has long been hypothesized that glia play important roles in directing neurons and their processes to appropriate locations and targets within the nervous system (Cajal, 1911; Chotard and Salecker, 2004). Pioneering work by Rakic (Rakic, 1971) based principally on static observations of granule cell migration in the developing cerebellum led to the hypothesis that granule neuron migration was guided by glia (Fig. 2). Similar observations suggested that within the cerebral cortex, radial glial cells, which extend processes from the subventricular zone to the pial surface, serve as tracks along which newly generated neurons migrate to reach their destinations (Rakic, 1988). It is now known that glia-guided migration is not the only mechanism of neuronal migration in developing nervous systems (reviewed in Hatten, 2002); nonetheless, it is a major aspect of nervous system development. Studies in which murine cerebellar glia and granule neurons were purified and plated together clearly showed both tight association of neurons with glial fibers and neuronal movement along these fibers (Edmondson and Hatten, 1987). Furthermore, a recent study in which mouse embryos were infected with a retrovirus encoding green fluorescent protein demonstrated not only that radial glia in the cortex give rise to neurons, and thus function as stem cells, but also that neurons generated by these glia proceed to migrate along radial glia fibers *in vivo* (Noctor *et al.*, 2001).

How glia and neurons establish affinity and how migration is executed are important questions for which answers are now emerging. To define the neuronal proteins involved in glial fiber recognition, postnatal cerebellar cells were used to raise antibodies recognizing cell surface moieties. One such immune activity blocked the formation of stable neuron–glia interactions in cultured cells, suggesting that it might recognize a neuronal epitope

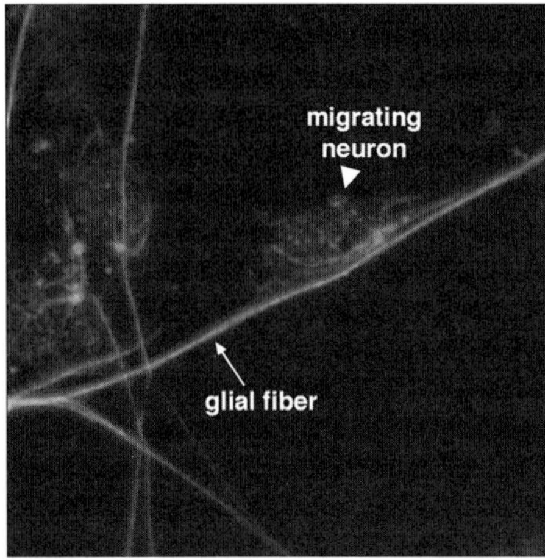

Figure 2   Cerebellar granule cell migrating on a Bergmann glia fiber. Green, ß-tubulin in a neuron, marking the glial fiber track; red, dynein intermediate chain in the nucleus and centrosome of the migrating neuron. (Image courtesy of David Solecki and Mary Beth Hatten.) (See Color Insert.)

essential for adhesion to glial fibers (Edmondson *et al.*, 1988; Fishell and Hatten, 1991). Further studies led to the cloning of a neuronal protein containing multiple extracellular protein-binding domains, termed astrotactin (Zheng *et al.*, 1996). The glial ligand for astrotactin has not yet been determined; however, the brain lipid-binding protein (BLBP) gene may be involved in supporting neuronal migration, since its expression in radial glia throughout the CNS is correlated with neuronal differentiation and migration along glial fibers (Feng and Heintz, 1995; Feng *et al.*, 1994). Additional neuronal components involved in adhesion and cytoskeleton organization have also been extensively defined (Hatten, 2002).

Some studies have suggested that neuronal processes can serve to guide migrating glial cells. For example, time-lapse imaging of migrating glia in zebrafish embryos revealed that these cells are guided by axons of the lateral line neurons. Ablation or misrouting of axons in these embryos prevented glial migration or caused abnormal misrouted migration, strongly indicating that glia follow neuronal tracks (Gilmour *et al.*, 2002). Similar effects on glial migration have also been documented in *Drosophila* embryos in mutants defective in sensory axon extension (Giangrande, 1994).

Glia also seem to be important for directing proper axon outgrowth and pathfinding. For example, the netrin guidance molecules are expressed in glia of *C. elegans* (Wadsworth *et al.*, 1996), *Drosophila* (Jacobs, 2000), and vertebrates (Serafini *et al.*, 1996), suggesting important roles in directing axon pathfinding. Expression of the chemorepellent Slit in midline glia of the *Drosophila* CNS is critical for preventing midline recrossing by axons (Kidd *et al.*, 1999). Guidance proteins produced by glia-like cells of the vertebrate floor plate also act to control midline crossing (Serafini *et al.*, 1996). In *Drosophila*, a subset of neurons in animals lacking the glial cells missing (gcm) gene fail to extend their normal processes (see also Hidalgo *et al.*, 1995), suggesting a possible role for their associated glia (whose processes are closely aligned with the neuronal processes) in process extension (Hosoya *et al.*, 1995; Jones *et al.*, 1995).

In *Drosophila* and vertebrates, once axons have reached the vicinity of their targets, extensive pruning takes place, whereby excessive processes are degraded. Interestingly, in the fruit fly, glia play an active role in this process. Glia infiltrate regions in which pruning will take place and engulf fragmented axonal processes. Temporary inactivation of these glia using a glial-targeted temperature-sensitive mutation in the *Drosophila* homolog of the vesicle pinching protein dynamin transiently prevented pruning, demonstrating an active role for glia in this process (Awasaki and Ito, 2004; Watts *et al.*, 2004).

Although a thorough molecular description of the roles of glia in neuronal and axonal migration is still lacking, recent studies have demonstrated the importance of glia in these processes. The identification of some proteins regulating neuronal movement should serve as an inroad to a more complete description of these glia–neuron interactions.

VII. Reciprocal Control of Cell Survival between Neurons and Glia

As described in the beginning of this chapter, glia are often required, both *in vitro* and *in vivo*, for the survival of the neurons with which they interact. Indeed, primary cultures of neurons are invariably mixed with glia, which presumably provide both trophic and nutritive support. Removal of glia without the addition of specific survival factors results in neuronal death (Meyer-Franke *et al.*, 1995). *In vivo*, loss of the glial sheaths associated with neurons can severely affect neuronal survival. For example, overexpression of a dominant negative form of the ErbB receptor in non-myelinating Schwann cells (cells that ensheath neurons but do not form a myelin sheath) of adult transgenic mice resulted in the death of these Schwann cells, followed by a loss of unmyelinated axons and the subsequent death of sensory neurons (Chen *et al.*, 2003).

In vitro, the ratio of glial cells to neurons required for neuronal survival is generally not stoichiometric. Thus, cultures with 5% glia are sufficient for robust neuronal survival (e.g., Fishell and Hatten, 1991). This observation supports previous assertions that glia-regulated neuronal survival is mediated, at least in part, by soluble factors.

Glial support of neuronal survival has been mainly explored in two areas: nutritive and trophic factor support. In vertebrates, glucose plays a key role as an energy source for cellular metabolism. Both glia and neurons possess the relevant glycolytic enzymes required to break down this sugar and use it for the production of ATP. Thus, it has long been assumed that in vertebrates, neuronal energy is supplied by sugar carried in the blood. Recent studies, however, have challenged this notion, suggesting instead that the primary energy source for active neurons is lactic acid, generated by glia. Interestingly, individual astrocytes in the CNS often contact both blood vessels and neurons and are thus perfectly situated to be energy mediators. The lactate shuttle hypothesis suggests that uptake of glutamate by astrocytes, following a bout of neuronal activity, stimulates astrocytic glycolysis and lactic acid production. Lactic acid, in turn, leaves the astrocyte and is taken up by the adjacent neuron, to be used as an energy source (Pellerin and Magistretti, 1994; Pellerin *et al.*, 1998). Although this hypothesis is attractive, since it allows neurons to tightly couple activity to energy utilization, it remains controversial. Objectors do not refute the idea that lactate could be used as a source of energy; however, they question whether it is the main energy source. Thus, it is possible that glucose is normally taken up directly by neurons, without astrocyte mediation (Chih *et al.*, 2001).

Trophic support of neurons was originally described by a series of now-classic papers by Rita Levi-Montalcini and Viktor Hamburger (Hamburger and Levi-Montalcini, 1949; Levi-Montalcini and Levi, 1943). Their pioneering work led to the discovery of nerve growth factor (NGF) and eventually to a host of other related neurotrophins. In culture, the requirement of glia for neuronal survival can be bypassed by incorporation of neurotrophins, including BDNF, in the culture medium. Furthermore, glia have been shown to express neurotrophin genes in culture (Condorelli *et al.*, 1995; Furukawa *et al.*, 1986; Gonzalez *et al.*, 1990; Yamakuni *et al.*, 1987), suggesting a role in survival of neurons. In *Drosophila*, neurons in animals carrying mutations in the gcm gene eventually die; however, whether death is a direct result of glial loss or a secondary consequence is unclear (Hosoya *et al.*, 1995; Jones *et al.*, 1995). Direct roles for glial-derived survival factors in neuronal survival *in vivo* have not been convincingly demonstrated in any organism.

Roles for neurons in promoting the survival of glia have been surprisingly well established both *in vivo* and *in vitro*. Many glial cell types express neurotrophin receptors (reviewed in Althaus and Richter-Landsberg, 2000), and signaling pathways within these cells in response to trophic factor stimulation have been elaborated (Althaus and Richter-Landsberg, 2000; Heumann, 1994). Perhaps the clearest example of trophic support provided by neurons to glia comes from studies of midline glia in *Drosophila* (Bergmann *et al.*, 2002; Hidalgo *et al.*, 2001). In the *Drosophila* embryo, the midline glia are important for separating and ensheathing commissural axons. Initially, about ten glia are generated in each segment. Eventually, most of these die by apoptosis, leaving approximately three glia per segment. Apoptosis of midline glia is initially prevented by the action of the neuronally generated transforming growth factor (TGF)-α-like ligand SPITZ. SPITZ activates the conserved epidermal growth factor (EGF) signaling pathway within midline glia, resulting in inhibition of the proapoptotic protein HID by phosphorylation. Midline glia apparently compete for a limited amount of SPITZ ligand, so that those receiving little signal eventually die in a head irritation defective (HID)-dependent fashion.

VIII. Genetic and Functional Studies of Glia in the Nematode *Caenorhabditis elegans*

Few functional studies of glia have been conducted in invertebrate animals. Recent studies, however, suggest that the nematode *C. elegans* may serve as an excellent organism from which both functional and molecular insights regarding glial roles in the nervous system may be gained.

Glia in *C. elegans* comprise a group of cells that conform to the criteria outlined in Section II of this chapter. *C. elegans* glia are closely associated with neurons and their processes (Perens and Shaham, submitted; Ward *et al.*, 1975), are not neurons themselves (as assessed by the absence of synaptic or gap junction connectivity to neighboring cells; Ward *et al.*, 1975; White *et al.*, 1986), and are lineally related to neurons. Although cells in the developing *C. elegans* embryo do not form germ layers, the lineage that gives rise to the 959 somatic cells in the adult hermaphrodite is essentially invariant. An examination of the lineal relatives of *C. elegans* glia reveals that sister cells of these are neurons, other glia, or epithelial cells (Sulston *et al.*, 1983), all cells of ectodermal origin in vertebrates. Furthermore, all glia in *C. elegans* extend processes that abut and also ensheath neurons with which they associate. Thus, these cells are highly reminiscent of vertebrate glia.

A. Anatomy

In the adult *C. elegans* hermaphrodite there are 56 glial cells that associate with specific subsets of the 302 neurons of the animal. The 56 glia can be divided into three classes: sheath glia, socket cells, and GLR cells. Sheath glia extend processes that associate with dendritic projections of sensory neurons. These cells ensheath sensory dendrites at the dendritic tip, where a specialized sensory apparatus is localized (Figs. 3 and 4) (Perkins *et al.*, 1986; Ward *et al.*, 1975). The roles of sheath glia in sensory function are discussed in detail below. Of the 24 sheath glia, four cells, the CEP neuron sheath cells, are bipolar, extending both dendrite-associated processes and processes that

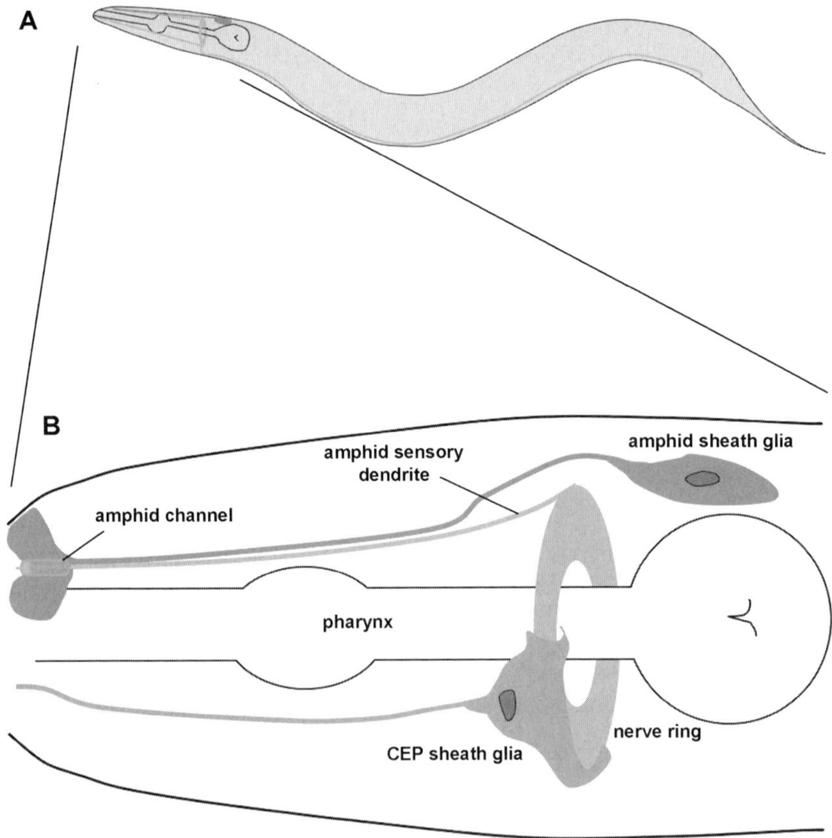

Figure 3   Glia in *C. elegans*. (A) Schematic of a *C. elegans* adult hermaphrodite. Anterior, left; dorsal, up. Major neural tracts (green) and an amphid sheath glia (red) are depicted. The outline of the pharynx of the animal is also shown. (B) Enlarged view of the anterior region. The nerve ring and an amphid channel neuron dendrite (green), the amphid sheath glia and channel (red), and the CEP sheath glia (pink) are depicted. (See Color Insert.)

envelop the nerve ring (Fig. 4), a discrete neuropil composed of many neuronal processes, that is generally viewed as the animal's brain. Most synaptic interactions between neurons occur in the nerve ring (Ware *et al.*, 1975). In addition to enveloping the nerve ring, the CEP sheath glia also send fine processes into the neuropil, where they can be found closely apposed to a small number of synapses (White *et al.*, 1986). The ventral CEP sheath cells express the *C. elegans* netrin UNC-6, suggesting that they may have roles in axon guidance within the nerve ring (Wadsworth *et al.*, 1996).

Twenty-six glia, termed socket cells, run along the terminal portion of sheath cell processes and surround the dendritic tips of a subset of sensory neurons,

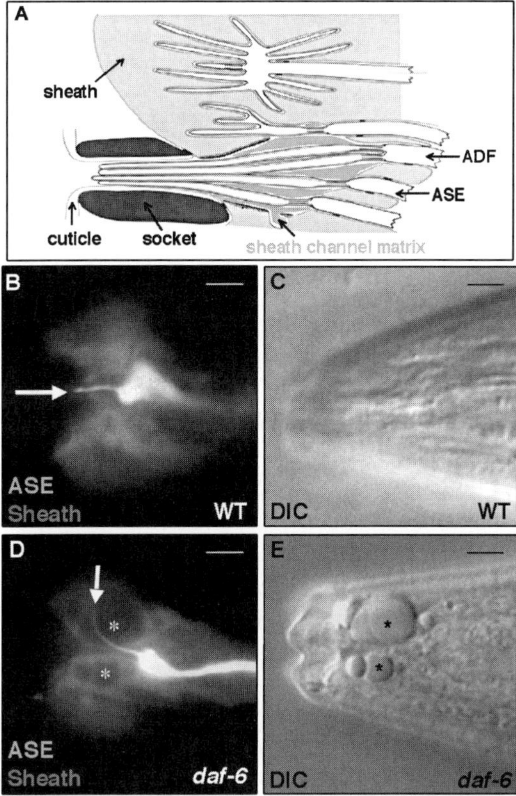

Figure 4 The glial channel of the *C. elegans* amphid sensory organ is closed in *daf-6* mutants. (A) Schematic showing the dendritic tip region of the amphid (adapted from Ward *et al.*, 1975). A representative neuron embedded within the sheath glia is shown. Two channel neurons, ADF and ASE, are labeled, as are the sheath and socket glia. The sheath secretes a matrix into the channel (green). The socket glia secrete cuticle, which is contiguous with the cuticle on the animal's exterior. This image is an enlarged view of the anterior of Figure 3B. (B) Fluorescence image of a wild-type amphid. Sheath (red) and ASE channel neuron (green) are shown. Arrow points to ASE cilium in the amphid channel. (C) differential interference contrast (DIC) image of animal in (B). (D) Fluorescence image of a *daf-6* mutant amphid. Note absence of exposed channel. (E) DIC image of animal in (D). Asterisks indicate vacuoles accumulating within the sheath glia. (See Color Insert.)

anterior to the sheath cell. In some cases socket cells make a pore through which sensory dendrites can access the animal's environment (Fig. 4).

Six glia, termed GLR cells, extend sheet-like projections that contact muscle arms in the head. In some cases GLR extensions have been seen at neuromuscular junctions in the head of the animal (White *et al.*, 1986).

An attractive model for studying the functions of sheath and socket glia is the amphid, the largest sensory organ in *C. elegans*. The amphid is composed of 12 neurons displaying sensory cilia at their dendritic tips, a single sheath glial cell, and a single socket glial cell (Ward *et al.*, 1975). The sheath glia envelopes all 12 neurons at the dendritic tip. Eight of these neurons (termed ADF, ASG, ASE, ASK, ASJ, ASI, ADL, and ASH) extend through a channel, made by the sheath glia, into the contiguous socket glia channel and are exposed to the outside environment through the socket glia pore. Four amphid neurons (AWA, AWB, AWC, and AFD) are fully embedded within the sheath glia at their dendritic tip. Numerous studies of amphid neurons have revealed roles for these cells in chemotaxis, odor sensation, thermotaxis, mechanosensation, avoidance of high osmolarity, and dauer pheromone sensation (Bargmann and Mori, 1997; Driscoll and Kaplan, 1997; Riddle and Albert, 1997). Thus, numerous molecular markers are available for these cells, and alterations in their functions can be easily scored by sensory behavior abnormalities or structural defects in the neurons.

B. Functional Studies

C. elegans has a small number of cells, each of which generally performs unique functions. Thus, eliminating single cells in this organism can be compared to the removal of entire tissues, or even organs, in vertebrates. Laser ablation of *C. elegans* cells has been an effective tool for studying cell function. Using this method, individual cells within live animals can be ablated at various times in embryos and larvae, and developmental and/or behavioral consequences can be assessed in operated animals. To examine the role of the sheath glia in amphid sensory functions, the bilateral sheath glia have been ablated. Anecdotal reports suggested that amphid sheath glia ablation could result in behavioral and developmental deficits (cited in Bargmann *et al.*, 1990; Vowels and Thomas, 1994). Indeed, recent comprehensive studies have conclusively shown that ablation of amphid sheath glia in animals in which the sensory organ has already formed impaired sensory functions of the organ (T. Bacaj and S. Shaham, unpublished results). Thus, amphid sheath glia are essential for proper neuronal sensory functions in this organism.

Examination of amphid neurons following sheath glia ablations revealed that the neurons did not die, but displayed stereotypic morphological abnormalities at the dendritic tip (T. Bacaj and S. Shaham, unpublished results). These observations are interesting in two respects. First, *C. elegans* amphid neurons can live normally in an intact organism in the absence of their associated glia. Thus, issues of neuronal viability, which have made the study of glia in vertebrates difficult (see Section I), are obviated. Second, the results suggest that glia are intimately involved in the maintenance and generation of dendritic ending structure. In vertebrates, dendritic spines are often associated with glia, and it is thus possible that the shape of these receptive ends is also controlled by glia. Indeed, a recent study examining spine morphology in the hippocampus demonstrated that mice lacking the ephrin receptor ephA4 have longer spine lengths than wild-type animals. Interestingly, hippocampal neurons express ephrin A3, and their associated astrocytes express ephA4. Furthermore, soluble ephrin A3 can result in spine length reduction (Murai *et al.*, 2003). Thus, as for *C. elegans* sensory endings, spine morphology may be regulated by glia in the hippocampus.

In addition to roles in morphogenesis of the amphid sensory organ, *C. elegans* glia seem to be important for proper nervous system assembly. Preliminary laser ablation studies of the CEP sheath cells suggest that these cells may be important for assembly and morphogenesis of the *C. elegans* nerve ring (S. Yoshimura and S. Shaham, unpublished observations). Similar studies in *Drosophila* suggest glial roles in axon guidance and fasciculation (Hidalgo *et al.*, 1995).

Studies in *C. elegans* have revealed interesting communication between sheath glia and their associated neurons. The extracellular space surrounding the ciliated endings of amphid neuron dendrites is composed of an electron-dense matrix that is housed in large vesicles within the surrounding sheath glial cell and secreted onto neurons (Ward *et al.*, 1975). Interestingly, mutations that affect cilia formation, such as mutations in genes encoding components of the intraflagellar transport (IFT) system (Sloboda, 2002) or mutations in the *daf-19* gene, which encodes a transcription factor required for cilia formation, result in the accumulation of numerous matrix-laden vesicles within the sheath glia (Perkins *et al.*, 1986). These observations suggest that the sheath glia can monitor and respond to the state of the neurons that they ensheath. Defects in the *che-12* gene suggest that the matrix secreted by sheath glia is important for neuronal function (Perkins *et al.*, 1986; Starich *et al.*, 1995). Animals carrying mutations in *che-12* possess fairly normal glia and neurons, as assessed by ultrastructural studies, yet these animals have profound chemosensory deficits. Furthermore, in wild-type animals, seven amphid neurons have the capacity to take up lipophilic dyes (such as DiI or fluorescein antibody search [FITC]) from the environment. Thus, these dyes can be used as indicators of exposure to the environment or neuronal function (Hedgecock *et al.*, 1985). In *che-12* mutants, these exposed channel neurons show reduced dye uptake (Perkins *et al.*, 1986; Starich *et al.*, 1995). These observations suggest that factors secreted by glia are important for sensory neuron properties and activities. The recent identification of a protein component of the matrix (Perens and Shaham, 2005; Sutherlin *et al.*, 2005) should allow a genetic dissection of this neuron–glia conversation.

Recent work on the amphid sensory organ has also identified important molecular players in generating the glial channel that ensheathes ciliated dendritic endings. Such a channel is formed by glia or specialized epithelia in sensory organs of many animals, including the olfactory, taste, and auditory organs of vertebrates (Burkitt *et al.*, 1993; Jan and Jan, 1993). In *C. elegans*, two genes, *daf-6* and *che-14*, act cooperatively to regulate channel formation (Michaux *et al.*, 2000; Perens and Shaham, 2005). Defects in either *daf-6* or *che-14* result in the inability of the sheath glia to form a channel contiguous with the socket glia channel. As a result, sensory endings of the amphid channel neurons are not exposed to the environment, and animals display profound sensory deficits (Fig. 4). DAF-6 protein is related to the Hedgehog receptor, Patched, and its sequence suggests that it is a member of a sub-family of sterol-sensing domain (SSD)-containing proteins of previously unknown function (Perens and Shaham, 2005). DAF-6 expression is restrict-ed to lumenal structures in *C. elegans*, and the protein is localized to the apical surfaces of all tube classes of the animal. In the amphid glia, DAF-6 function is required early during channel formation, and its expression lasts for only a short time during embryogenesis and the earliest larval stage.

CHE-14 protein is expressed in multiple epithelial cell types, including tubular epithelia. CHE-14 is related to the *Drosophila* Dispatched protein required for Hedgehog secretion (Michaux *et al.*, 2000). *che-14* mutants have abnormal amphid structure and display defects in cuticle structure, suggest-ing a role for *che-14* in the secretion of cuticle components by underlying hypodermal cells (the *C. elegans* equivalents of epidermal cells; Michaux *et al.*, 2000). Interestingly, animals harboring mutations in both *daf-6* and *che-14* exhibit synthetic defects in tube formation in several organs, suggest-ing that the proteins act together in this process. One attractive hypothesis for glia channel formation assigns a role for *daf-6* in inhibiting endo-cytosis from apical glial surfaces and for *che-14* in promoting exocytosis during channel formation (Michaux *et al.*, 2000; Perens and Shaham, 2005). These activities result in net membrane gain surrounding the channel, lead-ing to channel opening and expansion.

The studies of amphid channel formation have also revealed that neuron-ally expressed genes are required for proper channel morphogenesis. Mutants in *daf-19*, a neuronally expressed gene required for cilia formation, possess abnormal amphid channels, displaying irregular shape and size. Furthermore, the glial protein DAF-6 is not properly localized in *daf-19* mutants. Thus, *daf-19*, perhaps through its role in cilia formation, promotes normal glial cell shape.

Although still in its infancy, the study of glia in *C. elegans* has already revealed exciting and essential roles for these cells in the functioning of the nervous system. Continued functional studies using cell-specific ablations and further genetic studies to identify genes required for glia–neuron interactions should yield a rich understanding of the roles that glia play in the *C. elegans* nervous system. The remarkable conservation of many morphological and molecular features between the nervous systems of *C. elegans* and human beings suggests that glial genes and roles identified in the nematode may teach us much regarding glial genes and roles in humans.

IX. Summary

In the mammalian brain, there are roughly five times as many glia as there are neurons, yet glial functions and their mechanisms remain mysterious. The studies described in this chapter suggest that glia play essential and complex roles in regulating nervous system structure and function. Recent interest in these cells as active participants in nervous system behavior has led to the development of a number of important model systems to examine glia both *in vitro* and *in vivo*. Continued studies using these assay systems, as well as tractable *in vivo* genetic models, should help to elucidate the roles of these remarkable cells.

Acknowledgments

I thank members of my laboratory for helpful comments and discussions concerning this chapter and Mary Beth Hatten and Ben Barres for contributing images. I apologize to those whose work was not cited here due either to oversight on my part or to space constraints.

References

Adams, R. H., Sato, K., Shimada, S., Tohyama, M., Puschel, A. W., and Betz, H. (1995). Gene structure and glial expression of the glycine transporter GlyT1 in embryonic and adult rodents. *J. Neurosci.* **15,** 2524–2532.

Althaus, H. H., and Richter-Landsberg, C. (2000). Glial cells as targets and producers of neurotrophins. *Int. Rev. Cytol.* **197,** 203–277.

Alvarez-Buylla, A., Seri, B., and Doetsch, F. (2002). Identification of neural stem cells in the adult vertebrate brain. *Brain Res. Bull.* **57,** 751–758.

Alvarez-Maubecin, V., Garcia-Hernandez, F., Williams, J. T., and Van Bockstaele, E. J. (2000). Functional coupling between neurons and glia. *J. Neurosci.* **20,** 4091–4098.

Araque, A., Parpura, V., Sanzgiri, R. P., and Haydon, P. G. (1998). Glutamate-dependent astrocyte modulation of synaptic transmission between cultured hippocampal neurons. *Eur. J. Neurosci.* **10,** 2129–2142.

Araque, A., Carmignoto, G., and Haydon, P. G. (2001). Dynamic signaling between astrocytes and neurons. *Annu. Rev. Physiol.* **63,** 795–813.

Arroyo, E. J., Xu, T., Grinspan, J., Lambert, S., Levinson, S. R., Brophy, P. J., Peles, E., and Scherer, S. S. (2002). Genetic dysmyelination alters the molecular architecture of the nodal region. *J. Neurosci.* **22,** 1726–1737.

Awasaki, T., and Ito, K. (2004). Engulfing action of glial cells is required for programmed axon pruning during *Drosophila* metamorphosis. *Curr. Biol.* **14,** 668–677.

Bargmann, C. I. (1998). Neurobiology of the *Caenorhabditis elegans* genome. *Science* **282,** 2028–2033.

Bargmann, C. I., and Mori, I. (1997). Chemotaxis and thermotaxis. *In* "*C. elegans* II" (D. L. Riddle, T. Blumenthal, B. J. Meyer, and J. R. Priess, Eds.), pp. 717–737. Cold Spring Harbor Laboratory Press, Cold Spring Harbor, NY.

Bargmann, C. I., Thomas, J. H., and Horvitz, H. R. (1990). Chemosensory cell function in the behavior and development of *Caenorhabditis elegans. Cold Spring Harb. Symp. Quant. Biol.* **55,** 529–538.

Barlow, L. A. (2003). Toward a unified model of vertebrate taste bud development. *J. Comp. Neurol.* **457,** 107–110.

Bergles, D. E., and Jahr, C. E. (1997). Synaptic activation of glutamate transporters in hippocampal astrocytes. *Neuron* **19,** 1297–1308.

Bergmann, A., Tugentman, M., Shilo, B. Z., and Steller, H. (2002). Regulation of cell number by MAPK-dependent control of apoptosis: a mechanism for trophic survival signaling. *Dev. Cell* **2,** 159–170.

Bezzi, P., Carmignoto, G., Pasti, L., Vesce, S., Rossi, D., Rizzini, B. L., Pozzan, T., and Volterra, A. (1998). Prostaglandins stimulate calcium-dependent glutamate release in astrocytes. *Nature* **391,** 281–285.

Bezzi, P., Gundersen, V., Galbete, J. L., Seifert, G., Steinhauser, C., Pilati, E., and Volterra, A. (2004). Astrocytes contain a vesicular compartment that is competent for regulated exocytosis of glutamate. *Nat. Neurosci.* **7,** 613–620.

Bourque, M. J., and Robitaille, R. (1998). Endogenous peptidergic modulation of perisynaptic Schwann cells at the frog neuromuscular junction. *J. Physiol.* **512**(Pt. 1), 197–209.

Burkitt, H. G., Young, B., and Heath, J. W. (1993). "Wheater's Functional Histology," Special sense organs, pp. 374–399. Churchill Livingstone Press, Edinburgh, UK.

Cajal, R. S. (1911). "Histology of the Nervous System," Neuroglia, pp. 202–203. Oxford University Press, New York.

Cajal, R. S. (1913). Sobre un nuevo proceder de impregnacion de la neuroglia y sus resultados en los centros nervioses del hombre y animales. *Trab. Lab. Invest. Biol. Univ. Madr.* **14,** 155–162.

Charles, P., Tait, S., Faivre-Sarrailh, C., Barbin, G., Gunn-Moore, F., Denisenko-Nehrbass, N., Guennoc, A. M., Girault, J. A., Brophy, P. J., and Lubetzki, C. (2002). Neurofascin is a glial receptor for the paranodin/Caspr-contactin axonal complex at the axoglial junction. *Curr. Biol.* **12,** 217–220.

Chen, S., Rio, C., Ji, R. R., Dikkes, P., Coggeshall, R. E., Woolf, C. J., and Corfas, G. (2003). Disruption of ErbB receptor signaling in adult non-myelinating Schwann cells causes progressive sensory loss. *Nat. Neurosci.* **6,** 1186–1193.

Chih, C. P., Lipton, P., and Roberts, E. L., Jr. (2001). Do active cerebral neurons really use lactate rather than glucose? *Trends Neurosci.* **24,** 573–578.

Chotard, C., and Salecker, I. (2004). Neurons and glia: Team players in axon guidance. *Trends Neurosci.* **27,** 655–661.

Colomar, A., and Robitaille, R. (2004). Glial modulation of synaptic transmission at the neuromuscular junction. *Glia* **47,** 284–289.

Condorelli, D. F., Salin, T., Dell' Albani, P., Mudo, G., Corsaro, M., Timmusk, T., Metsis, M., and Belluardo, N. (1995). Neurotrophins and their trk receptors in cultured cells of the glial lineage and in white matter of the central nervous system. *J. Mol. Neurosci.* **6,** 237–248.

Couteaux, R., and Pecot-Dechavassine, M. (1974). Specialized areas of presynaptic membranes. *C R Acad. Sci. Hebd. Seances Acad. Sci. D* **278,** 291–293.

Dahl, D., Crosby, C. J., Sethi, J. S., and Bignami, A. (1985). Glial fibrillary acidic (GFA) protein in vertebrates: Immunofluorescence and immunoblotting study with monoclonal and polyclonal antibodies. *J. Comp. Neurol.* **239,** 75–88.

Doetsch, F. (2003). The glial identity of neural stem cells. *Nat. Neurosci.* **6,** 1127–1134.

Driscoll, M., and Kaplan, J. (1997). Mechanotransduction. *In* "*C. elegans* II," pp. 645–677. Cold Spring Harbor Laboratory Press, Cold Spring Harbor, NY.

Edmondson, J. C., and Hatten, M. E. (1987). Glial-guided granule neuron migration *in vitro*: A high-resolution time-lapse video microscopic study. *J. Neurosci.* **7,** 1928–1934.

Edmondson, J. C., Liem, R. K., Kuster, J. E., and Hatten, M. E. (1988). Astrotactin: A novel neuronal cell surface antigen that mediates neuron-astroglial interactions in cerebellar microcultures. *J. Cell Biol.* **106,** 505–517.

Eng, L., Gerstl, B., and Vanderhaeghen, J. (1970). A study of proteins in old multiple sclerosis plaques. *Trans. Am. Soc. Neurochem.* **1,** 42.

Eng, L. F., Vanderhaeghen, J. J., Bignami, A., and Gerstl, B. (1971). An acidic protein isolated from fibrous astrocytes. *Brain Res.* **28,** 351–354.

Fellin, T., Pascual, O., Gobbo, S., Pozzan, T., Haydon, P. G., and Carmignoto, G. (2004). Neuronal synchrony mediated by astrocytic glutamate through activation of extrasynaptic NMDA receptors. *Neuron* **43,** 729–743.

Feng, L., and Heintz, N. (1995). Differentiating neurons activate transcription of the brain lipid-binding protein gene in radial glia through a novel regulatory element. *Development* **121,** 1719–1730.

Feng, L., Hatten, M. E., and Heintz, N. (1994). Brain lipid-binding protein (BLBP): A novel signaling system in the developing mammalian CNS. *Neuron* **12,** 895–908.

Fishell, G., and Hatten, M. E. (1991). Astrotactin provides a receptor system for CNS neuronal migration. *Development* **113,** 755–765.

Francke, M., Pannicke, T., Biedermann, B., Faude, F., Wiedemann, P., Reichenbach, A., and Reichelt, W. (1997). Loss of inwardly rectifying potassium currents by human retinal glial cells in diseases of the eye. *Glia* **20,** 210–218.

Franklin, R. J. (2002). Why does remyelination fail in multiple sclerosis? *Nat. Rev. Neurosci.* **3,** 705–714.

Furukawa, S., Furukawa, Y., Satoyoshi, E., and Hayashi, K. (1986). Synthesis and secretion of nerve growth factor by mouse astroglial cells in culture. *Biochem. Biophys. Res. Commun.* **136,** 57–63.

Giangrande, A. (1994). Glia in the fly wing are clonally related to epithelial cells and use the nerve as a pathway for migration. *Development* **120,** 523–534.

Gilmour, D. T., Maischein, H. M., and Nusslein-Volhard, C. (2002). Migration and function of a glial subtype in the vertebrate peripheral nervous system. *Neuron* **34,** 577–588.

Gomeza, J., Hulsmann, S., Ohno, K., Eulenburg, V., Szoke, K., Richter, D., and Betz, H. (2003). Inactivation of the glycine transporter 1 gene discloses vital role of glial glycine uptake in glycinergic inhibition. *Neuron* **40,** 785–796.

Gonzalez, D., Dees, W. L., Hiney, J. K., Ojeda, S. R., and Saneto, R. P. (1990). Expression of beta-nerve growth factor in cultured cells derived from the hypothalamus and cerebral cortex. *Brain Res.* **511,** 249–258.

Goodman, M. B., Hall, D. H., Avery, L., and Lockery, S. R. (1998). Active currents regulate sensitivity and dynamic range in *C. elegans* neurons. *Neuron* **20,** 763–772.

Guastella, J., Brecha, N., Weigmann, C., Lester, H. A., and Davidson, N. (1992). Cloning, expression, and localization of a rat brain high-affinity glycine transporter. *Proc. Natl. Acad. Sci. USA* **89,** 7189–7193.

Hamburger, V., and Levi-Montalcini, R. (1949). Proliferation, differentiation and degeneration in the spinal ganglia of the chick embryo under normal and experimental conditions. *J. Exp. Zool.* **111,** 457–502.

Hatten, M. E. (2002). New directions in neuronal migration. *Science* **297,** 1660–1663.

Hatton, G. I. (2002). Glial-neuronal interactions in the mammalian brain. *Adv. Physiol. Educ.* **26,** 225–237.

Hedgecock, E. M., Culotti, J. G., Thomson, J. N., and Perkins, L. A. (1985). Axonal guidance mutants of *Caenorhabditis elegans* identified by filling sensory neurons with fluorescein dyes. *Dev. Biol.* **111,** 158–170.

Herrera, A. A., Qiang, H., and Ko, C. P. (2000). The role of perisynaptic Schwann cells in development of neuromuscular junctions in the frog (*Xenopus laevis*). *J. Neurobiol.* **45,** 237–254.

Heumann, R. (1994). Neurotrophin signalling. *Curr. Opin. Neurobiol.* **4,** 668–679.

Heumann, R., Villegas, J., and Herzfeld, D. W. (1981). Acetylcholine synthesis in the Schwann cell and axon in the giant nerve fiber of the squid. *J. Neurochem.* **36,** 765–768.

Hidalgo, A., Urban, J., and Brand, A. H. (1995). Targeted ablation of glia disrupts axon tract formation in the *Drosophila* CNS. *Development* **121,** 3703–3712.

Hidalgo, A., Kinrade, E. F., and Georgiou, M. (2001). The *Drosophila* neuregulin vein maintains glial survival during axon guidance in the CNS. *Dev. Cell* **1,** 679–690.

Hirata, K., Zhou, C., Nakamura, K., and Kawabuchi, M. (1997). Postnatal development of Schwann cells at neuromuscular junctions, with special reference to synapse elimination. *J. Neurocytol.* **26,** 799–809.

Hosoya, T., Takizawa, K., Nitta, K., and Hotta, Y. (1995). Glial cells missing: A binary switch between neuronal and glial determination in *Drosophila. Cell* **82,** 1025–1036.

Huang, Y. H., and Bergles, D. E. (2004). Glutamate transporters bring competition to the synapse. *Curr. Opin. Neurobiol.* **14,** 346–352.

Jacobs, J. R. (2000). The midline glia of *Drosophila*: A molecular genetic model for the developmental functions of glia. *Prog. Neurobiol.* **62,** 475–508.

Jahromi, B. S., Robitaille, R., and Charlton, M. P. (1992). Transmitter release increases intracellular calcium in perisynaptic Schwann cells *in situ. Neuron* **8,** 1069–1077.

Jan, Y. N., and Jan, L. Y. (1993). The peripheral nervous system. *In* "The Development of *Drosophila Melanogaster,*" pp. 1207–1244. Cold Spring Harbor Laboratory Press, Cold Spring Harbor, NY.

Jones, B. W. (2001). Glial cell development in the *Drosophila* embryo. *Bioessays* **23,** 877–887.

Jones, B. W., Fetter, R. D., Tear, G., and Goodman, C. S. (1995). Glial cells missing: A genetic switch that controls glial versus neuronal fate. *Cell* **82,** 1013–1023.

Kang, J., Jiang, L., Goldman, S. A., and Nedergaard, M. (1998). Astrocyte-mediated potentiation of inhibitory synaptic transmission. *Nat. Neurosci.* **1,** 683–692.

Kelly, A. M., and Zacks, S. I. (1969). The fine structure of motor endplate morphogenesis. *J. Cell Biol.* **42,** 154–169.

Kidd, T., Bland, K. S., and Goodman, C. S. (1999). Slit is the midline repellent for the robo receptor in *Drosophila. Cell* **96,** 785–794.

Lan, C. T., Shieh, J. Y., Wen, C. Y., Tan, C. K., and Ling, E. A. (1996). Ultrastructural localization of acetylcholinesterase and choline acetyltransferase in oligodendrocytes, glioblasts and vascular endothelial cells in the external cuneate nucleus of the gerbil. *Anat. Embryol. (Berl)* **194,** 177–185.

Le Douarin, N. M., and Dupin, E. (2003). Multipotentiality of the neural crest. *Curr. Opin. Genet. Dev.* **13,** 529–536.

Le Douarin, N., Dulac, C., Dupin, E., and Cameron-Curry, P. (1991). Glial cell lineages in the neural crest. *Glia* **4,** 175–184.

Levi-Montalcini, R., and Levi, G. (1943). Recherches quantitatives sur la marche du processus de différenciation des neurons dans les ganglions spinaux de l'embryon de poulet. *Arch. Biol. Liege* **54**, 183–206.

Liu, Q. R., Lopez-Corcuera, B., Mandiyan, S., Nelson, H., and Nelson, N. (1993). Cloning and expression of a spinal cord- and brain-specific glycine transporter with novel structural features. *J. Biol. Chem.* **268**, 22802–22808.

Liu, Q. R., Nelson, H., Mandiyan, S., Lopez-Corcuera, B., and Nelson, N. (1992). Cloning and expression of a glycine transporter from mouse brain. *FEBS Lett.* **305**, 110–114.

Liu, Q. S., Xu, Q., Arcuino, G., Kang, J., and Nedergaard, M. (2004). Astrocyte-mediated activation of neuronal kainate receptors. *Proc. Natl. Acad. Sci. USA* **101**, 3172–3177.

Lockery, S. R., and Goodman, M. B. (1998). Tight-seal whole-cell patch clamping of *Caenorhabditis elegans* neurons. *Methods Enzymol.* **293**, 201–217.

Mauch, D. H., Nagler, K., Schumacher, S., Goritz, C., Muller, E. C., Otto, A., and Pfrieger, F. W. (2001). CNS synaptogenesis promoted by glia-derived cholesterol. *Science* **294**, 1354–1357.

McBean, G. J. (1994). Inhibition of the glutamate transporter and glial enzymes in rat striatum by the gliotoxin, alpha aminoadipate. *Br. J. Pharmacol.* **113**, 536–540.

Meyer-Franke, A., Kaplan, M. R., Pfrieger, F. W., and Barres, B. A. (1995). Characterization of the signaling interactions that promote the survival and growth of developing retinal ganglion cells in culture. *Neuron* **15**, 805–819.

Michailov, G. V., Sereda, M. W., Brinkmann, B. G., Fischer, T. M., Haug, B., Birchmeier, C., Role, L., Lai, C., Schwab, M. H., and Nave, K. A. (2004). Axonal neuregulin-1 regulates myelin sheath thickness. *Science* **304**, 700–703.

Michaux, G., Gansmuller, A., Hindelang, C., and Labouesse, M. (2000). CHE-14, a protein with a sterol-sensing domain, is required for apical sorting in *C. elegans* ectodermal epithelial cells. *Curr. Biol.* **10**, 1098–1107.

Minchin, M. C., and Iversen, L. L. (1974). Release of (^{3}H)gamma-aminobutyric acid from glial cells in rat dorsal root ganglia. *J. Neurochem.* **23**, 533–540.

Minelli, A., Brecha, N. C., Karschin, C., DeBiasi, S., and Conti, F. (1995). GAT-1, a high-affinity GABA plasma membrane transporter, is localized to neurons and astroglia in the cerebral cortex. *J. Neurosci.* **15**, 7734–7746.

Minelli, A., DeBiasi, S., Brecha, N. C., Zuccarello, L. V., and Conti, F. (1996). GAT-3, a high-affinity GABA plasma membrane transporter, is localized to astrocytic processes, and it is not confined to the vicinity of GABAergic synapses in the cerebral cortex. *J. Neurosci.* **16**, 6255–6264.

Montana, V., Ni, Y., Sunjara, V., Hua, X., and Parpura, V. (2004). Vesicular glutamate transporter-dependent glutamate release from astrocytes. *J. Neurosci.* **24**, 2633–2642.

Murai, K. K., Nguyen, L. N., Irie, F., Yamaguchi, Y., and Pasquale, E. B. (2003). Control of hippocampal dendritic spine morphology through ephrin-A3/EphA4 signaling. *Nat. Neurosci.* **6**, 153–160.

Myers, A. J., and Goate, A. M. (2001). The genetics of late-onset Alzheimer's disease. *Curr. Opin. Neurol.* **14**, 433–440.

Nagler, K., Mauch, D. H., and Pfrieger, F. W. (2001). Glia-derived signals induce synapse formation in neurones of the rat central nervous system. *J. Physiol.* **533**, 665–679.

Nedergaard, M. (1994). Direct signaling from astrocytes to neurons in cultures of mammalian brain cells. *Science* **263**, 1768–1771.

Newman, E. A. (2003). Glial cell inhibition of neurons by release of ATP. *J. Neurosci.* **23**, 1659–1666.

Newswanger, D. L., and Warren, C. R. (2004). Guillain-Barre syndrome. *Am. Fam. Physician* **69**, 2405–2410.

Noctor, S. C., Flint, A. C., Weissman, T. A., Dammerman, R. S., and Kriegstein, A. R. (2001). Neurons derived from radial glial cells establish radial units in neocortex. *Nature* **409,** 714–720.

Oliet, S. H., Piet, R., and Poulain, D. A. (2001). Control of glutamate clearance and synaptic efficacy by glial coverage of neurons. *Science* **292,** 923–926.

Parpura, V., Basarsky, T. A., Liu, F., Jeftinija, K., Jeftinija, S., and Haydon, P. G. (1994). Glutamate-mediated astrocyte-neuron signalling. *Nature* **369,** 744–747.

Pellerin, L., and Magistretti, P. J. (1994). Glutamate uptake into astrocytes stimulates aerobic glycolysis: A mechanism coupling neuronal activity to glucose utilization. *Proc. Natl. Acad. Sci. USA* **91,** 10625–10629.

Pellerin, L., Pellegri, G., Bittar, P. G., Charnay, Y., Bouras, C., Martin, J. L., Stella, N., and Magistretti, P. J. (1998). Evidence supporting the existence of an activity-dependent astrocyte-neuron lactate shuttle. *Dev. Neurosci.* **20,** 291–299.

Peng, H. B., Yang, J. F., Dai, Z., Lee, C. W., Hung, H. W., Feng, Z. H., and Ko, C. P. (2003). Differential effects of neurotrophins and schwann cell-derived signals on neuronal survival/ growth and synaptogenesis. *J. Neurosci.* **23,** 5050–5060.

Perens, E., and Shaham, S. (2005). *C. elegans daf-6* encodes a Patched-related protein required for lumen formation. *Dev. Cell.* **8,** 893–906.

Perkins, L. A., Hedgecock, E. M., Thomson, J. N., and Culotti, J. G. (1986). Mutant sensory cilia in the nematode *Caenorhabditis elegans. Dev. Biol.* **117,** 456–487.

Peters, A., Palay, S. L., and Webster, H. D. (1991). The neuroglial cells. *In* "The Fine Structure of the Nervous System," pp. 273–311. Oxford University Press, New York.

Pfrieger, F. W., and Barres, B. A. (1997). Synaptic efficacy enhanced by glial cells *in vitro. Science* **277,** 1684–1687.

Plante-Bordeneuve, V., and Said, G. (2002). Dejerine-Sottas disease and hereditary demyelinating polyneuropathy of infancy. *Muscle Nerve* **26,** 608–621.

Rakic, P. (1971). Neuron-glia relationship during granule cell migration in developing cerebellar cortex. A Golgi and electronmicroscopic study in Macacus Rhesus. *J. Comp. Neurol.* **141,** 283–312.

Rakic, P. (1988). Specification of cerebral cortical areas. *Science* **241,** 170–176.

Reddy, L. V., Koirala, S., Sugiura, Y., Herrera, A. A., and Ko, C. P. (2003). Glial cells maintain synaptic structure and function and promote development of the neuromuscular junction *in vivo. Neuron* **40,** 563–580.

Reist, N. E., and Smith, S. J. (1992). Neurally evoked calcium transients in terminal Schwann cells at the neuromuscular junction. *Proc. Natl. Acad. Sci. USA* **89,** 7625–7629.

Riddle, D. L., and Albert, P. S. (1997). Genetic and environmental regulation of dauer larva development. *In* "*C. elegans* II," pp. 739–768. Cold Spring Harbor Laboratory Press, Cold Spring Harbor, NY.

Robitaille, R. (1998). Modulation of synaptic efficacy and synaptic depression by glial cells at the frog neuromuscular junction. *Neuron* **21,** 847–855.

Rochon, D., Rousse, I., and Robitaille, R. (2001). Synapse-glia interactions at the mammalian neuromuscular junction. *J. Neurosci.* **21,** 3819–3829.

Rothstein, J. D., Martin, L., Levey, A. I., Dykes-Hoberg, M., Jin, L., Wu, D., Nash, N., and Kuncl, R. W. (1994). Localization of neuronal and glial glutamate transporters. *Neuron* **13,** 713–725.

Rothstein, J. D., Dykes-Hoberg, M., Pardo, C. A., Bristol, L. A., Jin, L., Kuncl, R. W., Kanai, Y., Hediger, M. A., Wang, Y., Schielke, J. P., *et al.* (1996). Knockout of glutamate transporters reveals a major role for astroglial transport in excitotoxicity and clearance of glutamate. *Neuron* **16,** 675–686.

Serafini, T., Colamarino, S. A., Leonardo, E. D., Wang, H., Beddington, R., Skarnes, W. C., and Tessier-Lavigne, M. (1996). Netrin-1 is required for commissural axon guidance in the developing vertebrate nervous system. *Cell* **87,** 1001–1014.

Sloboda, R. D. (2002). A healthy understanding of intraflagellar transport. *Cell Motil. Cytoskeleton* **52**, 1–8.

Smit, A. B., Syed, N. I., Schaap, D., van Minnen, J., Klumperman, J., Kits, K. S., Lodder, H., van der Schors, R. C., van Elk, R., Sorgedrager, B., Brojc, K., Sixma, T. K., Suit, A. B., and Geraerts, W. P. (2001). A glia-derived acetylcholine-binding protein that modulates synaptic transmission. *Nature* **411**, 261–268.

Smith, K. E., Borden, L. A., Hartig, P. R., Branchek, T., and Weinshank, R. L. (1992). Cloning and expression of a glycine transporter reveal colocalization with NMDA receptors. *Neuron* **8**, 927–935.

Spacek, J. (1985). Three-dimensional analysis of dendritic spines. III. Glial sheath. *Anat. Embryol. (Berl)* **171**, 245–252.

Starich, T. A., Herman, R. K., Kari, C. K., Yeh, W. H., Schackwitz, W. S., Schuyler, M. W., Collet, J., Thomas, J. H., and Riddle, D. L. (1995). Mutations affecting the chemosensory neurons of *Caenorhabditis elegans*. *Genetics* **139**, 171–188.

Sulston, J. E., Schierenberg, E., White, J. G., and Thomson, J. N. (1983). The embryonic cell lineage of the nematode *Caenorhabditis elegans*. *Dev. Biol.* **100**, 64–119.

Sutherlin, M., Westlund, B., Burnam, L., Sluder, A., and Liu, L. (2001). Characterization of two Caenorhabditis elegans venum allergen-related proteins. *Mol. Thel.* Keystone Symp. Taos, New Mexico.

Tanaka, K., Watase, K., Manabe, T., Yamada, K., Watanabe, M., Takahashi, K., Iwama, H., Nishikawa, T., Ichihara, N., Kikuchi, T., Okuyama, S., Kawashima, N., Hori, S., Takimoto, M., and Wada, K. (1997). Epilepsy and exacerbation of brain injury in mice lacking the glutamate transporter GLT-1. *Science* **276**, 1699–1702.

Theodosis, D. T., and Poulain, D. A. (1993). Activity-dependent neuronal-glial and synaptic plasticity in the adult mammalian hypothalamus. *Neuroscience* **57**, 501–535.

Ullian, E. M., Sapperstein, S. K., Christopherson, K. S., and Barres, B. A. (2001). Control of synapse number by glia. *Science* **291**, 657–661.

Ullian, E. M., Christopherson, K. S., and Barres, B. A. (2004). Role for glia in synaptogenesis. *Glia* **47**, 209–216.

Ullian, E. M., Harris, B. T., Wu, A., Chan, J. R., and Barres, B. A. (2004). Schwann cells and astrocytes induce synapse formation by spinal motor neurons in culture. *Mol. Cell Neurosci.* **25**, 241–251.

Ulzheimer, J. C., Peles, E., Levinson, S. R., and Martini, R. (2004). Altered expression of ion channel isoforms at the node of Ranvier in P0-deficient myelin mutants. *Mol. Cell Neurosci.* **25**, 83–94.

Ventura, R., and Harris, K. M. (1999). Three-dimensional relationships between hippocampal synapses and astrocytes. *J. Neurosci.* **19**, 6897–6906.

Virchow, R. (1846). Ueber das granulierte Ansehen der Wandungen der Gehirnventrikel. *Allg. Z. Psychiatr.* **3**, 424–450.

Vowels, J. J., and Thomas, J. H. (1994). Multiple chemosensory defects in *daf-11* and *daf-21* mutants of *Caenorhabditis elegans*. *Genetics* **138**, 303–316.

Wadsworth, W. G., Bhatt, H., and Hedgecock, E. M. (1996). Neuroglia and pioneer neurons express UNC-6 to provide global and local netrin cues for guiding migrations in *C. elegans*. *Neuron* **16**, 35–46.

Ward, S., Thomson, N., White, J. G., and Brenner, S. (1975). Electron microscopical reconstruction of the anterior sensory anatomy of the nematode *Caenorhabditis elegans*. *J. Comp. Neurol.* **160**, 313–337.

Ware, R. S., Clark, D., Crossland, K., and Russell, R. L. (1975). The nerve ring of the nematode *Caenorhabditis elegans*: Sensory input and motor output. *J. Comp. Neur.* **162**, 71–110.

Watase, K., Hashimoto, K., Kano, M., Yamada, K., Watanabe, M., Inoue, Y., Okuyama, S., Sakagawa, T., Ogawa, S., Kawashima, N., Hori, S., Takimoto, M., Wada, K., and Tanaka, K. (1998). Motor discoordination and increased susceptibility to cerebellar injury in GLAST mutant mice. *Eur J. Neurosci.* **10,** 976–988.

Watts, R. J., Schuldiner, O., Perrino, J., Larsen, C., and Luo, L. (2004). Glia engulf degenerating axons during developmental axon pruning. *Curr. Biol.* **14,** 678–684.

White, J. G., Southgate, E., Thomson, J. N., and Brenner, S. (1986). The structure of the nervous system of the nematode *Caenorhabditis elegans*. *Phil. Trans. R. Soc. Lond. B* **314,** 1–340.

Wolff, J. R. (1976). The morphological organization of cortical neuroglia. *In* "Handbook of Electroencephalography and Clinical Neurophysiology," pp. 26–43. Elsevier, Amsterdam.

Yamakuni, T., Ozawa, F., Hishinuma, F., Kuwano, R., Takahashi, Y., and Amano, T. (1987). Expression of beta-nerve growth factor mRNA in rat glioma cells and astrocytes from rat brain. *FEBS Lett.* **223,** 117–121.

Zafra, F., Aragon, C., Olivares, L., Danbolt, N. C., Gimenez, C., and Storm-Mathisen, J. (1995). Glycine transporters are differentially expressed among CNS cells. *J. Neurosci.* **15,** 3952–3969.

Zafra, F., Gomeza, J., Olivares, L., Aragon, C., and Gimenez, C. (1995). Regional distribution and developmental variation of the glycine transporters GLYT1 and GLYT2 in the rat CNS. *Eur J. Neurosci.* **7,** 1342–1352.

Zhang, J. M., Wang, H. K., Ye, C. Q., Ge, W., Chen, Y., Jiang, Z. L., Wu, C. P., Poo, M. M., and Duan, S. (2003). ATP released by astrocytes mediates glutamatergic activity-dependent heterosynaptic suppression. *Neuron* **40,** 971–982.

Zheng, C., Heintz, N., and Hatten, M. E. (1996). CNS gene encoding astrotactin, which supports neuronal migration along glial fibers. *Science* **272,** 417–419.

Zucker, R. S., and Regehr, W. G. (2002). Short-term synaptic plasticity. *Annu. Rev. Physiol.* **64,** 355–405.

4

The Novel Roles of Glial Cells Revisited: The Contribution of Radial Glia and Astrocytes to Neurogenesis

Tetsuji Mori, Annalisa Buffo, and Magdalena Götz
Institute for Stem Cell Research, GSF-National Research Center
for Environment and Health, D-85764 Neuherberg/Munich, Germany

Astroglial cells are the most frequent cell type in the adult mammalian brain, and the number and range of their diverse functions are still increasing. One of their most striking roles is their function as adult neural stem cells and contribution to neurogenesis. This chapter discusses first the role of the ubiquitous glial cell type in the developing nervous system, the radial glial cells. Radial glial cells share several features with neuroepithelial cells, but also with astrocytes in the mature brain, which led to the name "radial glia." At the end of neurogenesis in the mammalian brain, radial glial cells disappear, and a subset of them transforms into astroglial cells. Interestingly, only some astrocytes maintain their neurogenic potential and continue to generate neurons throughout life. We discuss the current knowledge about the differences between the adult astroglial cells that remain neurogenic and act as neural stem cells and the majority of other astroglial cells that have apparently lost the capacity to generate neurons. Additionally, we review the changes in glial cells upon brain lesion, their dedifferentiation and recapitulation of radial glial properties, and the conditions under which reactive glia may reinitiate some neurogenic potential. Given that the astroglial cells are not only the most frequent cell type in an adult mammalian brain, but also the key cell type in the

Current Topics in Developmental Biology, Vol. 69
Copyright 2005, Elsevier Inc. All rights reserved.
0070-2153/05 $35.00
DOI: 10.1016/S0070-2153(05)69004-7

wound reaction of the brain to injury, it is essential to further understand their heterogeneity and molecular specification, with the final aim of using this unique source for neuronal replacement. Therefore, one of the key advances in the field of neurobiology is the discovery that astroglial cells can generate neurons not only during development, but also throughout adult life and potentially even after brain lesion. © 2005, Elsevier Inc.

I. Definition of Radial Glia

A. Similarities and Differences between Radial Glia and Astrocytes

The ubiquitous glial cells in the developing brain have a radial morphology: their somata are located in the ventricular zone (VZ) and extend two processes, one basally with attachment to the basal surface underneath the meninges and the other contacting the ventricle, the apical surface (Bentivoglio and Mazzarello, 1999; Cameron and Rakic, 1991) (Fig. 1), where cells are connected by adherence junctions (Aaku-Saraste *et al.*, 1996; Mollgard *et al.*, 1987; Shoukimas and Hinds, 1978) (Fig. 1). Radial glial cells express a variety of molecules that are characteristic for astrocytes at later stages, such as the astrocyte-specific L-glutamate/L-aspartate transporter (GLAST) (Hartfuss *et al.*, 2001; Malatesta *et al.*, 2000), the glutamine synthetase (GS) (Akimoto *et al.*, 1993), the β subunit of calcium-binding protein S100 (S100β) (Fig. 3), the glial fibrillary acidic protein (GFAP) (present in radial glia of the primate cortex [Levitt and Rakic, 1980], but not in the rodent brain [(Sancho-Tello *et al.*, 1995]), the extracellular matrix molecule Tenascin-C (TN-C) (Götz *et al.*, 1998), the intermediate filament protein vimentin (Schnitzer *et al.*, 1981), the brain lipid binding protein (BLBP) (Anthony *et al.*, 2004), and also contain glycogen granules like astrocytes in the adult brain (Götz *et al.*, 2002) (Table I and Fig. 3). None of these molecules is unique or specific for astroglial cells (for example, GS is largely expressed in astrocytes, but also in some subtypes of oligodendrocytes [Miyake and Kitamura, 1992]), but the combination of several of these molecules and ultrastructural features is unique to radial glia or astrocytes (see also Kimelberg, 2004), hence the name radial "glia" (Bentivoglio and Mazzarello, 1999; Rakic, 2003).

It is also important to note that there is heterogeneity among both astrocytes and radial glia (Hartfuss *et al.*, 2001; Kimelberg, 2004), so not all cells express all of these characteristics at the same time. For example, in the adult brain, only some astrocytes contain vimentin, BLBP, or TN-C (Owada *et al.*, 1996; Theodosis *et al.*, 1997; Young *et al.*, 1996), similar to a subset of radial glial cells expressing BLBP during development (Hartfuss *et al.*, 2001). To some extent the heterogeneity of some of these markers is due to their

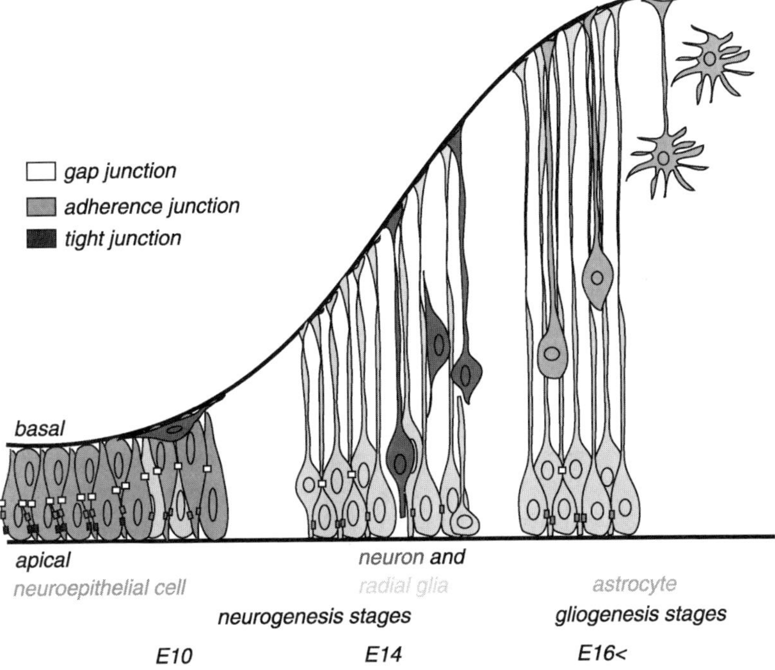

Figure 1 Schematic drawing of the developing mouse cerebral cortex. At early stages, all precursors are neuroepithelial cells depicted in orange to the left. Radial glial cells (yellow) appear around the onset of neurogenesis and exhibit many cell biological or molecular similarities with astrocytes. Postmitotic neurons are depicted in red, and the schematic drawing reflects the observation that when the first neurons appear some neuroepithelial precursors start to acquire the first radial glial features. At this time tight junctions (dark blue) are converted to adherens junctions (light blue). Radial glial cells generate neurons that either migrate by somal translocation, with the basal process attached to the basal surface, or they migrate basally along radial glial cells as depicted in the drawing. At the end of neurogenesis, radial glia transform into astrocytes (green). Tight junctions, adherence junctions, and gap junctions are dark blue, light blue, and white, respectively. (See Color Insert.)

regulation during development. For example, the number of BLBP-positive radial glial cells increases during development (Anthony *et al.*, 2004; Hartfuss *et al.*, 2001), and in the adult brain the number of BLBP-positive astrocytes increases after a brain lesion (Fig. 2). Indeed, as listed in Table I, several molecules present in radial glial cells during development are down-regulated in mature, quiescent astroglial cells in the adult brain, and are then re-expressed again in reactive astroglia. Notably, reactive astroglia, neural stem cells, and radial glia share all of these cell biological markers, implying a close similarity. Taken together, there is hardly any marker common to all astroglial cells, but the numerous features that radial glial cells and astroglia

Table I Glial Characters of Precursors and Astrocytes

	Neuroepithelial Cells	Radial Glia	Neural Stem Cells	Astrocytes after Injury	Astrocytes in the Adult Brain
Glycogen granules	−	+	+	+	+
GLAST	−	+	+	+	+
GS	−	+	+	+	+
S100β	−	+	+	+	+
GFAP	−	+ (not in rodents)	+	+	+/−
TN−C	−	+	+	+	+/−
Vimentin	−	+	+	+	+/−
BLBP	−	+	+	+	+/−
Nestin /RC2	+	+	+	+	−

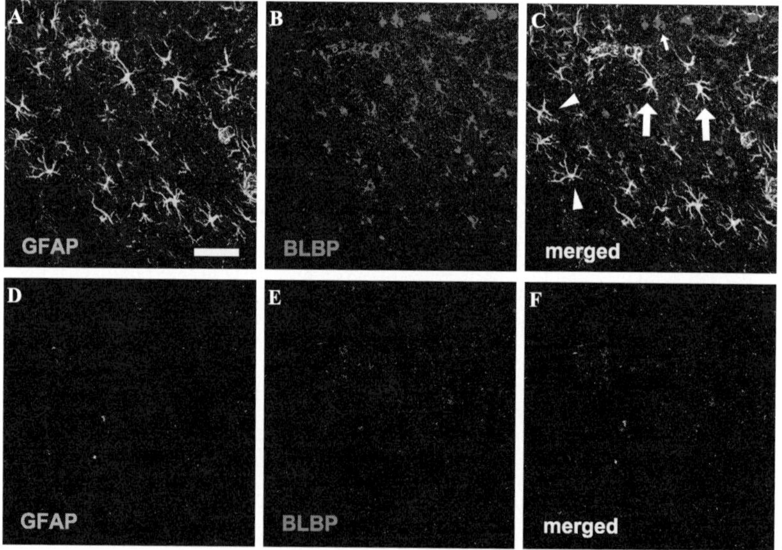

Figure 2 Heterogeneity of reactive astroglia of the adult cerebral cortex. Three days after a stab-wound lesion, strong GFAP and BLBP immunoreactivity is detectable in reactive astrocytes of the hemisphere subjected to the lesion (A, B, C), but not in astrocytes of the intact cortical parenchyma in the other hemisphere (D, E, F). Interestingly, the astroglia response to lesion is not homogeneous, as cells that express GFAP (arrowheads) or BLBP (small arrow) only can be identified, whereas others co-express the examined markers (large arrows). Scale bar: 50μm. (See Color Insert.)

share support their cellular and molecular kinship. This view is further supported by their lineage relation. It was long ago suggested that radial glial cells transform into astrocytes at the end of neurogenesis in mammals (Bentivoglio and Mazzarello, 1999; Choi, 1981; Pixley and de Vellis, 1984); Voigt *et al.* (Voigt, 1989) showed this directly in the cerebral cortex of the postnatal ferret by labeling radial glia from their basal endfeet at the pial surface with DiI (a lipophilic fluorescent dye) and thus following the transformation of the labeled radial glia into labeled astrocytes.

B. Similarities and Differences between Radial Glia and Neuroepithelial Cells

Intriguingly, not only the disappearance, but also the appearance of radial glial cells seem to be linked to the phase of neurogenesis (reviewed in Götz, 2003). At embryonic day (E) 10, neurogenesis has started in most regions of the mouse brain since some postmitotic neurons (βIII-tubulin immunopositive, but negative for the proliferation marker Ki67) can be detected underneath the pial surface (Fig. 3A). Just around this stage (E10–12), the first signs of radial glial differentiation, namely BLBP, GLAST (Fig. 3B and C), and vimentin immunoreactivity (Schnitzer *et al.*, 1981), become detectable and glycogen granules appear (Gadisseux and Evrard, 1985), suggesting that the first neurons must have been generated from precursors prior to the appearance of radial glial features. The ubiquitous cell type present in the developing neural tube prior to neurogenesis is called neuroepithelial cells (Fig. 1), named after their epithelial features with a pronounced apico-basal polarity and tight junctions separating the apical and basal parts of the cell membrane (reviewed in Huttner and Brand, 1997). In regard to their morphology, neuroepithelial and radial glial cells are very similar, with a bipolar cell shape. Neuroepithelial cells, radial glia, as well as reactive astroglia express nestin that also appears just prior to neurogenesis around E9 (Chanas-Sacre *et al.*, 2000; Edwards *et al.*, 1990; Frederiksen and McKay, 1988; Misson *et al.*, 1988a) (Table I). In addition, neuroepithelial cells and radial glial cells also share functional similarities, such as their connection to neighboring cells by gap junctions, a common feature among precursor cells and astroglia (Bennett *et al.*, 2003; Lo Turco and Kriegstein, 1991). Furthermore, both neuroepithelial cells and radial glial cells perform interkinetic nuclear migration within the VZ during the cell cycle, such that their nuclei move toward the basal side of the VZ at the phase of DNA synthesis, and then the nuclei move toward the apical side of the VZ during the mitotic phase (Fig. 1 and, e.g., Misson *et al.*, 1988b).

In contrast to radial glial cells, however, neuroepithelial cells do not express molecules shared with astroglia such as GLAST, GS, S100β,

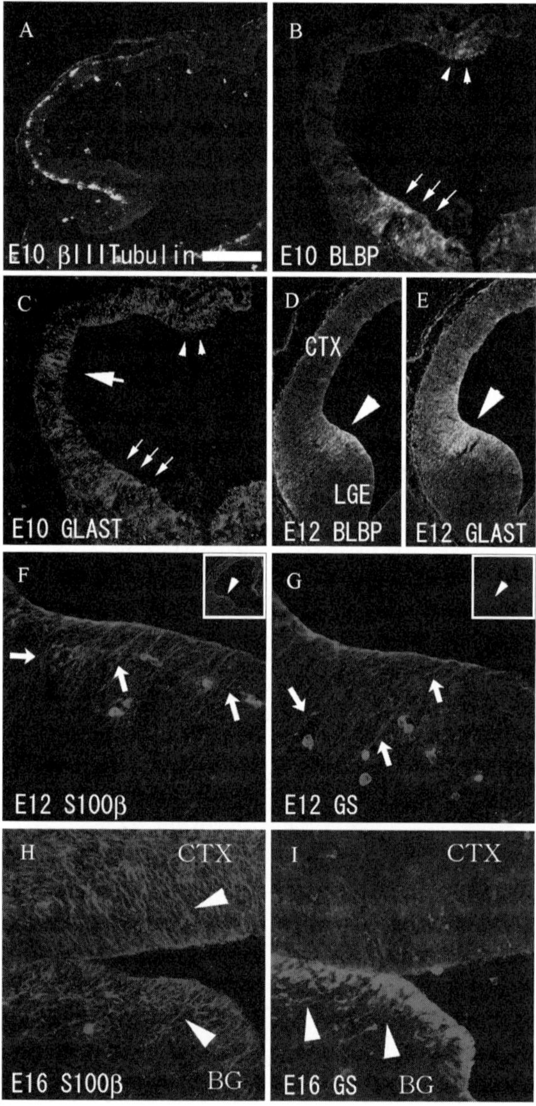

Figure 3 Astrocyte markers in radial glia. Fluorescent micrographs of coronal sections through the embryonic mouse cortex are depicted. At E10, postmitotic neurons immunolabeled for βIII-tubulin are already detectable in the basal part of the telencephalon (green cells in A). At this stage, BLBP (B) and GLAST (C) immunoreactivity become detectable in subsets of radial glia. Small arrows in (B) and (C) indicate the basal telencephalon; arrowheads indicate the cortical hem. Some radial glial cells express GLAST in the middle part of the telencephalon (large arrow in C). Panels (D) and (E) depict BLBP (D) and GLAST (E) immunoreactivity in the dorsal-most part of the lateral ganglionic eminence, the border between dorsal and ventral telencephalon, at E12 [arrowhead in (D) and (E)]. S100β (F) and GS (G) immunoreactivity

TN-C, BLBP, vimentin, or GFAP that are characteristic for radial glia and astrocytes as described above (Table I). Moreover, neuroepithelial cells do not yet contain glycogen granules (Gadisseux and Evrard, 1985). Thus, radial glial cells can be distinguished from neuroepithelial cells by all these features that are present in radial glia but absent in neuroepithelial cells. However, as mentioned above, these defining glial aspects do not all appear all simultaneously. It is important to note that vimentin and BLBP are among the first glial molecules to appear, immediately followed by GLAST (Anthony *et al.*, 2004; Malatesta *et al.*, 2003; Schnitzer *et al.*, 1981) (Fig. 3). These molecules are still contained in subsets of precursors at E12 but have spread to the majority of precursors in the cerebral cortex by E14 (Anthony *et al.*, 2004; Hartfuss *et al.*, 2001). S100β and GS immunoreactivities begin to be detectable still later in some radial glia, in the mouse telencephalon around E14–16 as depicted in Fig. 3F–I. Notably, the earliest appearance of many of these molecules characteristic for astroglia/radial glia occurs in the lateral ganglionic eminence (LGE, Fig. 3D, E, H, and I), the region corresponding to the origin of adult subventricular zone (SVZ), in which precursor cells continue neurogenesis throughout life, as discussed below.

A gradual transition of neuroepithelial cells toward a radial glial cell type is indicated by specific changes occurring in neuroepithelial cells, just before the appearance of the first radial glial markers. For example, neuroepithelial cells start to loose tight junctions around E9 and are solely connected by a special type of adherens junctions thereafter (Aaku-Saraste *et al.*, 1996; Mollgard *et al.*, 1987; Shoukimas and Hinds, 1978). A further intriguing change just prior to neurogenesis was recently identified by Hatakeyama and colleagues. These authors showed that the basic helix-loop-helix (bHLH) Hes transcription factors are required for the transition of neuroepithelial cells into a radial glial cell type (Hatakeyama *et al.*, 2004). Hes genes are downstream target genes of Notch signaling, and Notch signaling is involved in the maintenance of radial glia (Gaiano *et al.*, 2000). Mice deficient for Hes1 and Hes5 have normal neuroepithelial cells until E8.5, but the differentiation of radial glia is severely impaired after E9.5 (Hatakeyama *et al.*, 2004). These data suggest that precursor cells *in vivo* changed their properties from Hes-independent neuroepithelial cells to Hes-dependent transitory neuroepithelial cells, and finally to Hes-dependent radial glial cells. Notably,

starts to be weakly detectable at E12; arrows indicate radial glial processes. The insets in (F) and (G) show a low-power view of the telencephalon, and the arrowhead indicates where the high-power view is taken (at the border between the dorsal and the ventral telencephalon). Note that GS immunoreactivity is strongly upregulated at E16 with stronger signal in the ventral telencephalon generating the basal ganglia (BG) than the dorsal telencephalon, the future cortex (CTX). Arrowheads in (H) and (I) indicate radial glial processes. CTX; cortex, LGE; lateral ganglionic eminence, BG; basal ganglia. Scale bar: 200 μm (A, B, C, D, E); 50 μm (F, G, H, I). (See Color Insert.)

the loss of radial glial cells results in severe malformations and the failure of specific brain regions to form at all (Hatakeyama *et al.*, 2004). This study demonstrated that the proper differentiation of radial glia from neuroepithelial cells is essential for neuronal development and provided further functional evidence that radial glial cells derive from neuroepithelial cells.

These notions are further supported by studies on the differentiation of embryonic stem (ES) cells. Ying *et al.* (2003) succeeded in inducing the differentiation of ES cells into neurons by plating ES cells on gelatin-coated dishes and culturing them without leukemia inhibitory factor (LIF) and serum. This so-called monolayer protocol results in the differentiation of 60% ES cells into neuroectoderm/neuroepithelial cells as indicated by expression of Sox1, a transcription factor usually present in neuroepithelial cells (Wood and Episkopou, 1999). Interestingly, Bibel and colleagues recently showed that neuronal differentiation of ES cells also recapitulates a radial glial cell state (Bibel *et al.*, 2004; Plachta *et al.*, 2004). In their protocol achieving almost pure neuronal differentiation, ES cells formed embryoid bodies (EBs) in suspension culture without LIF following retinoic acid treatment and were then dissociated and cultured on an adherent surface. Remarkably, ES cells differentiated first almost completely into radial glia-like cells with elongated, spindle-shaped morphology that expressed nestin, RC2, BLBP, GLAST, and Pax6, and the majority of radial glial cells differentiated into glutamatergic neurons. This *in vitro* model provides a powerful tool for studying the mechanisms of radial glial differentiation at the molecular and biochemical levels due to the purity of the cell types. Thus, in ES cells different protocols result in the differentiation of neuroepithelial cells and radial glia and even the consecutive generation of first neurons and later glial cells (McKay, 2004). These data suggest that neuronal differentiation *in vitro* recapitulates the steps of neuronal differentiation *in vivo*.

In summary, radial glial cells appear around the onset of neurogenesis and gradually acquire most of the features characteristic of astrocytes, into which many of them transform at the end of neurogenesis (Voigt, 1989). The observation that they share many similarities with astrocytes while they are largely different from neuroepithelial cells is evidence in support of their glial identity.

II. Function of Radial Glia

A. Neurogenesis

Similar to astrocytes, radial glial cells were considered to be supporting cells exclusively. For example, radial glia support the migration of postmitotic neurons from the VZ to their final position in the outer, basal parts of the neural tube and developing brain (Rakic, 1988) (Fig. 1). Due to this idea of their role as stable support structure for migrating neurons, the evidence that radial glia are also proliferating throughout neurogenesis (Misson *et al.*, 1988b) has been neglected for a long time. However, Hartfuss and colleagues showed that all radial glial cells in the developing telencephalon were dividing from E12–E18, the phase of neurogenesis in this region (Hartfuss *et al.*, 2001, 2003). Since radial glial cells also compose the majority of all precursor cells during neurogenesis (Hartfuss *et al.*, 2001, 2003; Noctor *et al.*, 2002), these data started to suggest that radial glial cells may actively contribute to neurogenesis by generating neurons directly. The first direct evidence for the novel, neurogenic role of radial glia came from *in vitro* lineage analysis. Malatesta and colleagues isolated radial glia by FACS (fluorescence-activated cell sorting) using mice that contained the gene encoding for the green fluorescent protein (GFP) under control of the human GFAP (hGFAP) promoter (Malatesta *et al.*, 2000). As an alternative, independent method these authors also used the fluorescent tracers described above (Voigt, 1989) to label radial glial cells from the pial surface. The authors succeeded in isolating these fluorescently labeled radial glial cells by FACS, with the majority of the sorted cells being GLAST and BLBP immunoreactive, i.e., radial glial cells. These purified radial glial cells were then plated *in vitro* at such a low density that the progeny of a single radial glial cell could be identified as a cluster of cells distinct from other unlabeled cells (for details of this technique, see Anthony *et al.*, 2004; Malatesta *et al.*, 2000, 2003). Thus, the progeny of a single radial glial cell, a clone, was analyzed after 1 week *in vitro*, a time sufficient for further proliferation and differentiation. This analysis revealed a remarkable heterogeneity of radial glial cells, with the majority of them exclusively generating neurons when isolated at E14 or 16 and a significant proportion that were already specified to generate radial glia or astrocytes at these stages. Only a small proportion of radial glial cells were still bipotent, generating neurons and astrocytes, neurons and precursors, or neurons and oligodendrocytes (Malatesta *et al.*, 2000, 2003). These results suggest that there are at least three types of functionally distinct subtypes of radial glial cells in the cerebral cortex: a large fraction of neuronal precursors that appear to be restricted to the generation of neurons (see also Götz *et al.*, 2002), a smaller proportion of astroglial precursors, and a very small proportion of bi- or multipotent precursors. Notably, the population of neurogenic radial glial cells disappeared exclusively during development, and most radial glia isolated at later stages (E18 mouse cerebral cortex) generated cells of the astroglial phenotype (Malatesta *et al.*, 2000).

The next major breakthrough in this field came from the live analysis of radial glial cell division in cortical slice cultures (Miyata *et al.*, 2001, 2004; Noctor *et al.*, 2001, 2004), allowing the direct observation of the generation of neurons from radial glial cells. Noctor *et al.* (2001, 2002) monitored the cell division of single radial glial cells and followed their progeny in slice cultures isolated during the peak of neurogenesis from the rat cortex. To label dividing radial glia, they used a retroviral vector that incorporates the gene encoding for GFP only in the DNA of dividing cells and that is then inherited to their entire progeny (Price and Thurlow, 1988). Two days after viral infection many of the GFP-positive cells had the radial glial features of vimentin immunoreactivity and a bipolar morphology with a long radial process close to the previous pial surface. Note that the basement membrane—with all its potential influences—is absent in this *in vitro* system due to the removal of the meninges. Most of these labeled radial glial cells divided asymmetrically, with each of them giving rise to one new radial glial cell and one neuron (Fig. 4). Noctor and colleagues further suggested that radial processes might be inherited by the daughter radial glia (Noctor *et al.*, 2001, 2004) and might be used as a scaffold by postmitotic migrating neurons, as previously suggested by the model of Rakic, which proposed that clonally related cells establish the radial organization of the cerebral cortex (Rakic, 1988). Miyata and colleagues (Miyata *et al.*, 2001) also observed the cell division of single radial glial cells by time-lapse video microscopy using the fluorescent back-tracing technique discussed above

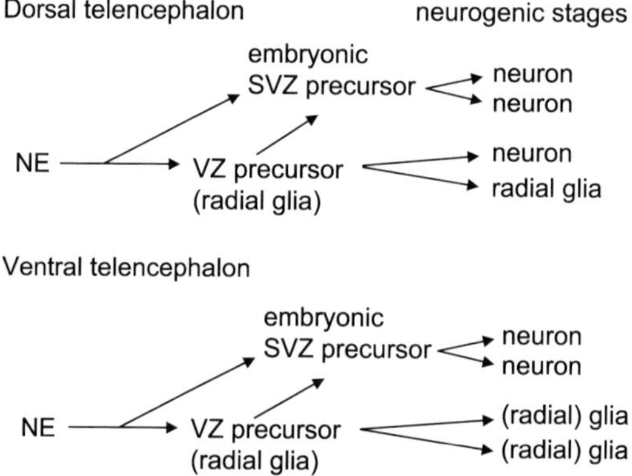

Figure 4   Model for the distinct cell lineages observed in the dorsal and ventral telencephalon during neurogenesis. First neuroepithelial precursors (NE) generate other ventricular zone precursors (VZ) that are composed mostly of radial glial cells in both regions; they also generate SVZ precursors that divide at some distance from the ventricle. SVZ precursors divide symmetrically and generate two neurons in both regions. Most radial glial cells in the dorsal telencephalon divide asymmetrically, generating one neuron and one radial glial cell, while most radial glial cells in the ventral telencephalon do not generate neurons. The latter may, however, generate neurons indirectly by generating SVZ precursors that are the majority of all precursors in the ventral telencephalon, while they are a minority in the dorsal telencephalon. Thus, while the dorsal telencephalon generates neurons in two modes, from radial glial cells directly and from SVZ precursors, mostly the latter mode of neurogenesis prevails in the ventral telencephalon.

(Malatesta *et al.*, 2000; Voigt, 1989). By retrograde labeling from the pial surface, radial glial fibers and their somata located in the VZ were fluorescently labeled (Miyata *et al.*, 2001). Furthermore, these authors showed that even if the radial glial cells generate two precursor cells, these two daughters differ in the inheritance of the radial process. One of them inherits radial process that become very thin during mitotic phase of the cell cycle, but another does not inherit it. In addition and in contrast to the observations described by Noctor and colleagues (Noctor *et al.*, 2001), Miyata observed many cases (about 50% of all radial glial cell divisions) in which the neurons generated by the radial glial cells inherited the radial processes from the mother radial glia by asymmetric cell division and then migrated toward the cortical plate by somal translocation (Morest, 1970; Nadarajah *et al.*, 2001). With the anchoring of the basal process to the basement membrane, neurons translocate to their final basal position by shortening their radial processes, as described by Morest and colleagues (reviewed in Fishell and Kriegstein, 2003). In parallel, the other, non-neuronal daughter cell entered the cell cycle again, started to extend its radial process to the pial surface, and became a new radial glia. These data are also consistent with the 3-D reconstruction of fluorescently labeled VZ cells, a subpopulation of cells with short radial processes and growth cones at the basally oriented tips (Hartfuss *et al.*, 2003).

While these beautiful experiments were performed *in vitro*—since live imaging of radial glial cells in the embryonic brain *in utero* was not yet feasible—Noctor and colleagues and Tamamaki and colleagues used viral labeling of radial glial cells *in vivo* by injection of the viral vectors in the lateral ventricle of rodent embryos *in utero* (Noctor *et al.*, 2001, 2004; Tamamaki *et al.*, 2001). Dividing radial glial cells were labeled either with a replication-deficient retrovirus or with an adenovirus expressing GFP. While this method is well established for retroviral vectors that incorporate their DNA into the host genome (see above and Price and Thurlow [1988]), adenoviral vectors remain episomal and should be diluted to half in each cell division, resulting in transient gene expression in the adenoviral vector-infected cells. However, Tamamaki and colleagues reported that the viral DNA was asymmetrically inherited by only one daughter cell. GFP protein was relatively stable for up to 1 week in the daughter cell inheriting the viral DNA, while the GFP protein was quickly lost in the other daughter cell that did not inherit the viral DNA. In examples of labeled cells examined shortly after cell division, these daughters had distinct morphologies, suggesting that radial glia gave rise to two daughter cells, one inheriting the radial process and the other with just a short process starting to extend toward the basal surface (Tamamaki *et al.*, 2001). Most notably, also *in vivo* many cells that inherited the radial process were neurons translocating to the cortical plate (Tamamaki *et al.*, 2001). In contrast, this was not reported in the work from

Noctor and colleagues (Noctor *et al.*, 2001, 2004). Thus, in addition to demonstrating by several independent lines of evidence that radial glial cells directly generate neurons, the live observations of radial glial cell division challenged another dogma, namely that dividing cells round up and retract all their processes. This finding had profound implications on the issue of asymmetric cell division, since the radial process is always inherited by only one daughter cell, be it the neuron or the precursor cell (reviewed in Fishell and Kriegstein, 2003). Finally, in these studies, the long-neglected mode of neuronal migration, the somal/perikaryal cell migration, was rediscovered (Morest and Silver, 2003).

B. Region-Specific Differences In Radial Glial Cell Fate

While these studies had focused on individual radial glial cells and their progeny, fate-mapping analysis revealed that radial glial cells not only contribute to neurogenesis, but also act as the major source for neurogenesis in the developing brain. To trace all the descendants of all radial glial cells, Malatesta and colleagues used the technique of *in vivo* recombination-based fate mapping (Zinyk *et al.*, 1998). Mice expressing the Cre recombinase under control of the hGFAP promoter (Zhuo *et al.*, 2001) contained Cre specifically in radial glia (Malatesta *et al.*, 2003). These mice were crossed with several reporter strains, such as the R26R reporter mice (Soriano, 1999), in which LacZ is expressed under the control of a ubiquitous promoter interrupted by a stop codon flanked by LoxP sequences. In this system, the stop cassette is deleted by Cre-mediated recombination in radial glial cells during development, followed by constitutive expression of the LacZ reporter gene in the recombined cells and all their progeny. With this technique, not just the progeny of some individual radial glia, but rather the progeny of all radial glial cells, at least in the telencephalon, can be followed. When the hGFAP promoter was used to drive Cre, the number of labeled neurons derived from Cre-positive radial glia in the dorsal telencephalon, the anlage of the cerebral cortex, was larger than those labeled in the ventral telencephalon, the anlage of the basal ganglia (Malatesta *et al.*, 2003). These data suggest that radial glia in the dorsal telencephalon generated most cortical neurons, while those in the basal telencephalon generated fewer neurons and were mostly gliogenic (Malatesta *et al.*, 2003).

In contrast, when the radial glial cell-specific element (RGE) of the BLBP promoter was used to mediate recombination, many labeled neurons were also detected in the ventral telencephalon (Anthony *et al.*, 2004). This discrepancy is best explained by the timing of Cre recombinase expression in the developing brain that differs between hGFAP and BLBP promoters. The BLBP promoter drives Cre expression as early as E10 in the ventral telencephalon, while hGFAP promoter mediates recombination at E14. As discussed above, the differentiation of radial glial cells is a gradual process, but most radial glial markers (except BLBP and vimentin) are not yet present at E10.

Interestingly, when GFP-positive radial glial cells of the ventral telenceph-alon were isolated from transgenic mice expressing GFP driven by the RGE of the BLBP promoter at E10 or E14, only the former generated neurons (Anthony *et al.*, 2004), further confirming the data from Malatesta and colleagues that radial glial cells in the ventral telencephalon are not neuro-genic at the peak of neurogenesis in this region. Thus, the two studies agree in the observation that radial glial cells in the ventral telencephalon are mostly not neurogenic except at very early stages. Since cells in the VZ undergo recombination as early as E10 in the RGE-Cre-based fate-mapping experi-ments, all their descendants will also be labeled, including the SVZ precur-sors that seem to act as the main source for neurons in the ventral telencephalon. Precursors located in the SVZ, basal to the VZ, form a secondary proliferating zone that is particularly prominent in the ventral telencephalon, and precursors in the SVZ of the ventral telencephalon reaches 60% of all proliferation precursors. Malatesta and colleagues sub-dissected the VZ and SVZ of the basal telencephalon at midneurogenesis (E14) and examined the progeny of these cells at clonal level *in vitro* (Malatesta *et al.*, 2003). This analysis confirmed that the VZ precursors of the E14 basal telencephalon, namely the radial glia, were largely not neuro-genic, while SVZ precursors of the basal telencephalon were mostly neuro-genic. Indeed, the neurogenic role of the SVZ precursors each generating two neurons has recently been demonstrated in live time-lapse video microscopy (Haubensak *et al.*, 2004; Miyata *et al.*, 2004; Noctor *et al.*, 2004). Thus, the prevailing model from these experiments is that early neurogenic cells gener-ate the first neurons as well as SVZ and VZ precursors. The latter are mainly radial glial cells that do not directly generate neurons in the ventral telen-cephalon (Fig. 4), in contrast to the radial glial cells in the dorsal telencepha-lon that directly generate neurons as observed by clonal analysis *in vitro* and the time-lapse analysis discussed above. Thus, direct neurogenesis from radial glial cells seems to be the predominant model of neurogenesis in the dorsal telencephalon. In contrast, except at early stages, neurogenesis in the ventral telencephalon originates from SVZ cells. SVZ cells are generated from early neuroepithelial cells and may persist as a mostly independent pool of precursors or be replenished constantly from the VZ precursors (Fig. 4). Thus, recombination in early precursors at E10 labels all of these cell types, and hence all cell types are labeled in the adult brain (Anthony *et al.*, 2004). In contrast to the dorsal telencephalon, the fact that VZ precursors and radial glia isolated from the ventral telencephalon at E14 do not generate neurons suggests that radial glial cells in this region can not generate neurons directly, but only via the generation of SVZ cells. Since proliferation stops *in vitro*, radial glial cell can not generate SVZ cells *in vitro* from the ventral telencephalon and hence no neurons are formed. Thus, taken together, both studies agree about the region-specific differences of radial glial cells at the peak of neurogenesis, with radial glial cells in the dorsal telencephalon directly generating numerous neurons, while the generation of SVZ precursors is required for neurogenesis in the ventral telencephalon (Fig. 4).

Indeed, this region-specific difference in radial glial cell fate is closely linked to the region-specific difference in the number of SVZ precursor cells. While SVZ precursors always constitute a minority of precursors during neurogenesis in the dorsal telencephalon of the mouse embryo (10% at E12; 30% at E16), they constitute the majority of precursors in the ventral telencephalon, reaching 60% of all precursors at E14 (Smart, 1976). An interesting and important difference between neurogenesis from SVZ or VZ cells is the mode of cell division. SVZ cells divide mostly symmetrically, generating two neurons, while the majority of direct neurogenesis from VZ/ radial glial cells in the dorsal telencephalon is asymmetric, with one neuron and one precursor generated. In addition, radial glial cells can also divide asymmetrically by generating a VZ and an SVZ precursor cell (Fig. 4). These modes of cell division may explain why the SVZ is specifically enlarged in the rodent basal ganglia anlage and the cerebral cortex of primates (Smart *et al.*, 2002), as a large SVZ can generate much larger numbers of neurons at the same time. Taken together, these different lines of evidence suggest a difference in the mode of neurogenesis in the dorsal and the ventral telencephalon in rodents. Thus, there are three different cell types that contribute to neurogenesis—early neuroepithelial cells for the first neurons, radial glial cells for most neurons in most brain regions, and SVZ precursors predominantly at later stages of neurogenesis and in the ventral telencephalon. One of the key questions to be addressed next is whether the molecular determinants of neurogenesis differ in these different cell types.

C. Functional Heterogeneity of Radial Glia

As a general rule, throughout the developing CNS neurogenesis always precedes gliogenesis. This may be achieved by two possible mechanisms. Radial glial cells might first generate neurons and then glia, in which case they would be bi- or multipotent. Alternately, distinct subtypes of radial glia might generate neurons while other radial glial subsets might give rise exclusively to radial glia first and later transform into astrocytes. The latter model would be consistent with the early fate restriction of radial glial precursors discussed previously. Clones derived from radial glial cells isolated from the telencephalon of hGFAP-GFP mice at midneurogenesis consisted mostly of a single cell type (Malatesta *et al.*, 2000, 2003), suggesting that at least *in vitro* radial glial cells do not generate first neurons and then glial cells, in which case the clones should be mixed, comprising both neurons and glia. However, when radial glial cells were sorted from E10 RGE-GFP mice and cultured differently, many more mixed clones comprising neurons and glia were observed, suggesting that these cells were not yet fate-restricted to the generation of a single cell type (Anthony *et al.*, 2004). These differences could be explained by either the different subtype of radial glial cells isolated by these authors or the culture conditions. The hGFAP-GFP transgene labels the GLAST-positive radial glial cells that comprise the majority of radial glial cells from midneurogenesis on (Hartfuss *et al.*, 2001; Malatesta *et al.*, 2000, 2003). In contrast, the BLBP-positive radial glial cells sorted by Anthony and colleagues are a subset of the GLAST-positive radial glia that had previously been hypothesized to be less fate-restricted (Hartfuss *et al.*, 2001). Moreover, Anthony *et al.* also added high concentrations of the fibroblast growth factor to their culture medium, which is well known to alter the fate of precursor cells (Hajihosseini and Dickson, 1999; Lillien, 1997; Qian *et al.*, 1998). We may tentatively conclude from these data that subsets of radial glial cells may indeed react differently to their environment and that a subset of radial glial cells—possibly with high levels of BLBP—may be less fate-restricted than others. However, clonal analysis *in vivo* as well as long-term time-lapse video microscopy of individual radial glial cells showed only a few examples of VZ/radial glial precursors generating both neurons and glial cells (Grove *et al.*, 1993; Luskin *et al.*, 1988; McCarthy *et al.*, 2001; Noctor *et al.*, 2004). Thus, most radial glial cells seemingly generate only a single cell type *in vivo* and in culture, even though it remains possible that their true potential may be broader.

Notably, when the progeny of neuroepithelial precursors labeled just prior to or at the onset of radial glial differentiation (around E9/10) was examined, a larger proportion of bi- or multipotent precursors was observed compared to later stages, but still a minority. For example, Qian *et al.* (1998, 2000) examined the lineage tree of single precursors isolated from E10 cortex, and observed only about 10–20% of the precursors as bi- or multipotent, while the majority of precursors generated neurons only (Qian *et al.*, 2000). Also, *in vivo* clonal analysis at these early stages (E9) revealed a surprisingly low proportion of clones (18%) containing both neurons and glial cells, while 34% of clones contained only neurons and 47% only glial cells (McCarthy *et al.*, 2001). These data not only suggest that most precursors exclusively generate a single cell type by this early stage of telencephalic development, but also showed that, more than 2 weeks prior to the phase of "gliogenesis," half of all precursors seem to be specified to the generation of glial cells. How is this possible?

Obviously, the ubiquitous glial cell type during development is the radial glial cells, so radial glial cells generating other radial glial cells exclusively are gliogenic. The only explanation for clones comprised exclusively of astrocytes in the adult brain is that radial glial cells generate first other radial glial cells that then transform into astrocytes at early postnatal stages. This scenario has also been demonstrated by experiments activating the Notch pathway (Gaiano *et al.*, 2000), which traps radial glial cells in the gliogenic mode. Upon overexpression of the intracellular domain of Notch, radial glial cells generate other radial glial cells exclusively and finally transform into astrocytes (Gaiano *et al.*, 2000). Taken together, several lines of evidence suggest that most, but not all, radial glial cells, both *in vitro* and *in vivo*, generate a single cell type exclusively. Thus, they are a heterogeneous population, comprising neuronal and glial precursors in distinct populations. Therefore, it appears that most neurogenic radial glia get depleted (theoretically by the generation of two postmitotic neurons [Noctor *et al.*, 2004; Qian *et al.*, 1998]) when gliogenesis starts, while only the gliogenic radial glia continue to survive and proliferate.

Even though multipotent or bipotent radial glial cells are a minority in the developing telencephalon, they are an important subset, since some of them may indeed persist into later stages. Consistent with the interpretation of BLBP-positive radial glial cells having a broader potential, the recent study of Li and colleagues suggests that this radial glial cell subset would be able to generate first neurons and then glial cells (Li *et al.*, 2004). Indeed, Qian *et al.* also found some precursor cells isolated from E10 cortex generating first neurons and then glial cells. Since this was observed in single cell cultures, the exciting possibility was raised that intrinsic fate determinants regulate this fate switch from neurogenesis to gliogenesis like an automatic clock mechanism (Qian *et al.*, 2000). Finally, long-term time-lapse analysis of radial glial cells *in vitro* also showed examples in which radial glial cells first generated neurons by asymmetric divisions, and then glial cells (Noctor *et al.*, 2004). This population may be particularly interesting, as it may be the source of adult neural stem cells.

III. Astrocytes and Neurogenesis

A. Transformation of Radial Glia into Astrocytes—The End of Neurogenesis?

As discussed previously, in correlation with the end of neurogenesis, radial glial cells transform into astrocytes in most of the avian and mammalian CNS regions (deAzevedo *et al.*, 2003; Marin-Padilla, 1995; Pixley and de Vellis, 1984). The adult mammalian brain contains no radial glia (astroglia with radial morphology) except Bergmann glia in the cerebellum and Müller glia in the retina. On the contrary, radial glial cells persist to a much larger extent in reptiles, amphibians, and fish (Garcia-Verdugo *et al.*, 2002; Naujoks-Manteuffel and Roth, 1989; Zupanc and Clint, 2003). As a potential reason for this phylogenetic development, it has been suggested that radial glial processes may approach their limits in supplying all the neurons in large brains for the K^+ buffering and glutamate uptake functions of astrocytes and that the larger number of evenly spread astrocytes may be better suited for this purpose (see Kimelberg, 2004).

Since adult neurogenesis is much more widespread in those species in which radial glial cells persist (Alvarez-Buylla *et al.*, 2001), this correlation leads to the question of whether neurogenesis comes to an end because radial glial cells transform into astrocytes, or whether that transformation may be a consequence of the end of neurogenesis. Alternatively, these two events may be merely coincidental and may not be related in any causal manner.

One answer to this question was provided by a series of recent studies showing that astrocytes can also be neurogenic, suggesting that the maintenance of radial glia is at least not a prerequisite for the maintenance of neurogenesis. For example, if astrocytes are cultured as floating spheres with the growth factors EGF and FGF2, they are dedifferentiated to such an extent that they can generate at least some neurons and oligodendrocytes (Laywell *et al.*, 2000). Notably, the same is the case for oligodendrocyte precursors that can be induced to a broader potential when cultured in such conditions (Kondo and Raff, 2000). However, this plasticity is still larger in young astrocytes isolated during the first 2 postnatal weeks from the rodent brain (Laywell *et al.*, 2000), a time period that corresponds with the disappearance of radial glia (Cameron and Rakic, 1991). Indeed, during the first postnatal week the morphological transformation of astrocytes into radial glial cells seems to still be bidirectional (Hunter and Hatten, 1995). If astrocytes are exposed *in vitro* or *in vivo* to cells of an embryonic cortex, they acquire a radial glial morphology (Hunter and Hatten, 1995), although it is not clear whether they also change their potential in correlation to the morphological transformation. One factor known to play a role in the morphological transformation is neuregulin (Schmid *et al.*, 2003). However, astrocytes in the adult CNS can still be induced to generate neurons and oligodendrocytes after culturing in the neurosphere conditions (Palmer *et al.*, 1999), suggesting that some glial precursor cells in the adult CNS can be dedifferentiated by environmental influences toward a more multipotent fate (for further discussion in the context of brain lesion see section III.C). Furthermore, several populations of glial precursors have been identified as "neurosphere-forming" cells, such as the A2B5-positive, 2', 3'-cyclic nucleotide 3'-phosphodiesterase (CNP) 2 promoter active cells or NG2-positive glial precursors in the white or gray matter of the cerebral cortex (Belachew *et al.*, 2003; Nunes *et al.*, 2003). There is also evidence that the latter may contribute to neurogenesis in the adult hippocampus *in vivo* (Aguirre *et al.*, 2004; Belachew *et al.*, 2003). Since NG2 and CNP have been suggested as oligodendrocyte precursor markers, these studies are reminiscent of the oligodendrocyte precursor formation of neurospheres described above (Kondo and Raff, 2000). These findings that different types of glial precursors have the capacity to form neurospheres and hence dedifferentiate into multipotent precursors are intriguing with regard to the possibility of evoking this potential also after brain lesion *in vivo* (see below).

Intrinsic fate determinants can also instruct a neurogenic fate in astrocytes. Upon expression of the paired type homeobox transcription factor Pax6, a high proportion of astrocytes isolated from the second week postnatal mouse cortex is induced to generate functional neurons (Heins *et al.*, 2002). Indeed, Pax6 expression, which is prominent in radial glial cells during neurogenesis in the embryonic cerebral cortex, is downregulated coincident with the end of neurogenesis (Götz *et al.*, 1998; Haubst *et al.*, 2004; Heins *et al.*, 2002), suggesting that Pax6 is a key neurogenic determinant for neurogenesis from radial glial/astroglial cells. Indeed, Pax6 is also necessary and sufficient to mediate neurogenesis in neurosphere cells (Hack *et al.*, 2004) that exhibit radial glial characteristics (Hartfuss *et al.*, 2001), and neuronal differentiation of ES cells apparently takes place via differentiation into Pax6-positive radial glial cells (Bibel *et al.*, 2004). Taken together, these results suggest that mammalian astrocytes may be able to generate neurons given a suitable environment and the necessary neurogenic fate determinants. Indeed, this is highlighted by a subset of astrocytes in specific regions of the mammalian telencephalon that continue to generate neurons throughout life.

B. Adult Neurogenesis in the Normal Mammalian Brain

Neurogenesis also continues throughout life in the adult brain of mammals (Alvarez-Buylla *et al.*, 2001; Gage, 2002), including humans (Bedard and Parent, 2004; Eriksson *et al.*, 1998; Sanai *et al.*, 2004), and astroglial cells act as the source of this adult neurogenesis. From the astrocytes in the SVZ lining the lateral wall of the lateral ventricle, interneurons of the olfactory bulb are newly generated throughout life and radial glial-like astrocytes in the subgranular layer (SGL) of the hippocampal dentate gyrus generate dentage granule neurons and possibly some GABAergic neurons throughout life (Belluzzi *et al.*, 2003; Carleton *et al.*, 2003; Doetsch *et al.*, 1999a; Liu *et al.*, 2003; Seri *et al.*, 2001).

Doetsch and colleagues (Doetsch *et al.*, 1997, 2002) classified cell types involved in adult neurogenesis in the adult mammalian SVZ by ultrastructural criteria and antigenic properties into four classes, the type A cells (PSA-NCAM and βIII-tubulin-positive neuroblasts), type B cells (GFAP-positive astrocytes), type C cells (Dlx2-positive transit amplifying cells), and type E cells (ependymal cells). The evidence that type B astrocytes are at the source of adult neurogenesis comes from lesion experiments using the antimitotic drug Ara-C (cytosine beta-arabinofuranoside) to kill the quickly dividing cells in the SVZ (Doetsch *et al.*, 1999b). Since type A and C cells are dividing faster than type B (ependymal cells do not proliferate), they are much more susceptible to Ara-C treatment, and only type B and E cells survive treatment with Ara-C for 6 days. However, 10 days after termination of this treatment, all SVZ cells types were fully reconstituted from type B cells because type E cells did not divide. These data along with further evidence from retroviral lineage analysis have made it clear that type B cells or a subset of them act as slowly dividing stem cells that then generate the quickly dividing transit-amplifying type C cells that then generate the neuroblasts, the type A cells. With the exception of some differences in cell cycle and ultrastructural characteristics of the transit-amplifying cells that are defined as type D cells in the SGL, the general lineage of adult neurogenesis in the SGL and the SVZ is similar (Seri *et al.*, 2001, 2004). Thus, in both of the adult neurogenic regions, precursors that are capable of regenerating all precursors for adult neurogenesis exhibit all the classical criteria of astrocytes (Table I and Fig. 5), including thick bundles of intermediate filaments and contact to blood vessels (Braun *et al.*, 2003; Doetsch *et al.*, 1997, 1999b). The definition of this astrocyte subset as adult neural stem cells is based on two main criteria: their ability to self-renew and their ability to fully reconstitute adult neurogenesis. Notably, however, it is not clear whether all astrocytes in these regions act as stem cells or only as a subtype of them. Neither is it known whether these cells are truly multipotent.

1. Adult Neural Stem Cells and Glial Cells *in Vitro*—The Neurosphere Assay

The evidence that neural stem cells may be multipotent is so far solely based on an *in vitro* system, the neurosphere cultures. Indeed, the entire field of adult neural stem cells was re-initiated when Reynolds and Weiss discovered in 1992 that some cells from the adult mammalian brain can be kept alive and even expanded free-floating in high concentrations of EGF and FGF2 (Reynolds and Weiss, 1992). These cells can then be passaged for a long time and hence have the capacity to self-renew. They can also be differentiated and generate some neurons, few oligodendrocytes, and many astrocytes (Gage, 1998; McKay, 1997), suggesting that they are multipotent. However, neurosphere-forming cells exist in many regions of the CNS, such as the spinal cord, where no neural stem cells are detectable *in vivo* (Palmer *et al.*,

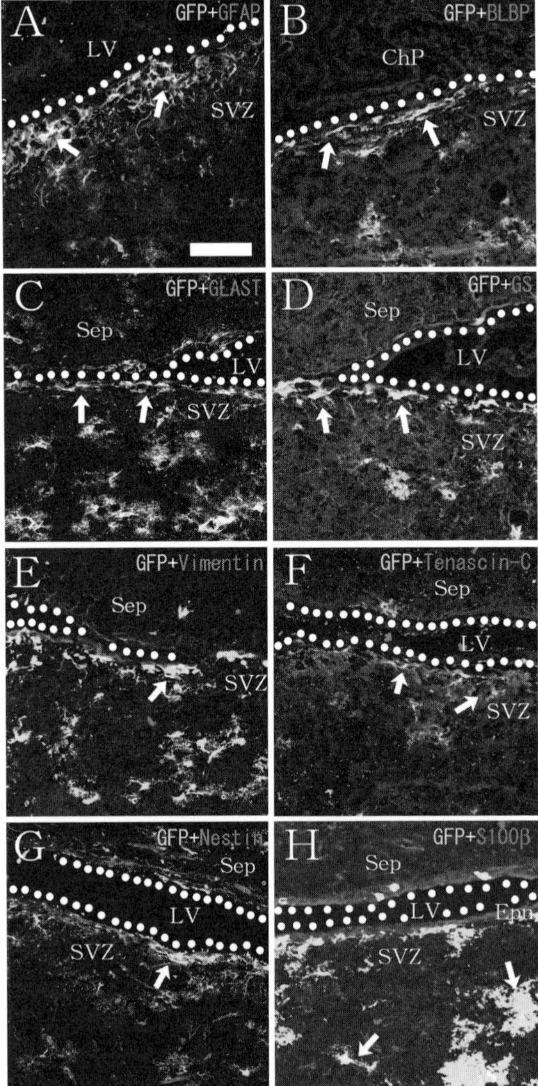

Figure 5 Adult SVZ cells express astrocyte markers. The panels depict single optical sections of confocal microscope micrographs taken from coronal sections of adult SVZ zone lining the lateral ventricle in transgenic mice expressing GFP under control of the human GFAP promoter. As indicated in the panels, GFP is depicted in green and the respective astrocyte markers are in red. Colocalization gives rise to yellow as indicated by arrows. The ventricle is indicated by the dotted lines. Note that GFP-positive astrocytes in the adult SVZ are also GFAP- (A), BLBP- (B), GLAST- (C), and GS- (D) immunopositive. Subsets of GFP-positive astrocytes were double-stained with antiserum directed against vimentin (E), Tenascin-C (F), or nestin (G), while S100β immunoreactivity was detected in ependymal cells and parenchymal astrocytes, but hardly in SVZ astrocytes (H). ChP, choroid plexus; LV, lateral ventricle; Sep, septum; SVZ, subventricular zone. Scale bar: 50 μm. (See Color Insert.)

1995; Weiss *et al.*, 1996). Conversely, neurosphere formation seems limited for cells isolated from the adult dentate gyrus (Seaberg and van der Kooy, 2002), despite the self-renewing properties of these cells *in vivo* (Seri *et al.*, 2001). Moreover, the neurosphere culture system is not merely an expansion system, but also severely alters the molecular expression characteristics of the precursor cells at the time of isolation and thereby affects their cell fate (Gabay *et al.*, 2003; Hack *et al.*, 2004). The expression of a variety of transcription factors involved in cell fate and regional specification of precursors, a prerequisite for the formation of specific types of neurons, is severely altered in the neurosphere culture system (Gabay *et al.*, 2003; Hack *et al.*, 2004). For example, cells that do not generate oligodendrocytes *in vivo* upregulate the transcription factor Olig2 in neurosphere cultures and thereby acquire the ability to generate oligodendrocytes. While there is so far no evidence that adult SVZ neural stem cell can generate oligodendrocytes *in vivo*, when these cells are cultured as neurospheres they can differentiate into neurons, astrocytes, and oligodendrocytes (Reynolds and Weiss, 1992, 1996). Conversely, neurogenesis is rather poor in neurosphere cultures obtained from adult neurogenic regions while neurogenesis is efficient *in vivo* (see, e.g., Hack *et al.*, 2004). Taken together, these data therefore suggest that the neurosphere culture system is necessarily a potent dedifferentiation system that can initiate multipotentiality in cells that do not possess or exhibit this property *in vivo*.

Thus, a neurosphere-forming cell *in vitro* is not necessarily a stem cell *in vivo*, but it is obviously a cell that can be dedifferentiated to acquire such properties upon exposure to the neurosphere culture conditions. In this context it is therefore also not surprising that both astrocytes that act as neural stem cells *in vivo* and the transit-amplifying precursors have been identified as neurosphere-forming cells from the adult SVZ (Doetsch *et al.*, 2002; Morshead *et al.*, 2003). Interestingly, the transit-amplifying cells *in vivo* express the bHLH transcription factor Olig2, which is important for neurosphere proliferation and self-renewal *in vitro* (Hack *et al.*, 2004).

2. Origin of Adult Neural Stem Cells, the SVZ Astrocytes

From the preceding discussion it becomes clear that most glial precursors lose their neurogenic potential at the end of neurogenesis when radial glia transform into astrocytes, while only those in the SVZ and SGL maintain their neurogenic potential *in vivo*. This prompts the crucial question of why only these and not other astrocytes maintain their neurogenic potential throughout life. To understand this, it may be helpful to know their developmental origin. It is known that SGL astrocytes derive from radial glia (Eckenhoff and Rakic, 1984), but the origin of SVZ astrocytes had been not clear until recently. Stenman and colleagues showed that in the SVZ of the LGE, there are two distinct precursor populations: the Dlx- and Isl-1-positive population generates striatal projection neurons, while the Dlx- and Er81-positive population, located in the dorsal-most LGE, generates olfactory bulb interneurons (Stenman *et al.*, 2003). The Dlx- and Er81-positive population expands at birth and persists into adult stages. Gsh1 and Gsh1/2 mutant mice, in which olfactory bulb interneurons are greatly reduced, have significantly fewer Er81-positive neurons, suggesting that Gsh2-, Dlx-, and Er81-positive precursors may contribute to the adult SVZ, at least in part (Stenman *et al.*, 2003; Toresson and Campbell, 2001; Toresson *et al.*, 2000). Notably, Gsh2-positive cells in the LGE are radial glia (Malatesta *et al.*, 2003), and the transformation of radial glial cells from this region has recently been directly demonstrated in fate-mapping experiments (Merkle *et al.*, 2004; Tramontin *et al.*, 2003). In several reporter mice (R26R, Z/EG [Lobe *et al.*, 1999]), Cre expressing adenovirus (Anton and Graham, 1995) was injected in the striatal region where the radial glial processes end. This resulted in retrograde labeling of radial glial cell bodies and a few striatal neurons (Merkle *et al.*, 2004; Tramontin *et al.*, 2003). Notably, when the progeny of these labeled cells were followed at later stages, not only were radial glial cells found to transform into ependymal cells, oligodendrocytes, and striatal astrocytes, but also many labeled cells were detected in the adult SVZ, contributing to adult neurogenesis as evident from the labeled olfactory bulb interneurons (Merkle *et al.*, 2004). Thus, adult neural stem cells are also derived from radial glial cells, apparently within the region of the dorsal LGE with potential contribution from the ventral pallium. Notably, precursors in the adult SVZ not only express transcription factors characteristic of the ventral telencephalon during development (Dlx [Doetsch *et al.*, 2002], Gsh2 [Stenman *et al.*, 2003], Mash1 [Parras *et al.*, 2004], Olig2 [Hack *et al.*, 2004]), but also contain transcription factors characteristic of the dorsal telencephalon such as Pax6 (Hack *et al.*, 2004) and Emx2 (Gangemi *et al.*, 2001). It will therefore be important to further examine whether dorsal and ventral telencephalic regions contribute to the adult SVZ, or whether it derives just from the boundary radial glia at this position (reviewed in Marshall *et al.*, 2003).

In further support of those observations, SVZ astrocytes still share several features with radial glial cells, as detailed in Table I. They contain not only the mature astroglial markers, but also the more immature markers such as nestin, vimentin, and RC2 (Table I) (Doetsch *et al.*, 1999a,b). Moreover, they share morphological features with radial glia, like the apical contact to the ventricle and the interkinetic nuclear migration (Tramontin *et al.*, 2003). From these observations it can be summarized that two adult neurogenic regions in the mammalian brain maintain the neurogenic potential in a subset of radial glia-derived astrocytes.

3. Neurogenesis from Astrocytes in the Adult Brain: Extrinsic Versus Intrinsic Factors

In contrast, most of the astrocytes in other regions of the adult brain are not neurogenic under normal conditions. It is not known why only some astrocytes maintain a neurogenic potential and others do not. Conceptually, this may be due to intrinsic differences or extrinsic differences in their local environment. In support of intrinsic differences between neurogenic and non-neurogenic astrocytes, distinct subtypes of astrocytes that have been discovered in several brain regions differ morphologically, electrophysiologically, and molecularly (Bachoo *et al.*, 2004; Walz, 2000). Alternatively or in addition, the local environment may differ within the adult SVZ and other regions of the adult mammalian brain. Indeed, a variety of extracellular cues persist from development exclusively in the adult SVZ (Gates *et al.*, 1995), but not in other regions of the adult mammalian brain. A particularly important signal may be noggin that is expressed in ependymal cells just beneath SVZ and that has been shown at the functional level as an important cue for the neurogenic niche in the SVZ (Lim *et al.*, 2000). Noggin functions as an inhibitor for the binding of bone morphogenic proteins (BMPs) to their receptors, and it is well known that BMPs inhibit neuronal specification and noggin induces neural tissue formation by blocking BMPs (reviewed in Mehler *et al.*, 1997). Consistent with this function of BMP, BMP over-expression in ependymal cells of the adult SVZ inhibits neurogenesis *in vivo*, while ectopic overexpression of noggin within the parenchyma of the striatum creates a more neurogenic environment for transplanted neurosphere cells (Lim *et al.*, 2000).

The importance of an environment permissive for neurogenesis is further supported by the results of many transplantation experiments. When cultured neural stem cells derived from embryonic or adult brains are transplanted into a neurogenic environment, such as in embryonic brains, the adult SVZ or hippocampus, they can differentiate into neurons (reviewed in Cao *et al.*, 2002). In contrast, when they are transplanted into non-neurogenic regions, such as the adult cerebellum, striatum or spinal cord, most of them differentiate into glia (Cao *et al.*, 2002). Even if multipotent ES cells were transplanted into an adult brain, only some regions, such as the hippocampus, cerebral cortex, and cerebellum, are permissive for the transplanted ES cells to differentiate into neurons (Harkany *et al.*, 2004). Thus, both intrinsic differences of the transplanted cells and the local environmental cues are important for the neuronal differentiation of transplanted cells, and signals supporting neurogenesis are obviously missing in the majority of regions in the adult mammalian brain. These observations raise the exciting possibility that the non-neurogenic environment can be modified to become neurogenic. As mentioned previously, blocking BMPs may be one way to convert a non-neurogenic into a neurogenic environment (Chmielnicki *et al.*, 2004). Conversely, it has been observed that factors supporting neurogenesis or neuronal differentiation such as brain-derived neurotrophic factor (BDNF) are helpful in supporting neurogenesis in normally non-neurogenic brain regions (Chmielnicki *et al.*, 2004). Moreover, infusion of growth factors into the SVZ can induce expansion of the adult SVZ precursor population (see, e.g., Craig *et al.*, 1996; Doetsch *et al.*, 2002; Zigova *et al.*, 1998).

Indeed, Nakatomi and colleagues demonstrated that growth factor infusion is sufficient for a striking degree of neuronal and functional recovery in a model of global ischemia (Nakatomi *et al.*, 2002). In the ischemic hippocampus, pyramidal cells in the CA1 region were newly generated from endogenous precursors in the posterior periventricle locating close to the hippocampus upon infusion of FGF2 and EGF. These new neurons migrated into the damaged CA1 region and replaced damaged neurons to a large extent (Nakatomi *et al.*, 2002). Interestingly, these authors also proposed that unidentified precursors in the hippocampal parenchyma might also contribute to the regeneration after ischemia. Taken together, these data suggest that manipulations of the environment in the adult brain not only may be helpful to reconstitute neurons from transplanted precursors but also may be sufficient to recover a neurogenic potential from endogenous precursors responding to injury and growth factor supply.

C. Adult Neurogenesis in the Injured Brain—The Role of Environmental Changes

Interestingly, injury itself induces endogenous growth factor upregulation (Wang *et al.*, 1998) that may be responsible for the instruction of some low degree of endogenous neurogenesis (Arvidsson *et al.*, 2002; Nakatomi *et al.*, 2002). The importance of environmental changes after injury is also indicated by transplantation experiments, since transplanted cells only integrate in regions exposed to lesions, not in regions of the intact adult brain (Cao *et al.*, 2002). Thus, injury itself obviously changes the environment, and it does so differently in different lesion paradigms. Macklis and colleagues established a unique method for inducing synchronous apoptotic degeneration of neurons in specific cortical layers by laser illumination. This then elicits apoptotic cell death of the back-labeled projection neurons in a single cortical layer with little inflammation, microglia activation, or gliotic reaction (Chen *et al.*, 2004; Macklis, 1993). In contrast to the poor neuronal differentiation of transplanted cells in other lesion paradigms (reviewed in Cao *et al.*, 2002), this specific lesion paradigm allows a highly efficient precursor maturation and neuronal replacement (Fricker-Gates *et al.*, 2002;

Shin *et al.*, 2000; Snyder *et al.*, 1997), suggesting that this special injury creates an environment permissive for neuronal differentiation. More interestingly, this holds true not only for transplanted cells, but also for endogenous precursors that are—even though in low numbers—recruited to neurogenesis in this lesion environment (Magavi *et al.*, 2000). Most strikingly, endogenously newly generated projection neurons in the cerebral cortex survive for a long time (more than a year) and even establish long-range axonal projections from the cortex to the spinal cord (Chen *et al.*, 2004). These data therefore further demonstrate that newly generated neurons are not subject to the cues inhibiting axonal regeneration in the adult mammalian nervous system, but are well able to extend long axons, as has previously been shown for embryonic neurons transplanted into the adult CNS (see, e.g., Li and Raisman, 1993; Wictorin *et al.*, 1990). Macklis and colleagues suggested that the new neurons differentiating in this specific lesion paradigm originate from at least two different precursor populations, the SVZ precursor population and endogenous precursors within the cortex (Magavi *et al.*, 2000). Newborn neurons detected by BrdU and Doublecortin (a marker of early migrating neurons) immunoreactivity with a migratory morphology oriented from the SVZ toward the cortex appeared to originate in the SVZ and migrate toward the lesion site in the cerebral cortex. Migration of new neurons from the neurogenic SVZ toward a lesion site has by now been observed in several lesion models, such as the stroke model of middle cerebral artery occlusion (MCAO) (Arvidsson *et al.*, 2002) or 6-hydroxydopamine (6-OHDA) lesion in the substantia nigra (Fallon *et al.*, 2000). Unfortunately, most of these new neurons then disappear again after some time, suggesting that the environment does not support their survival. This is notably different in the apoptotic lesion model of Macklis and colleagues, in which neurons survived for over a year. A further unique feature of this model is the apparent activation of endogenous neurogenic precursors. After lesion, precursors within the cortex started to proliferate, as indicated by pairs of BrdU-positive cells (Magavi *et al.*, 2000), and RC2-positive radial glial cells re-appeared in this lesion model (Leavitt *et al.*, 1999). Similarly, dedifferentiation of astrocytes into radial glial cells has been observed after chronic hypoxia (Ganat *et al.*, 2002). These observations suggest the possibility that radial glial cells dedifferentiating from astrocytes in the injured cortex may act as the endogenous neuronal progenitors. It is not yet known in which aspects the activated radial glia in the adult brain and radial glia in the embryonic brain differ during development, but these studies raise the exciting possibility that manipulation of extrinsic cues in the adult brain may be sufficient for converting astrocytes to an early stage of neurogenic radial glia after injury and thereby allowing repair from endogenous precursors. Therefore, further understanding of the radial glia–astroglia diversity and transition at the molecular and cellular levels will be key in directing astroglial cells that are already involved in the wound reaction to brain lesions (gliosis reaction) toward neuronal regeneration.

References

Aaku-Saraste, E., Hellwig, A., and Huttner, W. B. (1996). Loss of occludin and functional tight junctions, but not ZO-1, during neural tube closure—Remodeling of the neuroepithelium prior to neurogenesis. *Dev. Biol.* **180,** 664–679.

Aguirre, A. A., Chittajallu, R., Belachew, S., and Gallo, V. (2004). NG2-expressing cells in the subventricular zone are type C-like cells and contribute to interneuron generation in the postnatal hippocampus. *J. Cell Biol.* **165,** 575–589.

Akimoto, J., Itoh, H., Miwa, T., and Ikeda, K. (1993). Immunohistochemical study of glutamine synthetase expression in early glial development. *Brain Res. Dev. Brain Res.* **72,** 9–14.

Alvarez-Buylla, A., Garcia-Verdugo, J. M., and Tramontin, A. D. (2001). A unified hypothesis on the lineage of neural stem cells. *Nat. Rev. Neurosci.* **2,** 287–293.

Anthony, T. E., Klein, C., Fishell, G., and Heintz, N. (2004). Radial glia serve as neuronal progenitors in all regions of the central nervous system. *Neuron* **41,** 881–890.

Anton, M., and Graham, F. L. (1995). Site-specific recombination mediated by an adenovirus vector expressing the Cre recombinase protein: A molecular switch for control of gene expression. *J. Virol.* **69,** 4600–4606.

Arvidsson, A., Collin, T., Kirik, D., Kokaia, Z., and Lindvall, O. (2002). Neuronal replacement from endogenous precursors in the adult brain after stroke. *Nat. Med.* **8,** 963–970.

Bachoo, R. M., Kim, R. S., Ligon, K. L., Maher, E. A., Brennan, C., Billings, N., Chan, S., Li, C., Rowitch, D. H., Wong, W. H., and De Pinho, R. A. (2004). Molecular diversity of astrocytes with implications for neurological disorders. *Proc. Natl. Acad. Sci. USA* **101,** 8384–8349.

Bedard, A., and Parent, A. (2004). Evidence of newly generated neurons in the human olfactory bulb. *Brain Res. Dev. Brain Res.* **151,** 159–168.

Belachew, S., Chittajallu, R., Aguirre, A. A., Yuan, X., Kirby, M., Anderson, S., and Gallo, V. (2003). Postnatal NG2 proteoglycan-expressing progenitor cells are intrinsically multipotent and generate functional neurons. *J. Cell Biol.* **161,** 169–186.

Belluzzi, O., Benedusi, M., Ackman, J., and Lo Turco, J. J. (2003). Electrophysiological differentiation of new neurons in the olfactory bulb. *J. Neurosci.* **23,** 10411–10418.

Bennett, M. V., Contreras, J. E., Bukauskas, F. F., and Saez, J. C. (2003). New roles for astrocytes: Gap junction hemichannels have something to communicate. *Trends Neurosci.* **26,** 610–617.

Bentivoglio, M., and Mazzarello, P. (1999). The history of radial glia. *Brain Res. Bull.* **49,** 305–315.

Bibel, M., Richter, J., Schrenk, K., Tucker, K. L., Staiger, V., Korte, M., Goetz, M., and Barde, Y. A. (2004). Differentiation of mouse embryonic stem cells into a defined neuronal lineage. *Nat. Neurosci.* **7,** 1003–1009.

Braun, N., Sevigny, J., Mishra, S. K., Robson, S. C., Barth, S. W., Gerstberger, R., Hammer, K., and Zimmermann, H. (2003). Expression of the ecto-ATPase NTPDase2 in the germinal zones of the developing and adult rat brain. *Eur. J. Neurosci.* **17,** 1355–1364.

Cameron, R. S., and Rakic, P. (1991). Glial cell lineage in the cerebral cortex: A review and synthesis. *Glia* **4,** 124–137.

Cao, Q., Benton, R. L., and Whittemore, S. R. (2002). Stem cell repair of central nervous system injury. *J. Neurosci. Res.* **68,** 501–510.

Carleton, A., Petreanu, L. T., Lansford, R., Alvarez-Buylla, A., and Lledo, P. M. (2003). Becoming a new neuron in the adult olfactory bulb. *Nat. Neurosci.* **6,** 507–518.

Chanas-Sacre, G., Thiry, M., Pirard, S., Rogister, B., Moonen, G., Mbebi, C., Verdiere-Sahuque, M., and Leprince, P. (2000). A 295-kDA intermediate filament-associated protein in radial glia and developing muscle cells *in vivo* and *in vitro. Dev. Dyn.* **219,** 514–525.

Chen, J., Magavi, S. S., and Macklis, J. D. (2004). Neurogenesis of corticospinal motor neurons extending spinal projections in adult mice. *Proc. Natl. Acad. Sci. USA* **101,** 16357–16362.

Chmielnicki, E., Benraiss, A., Economides, A. N., and Goldman, S. A. (2004). Adenovirally expressed noggin and brain-derived neurotrophic factor cooperate to induce new medium spiny neurons from resident progenitor cells in the adult striatal ventricular zone. *J. Neurosci.* **24,** 2133–2142.

Choi, B. H. (1981). Radial glia of developing human fetal spinal cord: Golgi, immunohisto-chemical and electron microscopic study. *Brain Res.* **227,** 249–267.

Craig, C. G., Tropepe, V., Morshead, C. M., Reynolds, B. A., Weiss, S., and van der Kooy, D. (1996). *In vivo* growth factor expansion of endogenous subependymal neural precursor cell populations in the adult mouse brain. *J. Neurosci.* **16,** 2649–2658.

deAzevedo, L. C., Fallet, C., Moura-Neto, V., Daumas-Duport, C., Hedin-Pereira, C., and Lent, R. (2003). Cortical radial glial cells in human fetuses: Depth-correlated transformation into astrocytes. *J. Neurobiol.* **55,** 288–298.

Doetsch, F., Garcia-Verdugo, J. M., and Alvarez-Buylla, A. (1997). Cellular composition and three-dimensional organization of the subventricular germinal zone in the adult mammalian brain. *J. Neurosci.* **17,** 5046–5061.

Doetsch, F., Caille, I., Lim, D. A., Garcia-Verdugo, J. M., and Alvarez-Buylla, A. (1999a). Subventricular zone astrocytes are neural stem cells in the adult mammalian brain. *Cell* **97,** 703–716.

Doetsch, F., Garcia-Verdugo, J. M., and Alvarez-Buylla, A. (1999b). Regeneration of a germinal layer in the adult mammalian brain. *Proc. Natl. Acad. Sci. USA* **96,** 11619–11624.

Doetsch, F., Petreanu, L., Caille, I., Garcia-Verdugo, J. M., and Alvarez-Buylla, A. (2002). EGF converts transit-amplifying neurogenic precursors in the adult brain into multipotent stem cells. *Neuron* **36,** 1021–1034.

Eckenhoff, M. F., and Rakic, P. (1984). Radial organization of the hippocampal dentate gyrus: A Golgi, ultrastructural, and immunocytochemical analysis in the developing rhesus monkey. *J. Comp. Neurol.* **223,** 1–21.

Edwards, M. A., Yamamoto, M., and Caviness, V. S., Jr. (1990). Organization of radial glia and related cells in the developing murine CNS. An analysis based upon a new monoclonal antibody marker. *Neuroscience* **36,** 121–144.

Eriksson, P. S., Perfilieva, E., Bjork-Eriksson, T., Alborn, A. M., Nordborg, C., Peterson, D. A., and Gage, F. H. (1998). Neurogenesis in the adult human hippocampus. *Nat. Med.* **4,** 1313–1317.

Fallon, J., Reid, S., Kinyamu, R., Opole, I., Opole, R., Baratta, J., Korc, M., Endo, T. L., Duong, A., Nguyen, G., Karkehabadhi, M., Twardzik, D., Patel, S., and Loughlin, S. (2000). *In vivo* induction of massive proliferation, directed migration, and differentiation of neural cells in the adult mammalian brain. *Proc. Natl. Acad. Sci. USA* **97,** 14686–14691.

Fishell, G., and Kriegstein, A. R. (2003). Neurons from radial glia: The consequences of asymmetric inheritance. *Curr. Opin. Neurobiol.* **13,** 34–41.

Frederiksen, K., and McKay, R. D. (1988). Proliferation and differentiation of rat neuroepithelial precursor cells *in vivo. J. Neurosci.* **8,** 1144–1151.

Fricker-Gates, R. A., Shin, J. J., Tai, C. C., Catapano, L. A., and Macklis, J. D. (2002). Late-stage immature neocortical neurons reconstruct interhemispheric connections and form synaptic contacts with increased efficiency in adult mouse cortex undergoing targeted neurodegeneration. *J. Neurosci.* **22,** 4045–4056.

Gabay, L., Lowell, S., Rubin, L. L., and Anderson, D. J. (2003). Deregulation of dorsoventral patterning by FGF confers trilineage differentiation capacity on CNS stem cells *in vitro*. *Neuron* **40**, 485–499.

Gadisseux, J. F., and Evrard, P. (1985). Glial-neuronal relationship in the developing central nervous system. A histochemical-electron microscope study of radial glial cell particulate glycogen in normal and reeler mice and the human fetus. *Dev. Neurosci.* **7**, 12–32.

Gage, F. H. (1998). Stem cells of the central nervous system. *Curr. Opin. Neurobiol.* **8**, 671–676.

Gage, F. H. (2002). Neurogenesis in the adult brain. *J. Neurosci.* **22**, 612–613.

Gaiano, N., Nye, J. S., and Fishell, G. (2000). Radial glial identity is promoted by Notch1 signaling in the murine forebrain. *Neuron* **26**, 395–404.

Ganat, Y., Soni, S., Chacon, M., Schwartz, M. L., and Vaccarino, F. M. (2002). Chronic hypoxia up-regulates fibroblast growth factor ligands in the perinatal brain and induces fibroblast growth factor-responsive radial glial cells in the sub-ependymal zone. *Neuroscience* **112**, 977–991.

Gangemi, R. M., Daga, A., Marubbi, D., Rosatto, N., Capra, M. C., and Corte, G. (2001). Emx2 in adult neural precursor cells. *Mech Dev.* **109**, 323–329.

Garcia-Verdugo, J. M., Ferron, S., Flames, N., Collado, L., Desfilis, E., and Font, E. (2002). The proliferative ventricular zone in adult vertebrates: A comparative study using reptiles, birds, and mammals. *Brain Res. Bull.* **57**, 765–775.

Gates, M. A., Thomas, L. B., Howard, E. M., Laywell, E. D., Sajin, B., Faissner, A., Götz, B., Silver, J., and Steindler, D. A. (1995). Cell and molecular analysis of the developing and adult mouse subventricular zone of the cerebral hemispheres. *J. Comp. Neurol.* **361**, 249–266.

Götz, M. (2003). Glial cells generate neurons–master control within CNS regions: Developmental perspectives on neural stem cells. *Neuroscientist* **9**, 379–397.

Götz, M., Stoykova, A., and Gruss, P. (1998). Pax6 controls radial glia differentiation in the cerebral cortex. *Neuron* **21**, 1031–1044.

Götz, M., Hartfuss, E., and Malatesta, P. (2002). Radial glial cells as neuronal precursors: A new perspective on the correlation of morphology and lineage restriction in the developing cerebral cortex of mice. *Brain Res. Bull.* **57**, 777–788.

Grove, E. A., Williams, B. P., Li, D. Q., Hajihosseini, M., Friedrich, A., and Price, J. (1993). Multiple restricted lineages in the embryonic rat cerebral cortex. *Development* **117**, 553–561.

Hack, M. A., Sugimori, M., Lundberg, C., Nakafuku, M., and Götz, M. (2004). Regionalization and fate specification in neurospheres: The role of Olig2 and Pax6. *Mol. Cell Neurosci.* **25**, 664–678.

Hajihosseini, M. K., and Dickson, C. (1999). A subset of fibroblast growth factors (Fgfs) promote survival, but Fgf-8b specifically promotes astroglial differentiation of rat cortical precursor cells. *Mol. Cell Neurosci.* **14**, 468–485.

Harkany, T., Andang, M., Kingma, H. J., Gorcs, T. J., Holmgren, C. D., Zilberter, Y., and Ernfors, P. (2004). Region-specific generation of functional neurons from naive embryonic stem cells in adult brain. *J. Neurochem.* **88**, 1229–1239.

Hartfuss, E., Galli, R., Heins, N., and Götz, M. (2001). Characterization of CNS precursor subtypes and radial glia. *Dev. Biol.* **229**, 15–30.

Hartfuss, E., Forster, E., Bock, H. H., Hack, M. A., Leprince, P., Luque, J. M., Herz, J., Frotscher, M., and Götz, M. (2003). Reelin signaling directly affects radial glia morphology and biochemical maturation. *Development* **130**, 4597–4609.

Hatakeyama, J., Bessho, Y., Katoh, K., Ookawara, S., Fujioka, M., Guillemot, F., and Kageyama, R. (2004). Hes genes regulate size, shape and histogenesis of the nervous system by control of the timing of neural stem cell differentiation. *Development* **131**, 5539–5550.

Haubensak, W., Attardo, A., Denk, W., and Huttner, W. B. (2004). Neurons arise in the basal neuroepithelium of the early mammalian telencephalon: A major site of neurogenesis. *Proc. Natl. Acad. Sci. USA* **101**, 3196–3201.

Haubst, N., Berger, J., Radjendirane, V., Graw, J., Favor, J., Saunders, G. F., Stoykova, A., and Götz, M. (2004). Molecular dissection of Pax6 function: The specific roles of the paired domain and homeodomain in brain development. *Development* **131**, 6131–6140.

Heins, N., Malatesta, P., Cecconi, F., Nakafuku, M., Tucker, K. L., Hack, M. A., Chapouton, P., Barde, Y. A., and Götz, M. (2002). Glial cells generate neurons: The role of the transcription factor Pax6. *Nat. Neurosci.* **5**, 308–315.

Hunter, K. E., and Hatten, M. E. (1995). Radial glial cell transformation to astrocytes is bidirectional: Regulation by a diffusible factor in embryonic forebrain. *Proc. Natl. Acad. Sci. USA* **92**, 2061–2205.

Huttner, W. B., and Brand, M. (1997). Asymmetric division and polarity of neuroepithelial cells. *Curr. Opin. Neurobiol.* **7**, 29–39.

Kimelberg, H. K. (2004). The problem of astrocyte identity. *Neurochem. Int.* **45**, 191–202.

Kondo, T., and Raff, M. (2000). Oligodendrocyte precursor cells reprogrammed to become multipotential CNS stem cells. *Science* **289**, 1754–1757.

Laywell, E. D., Rakic, P., Kukekov, V. G., Holland, E. C., and Steindler, D. A. (2000). Identification of a multipotent astrocytic stem cell in the immature and adult mouse brain. *Proc. Natl. Acad. Sci. USA* **97**, 13883–13888.

Leavitt, B. R., Hernit-Grant, C. S., and Macklis, J. D. (1999). Mature astrocytes transform into transitional radial glia within adult mouse neocortex that supports directed migration of transplanted immature neurons. *Exp. Neurol.* **157**, 43–57.

Levitt, P., and Rakic, P. (1980). Immunoperoxidase localization of glial fibrillary acidic protein in radial glial cells and astrocytes of the developing rhesus monkey brain. *J. Comp. Neurol.* **193**, 815–840.

Li, H., Babiarz, J., Woodbury, J., Kane-Goldsmith, N., and Grumet, M. (2004). Spatiotemporal heterogeneity of CNS radial glial cells and their transition to restricted precursors. *Dev. Biol.* **271**, 225–238.

Li, Y., and Raisman, G. (1993). Long axon growth from embryonic neurons transplanted into myelinated tracts of the adult rat spinal cord. *Brain Res.* **629**, 115–127.

Lillien, L. (1997). Neural development: Instructions for neural diversity. *Curr. Biol.* **7**, R168–R171.

Lim, D. A., Tramontin, A. D., Trevejo, J. M., Herrera, D. G., Garcia-Verdugo, J. M., and Alvarez-Buylla, A. (2000). Noggin antagonizes BMP signaling to create a niche for adult neurogenesis. *Neuron* **28**, 713–726.

Liu, S., Wang, J., Zhu, D., Fu, Y., Lukowiak, K., and Lu, Y. M. (2003). Generation of functional inhibitory neurons in the adult rat hippocampus. *J. Neurosci.* **23**, 732–736.

Lo Turco, J. J., and Kriegstein, A. R. (1991). Clusters of coupled neuroblasts in embryonic neocortex. *Science* **252**, 563–566.

Lobe, C. G., Koop, K. E., Kreppner, W., Lomeli, H., Gertsenstein, M., and Nagy, A. (1999). Z/AP, a double reporter for cre-mediated recombination. *Dev. Biol.* **208**, 281–292.

Luskin, M. B., Pearlman, A. L., and Sanes, J. R. (1988). Cell lineage in the cerebral cortex of the mouse studied *in vivo* and *in vitro* with a recombinant retrovirus. *Neuron* **1**, 635–647.

Macklis, J. D. (1993). Transplanted neocortical neurons migrate selectively into regions of neuronal degeneration produced by chromophore-targeted laser photolysis. *J. Neurosci.* **13**, 3848–3863.

Magavi, S. S., Leavitt, B. R., and Macklis, J. D. (2000). Induction of neurogenesis in the neocortex of adult mice. *Nature* **405**, 951–955.

Malatesta, P., Hartfuss, E., and Götz, M. (2000). Isolation of radial glial cells by fluorescent-activated cell sorting reveals a neuronal lineage. *Development* **127**, 5253–5263.

Malatesta, P., Hack, M. A., Hartfuss, E., Kettenmann, H., Klinkert, W., Kirchhoff, F., and Götz, M. (2003). Neuronal or glial progeny: Regional differences in radial glia fate. *Neuron* **37**, 751–764.

Marin-Padilla, M. (1995). Prenatal development of fibrous (white matter), protoplasmic (gray matter), and layer I astrocytes in the human cerebral cortex: A Golgi study. *J. Comp. Neurol.* **357,** 554–572.

Marshall, C. A., Suzuki, S. O., and Goldman, J. E. (2003). Gliogenic and neurogenic progenitors of the subventricular zone: Who are they, where did they come from, and where are they going? *Glia* **43,** 52–61.

McCarthy, M., Turnbull, D. H., Walsh, C. A., and Fishell, G. (2001). Telencephalic neural progenitors appear to be restricted to regional and glial fates before the onset of neurogenesis. *J. Neurosci.* **21,** 6772–6781.

McKay, R. (1997). Stem cells in the central nervous system. *Science* **276,** 66–71.

McKay, R. D. (2004). Stem cell biology and neurodegenerative disease. *Philos. Trans. R Soc. Lond. B Biol. Sci.* **359,** 851–856.

Mehler, M. F., Mabie, P. C., Zhang, D., and Kessler, J. A. (1997). Bone morphogenetic proteins in the nervous system. *Trends Neurosci.* **20,** 309–317.

Merkle, F. T., Tramontin, A. D., Garcia-Verdugo, J. M., and Alvarez-Buylla, A. (2004). Radial glia give rise to adult neural stem cells in the subventricular zone. *Proc. Natl. Acad. Sci. USA* **101,** 17528–17532.

Misson, J. P., Edwards, M. A., Yamamoto, M., and Caviness, V. S., Jr. (1988a). Identification of radial glial cells within the developing murine central nervous system: Studies based upon a new immunohistochemical marker. *Brain Res. Dev. Brain Res.* **44,** 95–108.

Misson, J. P., Edwards, M. A., Yamamoto, M., and Caviness, V. S., Jr. (1988b). Mitotic cycling of radial glial cells of the fetal murine cerebral wall: A combined autoradiographic and immunohistochemical study. *Brain Res.* **466,** 183–190.

Miyake, T., and Kitamura, T. (1992). Glutamine synthetase immunoreactivity in two types of mouse brain glial cells. *Brain Res.* **586,** 53–60.

Miyata, T., Kawaguchi, A., Okano, H., and Ogawa, M. (2001). Asymmetric inheritance of radial glial fibers by cortical neurons. *Neuron* **31,** 727–741.

Miyata, T., Kawaguchi, A., Saito, K., Kawano, M., Muto, T., and Ogawa, M. (2004). Asymmetric production of surface-dividing and non-surface-dividing cortical progenitor cells. *Development* **131,** 3133–3145.

Mollgard, K., Balslev, Y., Lauritzen, B., and Saunders, N. R. (1987). Cell junctions and membrane specializations in the ventricular zone (germinal matrix) of the developing sheep brain: A CSF-brain barrier. *J. Neurocytol.* **16,** 433–444.

Morest, D. K. (1970). A study of neurogenesis in the forebrain of opossum pouch young. *Z. Anat. Entwicklungsgesch* **130,** 265–305.

Morest, D. K., and Silver, J. (2003). Precursors of neurons, neuroglia, and ependymal cells in the CNS: What are they? Where are they from? How do they get where they are going? *Glia* **43,** 6–18.

Morshead, C. M., Garcia, A. D., Sofroniew, M. V., and van Der Kooy, D. (2003). The ablation of glial fibrillary acidic protein-positive cells from the adult central nervous system results in the loss of forebrain neural stem cells but not retinal stem cells. *Eur. J. Neurosci.* **18,** 76–84.

Nadarajah, B., Brunstrom, J. E., Grutzendler, J., Wong, R. O., and Pearlman, A. L. (2001). Two modes of radial migration in early development of the cerebral cortex. *Nat. Neurosci.* **4,** 143–150.

Nakatomi, H., Kuriu, T., Okabe, S., Yamamoto, S., Hatano, O., Kawahara, N., Tamura, A., Kirino, T., and Nakafuku, M. (2002). Regeneration of hippocampal pyramidal neurons after ischemic brain injury by recruitment of endogenous neural progenitors. *Cell* **110,** 429–441.

Naujoks-Manteuffel, C., and Roth, G. (1989). Astroglial cells in a salamander brain (Salamandra salamandra) as compared to mammals: A glial fibrillary acidic protein immunohistochemistry study. *Brain Res.* **487,** 397–401.

Noctor, S. C., Flint, A. C., Weissman, T. A., Dammerman, R. S., and Kriegstein, A. R. (2001). Neurons derived from radial glial cells establish radial units in neocortex. *Nature* **409,** 714–720.

Noctor, S. C., Flint, A. C., Weissman, T. A., Wong, W. S., Clinton, B. K., and Kriegstein, A. R. (2002). Dividing precursor cells of the embryonic cortical ventricular zone have morphological and molecular characteristics of radial glia. *J. Neurosci.* **22,** 3161–3173.

Noctor, S. C., Martinez-Cerdeno, V., Ivic, L., and Kriegstein, A. R. (2004). Cortical neurons arise in symmetric and asymmetric division zones and migrate through specific phases. *Nat. Neurosci.* **7,** 136–144.

Nunes, M. C., Roy, N. S., Keyoung, H. M., Goodman, R. R., McKhann, G., 2nd, Jiang, L., Kang, J., Nedergaard, M., and Goldman, S. A. (2003). Identification and isolation of multipotential neural progenitor cells from the subcortical white matter of the adult human brain. *Nat. Med.* **9,** 439–447.

Owada, Y., Yoshimoto, T., and Kondo, H. (1996). Spatio-temporally differential expression of genes for three members of fatty acid binding proteins in developing and mature rat brains. *J. Chem NeuroaNat.* **12,** 113–122.

Palmer, T. D., Ray, J., and Gage, F. H. (1995). FGF-2-responsive neuronal progenitors reside in proliferative and quiescent regions of the adult rodent brain. *Mol. Cell Neurosci.* **6,** 474–486.

Palmer, T. D., Markakis, E. A., Willhoite, A. R., Safar, F., and Gage, F. H. (1999). Fibroblast growth factor-2 activates a latent neurogenic program in neural stem cells from diverse regions of the adult CNS. *J. Neurosci.* **19,** 8487–8497.

Parras, C. M., Galli, R., Britz, O., Soares, S., Galichet, C., Battiste, J., Johnson, J. E., Nakafuku, M., Vescovi, A., and Guillemot, F. (2004). Mash1 specifies neurons and oligodendrocytes in the postnatal brain. *EMBO J.* **23,** 4495–4505.

Pixley, S. K., and de Vellis, J. (1984). Transition between immature radial glia and mature astrocytes studied with a monoclonal antibody to vimentin. *Brain Res.* **317,** 201–209.

Plachta, N., Bibel, M., Tucker, K. L., and Barde, Y. A. (2004). Developmental potential of defined neural progenitors derived from mouse embryonic stem cells. *Development* **131,** 5449–5456.

Price, J., and Thurlow, L. (1988). Cell lineage in the rat cerebral cortex: A study using retroviral-mediated gene transfer. *Development* **104,** 473–482.

Qian, X., Goderie, S. K., Shen, Q., Stern, J. H., and Temple, S. (1998). Intrinsic programs of patterned cell lineages in isolated vertebrate CNS ventricular zone cells. *Development* **125,** 3143–3152.

Qian, X., Shen, Q., Goderie, S. K., He, W., Capela, A., Davis, A. A., and Temple, S. (2000). Timing of CNS cell generation: A programmed sequence of neuron and glial cell production from isolated murine cortical stem cells. *Neuron* **28,** 69–80.

Rakic, P. (1988). Specification of cerebral cortical areas. *Science* **241,** 170–176.

Rakic, P. (2003). Elusive radial glial cells: Historical and evolutionary perspective. *Glia* **43,** 19–32.

Reynolds, B. A., and Weiss, S. (1992). Generation of neurons and astrocytes from isolated cells of the adult mammalian central nervous system. *Science* **255,** 1707–1710.

Reynolds, B. A., and Weiss, S. (1996). Clonal and population analyses demonstrate that an EGF-responsive mammalian embryonic CNS precursor is a stem cell. *Dev. Biol.* **175,** 1–13.

Sanai, N., Tramontin, A. D., Quinones-Hinojosa, A., Barbaro, N. M., Gupta, N., Kunwar, S., Lawton, M. T., McDermott, M. W., Parsa, A. T., Manuel-Garcia Verdugo, J., Berger, M. S., and Alvarez-Buylla, A. (2004). Unique astrocyte ribbon in adult human brain contains neural stem cells but lacks chain migration. *Nature* **427,** 740–744.

Sancho-Tello, M., Valles, S., Montoliu, C., Renau-Piqueras, J., and Guerri, C. (1995). Developmental pattern of GFAP and vimentin gene expression in rat brain and in radial glial cultures. *Glia* **15,** 157–166.

Schmid, R. S., McGrath, B., Berechid, B. E., Boyles, B., Marchionni, M., Sestan, N., and Anton, E. S. (2003). Neuregulin 1-erbB2 signaling is required for the establishment of radial glia and their transformation into astrocytes in cerebral cortex. *Proc. Natl. Acad. Sci. USA* **100,** 4251–4256.

Schnitzer, J., Franke, W. W., and Schachner, M. (1981). Immunocytochemical demonstration of vimentin in astrocytes and ependymal cells of developing and adult mouse nervous system. *J. Cell Biol.* **90,** 435–447.

Seaberg, R. M., and van der Kooy, D. (2002). Adult rodent neurogenic regions: The ventricular subependyma contains neural stem cells, but the dentate gyrus contains restricted progenitors. *J. Neurosci.* **22,** 1784–1793.

Seri, B., Garcia-Verdugo, J. M., McEwen, B. S., and Alvarez-Buylla, A. (2001). Astrocytes give rise to new neurons in the adult mammalian hippocampus. *J. Neurosci.* **21,** 7153–7160.

Seri, B., Garcia-Verdugo, J. M., Collado-Morente, L., McEwen, B. S., and Alvarez-Buylla, A. (2004). Cell types, lineage, and architecture of the germinal zone in the adult dentate gyrus. *J. Comp. Neurol.* **478,** 359.

Shin, J. J., Fricker-Gates, R. A., Perez, F. A., Leavitt, B. R., Zurakowski, D., and Macklis, J. D. (2000). Transplanted neuroblasts differentiate appropriately into projection neurons with correct neurotransmitter and receptor phenotype in neocortex undergoing targeted projection neuron degeneration. *J. Neurosci.* **20,** 7404–7416.

Shoukimas, G. M., and Hinds, J. W. (1978). The development of the cerebral cortex in the embryonic mouse: An electron microscopic serial section analysis. *J. Comp. Neurol.* **179,** 795–830.

Smart, I. H. (1976). A pilot study of cell production by the ganglionic eminences of the developing mouse brain. *J. Anat.* **121,** 71–84.

Smart, I. H., Dehay, C., Giroud, P., Berland, M., and Kennedy, H. (2002). Unique morphological features of the proliferative zones and postmitotic compartments of the neural epithelium giving rise to striate and extrastriate cortex in the monkey. *Cereb. Cortex* **12,** 37–53.

Snyder, E. Y., Yoon, C., Flax, J. D., and Macklis, J. D. (1997). Multipotent neural precursors can differentiate toward replacement of neurons undergoing targeted apoptotic degeneration in adult mouse neocortex. *Proc. Natl. Acad. Sci. USA* **94,** 11663–11668.

Soriano, P. (1999). Generalized lacZ expression with the ROSA26 Cre reporter strain. *Nat. Genet.* **21,** 70–71.

Stenman, J., Toresson, H., and Campbell, K. (2003). Identification of two distinct progenitor populations in the lateral ganglionic eminence: Implications for striatal and olfactory bulb neurogenesis. *J. Neurosci.* **23,** 167–174.

Tamamaki, N., Nakamura, K., Okamoto, K., and Kaneko, T. (2001). Radial glia is a progenitor of neocortical neurons in the developing cerebral cortex. *Neurosci. Res.* **41,** 51–60.

Theodosis, D. T., Pierre, K., Cadoret, M. A., Allard, M., Faissner, A., and Poulain, D. A. (1997). Expression of high levels of the extracellular matrix glycoprotein, tenascin-C, in the normal adult hypothalamoneurohypophysial system. *J. Comp. Neurol.* **379,** 386–398.

Toresson, H., and Campbell, K. (2001). A role for Gsh1 in the developing striatum and olfactory bulb of Gsh2 mutant mice. *Development* **128,** 4769–4780.

Toresson, H., Potter, S. S., and Campbell, K. (2000). Genetic control of dorsal-ventral identity in the telencephalon: Opposing roles for Pax6 and Gsh2. *Development* **127,** 4361–4371.

Tramontin, A. D., Garcia-Verdugo, J. M., Lim, D. A., and Alvarez-Buylla, A. (2003). Postnatal development of radial glia and the ventricular zone (VZ): A continuum of the neural stem cell compartment. *Cereb. Cortex* **13,** 580–587.

Voigt, T. (1989). Development of glial cells in the cerebral wall of ferrets: Direct tracing of their transformation from radial glia into astrocytes. *J. Comp. Neurol.* **289,** 74–88.

Walz, W. (2000). Controversy surrounding the existence of discrete functional classes of astrocytes in adult gray matter. *Glia* **31,** 95–103.

Wang, Y., Sheen, V. L., and Macklis, J. D. (1998). Cortical interneurons upregulate neurotrophins *in vivo* in response to targeted apoptotic degeneration of neighboring pyramidal neurons. *Exp. Neurol.* **154,** 389–402.

Weiss, S., Dunne, C., Hewson, J., Wohl, C., Wheatley, M., Peterson, A. C., and Reynolds, B. A. (1996). Multipotent CNS stem cells are present in the adult mammalian spinal cord and ventricular neuroaxis. *J. Neurosci.* **16,** 7599–7609.

Wictorin, K., Brundin, P., Gustavii, B., Lindvall, O., and Bjorklund, A. (1990). Reformation of long axon pathways in adult rat central nervous system by human forebrain neuroblasts. *Nature* **347,** 556–558.

Wood, H. B., and Episkopou, V. (1999). Comparative expression of the mouse Sox1, Sox2 and Sox3 genes from pre-gastrulation to early somite stages. *Mech Dev.* **86,** 197–201.

Ying, Q. L., Stavridis, M., Griffiths, D., Li, M., and Smith, A. (2003). Conversion of embryonic stem cells into neuroectodermal precursors in adherent monoculture. *Nat. Biotechnol.* **21,** 183–186.

Young, J. K., Baker, J. H., and Muller, T. (1996). Immunoreactivity for brain-fatty acid binding protein in gomori-positive astrocytes. *Glia* **16,** 218–226.

Zhuo, L., Theis, M., Alvarez-Maya, I., Brenner, M., Willecke, K., and Messing, A. (2001). hGFAP-cre transgenic mice for manipulation of glial and neuronal function *in vivo*. *Genesis* **31,** 85–94.

Zigova, T., Pencea, V., Wiegand, S. J., and Luskin, M. B. (1998). Intraventricular administration of BDNF increases the number of newly generated neurons in the adult olfactory bulb. *Mol. Cell Neurosci.* **11,** 234–245.

Zinyk, D. L., Mercer, E. H., Harris, E., Anderson, D. J., and Joyner, A. L. (1998). Fate mapping of the mouse midbrain-hindbrain constriction using a site-specific recombination system. *Curr. Biol.* **8,** 665–668.

Zupanc, G. K., and Clint, S. C. (2003). Potential role of radial glia in adult neurogenesis of teleost fish. *Glia* **43,** 77–86.

5

Classical Embryological Studies and Modern Genetic Analysis of Midbrain and Cerebellum Development

Mark Zervas, Sandra Blaess,* and Alexandra L. Joyner*,†*
*Howard Hughes Medical Institute, Developmental Genetics Program, Skirball
Institute of Biomolecular Medicine, Department of Cell Biology, New York
University School of Medicine, New York, New York 10016
†Department of Physiology and Neuroscience, New York University School of
Medicine, New York, New York 10016

The brain is a remarkably complex anatomical structure that contains a
diverse array of subdivisions, cell types, and synaptic connections. It is equally
extraordinary in its physiological properties, as it constantly evaluates and
integrates external stimuli as well as controls a complicated internal
environment. The brain can be divided into three primary broad regions:
the forebrain, midbrain (Mb), and hindbrain (Hb), each of which contain
further subdivisions. The regions considered in this chapter are the Mb and
most-anterior Hb (Mb/aHb), which are derived from the mesencephalon
(mes) and rhombomere 1 (r1), respectively. The dorsal Mb consists of the
laminated superior colliculus and the globular inferior colliculus (Fig. 1A
and B), which modulate visual and auditory stimuli, respectively. The dorsal
component of the aHb is the highly foliated cerebellum (Cb), which is
primarily attributed to controlling motor skills (Fig. 1A and B). In contrast,

Mark Zervas and Sandra Blaess contributed equally to this chapter.

Current Topics in Developmental Biology, Vol. 69
Copyright 2005, Elsevier Inc. All rights reserved.

101

0070-2153/05 $35.00
DOI: 10.1016/S0070-2153(05)69005-9
"

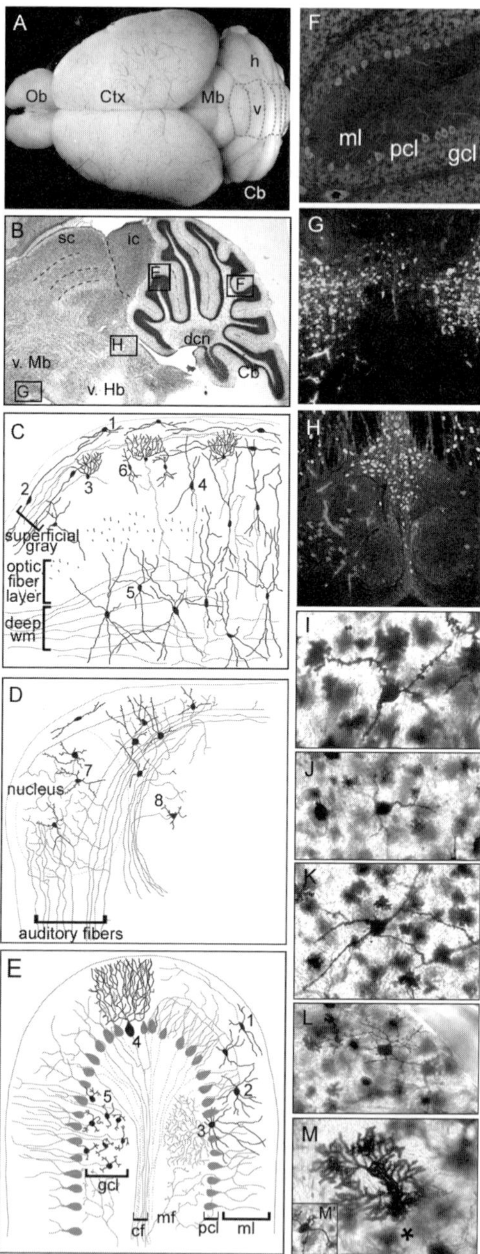

Figure 1 Midbrain and cerebellum morphology and cell types in the adult mouse. (A) The adult mouse brain showing the olfactory bulbs (Ob), cerebral cortex (Ctx), midbrain (Mb), as well as vermis (v) and hemispheres (h) of the cerebellum (Cb). (B) Nissl-stained sagittal section showing the superior (sc) and inferior (ic) colliculus of the dorsal Mb and Cb; also shown is the ventral Mb (v. Mb), hindbrain (v. Hb), and the deep cerebellar nuclei (dcn). The lettered

the ventral Mb/aHb (Fig. 1B) consists of distinct clusters of neurons that together comprise a network of nuclei and projections—notably, the Mb dopaminergic and Hb serotonergic and Mb/aHb cholinergic neurons (Fig. 1G and H), which modulate a collection of behaviors, including movement, arousal, feeding, wakefulness, and emotion.

Historically, the dorsal Mb and Cb have been studied using the chick as a model system because of the ease of performing both cell labeling and tissue transplants in the embryo *in ovo*; currently DNA electroporation techniques are also used. More recently the mouse has emerged as a powerful genetic system with numerous advantages to study events underpinning Mb/aHb development. There is a diverse array of spontaneous mutants with both Mb- and Cb-related phenotypes. In addition, numerous gene functions have been enumerated in mouse, gene expression is similar across vertebrates, and powerful genetic tools have been developed. Finally, additional insight into Mb/aHb function has been gained from studies of genetic diseases, such as Parkinson's disease, schizophrenia, cancer, and Dandy Walker syndrome, that afflict the Mb/aHb in humans and have genetic counterparts in mouse. Accordingly, this chapter discusses a spectrum of experiments, including classic embryology, *in vitro* assays, sophisticated genetic methods, and human diseases. We begin with an overview of Mb and aHb anatomy and physiology and mes/r1 gene expression patterns. We then provide a summary of fate-mapping studies that collectively demonstrate the complex cell behaviors that occur while the Mb and aHb primordia are established

---

marques indicate regions shown in illustrations or at higher magnification. (C–E) Semi-diagrammatic illustrations of the superior and inferior colliculi and the Cb as modified from Ramon y Cajal (Ramon y Cajal, 1995). (C, D) A diverse array of cell morphologies and the highly laminated versus globular cytoarchitecture of the sc (C) and ic (D) can be seen in these coronal views. Cell types: 1, marginal cell; 2, horizontal fusiform cell; 3, cell with complex dendritic bouquet; 4, large vertical fusiform cell; 5, large cells in transverse fiber layer; 6, radial fusiform cell; 7, spine laden triangular cells in ic nucleus; 8, multipolar cells in the central gray area. (E) The Cb folia displays a laminar arrangement and contain the following cell types: 1, stellate cells in molecular layer (ml); 2, basket cells in the molecular layer; 3, Golgi cells in Purkinje cell layer (pcl); 4, Purkinje cells in the pcl; 5, granule cells in granule cell layer (gcl); cf, climbing fibers; mf, mossy fibers. Note: solid and dashed red lines in cf region indicate cfs and Purkinje cell axons, respectively. (F) Calbindin-immunoreactive (IR, red) Purkinje cells in the Cb folium with dendrites in the ml; the gcl contains densely packed granule cells that can be observed with Hoechst staining (blue). (G, H) Horizontal section of v. Mb (G) or v. Hb (H). (G) Tyrosine hydroxylase-IR dopaminergic neurons of the v. Mb. (H) 5-hydroxy tryptophan-IR serotonergic (green) and choline acetyl transferase-IR cholinergic neurons (red) of the v. Hb. (I–M) Golgi impregnated neurons from adult mouse brain regions: (I) large triangular neuron with bulbous spines from the deep layer of sc; (J) small triangular neuron with few spines from superficial gray layer of ic; (K) pyramidal neuron from the substantia nigra of the v. Mb; (L) stellate cell from the Cb ml; (M, M′) Purkinje cell dendritic arbor (* indicates cell body that is out of focal plane) and granule cell, respectively. All cells in I-M were obtained at the same magnification for direct comparison. (See Color Insert.)

during embryogenesis and discuss the integration of both anterior-posterior (A-P) and dorsal-ventral (D-V) patterning. Finally, we describe some aspects of postnatal development and some of the insights gained from human diseases. © 2005, Elsevier Inc.

I. Cell Types and Projections of Mb and aHb

The most comprehensive details of cell types, morphology, and histology of the Mb and aHb can be found collectively in Ramon y Cajal's historic treatise, *Histology of the Nervous System* (Ramon y Cajal, 1995), and the wonderful series of descriptions by Altman and Bayer (Altman and Bayer, 1997). We have summarized many of their key histological and morphological descriptions in this section.

A. The Dorsal Mb and Cb

The dorsal Mb has two components: the superior and inferior colliculi (Fig. 1A and B). The superior colliculus is highly laminated; in chick it functions as the primary visual center and in mouse it processes and integrates visual stimuli largely as part of a reflex or attention module. The outermost layer of the superior colliculus is a fiber-rich marginal zone interspersed with horizontally oriented fusiform-shaped neurons (Fig. 1C). The intermediate layers contain a diverse collection of neurons, including small cells with complex dendritic arbors, radial fusiform cells, and triangular cells with large primary dendrites and bulbous spines (Fig. 1C and I). The deepest layer of both gray and white matter contains large cells with thick dendrites and axons that spread over long distances. The primary afferents (inputs) to the superior colliculus arise from the optic tract that traverses the lateral geniculate body and terminate in the more superficial layers. A deeper pathway of afferents innervating the superior colliculus originates from the cerebral cortex forming the corticotectal tract. The efferent (output) projections of the superior colliculus descend and innervate the trochlear, abducens, and oculomotor nuclei, which control eye muscles and pupilary reflexes.

The inferior colliculus, with a largely globular organization, is posterior to the superior colliculus and anterior to the Cb (Fig. 1A and B). The most superficial layer, again, is cell sparse (Fig. 1D). Mostly small to medium stellate-, fusiform-, and triangular-shaped cells populate the thin outer layer (Fig. 1J), while internally the inferior colliculus contains a layer of large multipolar neurons and a globular nucleus populated with spine-enriched neurons (Fig. 1D). The inferior colliculus in mouse is largely involved in the auditory reflex circuit and is extensively innervated. Axons relaying auditory stimuli course through the lateral lemniscus. Some end in the nucleus of the inferior colliculus, while others have terminals in both the inferior colliculus and medial geniculate nucleus. The primary efferent axons of the inferior colliculus project to the medial geniculate nucleus in the thalamus.

The avian and mammalian Cb has a foliated morphology that is species specific, although the cytoarchitecture within the folia is conserved. The mammalian Cb is subdivided into the medial vermis and the lateral paired hemispheres (Fig. 1A), whereas the avian cerebellum consists only of the vermis. Both the relative size of hemispheres to vermis and the complexity of foliation increase as the complexity of the mammal increases. The Cb contains only a few types of morphologically and physiologically distinct neurons that are arranged into three histologically distinct layers (Fig. 1B, E, and F): (1) the outer cell sparse molecular layer, which contains granule cell axons, Purkinje cell dendrites, and basket and stellate cells (Fig. 1L), (2) a monolayer of Purkinje cells (Fig. 1M) intercalated with Golgi cells, and (3) the internal granule cell layer containing small, densely packed granule cells (Fig. 1M). The Cb also contains deep cerebellar nuclei (Fig. 1B), which are organized in distinct medial-to-lateral clusters.

The Cb coordinates motion and proprioception, using sensory inputs from skin, joints, muscles, the vestibular apparatus, and the eye to fine-tune movements and balance. The two primary inputs into the Cb are the climbing fibers and mossy fibers (Fig. 1E). Climbing fibers originate in one of the pre-Cb nuclei (inferior olive) and terminate on spines of the large dendrites of Purkinje cells; they primarily mediate muscle proprioception. In contrast, mossy fibers originate in the spinal cord and in the pre-Cb nuclei located in the medulla and pons and form synapses with axons of Golgi cells and dendrites of granule cells in glomeruli. Mossy fibers from the pontine nuclei convey information from the cerebral cortex. The Golgi, basket, stellate, and granule cells are local circuit neurons making all of their connections within the Cb (Fig. 1E). The Purkinje cells integrate sensory information relayed to the Cb and project to the neurons of the deep cerebellar nuclei, which in turn project out of the Cb.

B. The Ventral Mb and aHb

The ventral Mb and aHb are a network of distinct nuclei that have highly specialized modulatory functions. Described in this section are a few of the well-characterized nuclei that encompass four primary neurotransmitter phenotypes: dopaminergic, serotonergic, cholinergic, and noradrenergic. Dopaminergic neurons express tyrosine hydroxylase (TH) and are present in the ventral Mb (Fig. 1G and K) but not in the ventral aHb. Mb dopaminergic neurons distributed medially in the ventral tegmental area project to the cerebral cortex and modulate cognitive processes, while the laterally positioned dopaminergic neurons of the substantia nigra pars compacta and pars reticularis innervate the striatum and control movement. Serotonergic neurons express 5-hydroxytryptophan (5-HT) and are dispersed along the A-P axis of the Hb (Fig. 1H), and in some adult species, such as mouse and human, they extend into the caudal Mb; collectively they modulate complex behaviors such as mood, arousal, and sleep. Cholinergic neurons express choline acetyl transferase (ChAT) and in the Mb are the most posterior and lateral nucleus: the parabigeminal nucleus (mouse) or the isthmic nucleus (chick). Cholinergic terminals are widely distributed throughout the brain, but the parabigeminal nucleus primarily has synaptic terminals in the superior colliculus and modulates attention to visual stimuli. Cholinergic neurons are also present in the trigeminal nucleus (motor component) and the facial nucleus of the ventral aHb (Fig. 1H). The locus coeruleus (LC) is located ventral to the Cb and is the major brain noradrenergic nucleus; its neurons project to the entire central nervous system and modulate behavioral and cognitive processes.

II. Gene Expression and Functional Analysis of the mes and r1

The Mb and aHb are an ideal system in which to study the complex mechanisms underpinning brain development because the two regions contain anatomically, histologically, and physiologically distinct structures despite their close regional proximity. Shortly after the neural tube closes, a series of morphologically distinct bulges, termed neuromeres, can be observed in the developing brain during early embryogenesis (Fig. 2A and B). The neuromeres that give rise to the Mb and aHb are the mesencephalon (mes) and rhombomere 1 (r1), respectively (Palmgren, 1921; Ramon y Cajal, 1995). Located between the posterior mes and r1 dorsally is a morphological constriction termed the isthmus (Altman and Bayer, 1997; Palmgren, 1921) (Fig. 2A and B); the morphological segregation of the mes and r1 ventrally is not as obvious, but can generally be demarcated by a small notch (Fig. 2B). In this section we describe mes/r1 gene expression patterns and the consequences of loss or gain of function of these genes. We also summarize the hierarchical yet interdependent cascade of transcription factors and secreted molecules that pattern the mes and r1 and that are critical for Mb/aHb development (Table I; Figs. 3 and 4).

A. Anterior-Posterior mes/r1

Otx2 and *Gbx2*, members of the homeobox family of transcription factors, are induced independently of each other and first subdivide the embryo into anterior and posterior regions during the early head fold stage of

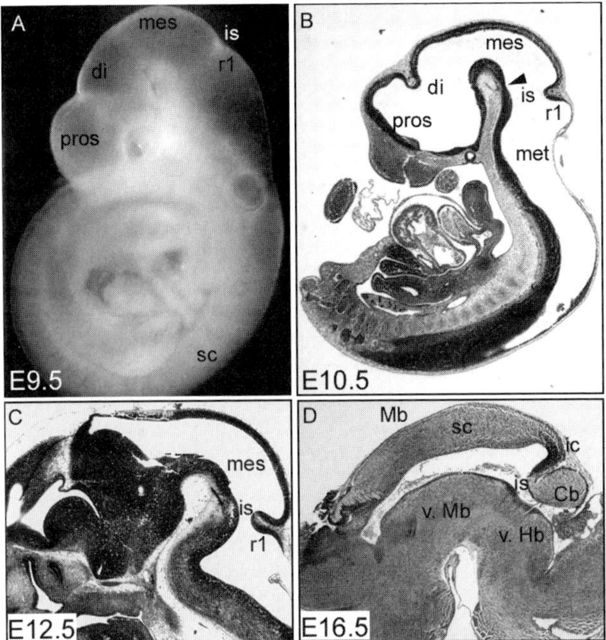

Figure 2 Mes/r1 morphogenesis in the developing mouse embryo. (A) E9.5 embryo showing the prosencephalon (pros), diencephalon (di), mesencephalon (mes), isthmus (is), rhombomere 1 (r1), and spinal cord (sc). (B–D) Hematoxylin/eosin-stained sagittal sections of an E10.5 embryo (B), E12.5 head (C), and E16.5 brain (D). Additional abbreviations: ic, inferior colliculus; sc, superior colliculus; v. Mb, ventral midbrain; v. Hb, ventral hindbrain. (See Color Insert.)

development. *Otx2* and *Gbx2* eventually become juxtaposed at a common interface located at the posterior mes/anterior isthmus (Li and Joyner, 2001; Wassarman *et al.*, 1997) (Figs. 3A and 4). Gene expression studies in loss-of-function mutants demonstrated that *Otx2* and *Gbx2* rapidly become dependent on each other to correctly delineate the limits of their expression domains (Broccoli *et al.*, 1999; Li and Joyner, 2001; Wassarman *et al.*, 1997). The removal of *Otx2* in the embryo causes *Gbx2* to expand anteriorly and a concomitant loss of Mb and an enlargement of the Cb. In contrast, the complete loss of Gbx2 results in a posterior expansion of *Otx2* and the loss of Cb at the expense of Mb expansion (Li and Joyner, 2001; Martinez-Barbera *et al.*, 2001). Although these two genes are critical for the development and positioning of the Mb/aHb, they are not required for the subsequent induction of additional mes/r1 genes (Li and Joyner, 2001).

Other key homeobox transcription factors include *Pax2/5*, *Engrailed1/2 (En1/2)*, and *Lmx1b,* which are members of the paired-rule, segmentation, and LIM homeodomain family of transcription factors, respectively, and are

Table I   Essential Genes for Mesencephalon and Rhombomere 1 Development[a]

Gene	Expression domains Domains E9.5 Mouse	Loss-of-function Mouse	Gain-of-function Mouse and Chick
Otx2	entire mes	Mb deletion	↑*Lmx1b, Wnt1* ↓*Gbx2* ectopic Mb induced in r1
Gbx2	r1	r1–r3 deletion	↓*Lmx1b, Wnt1, Otx2* ectopic Cb induced in mes
Pax2	mes/isthmus/ant. r1	post Mb deletion is variable, depends on allele	↑*Fgf8*
Pax5	mes/r1	mild post Mb deletion mild Cb foliation defect	
En1	mes/r1	Mb and Cb deletion	↑*Fgf8* (along with *Pax2*)
En2	mes/r1	mild Cb size reduction mild Cb foliation defect	
Lmx1b	v mes	substantia nigra depleted serotonergic nuclei absent	↑*Wnt1, Otx2* ↓*Fgf8*
Wnt1	post mes d mes-medial stripes v mes-medial stripes	entire Mb and Cb deletion	↑*En1, Fgf8* ectopic foliation in Cb (chick)
Fgf8	isthmus	Mb and Cb deletion	↑ *Lmx1b, Wnt1, Gbx2* ↑ *Pax5, En1/2* ↓*Otx2, Pax6* induces changes of tissue fate
Shh	v mes/r1 (floor plate)	v Mb and Hb deletion mes/r1 size reduction	↑*Gli1, Hnf3β,* ↓*Pax3/7, Gli3* induces changes of tissue fate
Bmps	d mes/r1 (roof plate)	d structures aberrant	

[a]Summary of essential genes in mes/r1 development. The details of gene expression, loss-of-function phenotypes, and gain-of-function changes in gene expression and associated phenotypes are discussed in the text. Gene expression patterns are listed for E9.5 mouse mes/r1 and are generally applicable to chick and zebrafish at an equivalent stage. Although many of these genes are expressed out of the mes/r1, their additional domains are not listed here. See text for relevant references and Section II.A and II.B for details of gene expression dynamics and loss-of-function phenotypes and Section II.C for gain-of-function experiments. Abbreviations: ventral (v), dorsal (d), posterior (post), anterior (ant), mesencephalon (mes), rhombomere 1 (r1), midbrain (Mb), cerebellum (Cb).

expressed in overlapping domains in the mes and r1. *Pax2* is initiated first, prior to somite formation, with *En1* following at the 1-somite stage and *Pax5* and *En2* by the 5-somite stage. These genes are first expressed broadly in the mes and r1 and then become refined to distinct domains of the

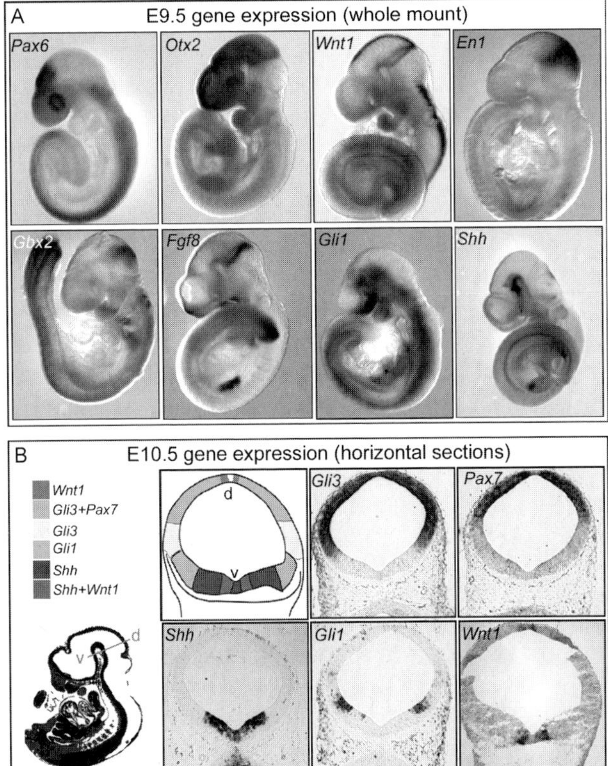

Figure 3 Gene expression in mes/r1 at embryonic stages. (A) Whole mount *in situ* hybridization with the indicated probes showing gene expression patterns at E9.5. *Pax6* is expressed in the prosencephalon, rhombomeres 2–7, and spinal cord. *Otx2* is expressed throughout the pros and mes. *Wnt1* is expressed in a semi-circle at the posterior limit of the mes and in a row at the dorsal and ventral midline of the diencephalon and mes as well as in r2–7 and sc. *En1* traverses both the posterior mes and r1. *Gbx2* is localized to r1 and the tail bud. *Fgf8* is expressed in signaling centers including r1 in a semi-circular pattern and in the anterior neural ridge, limb apical ectodermal ridge, branchial arches, and tail bud. *Gli1* is expressed adjacent to the floor plate and extends along the entire A-P axis. *Shh* is expressed in the floor plate at the ventral midline and becomes broader in the mes/r1 region. Figure 2A shows anatomical subdivisions and Figure 4B shows mes/r1 spatial relationships. (B) Gene expression along dorsal-ventral axes at E10.5. *In situ* hybridization on horizontal sections as shown by the red line through the embryo on the left. Dorsal (d) is at the top, ventral (v) at the bottom. The spatial relationship of the indicated genes is shown in the schematic in the upper left panel. (See Color Insert.)

posterior mes, isthmus, and r1 (Asano and Gruss, 1992; Davis and Joyner, 1988; Rowitch and McMahon, 1995; Urbanek *et al.*, 1997). Loss-of-function analysis of multiple *Pax2* mutant mice indicated that *Pax2* might play a critical role in Mb/Cb development: depending on the genetic background,

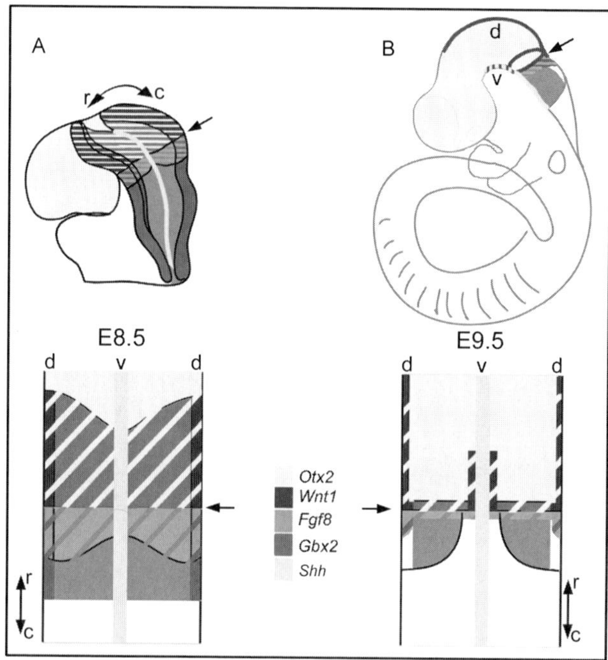

Figure 4 Illustrations of mouse embryos at E8.5 and E9.5 highlighting changes in morphology and gene expression patterns. Drawings depict whole mount embryos (top) or flat mount schematic (bottom). (A) Overlapping gene expression is shown at E8.5 when the neural tube is open; only the anterior region of the embryo is drawn. The bottom schematic shows the neural tube spread flat. The *Shh* domain is located in the midline (labeled as ventral, v) while *Wnt1* is strongest in the lateral neuroepithelium (labeled as dorsal, d) to indicate the M-L transition to D-V once the neural tube closes. *Otx2* and *Wnt1* are expressed in the mes versus *Gbx2* and *Fgf8*, which are expressed in r1; *Shh* is expressed in the floor plate. The rostral-caudal axis (r-c) is indicated; the arrow indicates the mes/r1 interface. (B) Shortly after neural tube closure, gene expression undergoes a dynamic change. *Wnt1* becomes restricted to a semi-circle at the posterior mes as well as the dorsal and ventral midline. *Fgf8* also becomes restricted to a semi-circle in the isthmus that is posterior and juxtaposed to *Wnt1*. The flat mount illustration depicts the neural tube cut along the dorsal midline; ventral and dorsal are therefore illustrated medially and laterally, respectively. This facilitates a direct comparison of the gene expression domains at E9.5 to E8.5. (See Color Insert.)

a targeted frameshift (null) mutation results either in no obvious mes/r1 phenotype or in Mb exencephaly; in contrast, a spontaneously occurring frameshift mutation causes a deletion of the posterior mes and r1 (Favor *et al.*, 1996; Schwarz *et al.*, 1997; Torres *et al.*, 1995). *Pax5* mutant mice display a slight reduction of the posterior Mb (inferior colliculus) and a mild Cb foliation defect (Urbanek *et al.*, 1994, 1997). Studies of *Pax2/5* double mutant mice revealed that the two genes function cooperatively in Mb and

Cb development (Schwarz *et al.*, 1997; Urbanek *et al.*, 1997). Consistent with this, *Pax5* knocked-in to the *Pax2* locus partially rescues the Mb/aHb phenotype, further demonstrating redundant functions and the necessity of the *Pax2/5* family in Mb/aHb development (Bouchard *et al.*, 2000). Loss of *En1*, the first *En* expressed, results in a loss of almost all Mb/aHb tissue (Wurst *et al.*, 1994). In contrast, *En2* loss of function causes a mild phenotype consisting of a smaller Cb with foliation defects (Joyner *et al.*, 1991; Millen *et al.*, 1994). Although *En2* can rescue the *En1* phenotype in knock-in experiments, indicating functional redundancy, the differences in their temporal expression and mutant phenotypes suggest that *En1* plays the primary role in early mes/r1 patterning (Hanks *et al.*, 1995). *Lmx1b* is expressed initially throughout the posterior mes, nested within the *Wnt1* domain (see below), and is also expressed in anterior r1 (overlapping with *Fgf8*) in chick. Eventually, chick *Lmx1b* becomes refined to the posterior mes coincident with *Wnt1* expression (Matsunaga *et al.*, 2002). In contrast, *Lmx1b* in E16 rat embryos, as shown by section *in situ* hybridization, is expressed in the ventral Mb/aHb. *Lmx1b* expression in mouse has not been described in detail, but a targeted mutation in murine *Lmx1b* results in the depletion of dopaminergic neurons of the Mb substantia nigra (see below) and a complete loss of all Hb serotonergic neurons (Ding *et al.*, 2003; Smidt *et al.*, 2000).

In addition to the diverse array of transcription factors expressed in the mes and r1, a number of key secreted molecules are critical for mes/r1 patterning. *Wnt1* is first expressed at the 1-somite stage, and by 6–8 somites (E8.5) is expressed throughout the mes neural plate in a mosaic pattern, although the lateral edges show the highest, most uniform expression (Echelard *et al.*, 1994; Wilkinson *et al.*, 1987) (Fig. 4A). As the neural tube closes, at E9.0–E9.5 in the mouse, the lateral edges come together to form the dorsal midline. At this stage, *Wnt1* undergoes a dramatic change in expression that results in *Wnt1* transcripts being restricted to a small, tight semi-circular domain at the posterior limit of *Otx2* expression in the mes and to a set of bilateral stripes along the dorsal and ventral midlines (Zervas *et al.*, 2004) (Figs. 3A and 4B). *Wnt1* continues to be expressed in this refined pattern until its cessation at E13.5 (Wilkinson *et al.*, 1987). *Wnt1* is a key player in Mb/aHb development, as evident by the complete loss of the Mb and almost the entire Cb in one knockout mouse line (McMahon and Bradley, 1990; McMahon *et al.*, 1992). A second targeted mutant allele of *Wnt1* causes the Mb and aHb to be significantly, but variably, disrupted (Thomas and Capecchi, 1990). The second allele is reminiscent of a spontaneous point mutation in *Wnt1* that was initially identified because of its characteristic swaying phenotype, referred to as the *Swaying* (*Wnt1^{sw}*) allele (Bronson and Higgins, 1967; Lane, 1967; Thomas *et al.*, 1991). Analysis of *Swaying* embryos in which ectopic clusters of *Wnt1* and *Otx2* are present in r1 suggests that *Wnt1* may play a role in maintaining the lineage restriction described below at the *Otx2/Gbx2* border (Bally-Cuif *et al.*, 1995). It has also been suggested that *Wnt1* regulates *En1* expression because a knock-in of *En1* into the *Wnt1* locus partially rescues the *Wnt1* Mb/aHb phenotype (Danielian and McMahon, 1996).

Fgf8 is first expressed during gastrulation and is eventually localized to a number of signaling centers important for embryonic patterning (Crossley and Martin, 1995). In the mes/r1 region, *Fgf8* is initially expressed in the neural plate in a band of cells located just posterior to the *Wnt1* expression domain (Fig. 4A). By E9.0–9.5 *Fgf8* becomes restricted to a semi-circular ring in the isthmus and acts as a signaling center that directs mes/r1 development (Crossley and Martin, 1995; Crossley *et al.*, 1996a,b; Meyers *et al.*, 1998; Reifers *et al.*, 1998) (Figs. 3A and 4B) (see Section IV for role in patterning). Loss-of-function analysis revealed that *Fgf8* is required for the formation of the entire Mb/aHb territory in a dose-dependent fashion (Chi *et al.*, 2003; Meyers *et al.*, 1998). Additional Fgfs such as *Fgf17* and *18* that are expressed in broader domains than *Fgf8* also have roles in mes/r1 development (Liu and Joyner, 2001a; Liu *et al.*, 2003; Xu *et al.*, 2000).

B. Dorsal-Ventral mes/r1

Sonic hedgehog (Shh) is a secreted signaling factor, and its role in the specification of the ventral spinal cord has been extensively studied (Briscoe and Ericson, 1999; Chiang *et al.*, 1996; Echelard *et al.*, 1993; Ericson *et al.*, 1996). Shh is expressed in the notochord that underlies the neural plate from the early headfold stage and induces Shh expression in the floor plate, which is located in the ventral midline of the neural tube (Echelard *et al.*, 1993). Within the mes/r1, Shh is expressed in a thin stripe of floor plate cells at E8.0 in the mouse and in a broader ventral-lateral domain of the mes/r1 at subsequent stages (E10.5–12.5) (Echelard *et al.*, 1993; Matise *et al.*, 1998; Zervas *et al.*, 2004) (Figs. 3 and 4). At similar stages in chick, Shh expression fans out from the ventral midline, occupying a broad expression domain in the mes and a more narrow domain in r1 (Agarwala *et al.*, 2001). In chick and mouse, Shh expression in the ventral midline is maintained at least up to late embryonic stages (Fu *et al.*, 2003). Throughout development, Shh-expressing cells are flanked laterally by domains that are positive for *Gli1* (Zervas *et al.*, 2004, and S. Blaess, unpublished data) (Fig. 3), a zinc finger transcription factor downstream of Shh signaling that is a transcriptional readout for positive Shh signaling (Bai *et al.*, 2002). It has been shown that Shh is essential for the induction and patterning of ventral neural tissue (Echelard *et al.*, 1993) (see Section IV for role in patterning). Accordingly, in *Shh* loss-of-function mouse mutants, all ventral neuronal subtypes are lost because the ventral domain is not specified and the dorsal markers *Pax3/7* expand throughout the entire D-V extent of the neural tube (Chiang *et al.*, 1996; Fedtsova and Turner, 2001).

The dorsal neural tube is characterized by the expression of several transcription factors. Secreted signals from the ectoderm and the roof plate, such as members of the bone morphogenetic protein (BMP) and Wnt families, specify dorsal cell types (Lee and Jessell, 1999). However, there is little experimental evidence demonstrating their role specifically in dorsal mes/r1 patterning. This is primarily because of redundancy in gene function, with two or more members of the same gene family being expressed in overlapping patterns. In addition, loss of function of a particular gene often results in the failure of the neural tube to close and/or early lethality (Lee and Jessell, 1999).

The paired box transcription factors *Pax3* and *Pax7* are initially expressed throughout the neural plate and are subsequently suppressed in ventral areas by Shh signaling from the notochord and floor plate (Ericson *et al.*, 1996; Goulding *et al.*, 1993). In the mes/r1, *Pax3/7* expression becomes restricted to the dorsal (alar) plate after neural tube closure (Fedtsova and Turner, 2001) (Fig. 3B). Defects in dorsal patterning of mes/r1 have not been observed in mice deficient for *Pax7*. However, *Pax3* (*Splotch*) mutants and *Pax3/7* double mutants display severe exencephaly and spina bifida, preventing the assessment of the mes/r1 phenotype (Auerbach, 1954; Mansouri and Gruss, 1998; Mansouri *et al.*, 1996). In chick gain-of-function studies, it has been reported that ectopic expression of *Pax3/7* in ventral mes causes the dorsal tectum to expand ventrally, indicating their role in dorsal specification (Matsunaga *et al.*, 2001). Expression of *Gli3*, a zinc finger transcription factor that is a repressor downstream of Shh signaling, is restricted to dorsal-lateral regions of the mes/r1 (Aoto *et al.*, 2002) (Fig. 3B). In *Gli3 extratoe* null mutants, the presumptive dorsal Mb/Cb anlage is present at E11.5 (Theil *et al.*, 1999), although both changes in gene expression and cell death in the mes/r1 have been described (Aoto *et al.*, 2002).

Evidence for the importance of roof plate signaling in dorsal mes/r1 development has come from mouse mutants with partial or complete loss of the roof plate. Defects in roof plate formation have been described in *dreher* mutant mice that have a spontaneous mutation in the LIM homeodomain protein *Lmx1a*, which is expressed in the roof plate. Mutant mice survive to adulthood and have an abnormal inferior colliculus, a loss of the vermis, and a severe perturbation of the foliation pattern in the Cb hemispheres (Millonig *et al.*, 2000). Genetic ablation of the roof plate by driving the expression of the diphtheria-toxin A subunit from the *Gdf7* locus results in the failure of the neural tube to close at the level of the mes/r1 and causes a reduction in the size of the mes/r1, but it is not clear if this is specific to the loss of dorsal mes/r1 (Lee *et al.*, 2000).

Members of the Bmp family, including *Bmp5/6/7* and *Gdf7,* are initially broadly expressed in the neural plate in zebrafish, chick, and mouse but are confined to the roof plate/epidermal ectoderm after neural tube closure and are involved in cell fate specification of dorsal r1 (Alder *et al.*, 1999; Alexandre and Wassef, 2003; Puelles *et al.*, 2003; Vogel-Hopker and Rohrer, 2002). In zebrafish, BMPs have been implicated in the generation of noradrenergic neurons of the LC. Even though the LC is located in the ventral aHb in the adult, the progenitors of these neurons are induced in dorsal/lateral r1. *Bmp5* is expressed in the dorsal neural tube, including the roof plate of r1, and *Bmp2b/Bmp7* are expressed in adjacent epidermal ectoderm. *Phox2a*-positive LC precursors are absent in *Bmp2b (swirl)* and in *Bmp7* (*snailhouse*) mutants, and the application of the BMP antagonist noggin leads to either the complete loss or dorsal relocation of these precursors (Guo *et al.*, 1999; Vogel-Hopker and Rohrer, 2002). However, inhibition of BMP signaling in zebrafish also leads to a loss of other roof plate markers and to a reduced size of the neural tube, leaving the possibility that other factors secreted from the roof plate could influence the induction of the LC precursors. In mouse, BMP6/7 or GDF7 added to E8.0 ventral mes/r1 explants induces *Math1/Zic2*-positive cerebellar granule cell precursors (GCPs) after 2 days *in vitro* (Alder *et al.*, 1999). Even though these experiments suggest a direct role for BMPs in specification of two distinct groups of dorsal cell types, it remains unclear whether BMPs are generally required for the induction of dorsal cells in mes/r1.

W*nt1/Wnt3a* are expressed in two bilateral stripes flanking the roof plate in the mes, with *Wnt3a* extending caudally into dorsal r1 (Parr *et al.*, 1993) (Fig. 3B). Wnts are also expressed in the ventral mes. *Wnt1* expression is observed in two ventral stripes in mouse that partially overlap with *Gli1* expression transiently at the 6–8 somite stage and is subsequently nested in the rapidly expanding Shh domain (Zervas *et al.*, 2004) (Fig. 3B). *Wnt5A* and *Wnt7A* are also expressed in stripes bilateral to the ventral midline in mouse and chick (Parr *et al.*, 1993; Sanders *et al.*, 2002). *Wnt1* mutant mice display severe A-P patterning defects, precluding direct functional analysis of D-V defects, but fate-mapping studies demonstrate that *Wnt1*-expressing cells themselves give rise to ventral Mb dopaminergic neurons, and it appears that at least a subset originates from the ventral midline (Zervas *et al.*, 2004). WNT-conditioned medium added to dissociated cells isolated from E14.5 rat ventral mes suggests specific roles for Wnt signaling in D-V patterning: WNT1 and WNT5a increase the number of TH-positive dopaminergic neurons by influencing cell proliferation and cell fate specification, respectively (Castelo-Branco *et al.*, 2003). No major defects in dorsal patterning have been reported for *Wnt3a* mutants (Takada *et al.*, 1994), possibly because of functional redundancy with *Wnt1*. *Wnt1/Wnt3a* double mutants, however, show a significant reduction in the number of dorsal interneurons in the spinal cord, and WNT3a can induce dorsal interneurons in medial neural plate explants in chick (Ikeya *et al.*, 1997; Muroyama *et al.*, 2002). These experiments provide some evidence for a potential general role for *Wnts* in D-V patterning of the mes/r1 that might be independent of *Wnt1* function in A-P patterning.

C. A Complex Genetic Cascade Regulates mes/r1 Development

Both loss- and gain-of-function experiments, as well as *in vitro* explants, have provided insight into the complex genetic hierarchy and entwined regulatory network that underlie Mb/aHb patterning. We have summarized many of the key gain-of-function experiments in Table I based primarily on certain experimental findings (Adams *et al.*, 2000; Liu and Joyner, 2001a,b; Liu *et al.*, 1999; Matsunaga *et al.*, 2002; Ye *et al.*, 2001). Loss-of-function studies tend to be the most difficult for interpreting epistatis among mes/r1 genes because the loss of almost any mes/r1 gene results in the subsequent decay of gene expression and/or loss of tissue, but we have listed many of the important loss-of-function findings discussed in Table I. Because of their early expression and role in subdividing the A-P domains of the embryo, *Otx2* and *Gbx2* may be placed near the top of the genetic mes/r1 hierarchy. The induction of ensuing mes/r1 genes proceeds independently of all known mes/r1 genes *in vivo,* precluding an obvious induction cascade. Gain-of-function experiments, however, demonstrate that many of the mes/r1 genes are sufficient to induce or repress mes/r1 gene expression. For example, *Otx2* can induce *Lmx1b* and *Wnt1* but represses *Gbx2.* In contrast, *Gbx2* represses *Otx2, Lmx1b,* and *Wnt1.* In addition, *Lmx1b* induces *Otx2* and *Wnt1* but represses *Fgf8. En1* induces the expression of *Fgf8,* but only in the presence of *Pax2,* although *Pax2* itself induces *Fgf8. Fgf8,* like *Otx2,* can induce *Lmx1b* and *Wnt1. Fgf8* also induces *Gbx2, Pax5,* and *En1/2* while repressing *Otx2* and *Pax6,* and *Wnt1* induces *En1* and *Fgf8.* The loss-of function and gain-of-function results indicate that mes/r1 genes are sufficient, but not necessary, for each other's induction and shows that gene induction proceeds independently of maintenance, but that the maintenance of mes/r1 gene expression rapidly becomes highly interdependent.

III. Cell Behaviors of the mes and r1

Gene expression is one mechanism involved in establishing the identity of an otherwise naïve developmental field. The emergence and refinement of gene expression must be coordinated with morphogenetic movements in order to generate the appropriate cell types arranged in characteristic histological organization along the A-P and D-V axes. We describe a wide spectrum of fate-mapping studies that collectively have elucidated the morphogenetic movements and cell behaviors that underlie mes/r1 development. A number of fate-mapping studies in chick suggested that the Mb is derived from the mes, but that the Cb was derived from both the mes and r1. In these classic quail-to-chick homotypic transplantations, donor quail tissue was micro-dissected and transplanted into similar host chick regions. After successful engraftment, quail tissue is readily detected using the QPCN antibody. Although quail and chick cells proliferate at different rates and grafting generally causes some tissue distortion, the method provided reproducible and insightful results. When Hamburger-Hamilton (HH) stage 12 (HH12) quail dorsal-posterior mes was transplanted into chick dorsal mes, it was observed that the quail-derived cells contributed to the chick Mb (Hallonet and Le Douarin, 1993; Hallonet *et al.*, 1990). Similarly, quail dorsal r1 transplants contributed to the Cb (Hallonet and Le Douarin, 1993; Hallonet *et al.*, 1990). An additional finding was that HH12 mes-derived tissue gave rise to an anterior-medial wedge in the Cb (Hallonet and Le Douarin, 1993; Hallonet *et al.*, 1990). Together, these results suggested that the Cb has a dual origin in both the mes and r1. This assertion was seemingly supported by short-term cell labeling and cell aggregation assays in HH11–12 chick, which showed that cells from the mes/r1 intermingle (Jungbluth *et al.*, 2001). An elegant transplant study in chick that relied on micro-dissecting mes/r1, in combination with genetic marker analysis, provided evidence that the dorsal mes contributes only to the dorsal Mb (tectum), while dorsal r1 contributes to the Cb (Millet *et al.*, 1996). Importantly, using *in situ* hybridization with *Otx2* as a genetic marker (Fig. 3A), Millet *et al.* determined that the caudal limit of *Otx2* expression delineates the posterior limit of the mes and hence the future Mb (Millet *et al.*, 1996). Furthermore, it was demonstrated that at HH10 there is a small Otx2-negative domain in the posterior mes that gives rise to the Cb, and by HH20 the Otx2-positive region is coincident with the isthmus. Therefore, the mes as defined by *Otx2* expression gives rise exclusively to the Mb and not Cb in chick (Millet *et al.*, 1996). The discrepancy of mes-derived tissue contributing to the Cb is apparently due to the caudal limit of the mes graft obtained at HH11–12 being in *Otx2*-negative territory. Short-term DiI labeling experiments also support the findings that cell movements between the mes and r1 in chick are restricted: labeled cells in lateral posterior mes tended to move medially and anteriorly, while labeled cells in medial posterior mes contributed to the isthmus, but not r1 (Alexandre and Wassef, 2003; Louvi *et al.*, 2003). The remarkable feature of the fate-mapping studies in chick was generally how accurate they were in determining the primordia of the adult Mb and Cb given the difficulty of precisely removing and replacing embryonic tissue *in ovo*.

More recently, a genetic inducible fate-mapping approach in mouse has been developed that allows small populations of cells to be marked with spatial resolution based on genetic lineage and temporal resolution based on inducible Cre recombinase (Danielian *et al.*, 1998; Kimmel *et al.*, 2000; Zervas *et al.*, 2004). In this tripartite system, regionally expressed promoters drive inducible Cre recombinase (CreER^T), tamoxifen induces CreER^T (Feil *et al.*, 1996), and recombination is monitored by the *R26R* (flox-stop-flox-lacZ) reporter allele (Soriano, 1999). Tamoxifen application induces recombination in embryos *in utero* and marks cells over 24–36 hr, allowing discrete populations of progenitors to be marked; *LacZ* is heritable and therefore provides a permanent record of marked cells and their descendants. This method thus allows developmental questions to be addressed noninvasively in mouse with precision unobtainable in transplantation studies, when appropriate promoters are available. This genetic fate-mapping approach was used to address the cell behaviors underpinning Mb/aHb development (Zervas *et al.*, 2004). In this study, using *Wnt1-CreER^T* to initially mark the entire mes at E8.5 or a small population at the posterior mes at E9.5, it was demonstrated that mes-derived cells do not migrate into r1, thus proving the presence of a lineage restriction boundary located at the dorsal-posterior mes (*Otx2* caudal border) in mouse (Fig. 5A and B). The small group of posterior cells marked at E9.5, however, expanded rapidly and underwent intracompartmental (mes) mixing and gave rise to the inferior colliculus (Fig. 5C). Interestingly, the boundary segregating the ventral-posterior mes from ventral r1 was established 1 day after the dorsal boundary, illustrating a unique temporal aspect to mes/r1 neuromere formation. Using an array of regional promoters (*En1, En2, Wnt1, Gli1*) to drive CreER^T in transgenic mice, two other observations were made. First, additional boundaries along the A-P axis at the anterior mes/diencephalon interface and r1/r2 interface partition the mes and r1 into distinct neuromeres that can be identified by genetic lineage: the mes is derived from a *Wnt1/En1* lineage, and r1 is derived from an *En1*, but not *Wnt1*, lineage. Second, cells originating from either the dorsal or ventral mes do not migrate into adjacent ventral or dorsal territories, respectively (Zervas *et al.*, 2004). Short-term studies of fluorescent cell labeling in zebrafish and chick indicated that the boundaries located at the anterior mes, anterior r1, and posterior r1 are bidirectional (Koster and Fraser, 2001; Larsen *et al.*, 2001; Wingate and Hatten, 1999).

Fate-mapping studies of the developing Cb have uncovered when and where cells are born and how they subsequently move to their final positions. During Cb development, most of the cell types are generated in distinct waves from the ventricular zone of dorsal r1 and contribute to the Cb primordium in a coordinated manner (Altman and Bayer, 1997). In contrast, GCPs are generated in the anterior rhombic lip that forms at the interface of the neural tube and the extended roof plate of the 4th ventricle of the

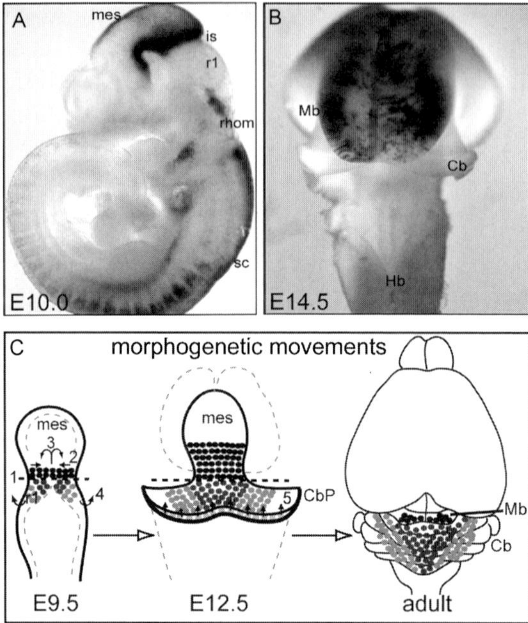

Figure 5 Morphogenetic movements in Mb and Cb development. (A) The initial population of *Wnt1*-derived cells marked by tamoxifen administration to Wnt1-CreERT; R*26R* mouse embryos at E8.5 is shown at E10.0. In this sagittal view, marked cells can be seen in the mes, but not in r1. (B) By E14.5 the mes and r1 have undergone growth and a dramatic change in morphology (compare A to B) to give rise to the Mb and Cb, respectively. During this time *Wnt1*-derived cells marked by tamoxifen at E8.5 are retained in mes and cannot migrate into r1, resulting in the Mb being marked, but not the Cb. (C) Schematic of dorsal views illustrating complex morphogenetic movements underlying Mb and Cb development. E9.5: A lineage boundary restricts mes-derived cells (blue circles) from posterior movement into r1 (1); mes-derived cells migrate from lateral to medial (2); but primarily expand anteriorly (3); r1 begins to undergo an orthogonal rotation (4). E12.5: Mes-derived cells have expanded to fill the posterior mes and r1 has rotated such that anterior r1 (red circles) becomes medial while posterior r1 (green circles) becomes lateral; granule cell precursors from the upper rhombic lip (orange circles) migrate over the surface of the Cb primordium (CbP) (5). Adult: The final distribution of the indicated cells is shown in the adult; granule cells have settled deep within the Cb during early postnatal development and are not shown (see Section III in text for details). (See Color Insert.)

hindbrain (Wingate, 2001). Furthermore, GCPs remain mitotically active as they move over the Cb primordium to form the external granule cell layer (Fig. 5C). During the first 2 postnatal weeks in mouse, GCPs differentiate and migrate through the Purkinje cell layer to form the internal granule cell layer (Hatten and Heintz, 1995). A recent genetic inducible fate-mapping study demonstrated that dorsal r1 first undergoes an orthogonal rotation such that the A-P axis of r1 at E9.5 becomes the M-L axis of dorsal r1 at

E12.5 (Fig. 5C) (Sgaier *et al.*, 2005). The M-L axis of ventricular-derived cells is then retained: Purkinje cells generated medially are located in the vermis while more lateral Purkinje cells populate the hemispheres (Fig. 5C). In contrast, the M-L generation of granule cells is only partially reflected in the adult Cb, with medial and more lateral derived granule cells of the Cb primordium being located in the anterior and posterior vermis, respectively (Sgaier *et al.*, 2005). Consistent with this, clonal analysis of Purkinje cells using LacZ mitotic recombination in mouse Cb showed that there is little M-L cell mixing since clonal dispersion in the Cb occurs primarily along the A-P axis (Mathis *et al.*, 1997). In addition, LacZ labeling of newly born Purkinje cells with replication-deficient adenovirus showed that Purkinje cells born on different days form distinct M-L segments (Hashimoto and Mikoshiba, 2003).

Fate mapping of r1 with chick-quail transplantation and DiI labeling demonstrated that GCPs are specifically derived from the part of the rhombic lip that is located in r1 (Fig. 5C). However, it seems that the same area also gives rise to ventrally migrating precursors that populate the lateral pontine nuclei (Wingate and Hatten, 1999) and might even contribute to additional ventral structures such as the LC, as shown with retroviral labeling (Lin *et al.*, 2001). The generation of these different cell types appears to be regulated in a temporal manner, with ventrally migrating neurons generated first (Gilthorpe *et al.*, 2002). It is not entirely clear, however, whether the LC is rhombic lip derived, since the cell labeling in the retroviral experiments might have occurred before the actual induction of the rhombic lip (HH10–12) (Lin *et al.*, 2001) and could have resulted in the labeling of a common pool of precursors that gives rise to both the rhombic lip and the LC (Wingate, 2001).

The pre-Cb system is a collection of six nuclei containing neurons that collectively make up the primary afferents to the Cb; the most rostral, pontine gray, nucleus is located in the ventral aHb. We mention its origin here because the pontine gray is located in the region where some putative r1-derivatives settle, two of the nuclei are located just ventral-posterior to the Cb, and the pre-Cb nuclei are intimately associated with the Cb. One study using quail-chick chimeras and two different genetic fate-mapping studies in mouse showed that in contrast to the origin of the Mb and Cb (mes and r1-derivatives, respectively), the pre-Cb system is derived from rhombomeres posterior to r1 (Rodriguez and Dymecki, 2000; Wingate, 2001; Wingate and Hatten, 1999; Zervas *et al.*, 2004). DiI labeling in chick suggests that some pre-Cb nuclei are derived from r2-r6, although their origin is not from any single rhombomere, indicating a multi-rhomobomere and lineage derivation. A noninducible genetic fate-mapping method in mouse using *Wnt1-Flp* recombinase and an *Frt-LacZ* reporter showed that *Wnt1*-derived neurons populate the five pre-Cb nuclei that project mossy fibers to the Cb while the inferior olive/climbing fibers do not derive from the *Wnt1* lineage (Rodriguez and Dymecki, 2000). In a second, inducible genetic fate-mapping approach using *Wnt1-CreER^T; R26R* mice, it was demonstrated that *Wnt1*-derived cells marked at E10.5–11.5, but not earlier, give rise to the pre-Cb system. By comparing these results with *En1-Cre* (cumulative) fate mapping, it is evident that the pre-Cb nuclei are not primarily derived from r1 (Zervas *et al.*, 2004).

Collectively, these fate-mapping studies demonstrate that the Mb/aHb are derived from distinct neuromeres (mes and r1) (Fig. 5C) that are established during early embryogenesis in a manner analogous to the more posterior rhombomeres and insect body/appendage segmentation plans (Crick and Lawrence, 1975; Garcia-Bellido *et al.*, 1979). The mes/r1 become further refined genetically, while complex morphogenetic movements sculpt the primordia into highly specialized anatomical and physiological structures (see VI).

IV. The mes and r1 are Patterned by Axis-Specific Organizers

An organizer is a signaling center that contains a morphogen, which instructs adjacent tissue to take on distinct fates. The presence of two primary mes/r1 organizers that function in coordination with the previously mentioned gene expression patterns and morphogenetic events illustrates the multi-tiered components involved in Mb/aHb development. The isthmic organizer specifically patterns the mes/r1 along the A-P axes, while the notochord and floor plate regulate D-V patterning but are not specific to the mes/r1, since they extend throughout the neural tube.

Evidence that the isthmic organizer (IsO) induces and patterns adjacent tissue to take on a specific fate comes from quail-to-chick chimeras, ectopic application of morphogens, and electroporation experiments (Fig. 6 and Table I). We present a historical look at some key experiments that provided insight into the organizing ability of the IsO and interpret them in light of current experimental findings. Early revealing experiments simply transplanted quail caudal prosencephalon (likely prosomeres 1 and 2) or rhombencephalon (hindbrain anlage) into the mes of HH9 (7–10 somites) chick (Nakamura *et al.*, 1986, 1988). The quail prosencephalon grafts underwent a fate change and were reprogrammed to mes, while rhombencephalon grafts gave rise to rhombencephalon derivatives or undetermined tissues (Nakamura *et al.*, 1986). This study illustrates an important point: at early stages, some tissues (prosencephalon, for example) are still malleable, whereas others are not. It was another 5 years before critical experiments demonstrated the importance of the isthmus region. Transplantation of grafts consisting primarily of isthmus into the diencephalon resulted in a new En-2 expression domain surrounding the graft in host tissue, which became an ectopic

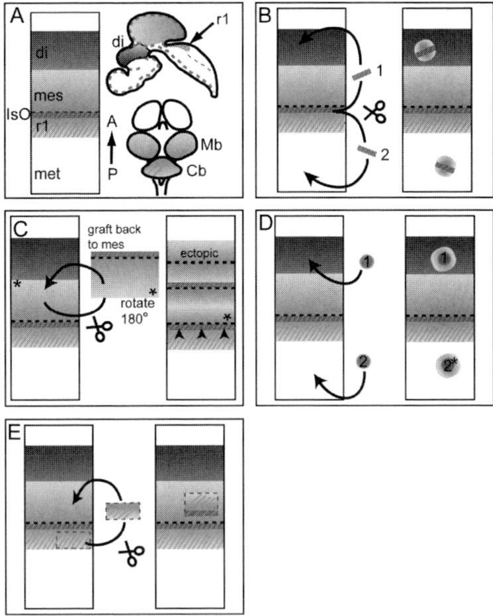

Figure 6 Schematic of IsO function in A-P patterning. (A) Schematic illustrating *Pax6* in the diencephalon (di) (green), *En2* in the mes/isthmus/r1 (blue), and *Fgf8* in the isthmus (red); r1 is indicated by yellow hatched lines and the isthmus constriction is depicted by black dashed line. Note that *En2* is expressed in two opposing gradients with highest levels at the isthmus and fading into the mes anteriorly and into r1 posteriorly. *Fgf8* is nested in the *En2* domain in the isthmus organizer (IsO). The chick brain is depicted at HH24 (top line drawing) and in the adult (bottom line drawing). (B) Ectopically transplanting small pieces of isthmus tissue into either diencephalon (1) or caudal metencephalon (met, 2) induces an ectopic *En2* gradient in host. The grafted isthmus induced host diencephalon to become an ectopic Mb and induces host metencephalon to become an ectopic Cb. (C) Removing the mes, rotating it 180°, and transplanting it back into mes region initially reverses the *En2* gradient in the mes, which is quickly reverted to a normal *En2* gradient. (* is for reference of graft orientation); presumably an *Fgf8* domain is re-established at the mes/r1 interface (arrowheads). The isthmus (*Fgf8*) and *En2* gradient of the transplanted graft is presumptively maintained in its new location. An ectopic *En2* gradient (ectopic) is established in the diencephalon and isthmus (black dashed line, ectopic). The presence of two organizers in the mes (*Fgf8*) results in the formation of a bi-caudal Mb. (D) Fgf8-soaked beads placed into the posterior diencephalon (1) induced En2 expression and an ectopic Mb, but when placed into metencephalon (2) did not induce an ectopic Cb (2*) although *En2* was induced; this is unlike isthmus transplants into metencephalon (B, 2). (E) Transplanting r1 into the mes results in the formation of a new mes/r1 interface complete with the induction of *Fgf8* and *En2*. (See Color Insert.)

Mb (Gardner and Barald, 1991; Martinez *et al.*, 1991) (Fig. 6B). In contrast, isthmus grafted into rhombomeres 2–4 also generated an ectopic En-2 expression domain in host tissue, but the induced tissue became an ectopic Cb (Martinez *et al.*, 1995) (Fig. 6B). In another study, two primary grafts were used: the first was an inversion of the mes excluding any putative isthmus region, while the second was an inversion of the mes plus anterior r1. The first graft was incorporated into host tissue, whereas the second type of graft induced a fate change in host tissue (Marin and Puelles, 1994). (Fig. 6C). When the mes/anterior r1 graft was inverted and transplanted back into the region that formerly housed the mes, the original anterior region of the mes/r1 graft (low En-2 expression) was reprogrammed to express high levels of En-2 (Fig. 6C, *). In addition, the posterior portion of the mes/r1 graft, which was rotated to an anterior position, maintained a high level of En-2 expression. The grafted tissue gave rise to two mirror-image caudal Mb structures (Marin and Puelles, 1994): one that presumably arose in response to the endogenous IsO that was repositioned anteriorly, and a second due to a newly induced IsO generated in its typical location at the mes/r1 interface (Fig. 6B, arrowheads). The presence of the repositioned IsO (anteriorly) re-patterned the host diencephalon, resulting in an ectopic En-2-expressing domain (Fig. 6C, ectopic), which developed into an ectopic Mb similar to when an isthmus was grafted into diencephalon (Fig. 6B). Importantly, ventral structures were also patterned according to the position of the presumptive IsO domains. Neither dorsal nor ventral transformations occurred when anterior r1 tissue was not included. Collectively, these results showed that isthmus/anterior r1 contains instructive properties and is therefore the location of an organizer.

The discovery of *Fgf8* and the localization of its expression to the IsO (Crossley and Martin, 1995) and subsequent experiments using *Fgf8*-soaked beads elucidated the role of *Fgf8* as the organizing molecule (Crossley *et al.*, 1996a; Liu *et al.*, 1999; Martinez *et al.*, 1999). *Fgf8*-soaked beads placed into host chick prosomeres 1 or 2 resulted in ectopic expression of *En1/2, Wnt1, Fgf8*, and the repression of *Otx2* (Fig. 6D). Subsequently, anterior-most diencephalon (prosomere 2) was locally transformed into an ectopic Mb/isthmus tissue. The placement of *Fgf8*-soaked beads in the caudal-most diencephalon (prosomere 1) or anterior mes (Fig. 6D) resulted in the formation of two ectopic mirror-image caudal Mb/isthmus structures—one located in the former anterior mes and one within prosomere 1 (Martinez *et al.*, 1999). A small ectopic Cb was often present in between the newly induced Mbs, and the ectopic mirror duplication also occurred ventrally. Collectively, this complex phenotype was reminiscent of the phenotype that resulted from the mes/r1 inversion (Fig. 6C). In both cases, the establishment of two competing IsO (*Fgf8* expression) domains partitioned the tissue between them to generate opposing mes domains. The *Fgf8*-soaked bead experiments were further supported by transgene misexpression of *Fgf8* in mouse embryos (Liu *et al.*, 1999) and by electroporation of *Fgf8* cDNA into HH10 chick, which resulted in the transformation of mes and diencephalon (Sato *et al.*, 2001). There are two *Fgf8* isoforms, *Fgf8a* and *Fgf8b*, both of which are expressed in the IsO and have distinct organizer activities relevant to Mb/aHb development (Liu *et al.*, 1999, 2003; Sato *et al.*, 2001). Fgf8b exerts a strong inductive effect that drives r1-expressing genes, represses *Otx2*, and directs tissue to acquire a Cb fate when misexpressed in mes/diencephalon (Liu *et al.*, 1999, 2003; Sato *et al.*, 2001). In contrast, *Fgf8a* drives mes fate as assessed by marker and morphological analysis (Liu *et al.*, 1999; Sato *et al.*, 2001); *Fgf17b/18* have inductive properties similar to *Fgf8a* (Liu *et al.*, 2003). Although the mes/r1 genes are induced independently of each other, transplantation studies in chick also demonstrated that the juxtaposition of mes and r1 can establish a new IsO (Irving and Mason, 1999). In one experiment, the isthmus was ablated and remarkably the IsO was regenerated, presumably as a result of mes and r1 tissue being juxtaposed. To determine if this interaction was specific to mes/r1, posterior r1 (devoid of *Fgf8* expression) or posterior rhombomeres (r2–r5) was grafted into host mes to generate a small ectopic mes/rhombomere interface. Only the juxtaposition of posterior r1 in the mes generated an ectopic *Fgf8* expression domain that was induced at the new mes/r1 border (Fig. 6E). The newly generated IsO caused the formation of ectopic mirror-image Mbs. Importantly, the transplantation studies in chick were extended to include tissue taken from mouse, which when grafted into chick gave very similar results as quail-chick transplantations (Martinez *et al.*, 1991, 1995). It was subsequently shown in mouse that genetically repositioning the IsO either anterior or posterior to its typical location by extending the *Gbx2* or *Otx2* domain, respectively, resulted in either an expansion or loss of Cb and Mb (Broccoli *et al.*, 1999; Brodski *et al.*, 2003; Millet *et al.*, 1999). These findings demonstrate that the IsO position is critical for determining the appropriate size of the Cb and Mb. Collectively, these experiments demonstrate that the juxtaposition of mes/r1 at a precise position results in the formation of an IsO that expresses *Fgf8a*, *Fgf8b*, and *Fgf17/18*, which in turn induces and patterns adjacent uncommitted tissue to take on a Mb or Cb fate.

The induction and patterning of the ventral neural tube is controlled by Shh, which is initially expressed in the notochord and subsequently in the floor plate (Echelard *et al.*, 1993). There are a number of studies demonstrating that Shh acts as an organizer to pattern ventral and possibly dorsal mes/r1 (Fig. 7). In chick, it has been shown that Shh ectopically induces the combinatorial expression of the transcription factors Pax2, Pax6, Evx1, and *Phox2a* in distinct domains in the lateral and dorsal mes in a concentration-dependent manner. These domains, which can be identified by E5, have been described as arcuate territories and give rise to specific ventral neuronal subpopulations (Agarwala and Ragsdale, 2002). Establishment of an ectopic point source of Shh by electroporation in dorsal or ventral-lateral mes results in the induction of precisely organized transcription factor domains around the Shh source (Agarwala *et al.*, 2001) (Fig. 7F). In addition, Shh signaling induces ventral cell fates in the mes/r1, including motor neurons and dopaminergic and serotonergic neurons. Ectopic sources of Shh placed

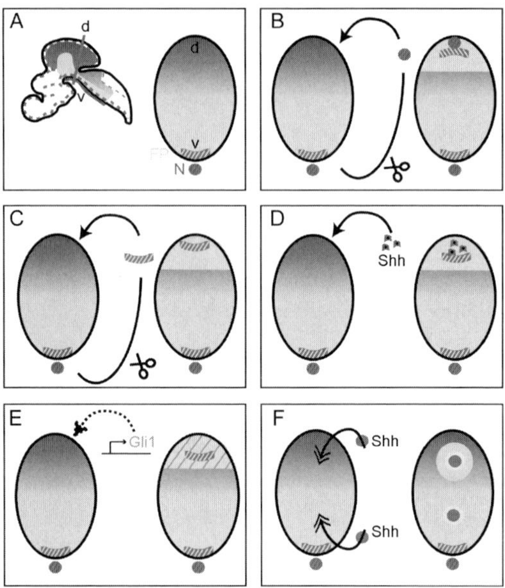

Figure 7 Schematic of organizer function in D-V patterning. (A) Schematic drawing of an embryonic brain (left) and a simplified cross-section (right) through the mes as depicted by the pink line. Dorsal (d) is at the top, ventral (v) at the bottom. The dorsal or alar plate of the mes/r1 is defined by the purple gradient, the ventral or basal plate by the orange gradient. Shh expression domains (red): notochord (N, dark red) and the floor plate (FP, yellow). Transplants of N (B) and FP (C), grafts of Shh-expressing cells (D), or ectopic expression of Gli1 in the dorsal mes (E) induce an ectopic floor plate as well as ventral markers and cell types. (F) Precisely localized expression of Shh by electroporation in either dorsal or ventral-lateral mes induces the graded expression of transcription factors normally observed along the D-V axis around the ectopic source of Shh. (See Color Insert.)

dorsally by transplanting floor plate and notochord, grafting Shh-expressing cell lines, or misexpressing the downstream transcriptional activator *Gli1* lead to the induction of ventral cell types in dorsal mes/r1 (Fig. 7B–E). This induction was accompanied by the loss of *Pax3/7* expression and roof plate markers, the concurrent ectopic expression of the floor plate marker *Hnf3β*, and the presence of ectopic ventral cell types, including dopaminergic and serotonergic neurons (Agarwala *et al.*, 2001; Fedtsova and Turner, 2001; Hynes *et al.*, 1997; Nomura and Fujisawa, 2000; Nomura *et al.*, 1998; Watanabe and Nakamura, 2000; Zhang *et al.*, 2000).

Shh appears to have an additional role in the overall expansion of the mes/r1. Loss of Shh signaling by a transient displacement of the notochord away from the mes in chick leads to reduced cell proliferation and increased cell death accompanied by collapse of the mes vesicle (Britto *et al.*, 2002). Consistent with this, in *Shh* mouse mutants the spinal cord and brain, including the mes/r1-derived region, are severely reduced in size, though this is not specific to ventral regions (Chiang *et al.*, 1996; Fedtsova and Turner, 2001). Accordingly, the ventral and dorsal diencephalon and anterior mes of *Shh* mutants have reduced proliferation and increased apoptosis at E9 (Ishibashi and McMahon, 2002). It has further been suggested that Shh normally regulates cell proliferation/apoptosis indirectly via FGF15 signaling, which affects the expression of *Bmp4* and *Tcf4*, a Wnt signaling component in the dorsal mes/diencephalon (Ishibashi and McMahon, 2002; Saitsu *et al.*, 2004). However, gain-of-function experiments produced contradictory results in chick and mouse; ectopic Shh signaling increases growth/proliferation in ventral chick mes and in dorsal mes/r1 in the mouse, but appears to inhibit proliferation and promote differentiation in dorsal chick mes (Agarwala and Ragsdale, 2002; Agarwala *et al.*, 2001; Hynes *et al.*, 1997; Lam *et al.*, 2003; Watanabe and Nakamura, 2000). These observed discrepancies might be due to differences in the time points chosen for the experiments or the level and extent of ectopic Shh expression/ pathway activation. Therefore, the details of the mechanism underlying the influence of Shh on the general expansion of the mes and whether it plays a similar role in r1 development have yet to be addressed. However, the general theme that emerges from these studies is that Shh has several distinct functions in mes/r1 development: (1) initial Shh signaling from the notochord is a prerequisite for the specification of a ventral (*Pax3/7* negative) domain and the floor plate, (2) a ventral-to-dorsal gradient of Shh signaling atterns the ventral (and possibly dorsal) domain by inducing specific transcription factors and neuronal precursors in a concentration-dependent manner, and (3) Shh signaling is required for the expansion of the mes vesicle by regulating proliferation and apoptosis in ventral and dorsal mes/r1 directly or indirectly.

As discussed previously, patterning of the dorsal neural tube is regulated primarily by members of the BMP and Wnt families that are secreted first from the ectoderm and then the roof plate at the dorsal midline of the neural tube (Lee and Jessell, 1999). The gain-of-function studies to date do not support a simple morphogen gradient action of these proteins. In addition, due to overlapping expression patterns, potential functional redundancy and/or early embryonic lethality loss-of-function experiments also have not demonstrated whether BMPs and/or Wnts act as actual organizing molecules in D-V patterning (Lee and Jessell, 1999).

V. The Intersection of A-P and D-V Signaling

The development of diverse mes/r1-derived structures and the generation of specific cell types at precise locations within the mes/r1 requires that D-V and A-P patterning be extremely well coordinated. Interactions between D-V and A-P signaling have been observed at three different levels: (1) induction of specific neuronal precursors requires signals from the isthmic organizer and the roof or floor plate, (2) transcription factors initially identified for their role in A-P patterning also influence D-V patterning and vice versa, and (3) changes in signals from one organizing center in the mes/r1 influence the normal formation of the other organizer.

Evidence for the requirement of combined signals for the induction of distinct neuronal cell types is provided by studies in zebrafish, which show that the normal induction of noradrenergic neurons of the LC requires a combination of BMP signals from the epidermal ectoderm and Fgf8 from the isthmic organizer (Guo *et al.*, 1999). In addition, elegant explant studies in rat demonstrated that the intersection of *Fgf8* and Shh signals are necessary and sufficient for the induction of Mb dopaminergic neurons in different brain regions. For example, co-culturing dorsal mes explants and the IsO *(Fgf8)* plus a source of Shh (floor plate or coated beads) results in the induction of dopaminergic neurons. Similarly, ventral diencephalon including the floor plate co-cultured with the IsO or *Fgf8*-coated beads induces dopaminergic neurons in explants (Ye *et al.*, 1998). Interestingly, addition of defined concentrations of *Fgf2* or *Fgf4* into the medium of mes/IsO explants is sufficient to cause dopaminergic neurons to take on a posterior, serotonergic fate (Ye *et al.*, 1998). This suggests that differentially expressed secreted factors are required for specification of distinct cell types. Induction of specific cell types in mes and r1 could also be explained by *Fgf8* and Shh signaling acting on prespecified tissue. The mes and r1 each express different transcription factors (*Otx2* versus *Gbx2*) and potentially different signaling components downstream of Fgf8/Shh/BMPs. In summary, the combinatorial effect of D-V and A-P organizers on the induction of dopaminergic, serotonergic, and noradrenergic neurons ensures that they are precisely positioned within the mes/r1.

The Otx homeoprotein transcription factors appear to regulate the intersection of D-V and A-P patterning by positioning the expression domains of both mes/r1 organizer signaling molecules. Both *Otx1* and *Otx2* are expressed in the mes in partially overlapping domains (Puelles *et al.*, 2003). It has been reported that in *Otx1* heterozygous mice, in which *Otx2* is specifically inactivated in the lateral mes, the expression of *Fgf8, Pax2,* and *En1* expands anteriorly. In addition, these mice show a lateral expansion of the *Shh* domain and alterations of the D-V expression domains of *Gli, Pax,* and *Nkx* transcription factors (Puelles *et al.*, 2003). In a follow-up study in which *Otx2* was inactivated throughout the mes/r1, the *Fgf8* expression domain was shifted dorsally and *Shh* and *Nkx2.2* expression expanded dorsally (Puelles *et al.*, 2004). Taken together, these studies suggest that *Otx2*, potentially in combination with *Otx1*, is required to position the correct domains of the two major organizer molecules in the mes/r1: *Fgf8* and Shh. This interpretation, however, is complicated by the fact that *Otx2* is expressed throughout the D-V axis of the neural tube, making it hard to assess why *Otx2* does not suppress Shh in its ventral floor plate domain.

Gain-of-function and heterotopic transplantation studies implicate the transcription factors *Pax3/7* in A-P patterning of the dorsal Mb and regulation of Fgf8 expression. *Pax3* and *Pax7* are expressed in dorsal mes/r1 by E9.5 in mouse (Fig. 3B) and E2.5 in chick and are also expressed in the posterior diencephalon, where they overlap with *Pax6* (Matsunaga *et al.*, 2001; Nomura *et al.*, 1998). Nomura *et al.* have reported that transplanting dorsal prosencephalon into dorsal mes causes the suppression of *Pax6* and induction of *Pax7,* resulting in the formation of an ectopic tectum (Nomura *et al.*, 1998). In addition, misexpression of *Pax3/7* in the dorsal diencephalon of chick leads to the repression of *Pax6*, induction of *Fgf8,* and subsequently *En2* expression and formation of an ectopic Mb (Matsunaga *et al.*, 2001). These data suggest that the dorsal genes *Pax3* and *Pax7* can also induce dorsal mes-like structures in the diencephalon, but this likely depends on the induction of *Fgf8*. Fgf8 levels in the transplants, however, are not sufficient to suppress *Otx2* expression and to induce Cb-like tissue.

Finally, there are a number of studies that indicate cross-regulation between signals from the IsO and the ventral organizer. It appears that changes in Shh signaling in the mes/r1 result in changes in both D-V and A-P gene expression, including *Fgf8*. For example, retroviral-mediated ectopic Shh expression induces ventral markers in dorsal mes/r1 but also leads to the A-P expansion of the *Pax2* and *En1* domains (Zhang *et al.*, 2000). In addition, misexpression of *Gli1* around the isthmus leads to the extension of the *En1, Fgf8,* and *Wnt1* domains (Hynes *et al.*, 1997). Finally, in *Gli3* mutants the *Fgf8* expression domain is broadened along the A-P axis (Aoto *et al.*, 2002). It remains to be shown whether Shh signaling regulates the described changes in gene expression directly or indirectly, for example, by altering cell proliferation and/or cell death.

VI. Maturation of Mb and Cb

Even though the genetic cascades and cell behaviors that regulate the formation of the embryonic Mb/aHb have been elucidated over the last decades, much less is known about how these primordia are further transformed into mature structures. This is especially true for the postnatal transition of the inferior and superior colliculus from a sheet-like structure into the layered and globular forms observed in the adult (compare Figs. 1A and B and 2A–D). It is, however, the Cb that undergoes the most dramatic changes in late embryonic and early postnatal development, including formation of three principal neuronal layers and a remarkable increase in size that is linked with the formation of the cerebellar folia (Altman and Bayer, 1997; Hatten and Heintz, 1995) (compare Figs. 1A and B and 2A–D). The mechanisms that control Cb foliation have not been identified, but it has been proposed that the high rate of GCP proliferation is disproportionate to the growth of the underlying white matter, resulting in buckling forces that lead to the formation of folia (Mares and Lodin, 1970). Furthermore, distinct rates of external granule cell layer expansion have been described in different areas of the Cb cortex; for example, the rate of proliferation is higher in the depth of the fissures (Allen *et al.*, 1981; Mares *et al.*, 1970) and in the anterior folia (Altman and Bayer, 1997; Charytoniuk *et al.*, 2002). Since the foliation patterns are conserved within a species, it has been suggested that the expansion of the GCP pool in the external granule cell layer is genetically controlled (Allen *et al.*, 1981).

After the external granule cell layer is established prenatally, Purkinje cells play an important role in regulating GCP proliferation. Ablation or loss of Purkinje cells in mutant mice results in a reduction in the number of granule cells and reduced foliation (Dussault *et al.*, 1998; Smeyne *et al.*, 1995; Steinmayr *et al.*, 1998). More recently, it was shown that GCP proliferation is regulated by Shh secreted from Purkinje cells. Shh can strongly stimulate GCP proliferation *in vitro* (Dahmane and Ruiz-i-Altaba, 1999; Wallace, 1999; Wechsler-Reya and Scott, 1999). More importantly, inhibition of Shh signaling *in vivo* by administration of a function-blocking antibody (Wechsler-Reya and Scott, 1999) or by gene inactivation in mouse (Corrales *et al.*, 2004; Lewis *et al.*, 2004) results in a significant decrease of GCPs, a smaller cerebellum, and a simplified foliation pattern. In addition, null mutations in the gene encoding the Shh receptor *Patched* in humans and mice, which lead to the constitutive activation of the Shh pathway, have been linked to the development of medulloblastomas, a tumor thought to be of GCP origin (Wechsler-Reya and Scott, 2001). While Shh seems to be the most important factor for GCP proliferation, there are a number of secreted molecules, cell surface receptors, and components of the cell cycle that modulate the response of GCPs to Shh (Blaess *et al.*, 2004; Kenney and Rowitch, 2000; Kenney *et al.*, 2003; Klein *et al.*, 2001; Pons *et al.*, 2001; Rios *et al.*, 2004.) However, even though our understanding of the mechanisms underlying GCP proliferation has grown over the last few years, it remains to be investigated how the establishment of the precise foliation pattern is influenced by events during embryonic development versus postnatal GCP proliferation.

VII. Human Diseases Affecting Mb and Cb

The identification of mutations causing human genetic diseases in the Mb and Cb is a potential source of further insight into the normal role of these genes during development. The most common human disease associated with the Mb is Parkinson's disease (PD). Sporadic PD is a progressive degenerative disorder in which Mb dopaminergic (MbDA) neurons are lost. It is characterized by an insidious onset in mid- or late adulthood, with muscular rigidity, postural instability, slowing of movements, and resting tremor. It is thought that it is caused by a combination of genetic predisposition and environmental influences (Huang *et al.*, 2004). A number of genes associated with PD have been identified in the rare familial forms of PD, including *α-synuclein, Parkin, Nurr-1,* and most recently *LRRK2 (leucine-rich repeat kinase 2)* (Huang *et al.*, 2004; Shen, 2004). Most of these genes seem to be involved in regulating the metabolism of mature MbDA. In humans, two heterozygous mutations in the orphan nuclear receptor *Nurr1* were found in several patients with familial PD (Le *et al.*, 2003). In mouse, *Nurr1* expression is first observed in the ventral mes at E10.5 and is maintained in MbDA neurons into adulthood (Zetterstrom *et al.*, 1996, 1997). In homozygous *Nurr1* mutant mice, TH, a key enzyme in dopamine synthesis, is not induced in MbDA precursors, resulting in their subsequent degeneration; the mice die soon after birth. Interestingly, it has further been reported that in behaviorally normal heterozygous *Nurr1* mice, dopamine levels are reduced, a phenotype that at least partially recapitulates PD (Saucedo-Cardenas *et al.*, 1998; Zetterstrom *et al.*, 1997). In addition, two transcription factors, Pitx3 and Lmx1b, have been implicated in the normal development of MbDA neurons in mouse, but have not been associated with PD in human. Expression of *Pitx3*, a bicoid-related homeobox gene, is confined to MbDA neurons in the brain in mouse and human (Smidt *et al.*, 1997). *Pitx3* is inactivated in the naturally occurring *aphakia* mouse (Semina *et al.*, 1997). In these mutants, only a subset of the MbDA neurons in the substantia nigra and ventral tegmental area appear to be lost during fetal and postnatal development, respectively (Smidt *et al.*, 2004; van den Munckhof *et al.*, 2003). It is not completely clear whether this is due to restricted expression of *Pitx3* in a subset of MbDA neurons or whether a subset is particularly susceptible to the loss of Pitx3 function. The LIM homeodomain transcription factor Lmx1b is localized to the MbDa neurons from late embryonic stages onward. In the absence of *Lmx1b*, TH and Nurr1-positive MbDa precursors are generated in the mouse, but Pitx3 is not induced, leading subsequently to the degeneration of a subset of MbDa (Smidt *et al.*, 2000). In addition to giving further insight into potential genetic cascades regulating MbDA neuron development, the subset-specific loss of these neurons in mutant mice resembles the degeneration of neuronal populations observed in PD quite accurately (Fearnley and Lees, 1991).

There are a number of genetic developmental malformations of the dorsal Mb and Cb in human. Most of these malformations seem to predominantly affect the Cb, and frequently the defects appear to be more severe in the vermis than in the hemispheres. This has been described, for example, in

Joubert syndrome, Dandy-Walker malformation, and cerebellar vermis hypoplasia (Parisi and Dobyns, 2003). Joubert syndrome is autosomal recessive and is characterized by hypoplasia/dysplasia of the cerebellar vermis, brainstem abnormalities, and abnormal axonal crossing in the corticospinal tract and superior cerebellar peduncles. Affected individuals are ataxic and have abnormal breathing patterns and cognitive problems. The first gene associated with Joubert syndrome was identified recently as *AHI1* (Abelson helper integration site gene) (Ferland *et al.*, 2004), which is a cytoplasmic adaptor protein highly expressed in human fetal brain. It is not clear how AHI1 causes the described defects, but it has been speculated that it might be downstream of transmembrane molecules that are involved in axonal path finding (Ferland *et al.*, 2004). In Dandy-Walker malformation (DWM), the hypoplastic vermis is rotated upward and the 4th ventricle is enlarged. Patients have hypotonia and ataxia and, less frequently, mental retardation and hydrocephalus. Recently, a heterozygous deletion of two Zinc finger genes, *ZIC1* and *ZIC4*, which are tightly linked, has been associated with DWM (Grinberg *et al.*, 2004). *Zic1* and *4* are both expressed in the developing mouse Cb. Whereas the *Zic4* null mutant phenotype has not been described, *Zic1* null mutants have a hypoplastic cerebellum. Furthermore, *Zic1* and *Zic4* heterozygous single or compound heterozygous mutants have defects in the foliation pattern and/or a smaller Cb, with variable severity (Aruga *et al.*, 1998; Grinberg *et al.*, 2004). Even though this demonstrates some role for *Zic1/4* in cerebellar development, the observed phenotype in mouse only partially resembles the human DWM.

In summary, while the analysis of human diseases has contributed significantly to our understanding of how the Mb and Cb function, the identification of the causative genes in diseases has so far given only limited insight into developmental processes. Similarly, genes that have been shown to play a role in Mb and Cb development based on genetic studies in model organisms have not yet been identified in any of the human diseases/malformations. This is most likely because homozygous mutations of these genes will result in embryonic or early postnatal lethality in humans.

References

Adams, K. A., Maida, J. M., Golden, J. A., and Riddle, R. D. (2000). The transcription factor Lmx1b maintains Wnt1 expression within the isthmic organizer. *Development* **127**, 1857–1867.

Agarwala, S., and Ragsdale, C. W. (2002). A role for midbrain arcs in nucleogenesis. *Development* **129**, 5779–5788.

Agarwala, S., Sanders, T. A., and Ragsdale, C. W. (2001). Sonic hedgehog control of size and shape in midbrain pattern formation. *Science* **291**, 2147–2150.

Alder, J., Lee, K. J., Jessell, T. M., and Hatten, M. E. (1999). Generation of cerebellar granule neurons *in vivo* by transplantation of BMP-treated neural progenitor cells. *Nat. Neurosci.* **2,** 535–540.

Alexandre, P., and Wassef, M. (2003). The isthmic organizer links anteroposterior and dorsoventral patterning in the mid/hindbrain by generating roof plate structures. *Development* **130,** 5331–5338.

Allen, C., Sievers, J., Berry, M., and Jenner, S. (1981). Experimental studies on cerebellar foliation. II. A morphometric analysis of cerebellar fissuration defects and growth retardation after neonatal treatment with 6-OHDA in the rat. *J. Comp. Neurol.* **203,** 771–783.

Altman, J., and Bayer, S. A. (1997). "Development of the Cerebellar System: In Relation to Its Evolution, Structure, and Functions." CRC Press Inc., New York.

Aoto, K., Nishimura, T., Eto, K., and Motoyama, J. (2002). Mouse GLI3 regulates Fgf8 expression and apoptosis in the developing neural tube, face, and limb bud. *Dev. Biol.* **251,** 320–332.

Aruga, J., Minowa, O., Yaginuma, H., Kuno, J., Nagai, T., Noda, T., and Mikoshiba, K. (1998). Mouse Zic1 is involved in cerebellar development. *J. Neurosci.* **18,** 284–293.

Asano, M., and Gruss, P. (1992). Pax-5 is expressed at the midbrain-hindbrain boundary during mouse development. *Mech. Dev.* **39,** 29–39.

Auerbach, R. (1954). Analysis of the developmental effects of a lethal mutation in the house mouse. *J. Exp. Zool.* **127,** 305–329.

Bai, C. B., Auerbach, W., Lee, J. S., Stephen, D., and Joyner, A. L. (2002). Gli2, but not Gli1, is required for initial Shh signaling and ectopic activation of the Shh pathway. *Development* **129,** 4753–4761.

Bally-Cuif, L., Cholley, B., and Wassef, M. (1995). Involvement of Wnt-1 in the formation of the mes/metencephalic boundary. *Mech. Dev.* **53,** 23–34.

Blaess, S., Graus-Porta, D., Belvindrah, R., Radakovits, R., Pons, S., Littlewood-Evans, A., Senften, M., Guo, H., Li, Y., Miner, J. H., Reichardt, L. F., and Muller, U. (2004). Beta1-integrins are critical for cerebellar granule cell precursor proliferation. *J. Neurosci.* **24,** 3402–3412.

Bouchard, M., Pfeffer, P., and Busslinger, M. (2000). Functional equivalence of the transcription factors Pax2 and Pax5 in mouse development. *Development* **127,** 3703–3713.

Briscoe, J., and Ericson, J. (1999). The specification of neuronal identity by graded Sonic Hedgehog signalling. *Semin. Cell Dev. Biol.* **10,** 353–362.

Britto, J., Tannahill, D., and Keynes, R. (2002). A critical role for sonic hedgehog signaling in the early expansion of the developing brain. *Nat. Neurosci.* **5,** 103–110.

Broccoli, V., Boncinelli, E., and Wurst, W. (1999). The caudal limit of Otx2 expression positions the isthmic organizer. *Nature* **401,** 164–168.

Brodski, C., Weisenhorn, D. M., Signore, M., Sillaber, I., Oesterheld, M., Broccoli, V., Acampora, D., Simeone, A., and Wurst, W. (2003). Location and size of dopaminergic and serotonergic cell populations are controlled by the position of the midbrain-hindbrain organizer. *J. Neurosci.* **23,** 4199–4207.

Bronson, R. T., and Higgins, D. C. (1967). The swaying (sw) mutation causes a midline sagittal fissure. *Mouse Genome* **89,** 851.

Castelo-Branco, G., Wagner, J., Rodriguez, F. J., Kele, J., Sousa, K., Rawal, N., Pasolli, H. A., Fuchs, E., Kitajewski, J., and Arenas, E. (2003). Differential regulation of midbrain dopaminergic neuron development by Wnt-1, Wnt-3a, and Wnt-5a. *Proc. Natl. Acad. Sci. USA* **100,** 12747–12752.

Charytoniuk, D., Porcel, B., Rodriguez Gomez, J., Faure, H., Ruat, M., and Traiffort, E. (2002). Sonic Hedgehog signalling in the developing and adult brain. *J. Physiol. Paris* **96,** 9–16.

Chi, C. L., Martinez, S., Wurst, W., and Martin, G. R. (2003). The isthmic organizer signal FGF8 is required for cell survival in the prospective midbrain and cerebellum. *Development* **130,** 2633–2644.

Chiang, C., Litingtung, Y., Lee, E., Young, K. E., Corden, J. L., Westphal, H., and Beachy, P. A. (1996). Cyclopia and defective axial patterning in mice lacking Sonic hedgehog gene function. *Nature* **383,** 407–413.

Corrales, J. D., Rocco, G. L., Blaess, S., Guo, Q., and Joyner, A. L. (2004). Spatial pattern of sonic hedgehog signaling through Gli genes during cerebellum development. *Development* **131,** 5581–5590.

Crick, F. H., and Lawrence, P. A. (1975). Compartments and polyclones in insect development. *Science* **189,** 340–347.

Crossley, P. H., and Martin, G. R. (1995). The mouse Fgf8 gene encodes a family of polypeptides and is expressed in regions that direct outgrowth and patterning in the developing embryo. *Development* **121,** 439–451.

Crossley, P. H., Martinez, S., and Martin, G. R. (1996). Midbrain development induced by FGF8 in the chick embryo. *Nature* **380,** 66–68.

Crossley, P. H., Minowada, G., Mac Arthur, C. A., and Martin, G. R. (1996). Roles for FGF8 in the induction, initiation, and maintenance of chick limb development. *Cell* **84,** 127–136.

Dahmane, N., and Ruiz-i-Altaba, A. (1999). Sonic hedgehog regulates the growth and patterning of the cerebellum. *Development* **126,** 3089–3100.

Danielian, P. S., and McMahon, A. P. (1996). Engrailed-1 as a target of the Wnt-1 signalling pathway in vertebrate midbrain development. *Nature* **383,** 332–334.

Danielian, P. S., Muccino, D., Rowitch, D. H., Michael, S. K., and McMahon, A. P. (1998). Modification of gene activity in mouse embryos *in utero* by a tamoxifen- inducible form of Cre recombinase. *Curr. Biol.* **8,** 1323–1326.

Davis, C. A., and Joyner, A. L. (1988). Expression patterns of the homeo box-containing genes En-1 and En-2 and the proto-oncogene int-1 diverge during mouse development. *Genes Dev.* **2,** 1736–1744.

Ding, Y. Q., Marklund, U., Yuan, W., Yin, J., Wegman, L., Ericson, J., Deneris, E., Johnson, R. L., and Chen, Z. F. (2003). Lmx1b is essential for the development of serotonergic neurons. *Nat. Neurosci.* **6,** 933–938.

Dussault, I., Fawcett, D., Matthyssen, A., Bader, J. A., and Giguere, V. (1998). Orphan nuclear receptor ROR alpha-deficient mice display the cerebellar defects of staggerer. *Mech. Dev.* **70,** 147–153.

Echelard, Y., Epstein, D. J., St-Jacques, B., Shen, L., Mohler, J., McMahon, J. A., and McMahon, A. P. (1993). Sonic hedgehog, a member of a family of putative signaling molecules, is implicated in the regulation of CNS polarity. *Cell* **75,** 1417–1430.

Echelard, Y., Vassileva, G., and McMahon, A. P. (1994). Cis-acting regulatory sequences governing Wnt-1 expression in the developing mouse CNS. *Development* **120,** 2213–2224.

Ericson, J., Morton, S., Kawakami, A., Roelink, H., and Jessell, T. M. (1996). Two critical periods of Sonic Hedgehog signaling required for the specification of motor neuron identity. *Cell* **87,** 661–673.

Favor, J., Sandulache, R., Neuhauser-Klaus, A., Pretsch, W., Chatterjee, B., Senft, E., Wurst, W., Blanquet, V., Grimes, P., Sporle, R., and Schughart, K. (1996). The mouse Pax2(1Neu) mutation is identical to a human PAX2 mutation in a family with renal-coloboma syndrome and results in developmental defects of the brain, ear, eye, and kidney. *Proc. Natl. Acad. Sci. USA* **93,** 13870–13875.

Fearnley, J. M., and Lees, A. J. (1991). Ageing and Parkinson's disease: Substantia nigra regional selectivity. *Brain* **114(5),** 2283–2301.

Fedtsova, N., and Turner, E. E. (2001). Signals from the ventral midline and isthmus regulate the development of Brn3.0-expressing neurons in the midbrain. *Mech. Dev.* **105,** 129–144.

Feil, R., Brocard, J., Mascrez, B., Le Meur, M., Metzger, D., and Chambon, P. (1996). Ligand-activated site-specific recombination in mice. *Proc. Natl. Acad. Sci. USA* **93**, 10887–10890.

Ferland, R. J., Eyaid, W., Collura, R. V., Tully, L. D., Hill, R. S., Al-Nouri, D., Al-Rumayyan, A., Topcu, M., Gascon, G., Bodell, A., Shugart, Y. Y., Ruvolo, M., and Walsh, C. A. (2004). Abnormal cerebellar development and axonal decussation due to mutations in AHI1 in Joubert syndrome. *Nat. Genet.* **36**, 1008–1013.

Fu, H., Cai, J., Rutledge, M., Hu, X., and Qiu, M. (2003). Oligodendrocytes can be generated from the local ventricular and subventricular zones of embryonic chicken midbrain. *Brain Res. Dev. Brain Res.* **143**, 161–165.

Garcia-Bellido, A., Lawrence, P. A., and Morata, G. (1979). Compartments in animal development. *Sci. Am.* **241**, 102–110.

Gardner, C. A., and Barald, K. F. (1991). The cellular environment controls the expression of engrailed-like protein in the cranial neuroepithelium of quail-chick chimeric embryos. *Development* **113**, 1037–1048.

Gilthorpe, J. D., Papantoniou, E. K., Chedotal, A., Lumsden, A., and Wingate, R. J. (2002). The migration of cerebellar rhombic lip derivatives. *Development* **129**, 4719–4728.

Goulding, M. D., Lumsden, A., and Gruss, P. (1993). Signals from the notochord and floor plate regulate the region-specific expression of two Pax genes in the developing spinal cord. *Development* **117**, 1001–1016.

Grinberg, I., Northrup, H., Ardinger, H., Prasad, C., Dobyns, W. B., and Millen, K. J. (2004). Heterozygous deletion of the linked genes ZIC1 and ZIC4 is involved in Dandy-Walker malformation. *Nat. Genet.* **36**, 1053–1055.

Guo, S., Brush, J., Teraoka, H., Goddard, A., Wilson, S. W., Mullins, M. C., and Rosenthal, A. (1999). Development of noradrenergic neurons in the zebrafish hindbrain requires BMP, FGF8, and the homeodomain protein soulless/Phox2a. *Neuron* **24**, 555–566.

Hallonet, M. E., and Le Douarin, N. M. (1993). Tracing neuroepithelial cells of the mesencephalic and metencephalic alar plates during cerebellar ontogeny in quail-chick chimaeras. *Eur. J. Neurosci.* **5**, 1145–1155.

Hallonet, M. E., Teillet, M. A., and Le Douarin, N. M. (1990). A new approach to the development of the cerebellum provided by the quail-chick marker system. *Development* **108**, 19–31.

Hanks, M., Wurst, W., Anson-Cartwright, L., Auerbach, A. B., and Joyner, A. L. (1995). Rescue of the En-1 mutant phenotype by replacement of En-1 with En-2. *Science* **269**, 679–682.

Hashimoto, M., and Mikoshiba, K. (2003). Mediolateral compartmentalization of the cerebellum is determined on the "birth date" of Purkinje cells *J. Neurosci.* **23**, 11342–11351.

Hatten, M. E., and Heintz, N. (1995). Mechanisms of neural patterning and specification in the developing cerebellum. *Annu. Rev. Neurosci.* **18**, 385–408.

Huang, Y., Cheung, L., Rowe, D., and Halliday, G. (2004). Genetic contributions to Parkinson's disease. *Brain Res. Brain Res. Rev.* **46**, 44–70.

Hynes, M., Stone, D. M., Dowd, M., Pitts-Meek, S., Goddard, A., Gurney, A., and Rosenthal, A. (1997). Control of cell pattern in the neural tube by the zinc finger transcription factor and oncogene Gli-1. *Neuron* **19**, 15–26.

Ikeya, M., Lee, S. M., Johnson, J. E., McMahon, A. P., and Takada, S. (1997). Wnt signalling required for expansion of neural crest and CNS progenitors. *Nature* **389**, 966–970.

Irving, C., and Mason, I. (1999). Regeneration of isthmic tissue is the result of a specific and direct interaction between rhombomere 1 and midbrain. *Development* **126**, 3981–3989.

Ishibashi, M., and McMahon, A. P. (2002). A sonic hedgehog-dependent signaling relay regulates growth of diencephalic and mesencephalic primordia in the early mouse embryo. *Development* **129**, 4807–4819.

Joyner, A. L., Herrup, K., Auerbach, B. A., Davis, C. A., and Rossant, J. (1991). Subtle cerebellar phenotype in mice homozygous for a targeted deletion of the En-2 homeobox. *Science* **251**, 1239–1243.

Jungbluth, S., Larsen, C., Wizenmann, A., and Lumsden, A. (2001). Cell mixing between the embryonic midbrain and hindbrain. *Curr. Biol.* **11**, 204–207.

Kenney, A. M., and Rowitch, D. H. (2000). Sonic hedgehog promotes G(1). Cyclin expression and sustained cell cycle progression in mammalian neuronal precursors. *Mol. Cell. Biol.* **20**, 9055–9067.

Kenney, A. M., Cole, M. D., and Rowitch, D. H. (2003). Nmyc upregulation by sonic hedgehog signaling promotes proliferation in developing cerebellar granule neuron precursors. *Development* **130**, 15–28.

Kimmel, R. A., Turnbull, D. H., Blanquet, V., Wurst, W., Loomis, C. A., and Joyner, A. L. (2000). Two lineage boundaries coordinate vertebrate apical ectodermal ridge formation. *Genes Dev.* **14**, 1377–1389.

Klein, R. S., Rubin, J. B., Gibson, H. D., De Haan, E. N., Alvarez-Hernandez, X., Segal, R. A., and Luster, A. D. (2001). SDF-1 alpha induces chemotaxis and enhances Sonic hedgehog-induced proliferation of cerebellar granule cells. *Development* **128**, 1971–1981.

Koster, R. W., and Fraser, S. E. (2001). Direct imaging of *in vivo* neuronal migration in the developing cerebellum. *Curr. Biol.* **11**, 1858–1863.

Lam, C. S., Sleptsova-Friedrich, I., Munro, A. D., and Korzh, V. (2003). SHH and FGF8 play distinct roles during development of noradrenergic neurons in the locus coeruleus of the zebrafish. *Mol. Cell Neurosci.* **22**, 501–515.

Lane, P. W. (1967). [Swaying]. *Mouse News Lett.* **36**, 40.

Larsen, C. W., Zeltser, L. M., and Lumsden, A. (2001). Boundary formation and compartition in the avian diencephalon. *J. Neurosci.* **21**, 4699–4711.

Le, W. D., Xu, P., Jankovic, J., Jiang, H., Appel, S. H., Smith, R. G., and Vassilatis, D. K. (2003). Mutations in NR4A2 associated with familial Parkinson disease. *Nat. Genet.* **33**, 85–89.

Lee, K. J., and Jessell, T. M. (1999). The specification of dorsal cell fates in the vertebrate central nervous system. *Annu. Rev. Neurosci.* **22**, 261–294.

Lee, K. J., Dietrich, P., and Jessell, T. M. (2000). Genetic ablation reveals that the roof plate is essential for dorsal interneuron specification. *Nature* **403**, 734–740.

Lewis, P. M., Gritli-Linde, A., Smeyne, R., Kottmann, A., and McMahon, A. P. (2004). Sonic hedgehog signaling is required for expansion of granule neuron precursors and patterning of the mouse cerebellum. *Dev. Biol.* **270**, 393–410.

Li, J. Y., and Joyner, A. L. (2001). Otx2 and Gbx2 are required for refinement and not induction of mid-hindbrain gene expression. *Development* **128**, 4979–4991.

Lin, J. C., Cai, L., and Cepko, C. L. (2001). The external granule layer of the developing chick cerebellum generates granule cells and cells of the isthmus and rostral hindbrain. *J. Neurosci.* **21**, 159–168.

Liu, A., and Joyner, A. L. (2001a). Early anterior/posterior patterning of the midbrain and cerebellum. *Annu. Rev. Neurosci.* **24**, 869–896.

Liu, A., and Joyner, A. L. (2001b). EN and GBX2 play essential roles downstream of FGF8 in patterning the mouse mid/hindbrain region. *Development* **128**, 181–191.

Liu, A., Losos, K., and Joyner, A. L. (1999). FGF8 can activate Gbx2 and transform regions of the rostral mouse brain into a hindbrain fate. *Development* **126**, 4827–4838.

Liu, A., Li, J. Y., Bromleigh, C., Lao, Z., Niswander, L. A., and Joyner, A. L. (2003). FGF17b and FGF18 have different midbrain regulatory properties from FGF8b or activated FGF receptors. *Development* **130**, 6175–6185.

Louvi, A., Alexandre, P., Metin, C., Wurst, W., and Wassef, M. (2003). The isthmic neuroepithelium is essential for cerebellar midline fusion. *Development* **130**, 5319–5330.

Mansouri, A., and Gruss, P. (1998). Pax3 and Pax7 are expressed in commissural neurons and restrict ventral neuronal identity in the spinal cord. *Mech. Dev.* **78,** 171–178.

Mansouri, A., Hallonet, M., and Gruss, P. (1996). Pax genes and their roles in cell differentiation and development. *Curr. Opin. Cell Biol.* **8,** 851–857.

Mares, V., and Lodin, Z. (1970). The cellular kinetics of the developing mouse cerebellum. II. The function of the external granular layer in the process of gyrification. *Brain Res.* **23,** 343–352.

Mares, V., Lodin, Z., and Srajer, J. (1970). The cellular kinetics of the developing mouse cerebellum. I. The generation cycle, growth fraction and rate of proliferation of the external granular layer. *Brain Res.* **23,** 323–342.

Marin, F., and Puelles, L. (1994). Patterning of the embryonic avian midbrain after experimental inversions: A polarizing activity from the isthmus. *Dev. Biol.* **163,** 19–37.

Martinez, S., Wassef, M., and Alvarado-Mallart, R. M. (1991). Induction of a mesencephalic phenotype in the 2-day-old chick prosencephalon is preceded by the early expression of the homeobox gene en. *Neuron* **6,** 971–981.

Martinez, S., Marin, F., Nieto, M. A., and Puelles, L. (1995). Induction of ectopic engrailed expression and fate change in avian rhombomeres: Intersegmental boundaries as barriers. *Mech. Dev.* **51,** 289–303.

Martinez, S., Crossley, P. H., Cobos, I., Rubenstein, J. L., and Martin, G. R. (1999). FGF8 induces formation of an ectopic isthmic organizer and isthmocerebellar development via a repressive effect on Otx2 expression. *Development* **126,** 1189–1200.

Martinez-Barbera, J. P., Signore, M., Boyl, P. P., Puelles, E., Acampora, D., Gogoi, R., Schubert, F., Lumsden, A., and Simeone, A. (2001). Regionalisation of anterior neuroectoderm and its competence in responding to forebrain and midbrain inducing activities depend on mutual antagonism between OTX2 and GBX2. *Development* **128,** 4789–4800.

Mathis, L., Bonnerot, C., Puelles, L., and Nicolas, J. F. (1997). Retrospective clonal analysis of the cerebellum using genetic laacZ/lacZ mouse mosaics. *Development* **124,** 4089–4104.

Matise, M. P., Epstein, D. J., Park, H. L., Platt, K. A., and Joyner, A. L. (1998). Gli2 is required for induction of floor plate and adjacent cells, but not most ventral neurons in the mouse central nervous system. *Development* **125,** 2759–2770.

Matsunaga, E., Araki, I., and Nakamura, H. (2001). Role of Pax3/7 in the tectum regionalization. *Development* **128,** 4069–4077.

Matsunaga, E., Katahira, T., and Nakamura, H. (2002). Role of Lmx1b and Wnt1 in mesencephalon and metencephalon development. *Development* **129,** 5269–5277.

McMahon, A. P., and Bradley, A. (1990). The Wnt-1 (int-1) proto-oncogene is required for development of a large region of the mouse brain. *Cell* **62,** 1073–1085.

McMahon, A. P., Joyner, A. L., Bradley, A., and McMahon, J. A. (1992). The midbrain-hindbrain phenotype of Wnt-1-/Wnt-1- mice results from stepwise deletion of engrailed-expressing cells by 9.5 days postcoitum. *Cell* **69,** 581–595.

Meyers, E. N., Lewandoski, M., and Martin, G. R. (1998). An Fgf8 mutant allelic series generated by Cre- and Flp-mediated recombination. *Nat. Genet.* **18,** 136–141.

Millen, K. J., Wurst, W., Herrup, K., and Joyner, A. L. (1994). Abnormal embryonic cerebellar development and patterning of postnatal foliation in two mouse Engrailed-2 mutants. *Development* **120,** 695–706.

Millet, S., Bloch-Gallego, E., Simeone, A., and Alvarado-Mallart, R. M. (1996). The caudal limit of Otx2 gene expression as a marker of the midbrain/hindbrain boundary: A study using *in situ* hybridisation and chick/quail homotopic grafts. *Development* **122,** 3785–3797.

Millet, S., Campbell, K., Epstein, D. J., Losos, K., Harris, E., and Joyner, A. L. (1999). A role for Gbx2 in repression of Otx2 and positioning the mid/hindbrain organizer. *Nature* **401,** 161–164.

Millonig, J. H., Millen, K. J., and Hatten, M. E. (2000). The mouse Dreher gene Lmx1a controls formation of the roof plate in the vertebrate CNS. *Nature* **403,** 764–769.

Muroyama, Y., Fujihara, M., Ikeya, M., Kondoh, H., and Takada, S. (2002). Wnt signaling plays an essential role in neuronal specification of the dorsal spinal cord. *Genes Dev.* **16,** 548–553.

Nakamura, H., Nakano, K. E., Igawa, H. H., Takagi, S., and Fujisawa, H. (1986). Plasticity and rigidity of differentiation of brain vesicles studied in quail-chick chimeras. *Cell Differ.* **19,** 187–193.

Nakamura, H., Takagi, S., Tsuji, T., Matsui, K. A., and Fujisawa, H. (1988). The prosencephalon has the capacity to differentiate into the optic tectum: Analysis by chick-specific monoclonal antibodies in quail-chick-chimeric brains. *Dev. Growth Differ.* **30,** 717–725.

Nomura, T., and Fujisawa, H. (2000). Alteration of the retinotectal projection map by the graft of mesencephalic floor plate or sonic hedgehog. *Development* **127,** 1899–1910.

Nomura, T., Kawakami, A., and Fujisawa, H. (1998). Correlation between tectum formation and expression of two PAX family genes, PAX7 and PAX6, in avian brains. *Dev. Growth Differ.* **40,** 485–495.

Palmgren, A. (1921). Embryoloical and morphological studies on the mid-brain and cerebellum of vertebrates. *Acta Zooligica* **2,** 1–94.

Parisi, M. A., and Dobyns, W. B. (2003). Human malformations of the midbrain and hindbrain: Review and proposed classification scheme. *Mol. Genet. Metab.* **80,** 36–53.

Parr, B. A., Shea, M. J., Vassileva, G., and McMahon, A. P. (1993). Mouse Wnt genes exhibit discrete domains of expression in the early embryonic CNS and limb buds. *Development* **119,** 247–261.

Pons, S., Trejo, J. L., Martinez-Morales, J. R., and Marti, E. (2001). Vitronectin regulates Sonic hedgehog activity during cerebellum development through CREB phosphorylation. *Development* **128,** 1481–1492.

Puelles, E., Acampora, D., Lacroix, E., Signore, M., Annino, A., Tuorto, F., Filosa, S., Corte, G., Wurst, W., Ang, S. L., and Simeone, A. (2003). Otx dose-dependent integrated control of antero-posterior and dorso-ventral patterning of midbrain. *Nat. Neurosci.* **6,** 453–460.

Puelles, E., Annino, A., Tuorto, F., Usiello, A., Acampora, D., Czerny, T., Brodski, C., Ang, S. L., Wurst, W., and Simeone, A. (2004). Otx2 regulates the extent, identity and fate of neuronal progenitor domains in the ventral midbrain. *Development* **131,** 2037–2048.

Ramon y Cajal, S. (1995). "Histology of the Nervous System." Oxford University Press Inc. New York, New York.

Reifers, F., Bohli, H., Walsh, E. C., Crossley, P. H., Stainier, D. Y., and Brand, M. (1998). Fgf8 is mutated in zebrafish acerebellar (ace) mutants and is required for maintenance of midbrain-hindbrain boundary development and somitogenesis. *Development* **125,** 2381–2395.

Rios, I., Alvarez-Rodriguez, R., Marti, E., and Pons, S. (2004). Bmp2 antagonizes sonic hedgehog-mediated proliferation of cerebellar granule neurones through Smad5 signalling. *Development* **131,** 3159–3168.

Rodriguez, C. I., and Dymecki, S. M. (2000). Origin of the precerebellar system. *Neuron* **27,** 475–486.

Rowitch, D. H., and McMahon, A. P. (1995). Pax-2 expression in the murine neural plate precedes and encompasses the expression domains of Wnt-1 and En-1. *Mech. Dev.* **52,** 3–8.

Saitsu, H., Komada, M., Suzuki, M., Nakayama, R., Motoyama, J., Shiota, K., and Ishibashi, M. (2004). Expression of the mouse Fgf15 gene is directly initiated by Sonic hedgehog signaling in the diencephalon and midbrain. *Dev. Dyn.* **232,** 282–292.

Sanders, T. A., Lumsden, A., and Ragsdale, C. W. (2002). Arcuate plan of chick midbrain development. *J. Neurosci.* **22,** 10742–10750.

Sato, T., Araki, I., and Nakamura, H. (2001). Inductive signal and tissue responsiveness defining the tectum and the cerebellum. *Development* **128,** 2461–2469.

Saucedo-Cardenas, O., Quintana-Hau, J. D., Le, W. D., Smidt, M. P., Cox, J. J., De Mayo, F., Burbach, J. P., and Conneely, O. M. (1998). Nurr1 is essential for the induction of the dopaminergic phenotype and the survival of ventral mesencephalic late dopaminergic precursor neurons. *Proc. Natl. Acad. Sci. USA* **95,** 4013–4018.

Schwarz, M., Alvarez-Bolado, G., Urbanek, P., Busslinger, M., and Gruss, P. (1997). Conserved biological function between Pax-2 and Pax-5 in midbrain and cerebellum development: Evidence from targeted mutations. *Proc. Natl. Acad. Sci. USA* **94,** 14518–14523.

Semina, E. V., Reiter, R. S., and Murray, J. C. (1997). Isolation of a new homeobox gene belonging to the Pitx/Rieg family: Expression during lens development and mapping to the aphakia region on mouse chromosome 19. *Hum. Mol. Genet.* **6,** 2109–2116.

Sgaier, S. K., Millet, S., Villanueva, M. P., Berenshteyn, F., Song, C., and Joyner, A. L. (2005). Morphogenetic and cellular movements that shape the mouse cerebellum: Insights from genetic fate mapping. *Neuron* **45,** 27–40.

Shen, J. (2004). Protein kinases linked to the pathogenesis of Parkinson's disease. *Neuron* **44,** 575–577.

Smeyne, R. J., Chu, T., Lewin, A., Bian, F., S, S. C., Kunsch, C., Lira, S. A., and Oberdick, J. (1995). Local control of granule cell generation by cerebellar Purkinje cells. *Mol. Cell Neurosci.* **6,** 230–251.

Smidt, M. P., van Schaick, H. S., Lanctot, C., Tremblay, J. J., Cox, J. J., van der Kleij, A. A., Wolterink, G., Drouin, J., and Burbach, J. P. (1997). A homeodomain gene Ptx3 has highly restricted brain expression in mesencephalic dopaminergic neurons. *Proc. Natl. Acad. Sci. USA* **94,** 13305–13310.

Smidt, M. P., Asbreuk, C. H., Cox, J. J., Chen, H., Johnson, R. L., and Burbach, J. P. (2000). A second independent pathway for development of mesencephalic dopaminergic neurons requires Lmx1b. *Nat. Neurosci.* **3,** 337–341.

Smidt, M. P., Smits, S. M., Bouwmeester, H., Hamers, F. P., van der Linden, A. J., Hellemons, A. J., Graw, J., and Burbach, J. P. (2004). Early developmental failure of substantia nigra dopamine neurons in mice lacking the homeodomain gene Pitx3. *Development* **131,** 1145–1155.

Soriano, P. (1999). Generalized lacZ expression with the ROSA26 Cre reporter strain. *Nat. Genet.* **21,** 70–71.

Steinmayr, M., Andre, E., Conquet, F., Rondi-Reig, L., Delhaye-Bouchaud, N., Auclair, N., Daniel, H., Crepel, F., Mariani, J., Sotelo, C., and Becker-Andre, M. (1998). staggerer phenotype in retinoid-related orphan receptor alpha-deficient mice. *Proc. Natl. Acad. Sci. USA* **95,** 3960–3965.

Takada, S., Stark, K. L., Shea, M. J., Vassileva, G., McMahon, J. A., and McMahon, A. P. (1994). Wnt-3a regulates somite and tailbud formation in the mouse embryo. *Genes Dev.* **8,** 174–189.

Theil, T., Alvarez-Bolado, G., Walter, A., and Ruther, U. (1999). Gli3 is required for Emx gene expression during dorsal telencephalon development. *Development* **126,** 3561–3571.

Thomas, K. R., and Capecchi, M. R. (1990). Targeted disruption of the murine int-1 proto-oncogene resulting in severe abnormalities in midbrain and cerebellar development. *Nature* **346,** 847–850.

Thomas, K. R., Musci, T. S., Neumann, P. E., and Capecchi, M. R. (1991). Swaying is a mutant allele of the proto-oncogene Wnt-1. *Cell* **67,** 969–976.

Torres, M., Gomez-Pardo, E., Dressler, G. R., and Gruss, P. (1995). Pax-2 controls multiple steps of urogenital development. *Development* **121,** 4057–4065.

Urbanek, P., Wang, Z. Q., Fetka, I., Wagner, E. F., and Busslinger, M. (1994). Complete block of early B cell differentiation and altered patterning of the posterior midbrain in mice lacking Pax5/BSAP. *Cell* **79,** 901–912.

Urbanek, P., Fetka, I., Meisler, M. H., and Busslinger, M. (1997). Cooperation of Pax2 and Pax5 in midbrain and cerebellum development. *Proc. Natl. Acad. Sci. USA* **94,** 5703–5708.

van den Munckhof, P., Luk, K. C., Ste-Marie, L., Montgomery, J., Blanchet, P. J., Sadikot, A. F., and Drouin, J. (2003). Pitx3 is required for motor activity and for survival of a subset of midbrain dopaminergic neurons. *Development* **130,** 2535–2542.

Vogel-Hopker, A., and Rohrer, H. (2002). The specification of noradrenergic locus coeruleus (LC) neurones depends on bone morphogenetic proteins (BMPs). *Development* **129,** 983–991.

Wallace, V. A. (1999). Purkinje-cell-derived Sonic hedgehog regulates granule neuron precursor cell proliferation in the developing mouse cerebellum. *Curr. Biol.* **9,** 445–448.

Wassarman, K. M., Lewandoski, M., Campbell, K., Joyner, A. L., Rubenstein, J. L., Martinez, S., and Martin, G. R. (1997). Specification of the anterior hindbrain and establishment of a normal mid/hindbrain organizer is dependent on Gbx2 gene function. *Development* **124,** 2923–2934.

Watanabe, Y., and Nakamura, H. (2000). Control of chick tectum territory along dorsoventral axis by Sonic hedgehog. *Development* **127,** 1131–1140.

Wechsler-Reya, R. J., and Scott, M. P. (1999). Control of neuronal precursor proliferation in the cerebellum by Sonic Hedgehog. *Neuron* **22,** 103–114.

Wechsler-Reya, R., and Scott, M. P. (2001). The developmental biology of brain tumors. *Annu. Rev. Neurosci.* **24,** 385–428.

Wilkinson, D. G., Bailes, J. A., and McMahon, A. P. (1987). Expression of the proto-oncogene int-1 is restricted to specific neural cells in the developing mouse embryo. *Cell* **50,** 79–88.

Wingate, R. J. (2001). The rhombic lip and early cerebellar development. *Curr. Opin. Neurobiol.* **11,** 82–88.

Wingate, R. J., and Hatten, M. E. (1999). The role of the rhombic lip in avian cerebellum development. *Development* **126,** 4395–4404.

Wurst, W., Auerbach, A. B., and Joyner, A. L. (1994). Multiple developmental defects in Engrailed-1 mutant mice: An early mid-hindbrain deletion and patterning defects in forelimbs and sternum. *Development* **120,** 2065–2075.

Xu, J., Liu, Z., and Ornitz, D. M. (2000). Temporal and spatial gradients of Fgf8 and Fgf17 regulate proliferation and differentiation of midline cerebellar structures. *Development* **127,** 1833–1843.

Ye, W., Shimamura, K., Rubenstein, J. L., Hynes, M. A., and Rosenthal, A. (1998). FGF and Shh signals control dopaminergic and serotonergic cell fate in the anterior neural plate. *Cell* **93,** 755–766.

Ye, W., Bouchard, M., Stone, D., Liu, X., Vella, F., Lee, J., Nakamura, H., Ang, S. L., Busslinger, M., and Rosenthal, A. (2001). Distinct regulators control the expression of the mid-hindbrain organizer signal FGF8. *Nat. Neurosci.* **4,** 1175–1181.

Zervas, M., Millet, S., Ahn, S., and Joyner, A. L. (2004). Cell behaviors and genetic lineages of the mesencephalon and rhombomere 1. *Neuron* **43,** 345–357.

Zetterstrom, R. H., Williams, R., Perlmann, T., and Olson, L. (1996). Cellular expression of the immediate early transcription factors Nurr1 and NGFI-B suggests a gene regulatory role in several brain regions including the nigrostriatal dopamine system. *Brain Res. Mol. Brain Res.* **41,** 111–120.

Zetterstrom, R. H., Solomin, L., Jansson, L., Hoffer, B. J., Olson, L., and Perlmann, T. (1997). Dopamine neuron agenesis in Nurr1-deficient mice. *Science* **276,** 248–250.

Zhang, X. M., Lin, E., and Yang, X. J. (2000). Sonic hedgehog-mediated ventralization disrupts formation of the midbrain-hindbrain junction in the chick embryo. *Dev. Neurosci.* **22,** 207–216.

6

Brain Development and Susceptibility to Damage; Ion Levels and Movements

Maria Erecinska, Shobha Cherian,[†] and Ian A. Silver**
*Department of Anatomy
School of Veterinary Science
Bristol BS2 8EJ, United Kingdom
[†]Department of Neonatal Medicine
University Hospital of Wales
Cardiff CF14 4XW, United Kingdom

Responses of immature brains to physiological and pathological stimuli often differ from those in the adult. Because CNS function critically depends on ion movements, this chapter evaluates ion levels and gradients during ontogeny and their alterations in response to adverse conditions. Total brain Na^+ and Cl^- content decreases during development, but K^+ content rises, reflecting shrinkage of the extracellular and increase in the intracellular water spaces and a reduction in total brain water volume. Unexpectedly, $[K^+]_i$ seems to fall during the first postnatal week, which should reduce $[K^+]_i/[K^+]_e$ and result in a lower V_m, consistent with experimental observations. Neuronal $[Cl^-]_i$ is high during early postnatal development, hence the opening of Cl^- conduction pathways may lead to plasma membrane depolarization. Equivalent loss of K_i^+ into a relatively large extracellular space leads to a smaller increase in $[K^+]_e$ in immature animals, while the larger reservoir of Ca_e^{2+} may result in a greater $[Ca^{2+}]_i$ rise. *In vivo* and *in vitro* studies show that compared with adult, developing brains are more resistant to hypoxic/ischemic ion leakage: increases in $[K^+]_e$ and decreases in $[Ca^{2+}]_e$ are slower and smaller, consistent with the known low level of energy utilization and better maintenance of [ATP]. Severe hypoxia/ischemia may,

Current Topics in Developmental Biology, Vol. 69
Copyright 2005, Elsevier Inc. All rights reserved.

0070-2153/05 $35.00
DOI: 10.1016/S0070-2153(05)69006-0

however, lead to large Ca_i^{2+} overload. Rises in $[K^+]_e$ during epileptogenesis *in vivo* are smaller and take longer to manifest themselves in immature brains, although the rate of K^+ clearance is slower. By contrast, *in vitro* studies suggest the existence of a period of enhanced vulnerability sometime during the developmental period. This chapter concludes that there is a great need for more information on ion changes during ontogeny and poses the question whether the rat is the most appropriate model for investigation of mechanisms of pathological changes in human neonates. © 2005, Elsevier Inc.

I. General Introduction

The *raison d'etre* of the nervous system is to generate and transmit impulses. To perform these functions, neural cells maintain high resting transmembrane gradients of sodium (Na^+), potassium (K^+), calcium (Ca^{2+}), and chloride (Cl^-). Neuronal activation increases the permeability of plasma membrane to these molecules and allows them to flow down their concentration gradients. The concomitant depolarization of the plasma membrane opens voltage-controlled, ion-conducting pathways (channels), which further accelerates ion movements. Membrane repolarization and restoration of the ionic disequilibria involve uphill movements of ions and hence require energy. Therefore, it is not surprising that a large proportion of cellular ATP in brain is consumed in the maintenance of excitability (Ames, 2000; Attwell and Laughlin, 2001; Erecinska and Silver, 1989; Erecinska *et al.*, 2004).

In contrast to other organs of the body, the brains of many mammals are very "immature" at birth and it takes weeks (mice, rats) to years (primates) to reach the adult level of performance. In physiological terms, postnatal "development" of the CNS signifies the formation and stabilization of neuronal networks (changes in cell numbers and size, growth of processes and formation and destruction of synapses) and involves alterations in numbers and properties of cellular ion-moving pathways, channels and transporters. Because their range of functions and complexities increase with age, neural cells require progressively more energy and exert a positive "pressure" on ATP-producing pathways. Therefore, proper growth and development of brain "activity" has to be accompanied by a parallel rise in glycolysis and oxidative phosphorylation (Erecinska *et al.*, 2004).

There are many studies on the ontogeny of cerebral energy metabolism (reviewed in Erecinska *et al.*, 2004; Jones, 1979; Jones and Rolph, 1985; Nehlig and Pereira de Vasconcelos, 1993; Volpe, 2000) as well as on the development and properties of neuronal channels and receptors. Some of the latter information has been reviewed very recently (Erecinska *et al.*, 2004). The behavior of energy metabolism under a variety of pathological conditions has received detailed scrutiny in an excellent textbook by Volpe (2000). However, while there is ample literature on ion levels and movements in the brains of adult mammals under physiological and pathological conditions (reviewed in, e.g., Erecinska and Silver, 1989, 1994; Hansen, 1985), much less attention has been paid to parallel situations in newborn and growing animals. Nevertheless, deeper knowledge in these areas would greatly enhance our understanding of the differences in electrical properties between mature and immature neurons and help explain the sometimes distinctive responses of immature brain to adverse conditions. The present chapter was designed to fill this gap. Its aim is to evaluate critically the literature on the subject, to pinpoint inconsistencies and disagreements in existing data, and to indicate venues for future investigations.

II. Ions and Membrane Potentials in Brains of Immature Mammals

A. Ion Levels and Gradients

1. Cations

It has been known for over 30 years (De Souza and Dobbing, 1971; Vernadakis and Woodbury, 1962) that in rat brain the content of key ions, Na^+, K^+, and Cl^-, changes during development (Fig. 1). The total amount of sodium remains relatively constant at 60–65 mmol/kg wet weight during the first few days after birth (until postnatal days [P] 7–12) but then steadily declines during the suckling period to reach the adult value of 40–45 mmol/kg at 3–4 weeks of age, i.e., shortly after weaning. Chloride follows a pattern similar to that of sodium (Vernadakis and Woodbury, 1962). Total brain water, which is 88–90% during the first postnatal week (Agrawal et al., 1968; De Souza and Dobbing, 1971; Vernadakis and Woodbury, 1962), falls somewhat more slowly than sodium and chloride and is still above the adult level (about 78% at 10 weeks of life; De Souza and Dobbing, 1971) at weaning (82% at P21 and 80% at P28; De Souza and Dobbing, 1971; Vernadakis and Woodbury, 1962). Total potassium content decreases from about 80 mmol/kg at birth to 70–75 mmol/kg at P5–8. It then rises to a maximum of 100–105 mmol/kg at P21–28 and declines very slowly over the next several weeks to reach the adult value of about 95 mmol/kg at 2–3 months (De Souza and Dobbing, 1971; Vernadakis and Woodbury, 1962). We were unable to find any data on the fate of total brain calcium.

Information on the total tissue content of ions does not translate easily into knowledge of their distribution between cells and their environment. Moreover, the physiologically relevant (i.e., biologically "active") amounts are those that are free in solution. Large fractions of sodium and potassium fulfill this criterion, while, by contrast, >99% of the intracellular calcium is either bound to various cell constituents or sequestered in subcompartments.

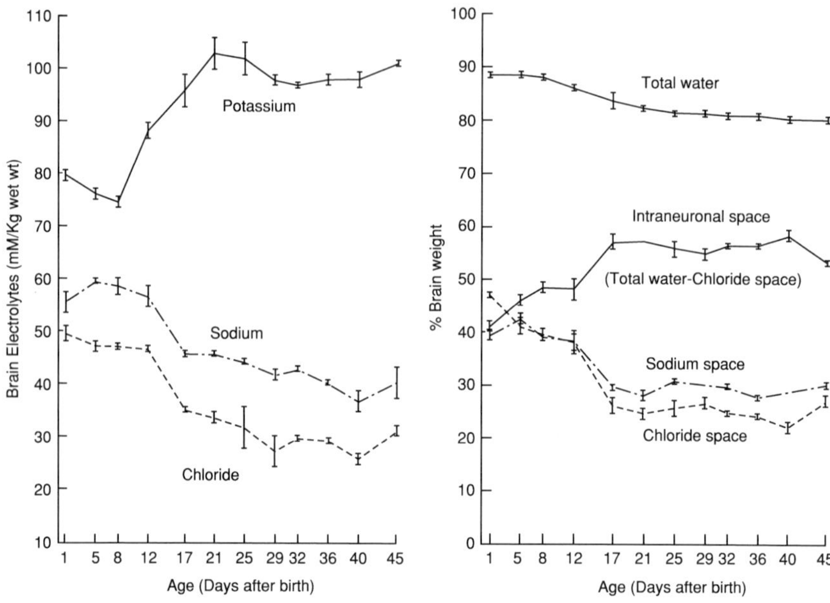

Figure 1 Changes in brain potassium, sodium, and chloride concentrations (left) and in brain total water content, chloride space, sodium space, and total water-chloride space (right) during maturation. Reproduced with permission of *Am. J. Physiol.* from Vernadakis and Woodbury (1962).

Estimates of the free ion levels require different methodologies that are more difficult to apply in brains of immature mammals because of their small size and relative fragility. Therefore, not surprisingly, the available data on the topic are rather meager. Table I summarizes most of the information we were able to find on the intra- and extracellular concentrations of Na^+, K^+, Ca^{2+}, H^+, and Cl^- in the CNS of newborn and developing mammals; these are the objects of our present scrutiny. The only "adult" values that are quoted in the table are those that were obtained as reference points and thus were inherent components of the individual studies. For a more extensive survey of ion levels in brain cells of mature mammals the interested reader is referred to earlier reviews (Erecinska and Silver, 1994; Hansen, 1985).

There is general agreement that the extracellular concentration of K^+, $[K^+]_e$, both *in vivo* and *in vitro* is 3–4 mM and is not species-, region-, or age-dependent (Ballanyi *et al.*, 1992; Hansen, 1977; Jiang *et al.*, 1992b; Mares *et al.*, 1976; Mutani *et al.*, 1974; Silver and Thoresen, unpublished data; Xia *et al.*, 1992) (Table I). (The somewhat higher mean of 4.4 ± 1.1 mM in the extracellular space of rabbit brain stem [Trippenbach *et al.*, 1990] is not significantly different.) By contrast, it has been found in several mammalian species, including the human, that the plasma potassium level is higher in the fetus and newborn (5–7 mM) than in adult (3.5–5 mM) (Anzai *et al.*, 2001; Dancis and Springer, 1970; Lelievre-Pegorier, *et al.*, 1983; Lorenz *et al.*, 1986, 1997; McCance and Widdowson, 1956; Rodriguez-Soriano *et al.*, 1981; Sulyok, *et al.*, 1979; Vernadakis and Woodbury, 1962). This indicates that early in ontogeny the blood–brain barrier is already essentially impermeable to potassium and that independent regulation of cerebral $[K^+]_e$ ensures that the brain is protected from fluctuations that may occur in the systemic circulation. Extracellular concentration of sodium, $[Na^+]_e$, in immature animals *in vivo* has not been quantified, while estimates in slices despite their dubious value (because of the large volume of the bathing medium and the high concentration of sodium in it) suggest a figure of 120–130 mM (Jiang *et al.*, 1992b). Determinations of extracellular calcium concentration, $[Ca^{2+}]_e$, *in vivo* in rat (P9–11, 1.11 mM; Puka-Sundvall *et al.*, 1994) and piglet (P1–3, 1.34 mM; Silver and Thoresen, unpublished data) cortex, and rabbit (P0–28) brain stem (mean 1.3 ± 0.4 mM; Trippenbach *et al.*, 1990) yielded very similar values (identical to those in adults), which indicates that $[Ca^{2+}]_e$, like $[K^+]_e$, is species-, region-, and age-independent. By contrast, it has been reported recently (Miyamoto *et al.*, 2004) that concentration of calcium in the cerebrospinal fluid of human babies younger than 11 months is 1.11–1.16 mM, which is slightly but significantly higher than the figure of 1.0 –1.07 mM seen in older children and adults. Why this should be is not immediately obvious, and the finding requires independent confirmation. Unfortunately, experimental measurements used in brains themselves do not have sufficient accuracy to detect differences of ≤ 0.1 mM.

Intracellular concentrations of Na^+ ($[Na^+]_i$) and K^+ ($[K^+]_i$) in brains of immature mammals have not yet been measured and are difficult to predict. Sodium is the most abundant extracellular cation in the brains of adult mammals, where it is present at a concentration of about 130 mM; its level inside cells is around 20 mM (Erecinska and Silver, 1994; Hansen, 1985). Given an extracellular water-space fraction of 0.23 in the brain of a mature rat (Bondareff and Pysh, 1968; Lehmenkuhler *et al.*, 1993; Levin, 1970; Vernadakis and Woodbury, 1965; Vorísek and Syková, 1997), one can calculate that two-thirds, or 66% (30 out of 45 mmol/kg), of the total CNS sodium is outside and only 33% is inside cells. During the first postnatal week, the total amount of sodium in the brain remains relatively constant and then begins to decline as total brain water falls to reach a new, lower steady state level at postnatal week 3/4 (Fig. 1). Using the value of 65 mmol/kg for the total Na^+ content (De Souza and Dobbing, 1971; Vernadakis and Woodbury, 1962) and a figure of 0.43 for the extracellular water-space fraction in the P4–6 rat (Vorísek and Syková, 1997), and assuming an $[Na^+]_e$ of 130 mM, one arrives at the conclusion that at this age, 56 out of 65 mmol/kg, or 86%, of total Na^+ is extracellular and only 14% is inside

Table I Ion Levels in Immature Brain

Species (preparation)	Age	K⁺ In (mM)	K⁺ Out (mM)	Na⁺ In (mM)	Na⁺ Out (mM)	Ca²⁺ In (mM)	Ca²⁺ Out (mM)	H⁺ (pH) In	H⁺ (pH) Out	Cl⁻ In (mM)	Cl⁻ Out (mM)	T (°C)	Reference
Rat													
Brain stem slices	P2 + 9 + 16		3.2									22–24	Jiang *et al.*
	Adult		3.8										(1992b)
	P5		3.1									35–36	Xia *et al.*
	Adult		3.2										(1992)
	P2–10				149.5					12.4	122.8	35–36	Jiang *et al.*
	Adult				150					11.4	128.3		(1992a)
Brain stem-spinal	P0–3		3–4										Ballanyi *et al.*
cord	Adult		3–4										(1992)
Cortical slices	P1–30					165							Bickler *et al.*
	Adult					165							(1993)
Synaptosomes	P5–15					~450							Keelan *et al.*
	P20–25					~400							(1996)
	P40					~300							
	P60					~250							
Hippocampal	P1–3					79							Marks *et al.*
neurons	Adult					79							(1996)
Cortical slices VZ	E16									37		21–23	Owens *et al.*
Cortical slices CP	E16									29.2			(1996)
	E19									23.8			
	P0									19.4			
	P1									18.8			
	P2									20.0			

	P4			18.8			
	P16			11.7			
Cortical slices VZ	P0–4			37.9			Shimizu-Okabe
Cortical slices CP				29.7			*et al.* (2002)
Cortical slices 1.V/VI				22.3			
Cortical slices, CP (or layer II/III)	P1–3			29.5	22–24		Yamada
	P5–7			~24			*et al.* (2004)
	P11–20			8.6			
Cortical layer V/VI	P1–3			~17			
	P5–7			~12			
Sensorimotor cortex (*in vivo*)	P5–19	3–3.7				Body	Mares *et al.* (1976)
Cortex (*in vivo*)	P4	4.6				Body	Hansen (1977)
	P7	2.9					
	P12	3.4					
	P16	3.1					
	P24	4.3					
	Adult	3.0					
Cortex (*in vivo*)	P9–11		1.11			Body	Puka-Sundvall *et al.* (1994)
Rabbit							
Cortex (*in vivo*)	P1–19	3–3.5				Body	Mutani *et al.* (1974)
Brain stem (*in vivo*)	4 hr	5.6				Body	Trippenbach *et al.* (1990)
	6 hr	4.4	0.95				
	12 hr	2.6					
	P3	6.3	1.83				
	P4	5.4	0.93				
	P9	5.0	1.52				
	P18	4.2	1.91				
	P21	3.3					

(*Continued*)

Table I *Continued*

Species (preparation)	Age	K⁺ In (mM)	K⁺ Out (mM)	Na⁺ In (mM)	Na⁺ Out (mM)	Ca^{2+} In (mM)	Ca^{2+} Out (mM)	H^+ (pH) In	H^+ (pH) Out	Cl⁻ In (mM)	Cl⁻ Out (mM)	T (°C)	Reference
	P21	4.6					1.00						
	P28	2.8											
Dog													
(*in vivo*)	P1–10								7.25*			Body	Young *et al.* (1985)
Pig													
(*in vivo*)	P7 ± 4								~7.0*			Body	Laptook *et al.* (1992)
Cortex (*in vivo*)	P0–3	3.4						1.34			7.2	Body	Silver and Thoresen, unpublished data
Human													
(*in vivo*)	42 weeks								~7.1*			Body	Azzopardi *et al.* (1989)

Values listed are means for which SDs and n can be found in the original publications. Unpublished data of Silver and Thoresen are means of 34 measurements for K^+ (SD = ± 0.07); 6 for Ca^{2+}_e (SD = ± 0.11), and 9 for pH_e (SD = ± 0.04). VZ = ventricular zone; CP = cortical plate; 1. = layer; ~ = values estimated by the present authors from figures in original publications; E = embryonic age in days; P = postnatal age in days. Ion concentrations were determined with ion-selective microelectrodes, except for pH_i (*), which was estimated by NMR from the chemical shift of the inorganic phosphate resonance peak relative to the phosphocreatine peak.

cells. This calculation seems to suggest that the average $[Na^+]_i$ in brain cells does not change during ontogeny, although it does not preclude the possibility that some decreases may occur in discrete brain regions and/or different populations of cells.

The situation may be different for potassium, which is the main intracellular cation in the CNS. In an adult rat, with a $[K^+]_e$ of 3–3.5 mM (see above) and an extracellular water-space fraction of 0.23, only 0.81 out of 95 mmol/kg, or about 1% of total K^+ content, is outside cells. This percentage is not much greater in brains of immature animals, which have a larger extracellular space but the same, low $[K^+]_e$ (Table I). Thus, one would expect that the postnatal rise in the intracellular water space (i.e., cell volume), which occurs at least in part at the expense of the diminishing extracellular space, would be accompanied by a parallel increase in total K^+ content. However, this is not the case as shown in Fig. 1, while the intraneuronal space rises continually, the total brain K^+ content initially decreases and only begins to increase sharply after P7, to reach a plateau at P21–28. The reasons for this pattern of K^+ behavior in rat brain are not clear. It is interesting that the total K^+ content is lowest at the age when many of the cerebral metabolic (e.g., activities of enzymes of energy producing pathways, the Na^+/K^+ ATPase; Erecinska *et al.*, 2004) and functional (e.g., growth of synapses; Aghajanian and Bloom, 1967) processes begin their most rapid "growth" and attains a stable, high level at the time when their development is almost complete and the brain acquires its adult properties (Erecinska *et al.*, 2004). The decrease in total brain K^+ in the presence of an unaltered $[K^+]_e$ implies that $[K^+]_i$ falls during the first week of postnatal life. However, in the absence of experimental measurements of the latter during brain ontogeny, this appealing suggestion remains a speculation. Moreover, if it is true, it would be important to establish whether the reduction in $[K^+]_i$ occurs in all cells and at all locations or only in particular cell types and at specific locations. Whether this unusual behavior of potassium is typical only of the rat brain or occurs in all growing animals at species-specific developmental times also deserves detailed scrutiny.

In dissociated rat hippocampal neurons, $[Ca^{2+}]_i$ (Table I, 79 nM; Marks *et al.*, 1996) was found to be half that in cells in cortical slices (165 ± 63 nM; Bickler *et al.*, 1993), and these values were the same in preparations from both newborn and adult animals. (The apparently large discrepancy between these two figures may be due either to differences in experimental material, slices vs. isolated cells, or to inadequacies of the methodology. Data acquired with fluorescent indicators are usually expressed as fluorescence intensity ratios and the concentrations sporadically calculated therefrom show a substantial spread, even for basal resting levels.) By contrast, a significant age-related decline in $[Ca^{2+}]_i$ has been described in synaptosomes (i.e., nerve-ending particles) from brains of developing rats (Keelan *et al.*,

1996). The higher level in preparations from P5–20 animals (400–450 nM vs about 250 nM in P60 rats) was explained by the immaturity of the calcium extrusion mechanisms. Although this explanation may hold true for the animals younger than P14, it does not for those at P20 because at the latter age most ion-handling processes in the brain, including the Na^+/K^+ ATPase, have attained their adult levels of activities (Erecinska *et al.*, 2004). Moreover, the fact that at 3 weeks of age rat brain possesses its full complement of synapses (Aghajanian and Bloom, 1967) while the $[Ca^{2+}]_i$ is still high seems to invalidate the postulate (Keelan *et al.*, 1996) that the elevated concentrations of the cation in younger animals may be necessary for normal growth and development of nerve terminals. Finally, the questions of why no such developmental differences were detected in slices from animals of the same species and age, and when the same technology was used for cation measurement (Bickler *et al.*, 1993) have to be answered. Although the results in slices may be dominated by events in cell bodies, one would nevertheless expect that the relatively large changes that occur in the very active neuronal compartment of nerve endings would yield noticeable effects in the whole slice preparations.

Using an even simpler system, Robertson *et al.* (2004) compared calcium uptake by isolated total (i.e., synaptosomal plus non-synaptosomal) brain mitochondria from P16–18 and 3-month-old rats. The authors found that the rates of accumulation in the presence of ATP at pH 6.5 were almost 5-fold lower in the organelles from young than from adult animals (645 vs 3110 nmol/mg protein). They proposed that this apparently low capacity for cation accumulation could lead to higher cytosolic $[Ca^{2+}]$ in an immature animal. Although the finding may have some significance, the conclusion seems rather far-fetched: pH 6.5 is not seen under physiological conditions, while the rates at pH 7.0 (which is still slightly lower than the pH *in vivo*) were not markedly different between the two age groups.

pH in the CNS is about 7.2, both internally (Azzopardi *et al.*, 1989; Laptook *et al.*, 1992; Young *et al.*, 1985) and externally (Silver and Thoresen, unpublished data), in dog, pig, and human (Table I); thus, there seems to be no transmembrane gradient for H^+ in cells from an immature brain.

A comparison between immature and mature animals (Table I) (Erecinska and Silver, 1994, 2001; Hansen, 1985) shows that there are no significant age-dependent differences in $[K^+]_e$, $[Ca^{2+}]_e$, and probably $[Na^+]_e$ and $[Na^+]_i$. $[Ca^{2+}]_e/[Ca^{2+}]_i$ may increase with age in synaptosomes due to a decline in $[Ca^{2+}]_i$, while the potassium gradient, $[K^+]_i/[K^+]_e$, appears to undergo a biphasic change due to alterations in $[K^+]_i$: it falls from birth to P6–7 and then rises to the adult value at P21–28. A lower potassium equilibrium potential, E_K (which is directly related to $[K^+]_i/[K^+]_e$), means a lesser efflux of potassium at a given membrane potential. Thus, after-hyperpolarizations, both calcium-dependent and -independent, will be reduced and the preceding depolarizing events, such as Ca^{2+} spikes, will be increased with a consequent increase in $[Ca^{2+}]_i$. Small membrane depolarizations under lowered E_K may also activate voltage-dependent conductances and enhance ion movements (including those of Ca^{2+}).

Changes with age in sizes of the extracellular and intracellular water spaces exert an independent influence on the relationships between intra- and extracellular cations. This may be particularly important for calcium, in which the total amount of the cation in the extracellular space is much greater early during development due to the larger size of the latter. Consequently, the amount of Ca^{2+} "available" per cell is greater in neonates and young mammals than in mature individuals despite the apparently identical extracellular concentrations and the magnitudes of the inward-directed driving force at all ages. This could lead to a higher $[Ca^{2+}]_i$ under some pathological conditions and a greater "loading" of cells with this cation. By contrast, more K^+ would have to leave brain cells of a P1 than of an adult rat to reach the same $[K^+]_i/[K^+]_e$, while identical leakage would result in a higher $[K^+]_i/[K^+]_e$ (and consequently smaller depolarization of the plasma membrane) in the younger animals.

The difference in the concentrations of free calcium between the cell interior (nM) and the extracellular space (mM) creates an over 10^5 gradient ($[Ca^{2+}]_e/[Ca^{2+}]_i$) across the cell plasma membrane that continuously drives the cation from outside to inside. This means that even a small amount of calcium entering a cell would make a large difference to its intracellular concentration. Among the mechanisms that "buffer" the internal calcium content are calcium-binding proteins (Baimbridge *et al.*, 1992). It has been noted that during the development of monkey and rat CNS (Endo *et al.*, 1985; Enderlin *et al.*, 1987; Hendrickson *et al.*, 1991; Nitsch *et al.*, 1990) the distribution of calbindin D-28K and parvalbumin, two such proteins that occur in GABAergic neurons, is complementary. Calbindin immunoreactivity was high at birth and increased shortly thereafter, and was followed by postnatal redistribution and overall decline. By contrast, parvalbumin immunoreactivity developed after birth and continued to rise into adulthood. However, in cat visual cortex (Stichel *et al.*, 1987) both proteins were present prenatally and increased in parallel after birth. The reasons for these species-specific patterns are not clear at present but evidence has emerged recently (Colin *et al.*, 2005) that calcium-binding macromolecules may play a role in neuronal plasticity and development and/or protection of neurons from hyperexcitability.

2. Anions

Intracellular concentration of chloride ($[Cl^-]_i$) in the CNS is determined by a balance between Cl^- efflux through the K^+/Cl^- cotransporter, KCC2, and the anion influx via the $Na^+/K^+/2Cl^-$ cotransporter NKCC1 (Mercado *et al.*, 2004; Payne *et al.*, 2003). The absence of K^+-coupled Cl^- transport in recordings from neonatal rat neurons (De Fazio *et al.*, 2000) correlates with low expression of KCC2 (Clayton *et al.*, 1998; DeFazio *et al.*, 2000; Lu *et al.*, 1999; Rivera *et al.*, 1999; Shimizu-Okabe *et al.*, 2002; Yamada *et al.*, 2004) during the first postnatal week. By contrast, NKCC1 expression is strong at birth and during the first few days of life but markedly decreases thereafter (Shimizu-Okabe *et al.*, 2002; Yamada *et al.*, 2004). Consistent with these results, it has been shown that *in vivo* $[Cl^+]_i$ decreases with age and that this reduction occurs earlier in the phylogenetically older parts of the brain (Table I) (Owens *et al.*, 1996; Shimizu-Okabe *et al.*, 2002; Yamada *et al.*, 2004). Because $[Cl^-]_e$ does not seem to change during development (Jiang *et al.*, 1992a), this means that the chloride gradient, $[Cl^-]_e/[Cl^-]_I$, increases with age. Consequently, together with the fact that permeability to the anion is high in immature neurons, opening of Cl^- conduction pathways leads to chloride efflux (and not influx) and membrane depolarization and not the usual hyperpolarization seen in adults.

B. Plasma Membrane Resting Potential and Input Resistance

Ionic gradients constitute one of two determinants of plasma membrane potential, the other being membrane permeability to ions. Representative values of plasma membrane resting potential measured during development in brain cells of different animal species, in various brain regions, and both *in vitro* and *in vivo* are presented in Table II. To allow critical analysis of the information, studies have been selected that investigated at least three different ages, one of which was adulthood or a point close to it. Measurements were taken using either sharp microelectrodes or patch-clamping (whole-cell or gramicidin-perforated) and involve temperatures between 20 and 37 °C.

Of the 18 *in vitro* studies listed in Table II, five (Schwartzkroin and Altschuler, 1977, cat; Schwartzkroin, 1981, rabbit; Fukuda and Prince, 1992; Psarropoulou and Descombes, 1999; Zhang *et al.*, 1991; all in rat) showed no changes in neuronal membrane potential with age. All of these results were obtained in the hippocampus; one was carried out at 23–24 °C (Zhang *et al.*, 1991) and the remainder at 32–37°C. Except for the experiments of Zhang *et al.* (1991), which utilized patch-clamping technology, all used microelectrodes. The first age "points" varied from P1 to P7–11.

Twelve studies noted a shift with increasing age in the membrane potential of neurons toward more negative values; five of those were by less than 10 mV (Mueller *et al.*, 1984, rabbit; Isagai *et al.*, 1999; Kriegstein *et al.*, 1987; Luhmann and Prince, 1991; McCormick and Prince, 1987; all in rat), five by 10–20 mV (Burgard and Hablitz, 1993; Cherubini *et al.*, 1989; Spigelman

Table II Resting Membrane Potentials (V_m) and Input Resistance (R_N) in Brains of Immature Mammals

Animal	Brain Region	Age	V_m (mV)	R_N (MΩ)	T (°C)	Reference
			IN VITRO (SLICES): A. NEURONS			
Cat						
	Hippocampus (CA1)	P2–3	−46	46	36.5	Schwartzkroin and Altschuler (1977)
		P5–6	−54	34		
		PW2	−48	33		
		PW4	−50	24		
		Adult	−52	23		
	Dorsal lateral	P1–3	−54	218	35	Pirchio *et al.* (1997)
	Geniculate	P5–7	−56	∼200		
	Nucleus	P9	−58	∼170		
		P12	∼−62	147		
		Adult	∼−65	137		
Rabbit						
	Hippocampus	E21		170	35	Schwartzkroin and Kunkel (1982)
		E23		87		
		E24		80		
		E26		76		
		E28		68		
	Hippocampus (CA1)	P1	−55 to −75	38.4	36.5	Schwartzkroin (1981)
		P3	−55 to −75	41.0		
		P5	−55 to −75	54.7		
		P8	−55 to −75	29.4		
		P14	−55 to −75	21.4		
		P21	−55 to −75	21.2		
		Adult	−55 to −75	22.7		
	Hippocampus (CA1)	P6–10	−53.1	52.2	35	Mueller *et al.* (1984)
		P11–16	−57.2	44.8		
		1 month	−59	36.8		

(Continued)

Table II *Continued*

Animal	Brain Region	Age	V_m (mV)	R_N (MΩ)	T (°C)	Reference
Rat						
	Cortex	<PW1	−68.7	64.1	35	Kriegstein *et al.* (1987)
		PW2–3	−73.6	35.2	37	
		Adult	−75.3	24.3		
	Cortex	P1	−71.8	150	35	McCormick and Prince (1987)
		P7		~85		
		P15		~50		
		P30	77.2	40		
	Hippocampus [CA1 (CA3)]	P1–4	−64.2 (−61.3)	76.5 (57.1)	34–35	Cherubini *et al.* (1989)
		P6–8	−66.3 (−64.7)	85.6 (82.7)		
		P10–14	−71.4	80.8		
		Adult	−75 (−70.3)	44.5 (60.7)		
	Cortex	P4–10	−68.0	113.8	20–25	Luhmann and Prince (1991)
		P11–16	−77.5	81.7		
		P28–41	−77.6	43.2		
	Hippocampus (CA1)	P2–5	−56.5	934	23–24	Zhang *et al.* (1991)
		P8–13	−53.5	358		
		P15–20	−54.8	186		
	Hippcampus (CA1)	P7–11	−68.4	44.8	37	Fukuda and Prince (1992)
		P21–25	−63.9	38.0		
		P35–39	−65.2	30.9		
	Hippocampus (CA1)	P1	~−60	~1100	20–25	Spigelman *et al.* (1992)
		P5	~−65	~800		
		P10	~−68	~400		
		P20	~−70	~150		
		P30	~−73	~150		
	Cortex	P3–5	~−51	~300	35	Burgard and Hablitz (1993)
		P6–9	~−53	~250		

Species	Region	Age				Reference
		P10–11	~−56			
		P12–15	~−61	~100		
	Cortex	P1	~−40	~3500	21–23	Zhou and Hablitz (1996)
		P5	~−50	~2500		
		P9	~−55	~2000		
		P13	~−60	~1000		
		P17	~−60	~600		
		P21	~−60	~600		
	Hippocampus (CA1)	P7	−66	69	36.5	Isagai *et al.* (1999)
		P14	−62	51		
		P21	−68	36		
		P28	−69	28		
		Adult	−73	31		
	Hippocampus (CA3)	P3–7	−64.8		32	Psarropoulou and Descombes (1999)
		P8–20	−69.1			
		P > 60	−63.2			
	Nucleus accumbens	P1	~−55	~3000	20–22	Belleau and Warren (2000)
		P10	~−70	~1000		
		P20	~−80	~500		
		>P21	84	305		
	Hippocampus	P0–2	−44/−58/−77	2600	20–22	Tyzio *et al.* (2003)
		P5	~−57			
		P10	~−65			
		P13–15	−67/nd/−77	250		
		P20–25	~72			
Mouse						
	Reticular nucleus	P3	~−50 (~−50)	~600 (~1200)	20–25	Warren and Jones (1997)
		P6	~−53 (~−53)	~600 (~900)		
	(ventral posterior nucleus)	P9	~−55 (~−55)	~500 (~700)		
		P14–29	−63 (−59)	~200 (~400)		
	Cortex	E14		17,065	24–26	Picken Bahrey and Moody (2003)
		E16		6396		

(Continued)

Table II *Continued*

Animal	Brain Region	Age	V_m (mV)	R_N (MΩ)	T (°C)	Reference
		E18		6923		
		P0		6000		
		P2		4901		
		P4		1107		
		P6		1772		
		P10		426		
		P12		524		
		IN VITRO (SLICES): B. ASTROCYTES				
Rat						
	Hippocampus	P5	−58 to −66	~500	22–24	Bordey and Sontheimer (1997)
		P50	−58 to −66	~100–200		
		IN VIVO (NEURONS)				
Rat						
	Neostriatum	P6–10	~−45	~52	Body	Tepper and Trent (1993); Tepper *et al.* (1998)
		P11–15	~−49	~55		
		P16–20	~−52	~45		
		P21–29	~−58	~41		
		P30–40	~−61	~31		
		Adult	~−63	~31		

Values are means. E = embryonic day (or gestational day); P = postnatal day; PW = postnatal week; ~ = numbers estimated from original graphs by the present authors. Methods: *Microelectrodes:* Cherubini *et al.* (1989), Fukuda and Prince (1992), Isagai *et al.* (1999), Kriegstein *et al.* (1987), Luhmann and Prince (1991), McCormick and Prince (1987), Mueller *et al.* (1984), Picken Bahrey and Moody (2003), Psarropoulou and Descombes (1999), Schwartzkroin (1981), Schwartzkroin and Altschuler (1977), Schwartzkroin and Kunkel (1982), Tepper and Trent (1993), and Tepper *et al.* (1998). *Whole-cell patch clamping:* Bellau and Warren (2000), Bordey and Sontheimer (1997), Burgard and Hablitz (1993), Spigelman *et al.* (1992), Warren and Jones (1997), Zhang *et al.* (1991), Zhou and Hablitz (1996). For data from Tyzio *et al.* (2003), where three numbers are given, the first refers to measurements on gramicidin-perforated patches, the second to whole-cell patch-clamping, and the third to estimates using cell-attached recordings of NMDA channels.

et al., 1992; all in rat; Warren and Jones, 1997, mouse; Pirchio *et al.*, 1997, cat), and two by 20 mV or more (Belleau and Warren, 2000; Zhou and Hablitz, 1996; both in rat). (The results reported by Tyzio *et al.*, 2003 were not included in any of the categories.) There was no correlation between the value of the membrane voltage and animal species, brain region, or method and temperature of measurement. Membrane potentials in astrocytes from hippocampal slices were -58 to -66 mV between P5 and P60 (Bordey and Sontheimer, 1997), independent of age. Since, unfortunately, these latter figures are well below the values routinely obtained in astrocytes from adult animals both *in vitro* and *in vivo* (reviewed in Silver and Erecinska, 1997) more work is needed to establish age dependency of plasma membrane voltage, or a lack thereof, in astrocytes.

In the two *in vivo* studies, a gradual but substantial hyperpolarization of neostriatal neurons was seen between P6–10 and adulthood (Table II) (Tepper and Trent, 1993; Tepper *et al.*, 1998).

Factors that determine the resting membrane potential in neurons during early development are not known. In mature cells, a high conductance for K^+ and an active Na^+/K^+ pump maintain the V_m at between -65 and -80 mV (values are higher *in vitro* than *in vivo* but are not very dependent on either animal species or brain region). Although these mechanisms may be similar at all ages, during early development the numbers of ion channels and transporters, including the Na^+/K^+ pump, are low, while chloride permeability and intracellular chloride concentration are high (Erecinska *et al.*, 2004); all could affect the plasma membrane voltage. Furthermore, at least in rat there appears to be a period between birth and P6–7 when $[K^+]_i$ and consequently $[K^+]_i/[K^+]_e$ are lower than in adults, which may lead to a somewhat depolarized plasma membrane. Thus, there are good potential reasons for a lowered V_m in developing cells, but they do not explain why the phenomenon was not seen by all investigators.

Another explanation for this apparently lower resting plasma membrane potential seen at some stages of brain development was put forward recently by Tyzio *et al.* (2003). These authors pointed out that there are technical difficulties in measuring membrane potentials in immature brain cells with either microelectrodes or patch-clamping technology (see also Ben-Ari *et al.*, 1989). They argued that the "depolarized" values obtained with perforated-patch and whole-cell recordings are due to a short circuit through the seal contact between the patch pipette and the membrane. In support of their proposal, they compared measurements obtained with three methods: (i) gramicidin-perforated patches, (ii) whole-cell recording, and (iii) cell-attached recordings of N-methyl-D-aspartate (NMDA) channels. They found that with gramicidin-perforated patches, V_m at P0–2 in hippocampal CA3 neurons was -44 mV and rose to -67 mV at P13–15. V_m determined by whole-cell recordings at P2–3 was -58 mV, while that from recordings of

NMDA channels was -77 mV and did not change with age. However, the technical difficulties do not explain why Lamsa *et al.* (2000), using the same material and animals of identical age, reported a membrane potential in gramicidin-perforated patches of -67.5 mV, i.e., over 25 mV higher than that of Tyzio *et al.* (2003).

The conclusion, hence, is that there is no obvious explanation for the discrepancy in experimental results for the values of membrane potential in immature neurons. Consequently, the crucial question of whether these voltages are lower early during development can not be answered definitively at the present time.

The question of whether the plasma membrane potential is indeed less negative early in postnatal life is important because it means that less depolarization should be needed to trigger an action potential. This would make immature nerve cells more excitable and could initiate a number of processes that influence intracellular events, such as a rise in $[Ca^{2+}]_i$. Entry of calcium secondary to depolarization occurs via two mechanisms: (i) activation of voltage-gated channels and (ii) ligand/receptor-mediated events. $[Ca^{2+}]_i$ can also increase upon its release from internal stores either through the 1,4,5-inositol triphosphate stimulation or by calcium-stimulated calcium release. The latter requires the presence of external calcium and provides a mechanism for amplifying the initial signal. Studies in dorsal root ganglion neurons (Kocsis *et al.*, 1994) showed that cells cultured for 1 day and possessing very few neuronal processes had much higher resting levels of fluorescence of a calcium indicator in their nucleus and cytosol than did 6-day-old neurons with a network of processes. Application of 60 mM KCl caused an 8-fold rise in nuclear and a 3-fold increase in the cytosolic $[Ca^{2+}]$ in the younger cells, while in the older, elevations were no greater than 2-fold and were more pronounced in the cytosol than in the nucleus. These observations led the authors to postulate that the increase in $[Ca^{2+}]_i$ may be of primary importance in initiating the cascade that leads to depolarization-induced gene expression and neuronal differentiation. Whether the resting V_m of the 1-day-old cells was lower than in more "mature" ones was not determined.

In contrast to the situation with resting plasma membrane potentials, all studies show that membrane input resistance, RN, decreases with age, during both the prenatal (Picken Bahrey and Moody, 2003; Schwartzkroin and Kunkel, 1982) and postnatal (Table II) period. Absolute values of resistance depend on the temperature and are lower at higher temperatures. This behavior is consistent with age-related increase in cell membrane surface and density of ion channels (ion currents). The laboratory-to-laboratory variations in actual values shown in the table for apparently the same material and conditions are more difficult to explain.

III. Changes in Ions under Pathological Conditions

A. Hypoxia/Ischemia

1. Survival in Hypoxia/Ischemia

As early as the 17th century, Boyle (1670) noted that young kittens were less sensitive to oxygen deprivation than adult cats. Since then, his observation has been confirmed repeatedly. It has also been found that resistance to hypoxia, as measured by survival time in nitrogen at room temperature (21–25°C), decreases with maturation, but at any given postnatal age varies greatly from one animal species to another (Fig. 2) (Fazekas *et al.*, 1941; Glass *et al.*, 1944; Jacobson and Windle, 1960; Kabat, 1940). At birth, the most resistant of the common laboratory species (rat, cat, dog, rabbit, guinea pig, and monkey) is the very immature rat (50 min survival) and the least is the rather mature guinea pig (6 min), the difference being over 8-fold. Most of these mammals, when adult, survive without oxygen for only 1.5 to 3 min, i.e., their sensitivity differs only by a factor of 2. This also means that the largest changes during development, 10-fold or greater from birth to adulthood, occur in rat, dog, and rabbit and the smallest in guinea pig.

2. Patterns of Damage

In a neonatal animal model of hypoxia/ischemia (unilateral common carotid artery ligation with hypoxia in a 7-day-old rat), it was found that the severe neuronal changes that occur in the ipsilateral hemisphere are distinct from those commonly encountered in adult animals: damage is not only faster but different in character (Rice *et al.*, 1981). The early stages from microvacuolation through ischemic cell damage, first without and then with incrustation, are not seen in the neonate. However, at 15–26 hr after insult, homogenizing cell change appears and is followed at 36–50 hr by a gliomesodermal reaction. In adult animals, visible damage evolution begins at 3 days and is most prominent at 5–7 days. Adult brain, previously damaged by hypoxia/ischemia, rarely undergoes calcification, whereas this is a prominent feature of healing injury in the developing brain (Stein and Vannucci, 1988). Regional sensitivity to hypoxia/ischemia also changes with age (Towfighi *et al.* 1997); e.g., hippocampus, which is amazingly resistant to damage at P2–3, becomes progressively more sensitive, and by P13 its vulnerability exceeds that of the cortex. Cortical lesions at P13 change from predominantly columnar cell death to laminar-selective death. Interestingly, in rats younger than P13, any residual perfusion during hypoxic/ischemic insult is characterized by columns of low flow, situated perpendicular to the pial surface, adjacent to columns of higher flow (Ringel *et al.*, 1991) and is accompanied

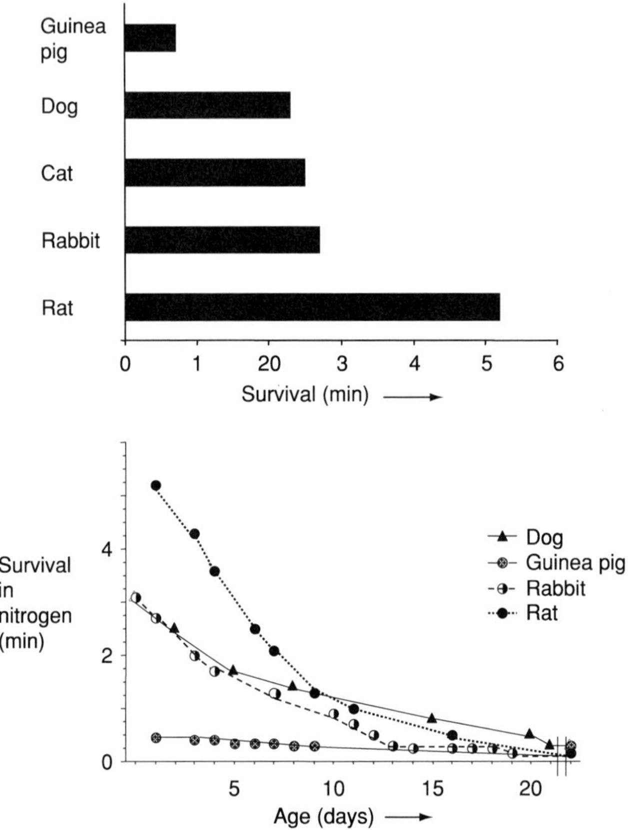

Figure 2 Species-dependent (top) and age-dependent (bottom) differences in susceptibility to hypoxia/ischemia. Susceptibility to hypoxia/ischemia was measured as survival in nitrogen. The results were taken from Fazekas *et al.* (1941) and Glass *et al.* (1944). The time to the last gasp for a newborn rhesus monkey placed in a jar with flowing nitrogen gas is 8–11 min (Jacobson and Windle, 1960).

by columnar alterations in NADH fluorescence (Welsh *et al.*,1982). Thus, the behavior of blood flow and metabolism corresponds to the pathologic pattern of injury seen within the cerebral cortex.

In human, as in rat, the main feature of cerebral hypoxic/ischemic injury is selective neuronal necrosis of a widespread but characteristic distribution (Johnston *et al.*, 2001; Volpe, 2000). In the term newborn, it involves neurons in the deep cortical layers, peri-Rolandic and calcarine cortex, putamen, thalamus, Purkinje cells of the cerebellum, cranial nerve nuclei of the brain stem, and anterior (ventral) horn cells of the spinal cord as well as cells in the supero-medial (particularly the parieto-occipital) parasagittal cortex. This pattern is similar to that in adults, in which the middle cortical laminae, especially over the vascular boundary zones, the CA1 and CA3 regions of the hippocampus, basal ganglia, Purkinje cells of the cerebellum, and the brain stem (Auer and Sutherland, 2002), are particularly prone to damage. In the premature infant, injury occurs in the subiculum of the hippocampus, the globus pallidus, the thalamus, internal granule cells of the cerebellum, the pons, and the cranial nerve nuclei of the brain stem (Friede, 1975; Larroche, 1977; Rorke, 1982).

The pathogenesis of the hypoxic/ischemic lesions is complex and involves both regional vascular and regional metabolic factors (Berger and Garnier, 1999; Volpe, 2000). The former include marked injury in vascular border zones, and the latter, differences in energy requirements, anaerobic glycolysis, lactate accumulation, and free radical formation and scavenging. Periventricular leucomalacia (necrosis of white matter dorsolateral to the external angles of the lateral ventricles), characteristically seen in preterm infants with evidence of hypoxia/ischemia and maternal chorioamnionitis (Gilles *et al.*, 1983; Larroche, 1977), has been attributed to the vulnerability of the developing oligodendrocytes to injury (Volpe, 2000).

3. Sensitivity to Hypoxia/Ischemia and Histological Damage in Experimental Animals

From their study on rats between P4 and P20, Ikonomidou *et al.* (1989) concluded that the sensitivity of the developing brain to hypoxic/ischemic injury peaks at 6 days of age. In the model these authors used, the only damage in P20 rats, which are already quite mature, was seen in the caudal caudate nucleus (80% frequency) and the rostral caudate and ventral thalamus (10% in each); there was no damage in nine other brain regions. These unprecedented findings, which are in direct conflict with the sensitivity to hypoxia measured by survival time in nitrogen (discussed above), have not received confirmation from investigations by other authors using the same or similar models. Grafe (1994) and Towfighi *et al.* (1997) reported a progressive increase in both the overall severity and frequency of hypoxic/ischemic damage in rats from P1 to P7 and P1 to P30, respectively, while Yager *et al.* (1996) showed that between the ages of 10–13 days and 6 months the most vulnerable were the 21- to 26-day-old animals. The patterns of damage in individual regions reported by Ikonomidou *et al.* (1989), Yager *et al.* (1996), and Towfighi *et al.* (1997) were also different. The discrepancies among the three groups in experimental findings are disconcerting and the reasons for their occurrence are baffling; the studies involved no differences in animal species, experimental model, or method of damage evaluation. Hence, unfortunately, even if there is increased sensitivity to hypoxia/ischemia during development, the lack of a consistent and uniform histological picture prevents any proposal being put forward that could designate the precise period and explain this enhanced vulnerability. Moreover, for the same reasons it can not be suggested with any degree of confidence that perinatal brain damage in humans is more profound or dangerous than that incurred at other stages of ontogeny.

4. Changes in Cerebral Ions during Hypoxia/Ischemia in Immature Mammals

Among the early responses of the CNS to hypoxia/ischemia are changes in ionic gradients (Erecinska and Silver, 1994; Hansen, 1985). These can be evaluated both *in vitro*, in brain slices or brain cells, and *in vivo*, in whole brain.

In vitro experiments using oxygen-free perfusion (or perifusion) fluids (Cherubini *et al.*, 1989; Haddad and Donnelly, 1990; Isagai *et al.*, 1999; Jiang *et al.*, 1992b; Krnjevic *et al.*, 1989; Luhmann and Kral, 1997; Nabetani *et al.*, 1997) either with or without glucose have shown that the time to anoxic depolarization (AD) is longer and the slope of the rise is smaller in nerve cells from immature than those from mature rats. The V_m of the plasma membrane in adult hypoglossal neurons exposed for 5 min to reduced (15–20 Torr) oxygen tensions declined by 32 mV, while it fell by only 10.4–11.2 mV in P3–16 cells (Haddad and Donnelly, 1990). The amplitude of the field potentials measured in neocortical slices at the time immediately preceding sudden anoxic depolarization declined by 3.1% in P5–8 rats, by 42% in P14–18 rats, and by 54% in >P28 animals (Luhmann and Kral, 1997). The frequency of AD increased from 39.1% to 87.5% and 90.5% in the same three groups, while the latency of onset decreased from 12.5 min to 8.7 min and 7.0 min, respectively. A similar observation was made in hippocampal slices, in which a 2-min hypoxic insult depressed excitatory postsynaptic potentials by 40–70% in preparations from P1–11 rats but by 90–100% in those from >P13 rats (Cherubini *et al.*, 1989). The latency to AD fell from 12.7 min at P14 to 5.8 min at P140 (Isagai *et al.*, 1999). Taken together, these results indicate that neurons from immature animals are resistant to short-term anoxia and relatively long but mild hypoxia. However, under more severe hypoxic conditions, the AD duration was found to be longer and its amplitude much greater in P5–8 and P14–18 groups of rats than in >P28 animals (Fig. 3; Luhmann and Kral, 1997).

It is well known that in adult neurons anoxia causes a substantial reduction in plasma membrane resistance due to an increased permeability of the membrane to ions. When measured in hippocampal slices, this decrease in R_N of CA1 and CA3 neurons was much smaller in P4 (by 7–11%) than P6–8 (by 35–38%) rats, while that in P10–14 animals (by 30–60%) was almost the same as in adults (by an average of 45.7%) (Cherubini *et al.*, 1989). No change was seen in young neonatal hypoglossal neurons during 5 min of

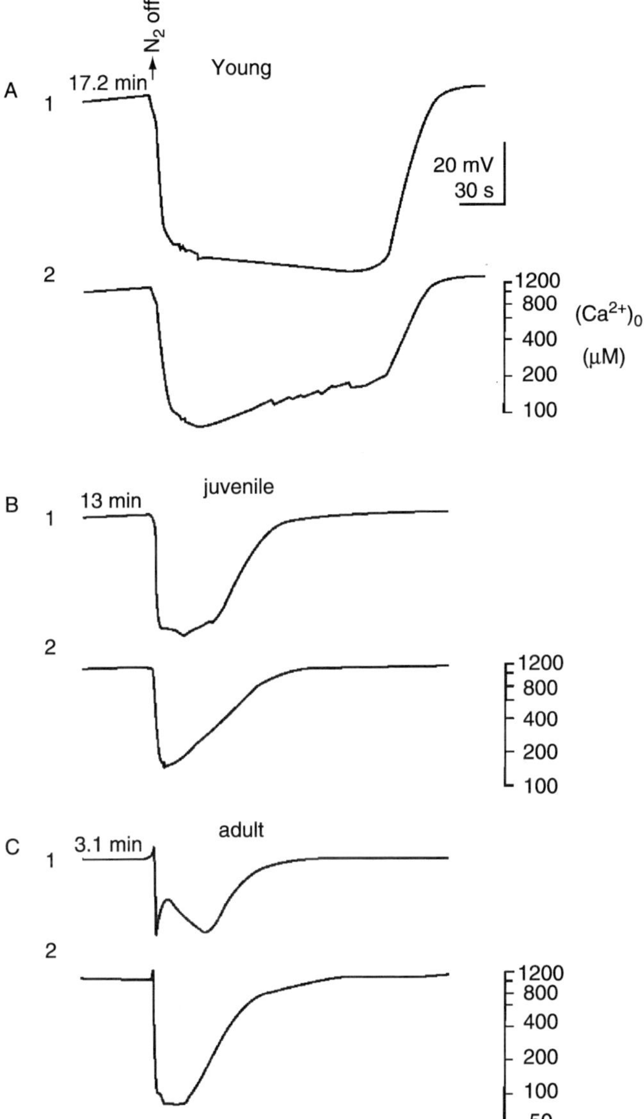

Figure 3 Developmental differences in responses of the rat neocortex to transient *in vitro* hypoxia. Simultaneous recordings of the extracellular DC potential (1) and extracellular Ca^{2+} concentration (2) were performed in a somatosensory cortical slice from a P7 (A), P17 (B), and adult (C) rat. Duration of hypoxia is given above trace 1 for each age group. Slices were reoxygenated at the onset of the anoxic depolarization (N₂ off). Reproduced with permission of *Am. J. Physiol.* from Luhmann and Kral (1997).

hypoxic exposure (Haddad and Donnelly, 1990). Consistent with these observations, the rises in extracellular potassium concentrations measured in slices from the brain stem of immature animals during short (4–10 min) periods of anoxia (Ballanyi *et al.*, 1992; Haddad and Donnelly, 1990; Jiang *et al.*, 1992b; Xia *et al.*, 1992) were also slower and much smaller in magnitude; after 4–5 min of oxygen deprivation, $[K^+]_e$ rose by <5 mM in preparations from P1–12 rats but by 30–40 mM in those from adults. Longer periods of anoxia and/or addition of iodoacetate, an inhibitor of glycolysis, induced much larger leakages of potassium, even in very young animals.

Anoxic changes in sodium, chloride, and calcium concentrations are also smaller and slower in immature than in mature animals. A period of 4 min of hypoxia reduced $[Cl^-]_e$ by 39 mM in adult brain stem slices but only by 5.3 mM in those from P2–10 pups (Jiang *et al.*, 1992a). Simultaneously, in the former, $[Cl^-]_e$ rose by 20.6 mM and in the latter by 4.5 mM. The corresponding decreases in $[Na^+]_e$ were 41.3 and 0 mM, respectively. It is interesting that 4–5 min of anoxia reduced extracellular water space by >50% in adult brain slices (Jiang *et al.*, 1992a,b; Pérez-Pinzón *et al.*, 1995) while very little (<5%) change was seen in brain tissue from immature animals (Jiang *et al.*, 1992a). However, longer periods of anoxia decreased extracellular space (ECS) in immature brain by about 30% (Jiang *et al.*, 1992a).

Hypoxia (95%N_2/5% CO_2) or 100 μM cyanide caused gradual elevation of $[Ca^{2+}]_i$ in neocortical slices from P1–15 rats but faster and larger changes in older animals (Bickler *et al.*, 1993). Surprisingly, no differences were found either between the P1–7 and P8–14 groups or among P15–21, P22–30, and adult groups. Perhaps measurements with fluorescent indicators are not accurate enough to detect small changes. By contrast to findings during either hypoxia alone or cyanide alone, a combination of cyanide and iodoacetate rapidly increased $[Ca^{2+}]_i$ in all age groups. It was later found (Bickler and Hansen, 1998) that after 5 min of anoxia followed by 6 hr of recovery CA1 neurons from hippocampal slices of P3–7 rats showed better survival, enhanced recovery, smaller increases in $[Ca^{2+}]_i$, less accumulation of glutamate, and less receptor-mediated Ca^{2+} influx than the same neurons from P18–22 animals. Survival of neurons in slices correlated with whole animal survival during oxygen deprivation (Fig. 2). In acutely isolated hippocampal CA1 neurons, the latency of the calcium rise during severe anoxia was five times longer (8.9 min vs 1.7 min) in P1–8 than in P21–40 rats (Friedman and Haddad, 1993). However, only in the former age group did $[Ca]_i$ continue to increase after anoxia ended. In hippocampal slices (Nabetani *et al.*, 1997), incubations in the absence of glucose and oxygen caused a rise in $[Ca^{2+}]_i$ with a lag time of 12.83 min at P7, 7.60 min at P10, and 3.94 min at P120. No increase in internal cation level was seen in isolated synaptosomes from P5–10 rats after 60 min of hypoxia/aglycemia, while 2-fold rises were seen at P20 and 3-fold rises at P60 (Keelan *et al.*, 1996). During perifusion with

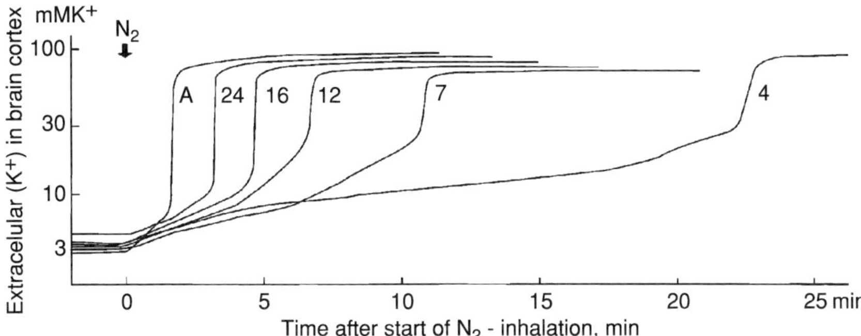

Figure 4 Changes in $[K^+]_e$ in the brain cortex of rats at different ages following exposure to nitrogen. Numbers indicate ages of animals in days. A denotes adult. Reproduced with permission of *Acta Physiol. Scand.* from Hansen (1977).

$95\%N_2/5\%$ CO_2, extracellular calcium was maintained at a basal level for longer periods in slices from young and juvenile than in those from mature rats (Luhmann and Kral, 1997). However, once the AD was triggered or during "long-term" hypoxia, decreases in extracellular calcium were significantly larger in P5–8 and P14–18 than in adult rats. Posthypoxic recoveries of calcium after ADs that follow short-term hypoxia were slowest in the youngest animals (see Fig. 3 in Luhmann and Kral, 1997) but those after long-term hypoxia were fastest (see Fig. 5 in Luhmann and Kral, 1997). These results are internally inconsistent, and the reasons for this difference in behavior are not immediately obvious.

In vivo studies in hypoxia/ischemia confirm results obtained in simpler preparations. As in adults (Hansen and Zeuthen, 1981; Silver and Erecinska 1990, 1992; reviewed in Erecinska and Silver, 1994; Hansen, 1985), changes in $[K^+]_e$ in brains of immature animals occur in four phases (Hansen, 1977; Mares *et al.*, 1976; Trippenbach *et al.*, 1990; Vorísek and Syková, 1997) (see also Fig. 4). After cerebral PO_2 falls to a very low value (<5 Torr), there is a small elevation in K_e^+ (phase I), followed by a second steeper increase (phase II) and an abrupt rise (phase III), culminating in a plateau (phase IV). The main differences between the developing and mature rats made hypoxic by inspiring N_2 (Hansen, 1977; Mares *et al.*, 1976) are the latency to the abrupt change in $[K^+]_e$ in phase III (which decreases from 25–35 min at P1 to <2 min in adulthood; [Hansen, 1977; Mares *et al.*, 1976] [Fig. 4]) and in its slope (which rises by 10-fold or more with age [Hansen, 1977; Mares *et al.*, 1976]). During decapitation-ischemia, changes in $[K^+]_e$ in the cerebral cortex of a P7 rat were much faster: phase II began at slightly more than 4 min and phase III at close to 6 min (Hansen and Nordstrom, 1979). Slower but qualitatively similar alterations were seen in the cardiac arrest model of ischemia in the same animal species (Vorísek and Syková, 1997). In rabbits, age 4 hr–28 days respiring for 4–5 min either 10% oxygen or N_2, tripartite changes were also observed: an initial slow phase, a secondary fast one, and a saturation at a level of 6–11 mM; the duration of the first two phases decreased with age, while their rates increased (Trippenbach *et al.*, 1990). The $[K^+]_e$ attained at the beginning of the abrupt phase III in hypoxic rats was shown either to be age-independent (about 20 mM; Mares *et al.*, 1976) or to decrease (about 25 mM at <P12 and about 10 mM at >P12; Hansen, 1977) or increase (Vorísek and Syková, 1997) with age. The final level of extracellular potassium was 52.6 mM at P1–5 and 75.6 mM at P19 (p < 0.01) in the study of Mares *et al.* (1976), 90 mM at all ages in the work of Hansen (1977), and 57–85 mM in that of Vorísek and Syková (1997).

Much less is known about alterations in calcium. No change in its external concentration was seen in rabbit brain stem during short-term, mild hypoxia (Trippenbach *et al.*, 1990), while large alterations were noted in P9–11 rats respiring nitrogen for 60 min (Puka-Sundvall *et al.*, 1994). These were biphasic in character; an increase during the first 10 min of anoxia from 1.11 mM to 1.53 mM was followed by a decline to 0.29 mM after 50 min. The decline phase was slow and protracted in 50% of animals but abrupt (after 14 ± 2.1 min of anoxia) in the remaining 50%. The initial, large rise is puzzling, since it is not observed in brains of adult rats (Silver and Erecinska, 1990, 1992). It was explained by the authors as being due either to shrinkage of the ECS or to movement of calcium from cells to the outside. However, ECS in brains of P9–11 rats is considerably larger than that in mature animals, and moreover, its shrinkage early during anoxia is very small (Vorísek and Syková, 1997). Intracellular free $[Ca^{2+}]$ is so low that its hypothetical movement (which moreover requires energy) would make very little, if any, difference to the high $[Ca^{2+}]_e$ already present in the ECS. Without a good explanation, these findings of Puka-Sundvall *et al.* (1994) require independent confirmation. The diversity of responses in the decline phase that the authors recorded probably indicates individual sensitivities of pups to anoxia, with about 50% of animals being more resistant and not experiencing rapid anoxic depolarization. Anoxic calcium changes were not sensitive to MK-801, an inhibitor of the NMDA receptor, at any phase, which indicates that this subtype of glutamate receptors was not involved in the cation movements.

In a 7-day rat model of unilateral common carotid occlusion plus hypo-xia (Rice *et al.*, 1981), prominent increases in $^{45}Ca^{2+}$ accumulation were observed during insult in the neocortex, hippocampus, striatum, and thalamus of the ipsilateral side, while only small rises were noted in about 40% of animals in the contralateral hemisphere (Stein and Vannucci, 1988). Further and progressive accumulation of calcium occurred in the ipsilateral hemisphere during a 15-day recovery period, while no increase in radioactivity was noted in the contralateral side. Studies in the same model on subcellular distribution combined with ultrastructural evaluation after 30 min and 3 and 24 hours of reperfusion showed deposits of calcium in the endoplasmic reticulum, cytoplasm, and nucleus and, predominantly, in the mitochondrial matrix (Puka-Sundvall *et al.*, 2000) This means that in immature animals as the reaction to injury proceeds, there is a continuing deposition of calcium into the damaged tissue. Mitochondrial calcium loading impairs cell function and eventually leads to its death. Calcium may also become incorporated into the structural elements of the brain and result in calcification. It is well known that in humans, mature brain previously damaged by hypoxia/ischemia seldom undergoes calcification of the necrotic areas, while such depositions are a prominent feature of the developing brain after injury (Friede, 1975; Larroche, 1977; Rorke, 1982; Rutherford, 2002).

There are no comparative studies on the age dependence of changes in brain pH during limitations in oxygen supply. Vorísek and Syková (1997) noted progressive acidosis during the evolution of brain ischemia after cardiac arrest, but no attempt was made to discern any possible difference between mature and immature rats. During partial ischemia induced by reduction in cerebral blood flow to $<50\%$ of control, pH_i in fed 7- $\pm$ 4-day-old piglets fell to 6.30 ± 0.22 and in fasted animals of the same age to 6.51 ± 0.06 (Laptook *et al.*, 1992). Moreover, no difference was seen either in the time of appearance of maximal acidosis following ischemia or in the half-lives of acid clearance between newborn (1–3 days) and 1-month-old piglets (Corbett *et al.*, 1999). The only conclusion that can be drawn from this very limited data is that pH in immature CNS responds to hypoxia/ischemia qualitatively in the same manner as it does in the adult brain.

Changes in diffusion parameters in the brain ECS during anoxia/ischemia were also found to be age dependent and slower in unmyelinated white matter than in gray matter (Vorísek and Syková, 1997). Final changes in P4–6 rats were reached at 37 min in cortical layer V and at about 54 min in the white matter; at P10–12, the values were 24 min and 27 min, respectively, and those at P21–23 were 15 and 17 min, respectively. Since in the youngest group the ECS fraction was 0.43 in gray matter and 0.45 in the white matter, while in the oldest group the corresponding figures were 0.23 and 0.23, the observed patterns mean that the larger the extracellular space volume, the longer the time course of anoxic changes. However, the final value to which ECS shrinks in anoxia/ischemia is the same at all ages.

Taken together, the information obtained *in vitro* and *in vivo* shows that immature animals are more resistant to hypoxic ion leakage than their mature counterparts: the increase in $[K^+]_e$ and the decrease in $[Ca^{2+}]_e$ are slower and initially smaller, and cell swelling, as manifested by alterations in the diffusion parameters, occurs at a later time. This behavior is consistent with the known low level of energy utilization in early development, which allows better maintenance of high energy phosphate compounds, ATP in particular, during limitations in oxygen delivery both *in vivo* (Duffy *et al.*, 1975; Lolley *et al.*, 1961, Lowry *et al.*, 1964; Thurston and McDougal, 1969) and *in vitro* (Bickler *et al.*, 1993; Kawai *et al.*, 1989; Nabetani *et al.*, 1997) in neonates compared with adults. Better maintenance of ion gradients should also slow leakage of glutamate via the reversal of excitatory amino acid transporters.

The mechanisms of hypoxic/ischemic changes in ions deserve a comment. The initial, small rise in $[K^+]_e$ is most likely caused by opening of Ca-dependent K^+ channels because based on experiments in adult rats the very early event that occurs after oxygen tension falls very close to zero is a slight elevation in $[Ca^{2+}]_i$ (Silver and Erecinska, 1990). Opening of ATP-controlled K^+ channels is another possibility, although reductions in cerebral [ATP] at early time points of anoxia are very small, particularly in immature animals. Moreover, since it has been shown that the increase in numbers of both types of these channels occurs later in ontogeny (reviewed in Erecinska *et al.*, 2004), one would expect a prolongation in early phases of the $[K^+]_e$ rise relative to adults. The elevation of extracellular potassium and its consequent decrease inside cells lead to depolarization of the plasma membrane and an increase in its permeability to ions via opening of voltage-controlled channels. An abrupt leakage of K^+ from, as well as entry of Na^+ and Ca^{2+} into, cells follows. It was noted in some (Hansen, 1977) studies that $[K^+]_e$ at the beginning of its steep rise (phase III) was higher in rats <P12 than in older animals. To account for this finding, Hansen (1977) postulated that because immature neurons are less permeable to K^+, a sufficient depolarization for a rise in potassium permeability can only be reached with a higher $[K^+]_e$. This supposition is consistent with the reported low densities of K^+ channels in immature animals (Erecinska *et al.*, 2004), but it does not explain why not all investigators report higher $[K^+]_e$ at the beginning of phase III. There is also no agreement on whether the final anoxic potassium levels are the same at all ages. The decrease in total K brain content in P4–6 rats and a peak at around the third postnatal week (Fig. 1) is consistent with the finding of Mares *et al.* (1976) that $[K^+]_e$ was significantly higher in P19 than in P1–5 rats. The issue of age-related differences in the final $[K^+]_e$ was not specifically addressed by the study of Vorísek and Syková (1997), while the 90 mM concentrations at P4 and P7 measured by Hansen (1977) seem to be disproportionately high as compared to the total brain content of 70–75 mmol/kg (see above).

Despite the dearth of observations on the effects of either prolonged or severe hypoxia/ischemia during brain development, two lines of experimental evidence suggest that they may be profound. The first is that during protracted anoxic depolarizations, reductions in $[Ca^{2+}]_e$ are larger (Luhmann and Kral, 1997) in slices from immature than in those from adult rats. This suggests that $[Ca^{2+}]_i$ may rise to higher levels in neurons from younger as compared to older animals. Consistent with this suggestion is the second line of evidence, which is that calcification of hypoxia/ischemia-induced necrotic areas is a feature of immature but not adult mammals (Stein and Vannucci, 1988). Both observations indicate that during pathological situations, once the "front line" defense mechanisms (fewer pathways of calcium or sodium entry, larger extracellular space, etc.) break down, excessive loading of cells, primarily neurons, with calcium (a common causative factor for a host of deleterious consequences) results in a high degree of damage. This includes calcium overload of mitochondria and consequent impairment of oxidative phosphorylation. Immaturity of the mechanisms that extrude calcium, such as the calcium pump (Brandt and Neve, 1992; Guerini et $al.$, 1999) and the Na^+/Ca^{2+} exchange (Gibney et $al.$, 2002; Sakaue et $al.$, 2000), may further accentuate the injury. The inconsistencies in the existing literature (see above) do not at present allow us to distinguish whether this apparently enhanced vulnerability simply diminishes as a function of age or whether a combination of positive and negative factors creates a developmental window, such as that around birth in humans, in which susceptibility of the CNS to limitations in oxygen supply is at its peak.

B. Epileptogenesis

Early experimental studies in kittens in $vivo$ indicated that the immature brain cortex has a relatively low susceptibility to epileptogenesis (Prince and Gutnick, 1972; Purpura et $al.$, 1968). By contrast, hyperthermia-induced seizure activity in rat pups was reported to increase between P2–3 and P10, at which age a plateau was reached (Holtzman et $al.$, 1981). Similarly, a number of in $vitro$ investigations in slices from either cerebral cortex or hippocampus of both rat and rabbit, and carried out with different models, showed a period of increased sensitivity to epileptiform discharges at a certain stage of development, usually during the second/third postnatal week (Hablitz, 1987; Hablitz and Heinemann, 1987, 1989; Haglund and Schwartzkroin, 1984, 1990; Moshé et $al.$, 1983; Psarropoulou and Avoli, 1993; Swann and Brady, 1984; Világi et $al.$, 1991; Wong and Yamada, 2001). In $vivo$ models of epileptogenesis devised mainly in rats (kindling or applications either systemically or locally of one of the following chemical compounds: flurothyl, bicuculline, penicillin, pentylenetetrazol, kainic acid, pilocarpine, 4-aminopyridine) show that compared with adult brains, those of immature animals are more prone to seizures. However, although the mechanisms of epileptogenesis are multiple and depend on the model used (e.g., Sankar et $al.$, 2000), in general much less histological damage and fewer disturbances in long-term cognition result from seizures in the immature brain than from seizures of similar duration and intensity in the CNS of mature animals (Riviello *et al.*, 2002; reviewed in Holmes, 1997, 2002; Holmes and Ben-Ari, 1998, 2001; Holmes *et al.*, 2002; Scher, 2003; Wasterlain, 1997). Patterns of injury are age dependent, and resistance to damage seems to be greatest at <P10 (Sankar *et al.*, 1998). Nevertheless, repetitive seizures induced in rats during the first consecutive postnatal days can cause impaired learning, decreased activity level, and increased sensitivity to epileptic agents later in life (Holmes *et al.*, 1998, 1999; McCabe *et al.*, 2001; Schmid *et al.*, 1999). The absence of visible cell injury (evaluated by histological methods) in the overwhelming majority of studies has resulted in a consensus that long-term functional impairments caused by seizures occurring during development are not due to overt neuronal death but to altered synaptogenesis and connectivity, which may subsequently influence brain growth and development.

Metabolic studies show that during seizures local cerebral blood flow (Nehlig and Pereira de Vasconcelos, 1996) and brain energy metabolism (cerebral metabolic rate for glucose) increase in various areas of the brain (Fujikawa *et al.*, 1989; Kato *et al.*, 1980; Nehlig and Pereira de Vasconcelos, 1996; Tremblay *et al.*, 1984), while the levels of high energy phosphate compounds in whole brain fall (Fujikawa *et al.*, 1988; Holtzman *et al.*, 1998; Young *et al.*, 1985). In term human newborns, seizures that accompany perinatal asphyxia have been shown to lead to a substantial rise in the level of brain lactate and lactate/choline, in particular in the subcortical intervascular boundary zone (Miller *et al.*, 2002). This indicates that epileptogenesis enhances glucose utilization not only in model studies but also in relatively common pathological situations.

Sixty days after pentylenetetrazol-induced status epilepticus in P10 and P21 rats, long-term decreases in the regional rates of glucose utilization were observed with patterns somewhat different in the two age groups (Nehlig and Pereira de Vasconcelos, 1996). However, in the lithium-pilocarpine model of epilepsy, no metabolic consequences were seen after 60 days in rats subjected to status epilepticus at P10, while in experimental P21 pups, both increases and decreases were seen in various areas of the brain (Dubé *et al.*, 2000). The latter were accompanied by minor and often transient changes in the activity of cytochrome *c* oxidase (Raffo *et al.*, 2004). These results indicate that seizures may trigger persistent metabolic alterations, but these are both age- and model-dependent and the outcome can not be extrapolated from one system to another.

Changes in $[K^+]_e$ that accompany increased activity induced either by electrical stimulation or by administration of "epileptogenic" compounds have been measured both *in vitro* and *in vivo*. Cortical slices from three age groups of rats, P8–15 (Hablitz and Heinemann, 1987), P16–30, and >P40 (Hablitz and Heinemann, 1989), were analyzed in some detail. Following picrotoxin administration to slices from P8–15 animals, $[K^+]_e$ rose to 12 mM during interictal spikes and to 20 mM during ictal-like discharges, but, in contrast to an earlier study by Hablitz (1987), episodes of spreading depression were very rare. The increases in $[K^+]_e$ declined rapidly, which seems to indicate that K^+-clearing mechanisms were fully developed at this age. However, the authors argued that a low activity of the Na^+/K^+ ATPase was in part responsible for the large increase in $[K^+]_e$ in brains of P8–15 rats. In the P16–30 animals, potassium rose to 12.77 mM after repetitive stimulation and to 14.8 mM after picrotoxin addition. Corresponding changes in $[K^+]_e$ in the eldest (>P40) group was to 8.85 mM and 11.3 mM, respectively. In the model used, spontaneous epileptiform activity and spreading depression episodes were not typically observed in animals over P40 but were common in the P16–30 group. During the repetitive discharges that preceded the onset of spreading depression, K^+ rose by 5.2 mM in P16–30 animals (and attained a value of 30.3 ± 18.5 mM during the episode) but rose only by 1.9 mM in the mature rats. Hablitz and Heinemann (1989) concluded that the level of $[K^+]_e$ reached during epileptiform discharges decreases with increase in maturity and postulated that large increases in extracellular potassium alone are not sufficient to generate spreading depression.

Similar observations were made in the hippocampus. In the CA3 area of P9–16 rats, the "ceiling" level of extracellular potassium after discharges caused by penicillin was 16.9 mM, while it was no higher than 12 mM in the same area of P30–35 animals (Swann *et al.*, 1986). ΔK_e^+ following the burst of activity was 4.31 mM in the former and 0.97 mM in the latter. A comparison between the CA1 and CA3 area in slices from P8–12 rabbits showed that the CA1 region was much more prone to spreading depression than the CA3 area: such events were more common (72% vs 11%), were longer in duration (69.8 s vs 51.9 s), and led to larger rises in $[K^+]_e$ (60.3 mM vs 21.4 mM) (Haglund and Schwartzkroin, 1990). During repetitive stimulation at 10 Hz, $[K^+]_e$ reached a plateau of 18.3 mM in the CA1 and 11.4 mM in CA3. Based on their earlier finding of regional differences in the localization of Na^+/K^+-ATPase activity (Haglund *et al.*, 1985), the authors postulated that the low, or inhibited, activity of this enzyme in the CA1 area of the hippocampus is responsible for its greater sensitivity to seizures. In support of their suggestion was the larger basal $[K^+]_e$ in the same region. However, the value of 6.1 mM was higher than that in the medium (5 mM) and almost twice that seen *in vivo* (see above). Moreover, no regional differences in immunostaining for the Na^+/K^+-ATPase within the hippocampus were seen in other studies (e.g., Fukuda and Prince, 1992). Hence, other factors must contribute to the greater vulnerability of the CA1 area at this age.

In a model of epileptogenesis induced in rat cortical and hippocampal slices by withdrawal of Mg^{2+} (Gloveli *et al.*, 1995) the epileptogenic activity was greatest in P15–17 slices but was difficult to recognize in slices <P7. The single discharges in the hippocampal slices were associated with small fluctuations of K^+, in the order of 0.1 mM in P7–8, 0.15 mM in P15–17, and 0.5 mM in P23–25. During seizure-like events, $[K^+]_e$ rose to 5–6 mM in P7 rats, 12 mM in P15–17 rats, and 13 mM in P23–25 animals. Thus, in this model of seizures, in contrast to that involving the use of epileptogenic compounds, the ceiling level of potassium was not higher in younger animals.

In the developing rat optic nerve (Connors *et al.*, 1982), the maximum level of $[K^+]_e$ measured with optimal frequencies of stimulation was 17.2 mM in unmyelinated P1–3 nerves but only 9.8 mM in adult, myelinated nerves. The rise in $[K^+]_e$ produced by a single stimulus was also much larger in early development. The authors concluded that extracellular K^+ accumulation was greater and persisted longer in neonates than in adults, regardless of stimulus frequency. This is somewhat surprising in view of the plethora of results that show that the numbers of potassium channels (which were presumably the routes for cation release) increase and not decrease with age. The rate of dissipation of comparable concentrations of the accumulated K^+ was also slightly greater in immature optic nerves, which led the authors to suggest that the reuptake mechanisms do not show any developmental change. However, other substantial experimental evidence indicates that this is not the case (Erecinska *et al.*, 2004). There was no noticeable shrinkage of extracellular space in nerves from rats <P5.

Epileptiform activity and spreading depression (SD) episodes are also accompanied by changes in concentrations of other ions. In P16–30 rats (Hablitz and Heinemann, 1989), $[Na^+]_e$ fell by 5–10 mM during repetitive discharges and in SDs by 56 mM; chloride declined in the latter by 41 mM. In P8–15 pups $[Ca^{2+}]_e$ fell by 0.1–0.2 mM during interictal spikes and by 0.3–0.4 mM during ictal-like discharges (Hablitz and Heinemann, 1987). After the onset of the large amplitude, negative slow potential characteristic of SD (Hablitz and Heinemann, 1989), alterations were much greater. Laminar analyses of their profiles showed the largest decrease, by 1.12 mM, in the layers 800–1100 μm below the cortex surface. Changes in calcium concentration in P8–15 and P16–30 pups were not different from those in mature animals, although smaller reductions might have been expected owing to the greater fraction of the ECS water space, but the same $[Ca^{2+}]_e$, in the former. This may suggest that enhanced CNS activity and seizures elevate $[Ca^{2+}]_i$ to a much higher value in immature than in mature brain.

There are very few *in vivo* studies on the movements of ions during increased activity; they have concentrated on the behavior of potassium and seem to paint a somewhat different picture of events. In early work on immature rabbits of ages P1–4, P5–12, and P19, epileptogenic activity was induced by penicillin (Mutani *et al.*, 1974). During interictal discharges, for a particular baseline $[K^+]_e$, the increases in the level of the cation were smaller in the younger animals. At a baseline of 4 mM, $[K^+]_e$ rose by 0.3–0.8 mM at P1–4, by 0.5–1.6 mM at P5–12, and by 1.0–2.6 mM at P19. The time to the peak of $[K^+]_e$ was 0.6–1.1 s at P1–4 and 0.15–0.3 s at P19. The decay time of the potassium responses showed a progressive decrease with age. During ictal episodes, $[K^+]_e$ rose by 3–7 mM, being least in youngest rabbits. Again, the rise time was slower and the decay phase longer in the less mature animals.

The fate of $[K^+]_e$ during increased activity was also studied in the hippocampus. Using trains of stimulation from an electrode placed in the CA3 area, Stringer and Lothman (1996) were unable to find any significant age-dependent differences in CA1 $[K^+]_e$ levels; the latter rose to 10.7–11.2 mM in both the P10–27 and adult rats. A later, more detailed study (Stringer, 1998) on several groups of immature (P9–11, P14–15, P17–26) and mature rats used multiple trains of stimuli (every 10 min) for 3–4 hr and analyzed $[K^+]_e$ both during peak stimulation and during after-discharges. The peak $[K^+]_e$ during stimulation was 8–9 mM in P9–11 rats and 11–13 mM in all older animals. Half-time of recovery was 11–12 s in the youngest group and 4.5 s in adults. However, during after-discharges, a secondary peak of increased $[K^+]_e$ (mean of max. 12.9 mM) appeared in animals of the youngest group (P9–11). This peak was seen only in three out of seven rats of the P14–15 group, while in all others the presence of after-discharges did not alter the maximum of $[K^+]_e$. The responses of CA1 and CA3 were identical at all ages, which is at variance with observations on rabbit hippocampal slices (Haglund and Schwartzkroin, 1990).

Based on the analysis above, one can conclude that there are important differences in findings *in vivo* and *in vitro*. The former show that during enhanced CNS activity in rabbit as well as rat, both the rate and magnitude of potassium release increase with age, as does the rate of $[K^+]_e$ clearance. No differences in patterns seem to exist between the cortex and the hippocampus, or the hippocampal CA1 and CA3 areas. No periods of increased sensitivity can be seen. By contrast, *in vitro* experiments suggest a period of enhanced susceptibility, which in the rat CNS is between P10 and P20, i.e., the age at which most brain properties (numbers and characteristics of ion channels, activities of enzymes, etc.) are changing extremely rapidly in this species, to attain an almost adult level by postnatal week 3/4. The discrepancies between data on slices and in whole brain can not be explained simply on the basis of differences in species, models, or techniques.

Despite these differences, one cannot escape the conclusion that during excessive stimulation an immature brain is unable to regulate the extracellular concentration of potassium properly. The longer time to peak and its lower value in very young animals are consistent with smaller numbers of K^+ channels and perhaps even the numbers of firing neurons. The most likely explanation for the longer recovery times is the relatively low activity of glial and neuronal clearing/uptake mechanisms, including the Na^+/K^+-ATPase. Although an inability to regulate $[K^+]_e$ is not the only reason for the increased propensity of immature animals to seize, it undoubtedly is a contributory factor.

A second reason for the enhanced tendency to seize is an imbalance between excitation and inhibition. A likely proconvulsant factor is the paradoxical action of GABA, which in the early postnatal period depolarizes the neuronal plasma membrane via its action on the $GABA_A$ receptor (Ben-Ari *et al.*, 1989; Gao and Van Den Pol, 2001; Leinekugel *et al.*, 1995; Lin *et al.*, 1994; LoTurco *et al.*, 1995; Luhmann and Prince, 1991; Owens *et al.*, 1996, 1999). Recent *in vitro* studies in rat hippocampus (Dzhala and Staley, 2003) have demonstrated that endogenously released GABA is indeed excitatory and that it contributes to the initiation of ictal activity throughout the developmental window (between P1 and P12 in control conditions and up to P23 when $[K^+]_e$ was elevated) in which $GABA_A$ receptors trigger action potentials.

Seizures are often a complication of perinatal or early postnatal hypoxia/ischemia (Jensen *et al.*, 1991; Romijn *et al.*, 1994; Owens *et al.*, 1997; Volpe, 2000). It has been shown in an *in vitro* model that epileptiform activities generated during anoxic/aglycemic episodes accelerate the development of anoxia-induced depolarization and associated neuronal death in the neonatal rat hippocampus (Dzhala *et al.*, 2000). By contrast, seizures produced by kainic acid or inhalation of flurothyl vapors in P7 and P13 rats 24 and 6 hr before hypoxia/ischemia conferred protection against damage, while those induced 2 and 24 hr after the same insult had no significant effect on the pathogenesis of the injury (Towfighi *et al.*, 1999).

Discrepancies in experimental findings even in the same species and same brain region as well as between the *in vivo* and *in vitro* results make it difficult to produce general conclusions that would be applicable to several mammalian species. Perhaps the safest conclusion is that differences exist in mechanisms for excitability at various ages and in different brain regions, although it is not clear whether they are species dependent. The existence, or persistence, of enhanced susceptibility to seizures in rats between P10 and P20 can bear little or no relation to perinatal CNS damage in newborn humans, because the brains of rats at this age interval undergo almost the full gamut of developmental changes and even at P10 are more "mature" than human newborns.

C. Excitotoxicity

It is now generally believed that brain damage produced by hypoxia/ischemia and intense seizures in mature as well as immature mammals is caused to some extent by the action of endogenous excitatory amino acids, in particular glutamate, released during insults into the external environment. Glutamate binding to its receptors results in increased membrane permeability to Na^+, K^+, and Ca^{2+} and collapses their gradients. There is quite extensive literature on the excitatory amino acids and their receptors in developing brain, as well as their involvement in pathology, and an interested reader should consult specialized reviews and textbooks. This section evaluates only studies that specifically explored changes in ions induced by glutamate and/or its agonists in the brains of immature animals. The important issue is not only whether such alterations do occur naturally but also whether there is a period during development when ionic responses to excitatory amino acids are exaggerated and might increase the sensitivity of immature CNS to the action of these compounds.

In neonatal rats, *in vivo* hypoxia/ischemia leads to a rise in the level of extracellular glutamate (but not aspartate) in various CNS regions (Gordon *et al.*, 1991; Silverstein *et al.*, 1991), although the changes are much smaller than those in mature animals subjected to insults of the same severity. In hippocampal slices, anoxia (Bickler and Hansen, 1998) and simulated ischemia (Cherici *et al.*, 1991; Minc-Golomb *et al.*, 1987) cause very little, if any, glutamate release during the first 2 weeks of life. Moreover, K^+-stimulated efflux both *in vivo* (Silverstein and Naik, 1991) and in cortical and hippocampal synaptosomes (Collard *et al.*, 1993) is very sluggish in rats younger than 15 days. Taken together, these results indicate that during early development the level to which brain [glutamate]$_e$ rises during pathological insults is relatively low, probably due to the larger ECS at this time. Moreover, the immaturity of components of the vesicular release machinery limits exocytosis, while better maintained ion gradients minimize efflux via the reversal of the amino acid transporters (reviewed in Erecinska *et al.*, 2004).

In 1988 McDonald and co-workers showed that direct injection of NMDA, an agonist of one of the subclasses of glutamate receptors, into corpus striatum of rats at P7 resulted in major brain damage that extended into dorsal hippocampus and neocortex. It produced an area of destruction 16-fold larger than that observed after direct injection of three times the amount of the same toxin into the hippocampus of adult animals. Moreover, Ikonomidou *et al.* (1989) reported parallels between the sensitivity of the developing brain to hypobaric/ischemic injury and the degree of neurotoxicity caused by intrastriatal injection of NMDA, both peaking at 6 days of age. These observations provided a basis for the suggestion that during early postnatal development the brain passes through a period of particularly great vulnerability to excitatory amino acids (McDonald and Johnston, 1990). However, Liu and co-workers (1996) demonstrated that *in vivo* administration of glutamate, the physiological neurotransmitter and an excitotoxin, into the CA1 region of the hippocampus did not result in any brain damage in P10 rats but caused injuries in older animals, the severity of which increased progressively with age.

There is also no unequivocal evidence that immature rat brain, at any stage of its early development, shows enhanced sensitivity to another neurotoxin, kainic acid, which has affinity for another subclass of glutamate receptors. A detailed study of the profile of toxicity after intrastriatal injection of this compound (Campochiaro and Coyle, 1978) found no vulnerability in P7 rats but an increase thereafter, which reached a plateau equal to adult sensitivity at P21. The authors suggested that this pattern parallels the marked rise of dendritic arborization and the formation of spines on the intrinsic neurons of rat striatum (Lu and Brown, 1977) as well as the increase in density of synaptic profiles (Hattori and McGeer, 1973). Kainic acid given systemically in rats younger than P18 provoked severe tonico-clonic convulsions and electrical seizures as well as enhanced metabolic activity in the hippocampus (Tremblay *et al.* 1984) but did not produce even the slightest histological brain damage (Ben-Ari *et al.*, 1984; Nitecka *et al.*, 1984). However, morphological injury after the same treatment appeared and progressively increased between P18 and P35 (Albala *et al.*, 1984; Nitecka *et al.*, 1984; Stafstrom *et al.*, 1992). No neuronal death was seen 2 days after systemic administration of kainic acid to either P6–7 or P10–12 rabbits (Towfighi *et al.*, 2004). However, damage could be produced even in P6–9 rats after direct injection of high doses of the toxin into the hippocampus (Cook and Crutcher, 1986).

Analyses of the consequences of actions of excitatory amino acids in simpler systems such as slices or cells are extremely rare. Hamon and Heinemann (1988) compared responses of hippocampal slices from two groups of rats, P5–9 and P12–30, to electrical stimulation and administration of NMDA and quisqualate. No differences in slow field potentials or decreases in $[Ca^{2+}]_e$ between the two groups were seen after electrical stimulation or administration of quisqualate. By contrast, the authors claimed that the laminar pattern of standardized responses to NMDA showed differences between the two age groups: while maximum changes in stratum radiatum were seen in P12–30 rats, reduction in $[Ca^{2+}]_e$ in stratum pyramidale were greatest at P5–9. It was concluded that during the second postnatal week apical dendrites become more sensitive to NMDA, which is expressed by large influxes of calcium. Although, the changes produced by NMDA were statistically significant, the actual data presented by Hamon and Heinemann (see Fig. 3 in Hamon and Heinemann, 1988) do not show very convincing differences between the patterns seen with this compound and those obtained with quisqualate.

In acutely dissociated CNS neurons from 1- to 25-day-old rats, Marks *et al.* (1996) saw a continuing increase in susceptibility to glutamate, as evaluated by morphological changes, rise in intracellular calcium, and cell death. Bickler and Hansen (1998) found no release of glutamate during anoxia in slices from P4 rats, although it was readily seen in those from P20 animals. Moreover, MK-801, an antagonist of the NMDA receptor, decreased the viability index of preparations from P3–7 but enhanced survival in P18–22 slices.

The inescapable conclusion from these *in vivo* and *in vitro* studies is that there is no consistent functional or morphological evidence which would indicate that there is a period in the life of a developing rat when the brain shows transient, markedly enhanced sensitivity to glutamate or other excitatory amino acids. No information is currently available on the behavior of other species. NMDA is not an endogenous neurotransmitter in the mammalian CNS, but it is possible that immature glutamate receptors of the NMDA subtype, which continually change their composition and/or structure during development (Erecinska *et al.*, 2004; McDonald and Johnston, 1990), may be particularly sensitive to that compound at some stage during the first 2 weeks of life, consistent with the observation in the brain of 10-day-old mice that glutamate is 100-fold less toxic than NMDA (Olney *et al.*, 1971).

IV. Conclusions

Developing brain is a dynamic system that continuously undergoes morphological and functional changes; the latter take only weeks to complete in mice and rats, but years in primates. This means that developing rat brain, the most commonly used system in experimental studies, is neither an easy nor convenient model for mimicking and investigating pathologies that occur in developing immature human CNS. An even more serious question has to be asked, whether or not it is an appropriate one. There is widespread interest in diseases of, and injuries to, the brain of human newborns, and there is a great need to elucidate alterations that take place at a cellular level, in particular, changes in ion levels and gradients. It is commonly agreed that the neonatal period in humans encompasses the first 4 weeks of life, which in the time scale of the rat CNS development corresponds to minutes or hours at some stage of this animal's early postnatal ontogeny. Hence, for example, repeated seizures induced for 5 days, from P0 to P4, in newborn rats, a period that correspond to 20–25% of the brain developmental period in the rodent, can hardly be considered equivalent to epileptic fits occurring in the first month of human life. Although application of rapidly developing noninvasive methods, such as the imaging techniques, to human studies opens very promising avenues for investigations, for the near future at least there are still no useful substitutes for experiments in animal models, in particular those *in vivo*, that can probe responses at the cellular level.

From the *in vitro* and *in vivo* experimental studies summarized in this chapter, there is consistent evidence that immature brain is less sensitive to hypoxia (and to some extent to hypoxia/ischemia) of the same duration and intensity than the mature organ: ATP is better maintained, histological damage is smaller (albeit different), there is less leakage of ions, and animals survive longer. This is due mainly to the overall low rate of cerebral ATP utilization in the very young. A much larger extracellular space in the brains of immature mammals slows and reduces hypoxic/ischemic rises in $[K^+]_e$. Perhaps this relative resistance to the lack of oxygen constitutes physiological adaptation or a defense mechanism to the transient hypoxia that almost inevitably occurs at birth. Susceptibility to hypoxia/ischemia increases with age, but no consistent evidence that there is a window of greatly enhanced transient vulnerability during postnatal development has been found.

Increased susceptibility of newborn and young animals to epileptic seizures may be caused by the immaturity of mechanisms that clear extracellular potassium, rather than to leakages of ions greater than in adults, which are not found reproducibly *in vivo*. There is also no evidence from either *in vivo* or *in vitro* studies that immature rat brain at P6–8 is unusually susceptible to the action of glutamate. The nature of the enhanced response reported by some investigators to the nonphysiological ligand NMDA during that period needs clarification. The relevance of this phenomenon to the effects of the native neurotransmitter, glutamate, has yet to be established. It is particularly worrying that results obtained *in vivo* often do not agree with those *in vitro*. Discrepancies of this kind are usually glossed over and/or not even mentioned in the relevant literature.

Two more points deserve comment. The first is that the mechanisms involved in both hypoxic/ischemic and excitotoxic damage are multiple and, in addition, are likely to differ in the immature and mature brain. The same is true about the cerebral regions vulnerable to these insults. The second point is that homeostatic mechanisms that reconstitute the "original" state (ion gradients) after insult may not be fully developed in growing brain, and this may slow the rate of damage repair. This is particularly relevant in the case of calcium, a known initiator of multiple "destruction" pathways. Therefore, while greater resistance to ion leakage may render an immature brain almost immune to short harmful events, longer periods of stressful situations may lead to damage greater than that seen in adults.

References

Aghajanian, G. K., and Bloom, F. E. (1967). The formation of synaptic junctions in developing rat brain: A quantitative electron microscopic study. *Brain Res.* **6**, 716–727.

Agrawal, H. C., Davis, J. M., and Himwich, W. A. (1968). Water content of dog brain parts in relation to maturation of the brain. *Am. J. Physiol.* **215,** 846–848.

Albala, B. J., Moshé, S. L., and Okada, R. (1984). Kainic-acid-induced seizures: A developmental study. *Brain Res.* **315,** 139–148.

Ames, A., 3rd. (2000). CNS energy metabolism as related to function. *Brain Res. Rev.* **34,** 42–68.

Anzai, N., Suzuki, Y., Nishikitani, M., Izumida-Moriguchi, I., Kokubo, A., and Kawahara, K. (2001). Development of renal potassium excretion capacity in the neonatal rat. *Jpn. J. Physiol.* **51,** 745–752.

Auer, R. L., and Sutherland, G. R. (2002). Hypoxia and related conditions. *In* "Greenfield's Neuropathology" (B. Graham and P. L. Lantos, Eds.), pp. 233–280. Arnold, London.

Azzopardi, D., Wyatt, J. S., Hamilton, P. A., Cady, E. B., Delpy, D. T., Hope, P. L., and Reynolds, E. O. (1989). Phosphorus metabolites and intracellular pH in the brains of normal and small for gestational age infants investigated by magnetic resonance spectroscopy. *Pediatr. Res.* **25,** 440–444.

Baimbridge, K. G., Celio, M. R., and Rogers, J. H. (1992). Calcium-binding proteins in the nervous system. *Trends Neurosci.* **15,** 303–308.

Ballanyi, K., Kuwana, S., Volker, A., Morawietz, G., and Richter, D. W. (1992). Developmental changes in the hypoxia tolerance of the *in vitro* respiratory network of rats. *Neurosci. Lett.* **148,** 141–144.

Belleau, M. L., and Warren, R. A. (2000). Postnatal development of electrophysiological properties of nucleus accumbens neurons. *J. Neurophysiol.* **84,** 2204–2216.

Ben-Ari, Y., Tremblay, E., Berger, M., and Nitecka, L. (1984). Kainic acid seizure syndrome and binding sites in developing rats. *Dev. Brain Res.* **14,** 284–288.

Ben-Ari, Y., Cherubini, E., Corradetti, R., and Gaiarsa, J. L. (1989). Giant synaptic potentials in immature rat CA3 hippocampal neurones. *J. Physiol.* **416,** 303–325.

Berger, R., and Garnier, Y. (1999). Pathophysiology of perinatal brain damage. *Brain Res. Rev.* **30,** 107–134.

Bickler, P. E., and Hansen, B. M. (1998). Hypoxia-tolerant neonatal CA1 neurons: Relationship of survival to evoked glutamate release and glutamate receptor-mediated calcium changes in hippocampal slices. *Dev. Brain Res.* **106,** 57–69.

Bickler, P. E., Gallego, S. M., and Hansen, B. M. (1993). Developmental changes in intracellular calcium regulation in rat cerebral cortex during hypoxia. *J. Cereb. Blood Flow Metab.* **13,** 811–819.

Bondareff, W., and Pysh, J. J. (1968). Distribution of the extracellular space during postnatal maturation of rat cerebral cortex. *Anat. Rec.* **160,** 773–780.

Bordey, A., and Sontheimer, H. (1997). Postnatal development of ionic currents in rat hippocampal astrocytes *in situ. J. Neurophysiol.* **78,** 461–477.

Boyle, R. (1670). New pneumatical experiments about respiration. *Philos. Trans. Roy. Soc. London* **5,** 2011–2131.

Brandt, P., and Neve, R. L. (1992). Expression of plasma membrane calcium-pumping ATPase mRNAs in developing rat brain and adult brain subregions: Evidence for stage-specific expression. *J. Neurochem.* **59,** 1566–1569.

Burgard, E. C., and Hablitz, J. J. (1993). Developmental changes in NMDA and non-NMDA receptor-mediated synaptic potentials in rat neocortex. *J. Neurophysiol.* **69,** 230–240.

Campochiaro, P., and Coyle, J. T. (1978). Ontogenetic development of kainate neurotoxicity: Correlates with glutamatergic innervation. *Proc. Natl. Acad. Sci. USA* **75,** 2025–2029.

Cherici, G., Alesiani, M., Pellegrini-Giampietro, D. E., and Moroni, F. (1991). Ischemia does not induce the release of excitotoxic amino acids from the hippocampus of newborn rats. *Dev. Brain Res.* **60,** 235–240.

Cherubini, E., Ben-Ari, Y., and Krnjevic, K. (1989). Anoxia produces smaller changes in synaptic transmission, membrane potential, and input resistance in immature rat hippocampus. *J. Neurophysiol.* **62,** 882–895.

Clayton, G. H., Owens, G. C., Wolff, J. S., and Smith, R. L. (1998). Ontogeny of cation-Cl⁻ cotransporter expression in rat neocortex. *Dev. Brain Res.* **109,** 281–292.

Colin, T., Chat, M., Lucas, M. G., Moreno, H., Recay, P., Schwaller, B., Morti, A., and Llano, I. (2005). Developmental changes in parvalbumin regulate presynaptic Ca^{2+} signaling. *J. Neurosci.* **25,** 96–107.

Collard, K. J., Edwards, R., and Liu, Y. (1993). Changes in synaptosomal glutamate release during postnatal development in the rat hippocampus and cortex. *Dev. Brain Res.* **71,** 37–43.

Connors, B. W., Ransom, B. R., Kunis, D. M., and Gutnick, M. J. (1982). Activity-dependent K^{+} accumulation in the developing rat optic nerve. *Science* **216,** 1341–1343.

Cook, T. M., and Crutcher, K. A. (1986). Intrahippocampal injection of kainic acid produces significant pyramidal cell loss in neonatal rats. *Neuroscience* **18,** 79–92.

Corbett, R., Laptook, A., Kim, B., Tollefsbol, G., Silmon, S., and Garcia, D. (1999). Maturational changes in cerebral lactate and acid clearance following ischemia measured *in vivo* using magnetic resonance spectroscopy and microdialysis. *Dev. Brain Res.* **113,** 37–46.

Dancis, J., and Springer, D. (1970). Fetal homeostasis in maternal malnutrition: Potassium and sodium deficiency in rats. *Pediatr. Res.* **4,** 345–351.

DeFazio, R. A., Keros, S., Quick, M. W., and Hablitz, J. J. (2000). Potassium-coupled transport controls intracellular chloride in rat pyramidal neurons. *J. Neurosci.* **20,** 8069–8076.

De Souza, S. W., and Dobbing, J. (1971). Cerebral edema in developing brain. I. Normal water and cation content in developing rat brain and postmortem changes. *Exp. Neurol.* **32,** 431–438.

Dubé, C., Boyet, S., Marescaux, C., and Nehlig, A. (2000). Progressive metabolic changes underlying the chronic reorganization of brain circuits during the silent phase of the lithium-pilocarpine model of epilepsy in the immature and adult Rat. *Exp. Neurol.* **162,** 146–157.

Duffy, T. E., Kohle, S. J., and Vannucci, R. C. (1975). Carbohydrate and energy metabolism in perinatal rat brain: Relation to survival in anoxia. *J. Neurochem.* **24,** 271–276.

Dzhala, V., Ben-Ari, Y., and Khazipov, R. (2000). Seizures accelerate anoxia-induced neuronal death in the neonatal rat hippocampus. *Ann. Neurol.* **48,** 632–640.

Dzhala, V. I., and Staley, K. J. (2003). Excitatory actions of endogenously released GABA contribute to initiation of ictal epileptiform activity in the developing hippocampus. *J. Neurosci.* **23,** 1840–1846.

Enderlin, S., Norman, A. W., and Celio, M. R. (1987). Ontogeny of the calcium binding protein calbindin D-28k in the rat nervous system. *Anat. Embryol. (Berl.)* **177,** 15–28.

Endo, T., Kobayashi, S., and Onaya, T. (1985). Parvalbumin in rat cerebrum, cerebellum and retina during postnatal development. *Neurosci. Lett.* **60,** 279–282.

Erecinska, M., and Silver, I. A. (1989). ATP and brain function. *J. Cereb. Blood Flow Metab.* **9,** 2–19.

Erecinska, M., and Silver, I. A. (1994). Ions and energy in mammalian brain. *Prog. Neurobiol.* **43,** 37–71.

Erecinska, M., and Silver, I. A. (2001). Tissue oxygen tension and brain sensitivity to hypoxia. *Respir. Physiol.* **128,** 263–276.

Erecinska, M., Cherian, S., and Silver, I. A. (2004). Energy metabolism in mammalian brain during development. *Prog. Neurobiol.* **73,** 397–445.

Fazekas, J. F., Alexander, F. A. D., and Himwich, H. E. (1941). Tolerance of the newborn to anoxia. *Am. J. Physiol.* **134,** 281–287.

Friede, R. L. (1975). "Developmental Neuropathology." Springer-Verlag, New York, NY.

Friedman, J. E., and Haddad, G. G. (1993). Major differences in Ca_i^{2+} response to anoxia between neonatal and adult rat CA1 neurons: Role of Ca_o^{2+} and Na_o^+. *J. Neurosci.* **13**, 63–72.

Fujikawa, D. G., Vannucci, R. C., Dwyer, B. E., and Wasterlain, C. G. (1988). Generalized seizures deplete brain energy reserves in normoxemic newborn monkeys. *Brain Res.* **454**, 51–59.

Fujikawa, D. G., Dwyer, B. E., Lake, R. R., and Wasterlain, C. G. (1989). Local cerebral glucose utilization during status epilepticus in newborn primates. *Am. J. Physiol.* **256**, C1160–1167.

Fukuda, A., and Prince, D. A. (1992). Postnatal development of electrogenic sodium pump activity in rat hippocampal pyramidal neurons. *Dev. Brain Res.* **65**, 101–114.

Gao, X. B., and Van Den Pol, A. N. (2001). GABA, not glutamate, a primary transmitter driving action potentials in developing hypothalamic neurons. *J. Neurophysiol.* **85**, 425–434.

Gibney, G. T., Zhang, J. H., Douglas, R. M., Haddad, G. G., and Xia, Y. (2002). Na^+/Ca^{2+} exchanger expression in the developing rat cortex. *Neuroscience* **112**, 65–73.

Gilles, F. H., Leviton, A., and Dooling, E. C. (1983). "The Developing Human Brain: Growth and Epidemiologic Neuropathology." John Wright Inc., Boston, MA.

Glass, H. G., Snyder, F. F., and Webster, E. (1944). The rate of decline in resistance to anoxia of rabbits, dogs and guinea pigs from the onset of viability to adult life. *Am. J. Physiol.* **140**, 609–614.

Gloveli, T., Albrecht, D., and Heinemann, U. (1995). Properties of low Mg^{2+} induced epileptiform activity in rat hippocampal and entorhinal cortex slices during adolescence. *Dev. Brain Res.* **87**, 145–152.

Gordon, K. E., Simpson, J., Statman, D., and Silverstein, F. S. (1991). Effects of perinatal stroke on striatal amino acid efflux in rats studied with *in vivo* microdialysis. *Stroke* **22**, 928–932.

Grafe, M. R. (1994). Developmental changes in the sensitivity of the neonatal rat brain to hypoxic/ischemic injury. *Brain Res.* **653**, 161–166.

Guerini, D., Garcia-Martin, E., Gerber, A., Volbracht, C., Leist, M., Merino, C. G., and Carafoli, E. (1999). The expression of plasma membrane Ca^{2+} pump isoforms in cerebellar granule neurons is modulated by Ca^{2+}. *J. Biol. Chem.* **274**, 1667–1676.

Hablitz, J. J. (1987). Spontaneous ictal-like discharges and sustained potential shifts in the developing rat neocortex. *J. Neurophysiol.* **58**, 1052–1065.

Hablitz, J. J., and Heinemann, U. (1987). Extracellular K^+ and Ca^{2+} changes during epileptiform discharges in the immature rat neocortex. *Dev. Brain Res.* **433**, 299–303.

Hablitz, J. J., and Heinemann, U. (1989). Alterations in the microenvironment during spreading depression associated with epileptiform activity in the immature neocortex. *Dev. Brain Res.* **46**, 243–252.

Haddad, G. G., and Donnelly, D. F. (1990). O_2 deprivation induces a major depolarization in brain stem neurons in the adult but not in the neonatal rat. *J. Physiol.* **429**, 411–428.

Haglund, M. M., and Schwartzkroin, P. A. (1984). Seizure-like spreading depression in immature rabbit hippocampus *in vitro*. *Brain Res.* **316**, 51–59.

Haglund, M. M., and Schwartzkroin, P. A. (1990). Role of Na-K pump potassium regulation and IPSPs in seizures and spreading depression in immature rabbit hippocampal slices. *J. Neurophysiol.* **63**, 225–239.

Haglund, M. M., Stahl, W. L., Kunkel, D. D., and Schwartzkroin, P. A. (1985). Developmental and regional differences in the localization of Na,K-ATPase activity in the rabbit hippocampus. *Brain Res.* **343**, 198–203.

Hamon, B., and Heinemann, U. (1988). Developmental changes in neuronal sensitivity to excitatory amino acids in area CA1 of the rat hippocampus. *Dev. Brain Res.* **38**, 286–290.

Hansen, A. J. (1977). Extracellular potassium concentration in juvenile and adult rat brain cortex during anoxia. *Acta Physiol. Scand.* **99**, 412–420.

Hansen, A. J. (1985). Effect of anoxia on ion distribution in the brain. *Physiol. Rev.* **65,** 101–148.

Hansen, A. J., and Nordstrom, C. H. (1979). Brain extracellular potassium and energy metabolism during ischemia in juvenile rats after exposure to hypoxia for 24 h. *J. Neurochem.* **32,** 915–920.

Hansen, A. J., and Zeuthen, T. (1981). Extracellular ion concentration during spreading depression and ischemia in the rat cortex. *Acta Physiol. Scand.* **113,** 437–445.

Hattori, T., and McGeer, P. L. (1973). Synaptogenesis in the corpus striatum of infant rat. *Exp. Neurol.* **38,** 70–79.

Hendrickson, A. E., Van Brederode, J. F., Mulligan, K. A., and Celio, M. R. (1991). Development of the calcium-binding protein parvalbumin and calbindin in monkey striate cortex. *J. Comp. Neurol.* **307,** 626–646.

Holmes, G. L. (1997). Epilepsy in the developing brain: Lessons from the laboratory and clinic. *Epilepsia* **38,** 12–30.

Holmes, G. L. (2002). Seizure-induced neuronal injury: Animal data. *Neurology* **59,** S3–6.

Holmes, G. L., and Ben-Ari, Y. (1998). Seizures in the developing brain: Perhaps not so benign after all. *Neuron* **21,** 1231–1234.

Holmes, G. L., and Ben-Ari, Y. (2001). The neurobiology and consequences of epilepsy in the developing brain. *Pediatr. Res.* **49,** 320–325.

Holmes, G. L., Gairsa, J. L., Chevassus-Au-Louis, N., and Ben-Ari, Y. (1998). Consequences of neonatal seizures in the rat: Morphological and behavioral effects. *Ann. Neurol.* **44,** 845–857.

Holmes, G. L., Khazipov, R., and Ben-Ari, Y. (2002). New concepts in neonatal seizures. *Neuroreport* **13,** A3–8.

Holmes, G. L., Sarkisian, M., Ben-Ari, Y., and Chevassus-Au-Louis, N. (1999). Mossy fiber sprouting after recurrent seizures during early development in rats. *J. Comp. Neurol.* **404,** 537–553.

Holtzman, D., Obana, K., and Olson, J. (1981). Hyperthermia-induced seizures in the rat pup: A model for febrile convulsions in children. *Science* **213,** 1034–1036.

Holtzman, D., Mulkern, R., Meyers, R., Cook, C., Allred, E., Khait, I., Jensen, F., Tsuji, M., and Laussen, P. (1998). *In vivo* phosphocreatine and ATP in piglet cerebral gray and white matter during seizures. *Brain Res.* **783,** 19–27.

Ikonomidou, C., Mosinger, J. L., Salles, K. S., Labruyere, J., and Olney, J. W. (1989). Sensitivity of the developing rat brain to hypobaric/ischemic damage parallels sensitivity to N-methyl-aspartate neurotoxicity. *J. Neurosci.* **9,** 2809–2818.

Isagai, T., Fujimura, N., Tanaka, E., Yamamoto, S., and Higashi, H. (1999). Membrane dysfunction induced by *in vitro* ischemia in immature rat hippocampal CA1 neurons. *J. Neurophysiol.* **81,** 1866–1871.

Jacobson, H. N., and Windle, W. F. (1960). Responses of foetal and new-born monkeys to asphyxia. *J. Physiol.* **153,** 447–456.

Jensen, F. E., Applegate, C. D., Holtzman, D., Belin, T. R., and Burchfiel, J. L. (1991). Epileptogenic effect of hypoxia in the immature rodent brain. *Ann. Neurol.* **29,** 629–637.

Jiang, C., Agulian, S., and Haddad, G. G. (1992). Cl^- and Na^+ homeostasis during anoxia in rat hypoglossal neurons: Intracellular and extracellular *in vitro* study. *J. Physiol.* **448,** 697–708.

Jiang, C., Xia, Y., and Haddad, G. G. (1992). Role of ATP-sensitive K^+ channels during anoxia: Major differences between rat (newborn and adult) and turtle neurons. *J. Physiol.* **448,** 599–612.

Johnston, M. V., Trescher, W. H., Ishida, A., and Nakajima, W. (2001). Neurobiology of hypoxic-ischemic injury in the developing brain. *Pediatr. Res.* **49,** 735–741.

Jones, C. T., and Rolph, T. P. (1985). Metabolism during fetal life: A functional assessment of metabolic development. *Physiol. Rev.* **65,** 357–430.

Jones, M. D., Jr. (1979). Energy metabolism in the developing brain. *Semin. Perinatol.* **3**, 121–129.

Kabat, H. (1940). The greater resistance of very young animals to arrest of the brain circulation. *Am. J. Physiol.* **130**, 588–599.

Kato, M., Malamut, B. L., Caveness, W. F., Hosokawa, S., Wakisaka, S., and O'Neill, R. R. (1980). Local cerebral glucose utilization in newborn and pubescent monkeys during focal motor seizures. *Ann. Neurol.* **7**, 204–212, 230–237.

Kawai, S., Yonetani, M., Nakamura, H., and Okada, Y. (1989). Effects of deprivation of oxygen and glucose on the neural activity and the level of high energy phosphates in the hippocampal slices of immature and adult rat. *Dev. Brain Res.* **48**, 11–18.

Keelan, J., Bates, T. E., and Clark, J. B. (1996). Intrasynaptosomal free calcium concentration during rat brain development: Effects of hypoxia, aglycaemia, and ischaemia. *J. Neurochem.* **66**, 2460–2467.

Kocsis, J. D., Rand, M. N., Lankford, K. L., and Waxman, S. G. (1994). Intracellular calcium mobilization and neurite outgrowth in mammalian neurons. *J. Neurobiol.* **25**, 252–264.

Kriegstein, A. R., Suppes, T., and Prince, D. A. (1987). Cellular and synaptic physiology and epileptogenesis of developing rat neocortical neurons *in vitro*. *Brain Res.* **431**, 161–171.

Krnjevic, K., Cherubini, E., and Ben-Ari, Y. (1989). Anoxia on slow inward currents of immature hippocampal neurons. *J. Neurophysiol.* **62**, 896–906.

Lamsa, K., Palva, J. M., Ruusuvuori, E., Kaila, K., and Taira, T. (2000). Synaptic $GABA_A$ activation inhibits AMPA-kainate receptor-mediated bursting in the newborn (P0-P2) rat hippocampus. *J. Neurophysiol.* **83**, 359–366.

Laptook, A. R., Corbett, R. J., Arencibia-Mireles, O., and Ruley, J. (1992). Glucose-associated alterations in ischemic brain metabolism of neonatal piglets. *Stroke* **23**, 1504–1511.

Larroche, J. C. (1977). Developmental Pathology of the Neonate. Exerpta Medica, New York.

Lehmenkuhler, A., Syková, E., Svoboda, J., Zilles, K., and Nicholson, C. (1993). Extracellular space parameters in the rat neocortex and subcortical white matter during postnatal development determined by diffusion analysis. *Neuroscience* **55**, 339–351.

Leinekugel, X., Tseeb, V., Ben-Ari, Y., and Bregestovski, P. (1995). Synaptic $GABA_A$ activation induces Ca^{2+} rise in pyramidal cells and interneurons from rat neonatal hippocampal slices. *J. Physiol.* **487**, 319–329.

Lelievre-Pegorier, M., Merlet-Benichou, C., Roinel, N., and de Rouffignac, C. (1983). Developmental pattern of water and electrolyte transport in rat superficial nephrons. *Am. J. Physiol.* **245**, F15–21.

Levin, V. A., Fenstermacher, J. D., and Patlak, C. S. (1970). Sucrose and inulin space measurements of cerebral cortex in four mammalian species. *Am. J. Physiol.* **219**, 1528–1533.

Lin, M. H., Takahashi, M. P., Takahashi, Y., and Tsumoto, T. (1994). Intracellular calcium increase induced by GABA in visual cortex of fetal and neonatal rats and its disappearance with development. *Neurosci. Res.* **20**, 85–94.

Liu, Z., Stafstrom, C. E., Sarkisian, M., Tandon, P., Yang, Y., Hori, A., and Holmes, G. L. (1996). Age-dependent effects of glutamate toxicity in the hippocampus. *Dev. Brain Res.* **97**, 178–184.

Lolley, R. N., Balfour, W. M., and Samson, F. E., Jr. (1961). The high-energy phosphates in developing brain. *J. Neurochem.* **7**, 289–297.

Lorenz, J. M., Kleinman, L. I., and Disney, T. A. (1986). Renal response of newborn dog to potassium loading. *Am. J. Physiol.* **251**, F513–519.

Lorenz, J. M., Kleinman, L. I., and Markarian, K. (1997). Potassium metabolism in extremely low birth weight infants in the first week of life. *J. Pediatr.* **131**, 81–86.

Lo Turco, J. J., Owens, D. F., Heath, M. J. S., Davis, M. B. E., and Kriegstein, A. R. (1995). GABA and glutamate depolarize cortical progenitor cells and inhibit DNA synthesis. *Neuron* **15**, 1287–1298.

Lu, E. J., and Brown, W. J. (1977). The developing caudate nucleus in the euthyroid and hypothyroid rat. *J. Comp. Neurol.* **171,** 261–284.

Lu, J., Karadsheh, M., and Delpire, E. (1999). Developmental regulation of the neuronal-specific isoform of K-Cl cotransporter KCC2 in postnatal rat brains. *J. Neurobiol.* **39,** 558–568.

Luhmann, H. J., and Kral, T. (1997). Hypoxia-induced dysfunction in developing rat neocortex. *J. Neurophysiol.* **78,** 1212–1221.

Luhmann, H. J., and Prince, D. A. (1991). Postnatal maturation of the GABAergic system in rat neocortex. *J. Neurophysiol.* **65,** 247–263.

Mares, P., Kríz, N., Brozek, G., and Bures, J. (1976). Anoxic changes of extracellular potassium concentration in the cerebral cortex of young rats. *Exp. Neurol.* **53,** 12–20.

Marks, J. D., Friedman, J. E., and Haddad, G. G. (1996). Vulnerability of CA1 neurons to glutamate is developmentally regulated. *Dev. Brain Res.* **97,** 194–206.

McCabe, B. K., Silveira, D. C., Cilio, M. R., Cha, B. H., Liu, X., Sogawa, Y., and Holmes, G. L. (2001). Reduced neurogenesis after neonatal seizures. *J. Neurosci.* **21,** 2094–2103.

McCance, R. A., and Widdowson, E. M. (1956). The effect of development on the composition of the serum and extracellular fluids. *Clin. Sci. (Lond)* **15,** 361–365.

McCormick, D. A., and Prince, D. A. (1987). Post-natal development of electrophysiological properties of rat cerebral cortical pyramidal neurones. *J. Physiol.* **393,** 743–762.

McDonald, J. W., and Johnston, M. V. (1990). Physiological and pathophysiological roles of excitatory amino acids during central nervous system development. *Brain Res. Rev.* **15,** 41–70.

McDonald, J. W., Silverstein, F. S., and Johnston, M. V. (1988). Neurotoxicity of N-methyl-D-aspartate is markedly enhanced in developing rat central nervous system. *Brain Res.* **459,** 200–203.

Mercado, A., Mount, D. B., and Gamba, G. (2004). Electroneutral cation-chloride cotransporters in the central nervous system. *Neurochem. Res.* **29,** 17–25.

Miller, S. P., Weiss, J., Barnwell, A., Ferriero, D. M., Latal-Hajnal, B., Ferrer-Rogers, A., Newton, N., Partridge, J. C., Glidden, D. V., Vigneron, D. B., and Barkovich, A. J. (2002). Seizure-associated brain injury in term newborns with perinatal asphyxia. *Neurology* **58,** 542–548.

Minc-Golomb, D., Levy, Y., Kleinberger, N., and Schramm, M. (1987). D-[^{3}H]aspartate release from hippocampus slices studied in a multiwell system: Controlling factors and postnatal development of release. *Brain Res.* **402,** 255–263.

Miyamoto, Y., Yamamoto, H., Murakami, H., Kamiyama, N., and Fukuda, M. (2004). Studies on cerebrospinal fluid ionized calcium and magnesium concentrations in convulsive children. *Pediatr. Int.* **46,** 394–397.

Moshé, S. L., Albala, B. J., Ackermann, R. F., and Engel, J., Jr. (1983). Increased seizure susceptibility of the immature brain. *Dev. Brain Res.* **7,** 81–85.

Mueller, A. L., Taube, J. S., and Schwartzkroin, P. A. (1984). Development of hyperpolarizing inhibitory postsynaptic potentials and hyperpolarizing response to γ-aminobutyric acid in rabbit hippocampus studied *in vitro*. *J. Neurosci.* **4,** 860–867.

Mutani, R., Futamachi, K. J., and Prince, D. A. (1974). Potassium activity in immature cortex. *Brain Res.* **75,** 27–39.

Nabetani, M., Okaba, Y., Takata, T., Takada, S., and Nakamura, H. (1997). Neural activity and intracellular Ca^{2+} mobilization in the CA1 area of hippocampal slices from immature and mature rats during ischemia or glucose deprivation. *Brain Res.* **769,** 158–162.

Nehlig, A., and Pereira de Vasconcelos, A. (1993). Glucose and ketone body utilization by the brain of neonatal rats. *Prog. Neurobiol.* **40,** 163–221.

Nehlig, A., and Pereira de Vasconcelos, A. (1996). The model of pentylenetetrazol-induced status epilepticus in the immature rat: Short- and long-term effects. *Epilepsy Res.* **26,** 93–103.

Nitecka, L., Tremblay, E., Charton, G., Bouillot, J. P., Berger, M. L., and Ben-Ari, Y. (1984). Maturation of kainic acid seizure-brain damage syndrome in the rat. II. Histopathological sequelae. *Neuroscience* **13**, 1073–1094.

Nitsch, R., Bergmann, I., Kuppers, K., Mueller, G., and Frotscher, M. (1990). Late appearance of parvalbumin-immunoreactivity in the development of GABAergic neurons in the rat hippocampus. *Neurosci. Lett.* **118**, 147–150.

Olney, J. W., Ho, O. W., and Rhee, V. (1971). Cytotoxic effects of acidic and sulphur-containing amino acids on the infant mouse central nervous system. *Exp. Brain Res.* **14**, 61–76.

Owens, D. F., Boyce, L. H., Davis, M. B., and Kriegstein, A. R. (1996). Excitatory GABA responses in embryonic and neonatal cortical slices demonstrated by gramicidin perforated-patch recordings and calcium imaging. *J. Neurosci.* **16**, 6414–6423.

Owens, D. F., Liu, X., and Kriegstein, A. R. (1999). Changing properties of $GABA_A$ receptor-mediated signaling during early neocortical development. *J. Neurophysiol.* **82**, 570–583.

Owens, J., Jr., Robbins, C. A., Wenzel, H. J., and Schwartzkroin, P. A. (1997). Acute and chronic effects of hypoxia on the developing hippocampus. *Ann. Neurol.* **41**, 187–199.

Payne, J. A., Rivera, C., Voipio, J., and Kaila, K. (2003). Cation-chloride co-transporters in neuronal communication, development and trauma. *Trends Neurosci.* **26**, 199–206.

Picken Bahrey, H. L., and Moody, W. J. (2003). Early development of voltage-gated ion currents and firing properties in neurons of the mouse cerebral cortex. *J. Neurophysiol.* **89**, 1761–1773.

Pirchio, M., Turner, J. P., Williams, S. R., Asprodini, E., and Crunelli, V. (1997). Postnatal development of membrane properties and delta oscillations in thalamocortical neurons of the cat dorsal lateral geniculate nucleus. *J. Neurosci.* **17**, 5428–5444.

Prince, D. A., and Gutnick, M. J. (1972). Neuronal activities in epileptogenic foci of immature cortex. *Brain Res.* **45**, 455–468.

Psarropoulou, C., and Avoli, M. (1993). 4-Aminopyridine-induced spreading depression episodes in immature hippocampus: Developmental and pharmacological characteristics. *Neuroscience* **55**, 57–68.

Psarropoulou, C., and Descombes, S. (1999). Differential bicuculline-induced epileptogenesis in rat neonatal, juvenile and adult CA3 pyramidal neurons *in vitro*. *Dev. Brain Res.* **117**, 117–120.

Puka-Sundvall, M., Hagberg, H., and Andiné, P. (1994). Changes in extracellular calcium concentration in the immature rat cerebral cortex during anoxia are not influenced by MK-801. *Dev. Brain Res.* **77**, 146–150.

Puka-Sundvall, M., Gajkowska, B., Cholewinski, M., Blomgren, K., Lazarewicz, J. W., and Hagberg, H. (2000). Subcellular distribution of calcium and ultrastructural changes after cerebral hypoxia-ischemia in immature rats. *Dev. Brain Res.* **125**, 31–41.

Purpura, D. P., Prelevic, S., and Santini, M. (1968). Postsynaptic potentials and spike variation in the feline hippocampus during postnatal ontogenesis. *Exp. Neurol.* **22**, 408–422.

Raffo, E., Koning, E., and Nehlig, A. (2004). Postnatal maturation of cytochrome oxidase and lactate dehydrogenase activity and age-dependent consequences of lithium-pilocarpine status epilepticus in the rat: A regional histoenzymology study. *Pediatr. Res.* **56**, 647–655.

Rice, J. E., 3rd., Vannucci, R. C., and Brierley, J. B. (1981). The influence of immaturity on hypoxic-ischemic brain damage in the rat. *Ann. Neurol.* **9**, 131–141.

Ringel, M., Bryan, R. M., and Vannucci, R. C. (1991). Regional cerebral blood flow during hypoxia-ischemia in the immature rat: Comparison of iodoantipyrine and iodoamphetamine as radioactive tracers. *Dev. Brain Res.* **59**, 231–235.

Rivera, C., Voipio, J., Payne, J. A., Ruusuvuori, E., Lahtinen, H., Lamsa, K., Pirvola, U., Saarma, M., and Kaila, K. (1999). The K^+/Cl^- co-transporter KCC2 renders GABA hyperpolarizing during neuronal maturation. *Nature* **397**, 251–255.

Riviello, P., de Rogalski Landrot, I., and Holmes, G. L. (2002). Lack of cell loss following recurrent neonatal seizures. *Dev. Brain Res.* **135**, 101–104.

Robertson, C. L., Bucci, C. J., and Fiskum, G. (2004). Mitochondrial response to calcium in the developing brain. *Dev. Brain Res.* **151,** 141–148.

Rodriguez-Soriano, J., Vallo, A., Castillo, G., and Oliveros, R. (1981). Renal handling of water and sodium in infancy and childhood: A study using clearance methods during hypotonic saline diuresis. *Kidney Int.* **20,** 700–704.

Romijn, H. J., Voskuyl, R. A., and Coenen, A. M. (1994). Hypoxic-ischemic encephalopathy sustained in early postnatal life may result in permanent epileptic activity and an altered cortical convulsive threshold in rat. *Epilepsy Res.* **17,** 31–42.

Rorke, L. B. (1982). "Pathology of Perinatal Brain Injury." Raven Press, New York, NY.

Rutherford, M. (2002). "MRI of the Neonatal Brain." W.B. Saunders, London, UK.

Sakaue, M., Nakamura, H., Kaneko, I., Kawasaki, Y., Arakawa, N., Hashimoto, H., Koyama, Y., Baba, A., and Matsuda, T. (2000). Na^+-Ca^{2+} exchanger isoforms in rat neuronal preparations: Different changes in their expression during postnatal development. *Brain Res.* **881,** 212–216.

Sankar, R., Shin, D. H., Liu, H., Mazarati, A., Pereira de Vasconcelos, A., and Wasterlain, C. G. (1998). Patterns of status epilepticus-induced neuronal injury during development and long-term consequences. *J. Neurosci.* **18,** 8382–8393.

Sankar, R., Shin, D., Mazarati, A. M., Liu, H., Katsumori, H., Lezama, R., and Wasterlain, C. G. (2000). Epileptogenesis after status epilepticus reflects age- and model-dependent plasticity. *Ann. Neurol.* **48,** 580–589.

Scher, M. S. (2003). Neonatal seizures and brain damage. *Pediatr. Neurol.* **29,** 381–390.

Schmid, R., Tandon, P., Stafstrom, C. E., and Holmes, G. L. (1999). Effects of neonatal seizures on subsequent seizure-induced brain injury. *Neurology* **53,** 1754–1761.

Schwartzkroin, P. A. (1981). Development of rabbit hippocampus: Physiology. *Dev. Brain Res.* **2,** 469–486.

Schwartzkroin, P. A., and Altschuler, R. J. (1977). Development of kitten hippocampal neurons. *Brain Res.* **134,** 429–444.

Schwartzkroin, P. A., and Kunkel, D. D. (1982). Electrophysiology and morphology of the developing hippocampus of fetal rabbits. *J. Neurosci.* **2,** 448–462.

Shimizu-Okabe, C., Yokokura, M., Okabe, A., Ikeda, M., Sato, K., Kilb, W., Luhmann, H. J., and Fukuda, A. (2002). Layer-specific expression of Cl^- transporters and differential $[Cl^-]_i$ in newborn rat cortex. *Neuroreport* **13,** 2433–2437.

Silver, I. A., and Erecinska, M. (1990). Intracellular and extracellular changes of $[Ca^{2+}]$ in hypoxia and ischemia in rat brain *in vivo*. *J. Gen. Physiol.* **95,** 837–866.

Silver, I. A., and Erecinska, M. (1992). Ion homostasis in rat brain *in vivo*: Intra- and extracellular $[Ca^{2+}]$ and $[H^+]$ in the hippocampus during recovery from short-term transient ischemia. *J. Cereb. Blood Flow Metab.* **12,** 759–772.

Silver, I. A., and Erecinska, M. (1997). Energetic demands of the Na^+/K^+ ATPase in mammalian astrocytes. *Glia* **21,** 35–45.

Silverstein, F. S., and Naik, B. (1991). Effect of depolarization on striatal amino acid efflux in perinatal rats: An *in vivo* microdialysis study. *Neurosci. Lett.* **128,** 133–136.

Silverstein, F. S., Naik, B., and Simpson, J. (1991). Hypoxia-ischemia stimulates hippocampal glutamate efflux in perinatal rat brain: An *in vivo* microdialysis study. *Pediatr. Res.* **30,** 587–590.

Spigelman, I., Zhang, L., and Carlen, P. L. (1992). Patch-clamp study of postnatal development of CA1 neurons in rat hippocampal slices: Membrane excitability and K^+ currents. *J. Neurophysiol.* **68,** 55–69.

Stafstrom, C. E., Thompson, J. L., and Holmes, G. L. (1992). Kainic acid seizures in the developing brain: Status epilepticus and spontaneous recurrent seizures. *Dev. Brain Res.* **65,** 227–236.

Stein, D. T., and Vannucci, R. C. (1988). Calcium accumulation during the evolution of hypoxic-ischemic brain damage in the immature rat. *J. Cereb. Blood Flow Metab.* **8,** 834–842.

Stichel, C. C., Singer, W., Heizmann, C. W., and Norman, A. W. (1987). Immunohistochemical localization of calcium-binding proteins, parvalbumin and calbindin-D 28k, in the adult and developing visual cortex of cats: A light and electron microscopic study. *J. Comp. Neurol.* **262,** 563–577.

Stringer, J. L. (1998). Regulation of extracellular potassium in the developing hippocampus. *Dev. Brain Res.* **110,** 97–103.

Stringer, J. L., and Lothman, E. W. (1996). During afterdischarges in the young rat *in vivo* extracellular potassium is not elevated above adult levels. *Dev. Brain Res.* **91,** 136–139.

Sulyok, E., Nemeth, M., Tenyi, I., Csaba, I. F., Varga, F., Gyory, E., and Thurzo, V. (1979). Relationship between maturity, electrolyte balance and the function of the renin-angiotensin-aldosterone system in newborn infants. *Biol. Neonate* **35,** 60–65.

Swann, J. W., and Brady, R. J. (1984). Penicillin-induced epileptogenesis in immature rat CA3 hippocampal pyramidal cells. *Dev. Brain Res.* **12,** 243–254.

Swann, J. W., Smith, K. L., and Brady, R. J. (1986). Extracellular K^+ accumulation during penicillin-induced epileptogenesis in the CA3 region of immature rat hippocampus. *Dev. Brain Res.* **30,** 243–255.

Tepper, J. M., and Trent, F. (1993). *In vivo* studies of the postnatal development of rat neostriatal neurons. *Prog. Brain Res.* **99,** 35–50.

Tepper, J. M., Sharpe, N. A., Koos, T. Z., and Trent, F. (1998). Postnatal development of the rat neostriatum: Electrophysiological, light- and electron-microscopic studies. *Dev. Neurosci.* **20,** 125–145.

Thurston, J. H., and McDougal, D. B., Jr. (1969). Effect of ischemia on metabolism of the brain of the newborn mouse. *Am. J. Physiol.* **216,** 348–352.

Towfighi, J., Mauger, D., Vannucci, R. C., and Vannucci, S. J. (1997). Influence of age on the cerebral lesions in an immature rat model of cerebral hypoxia-ischemia: A light microscopic study. *Dev. Brain Res.* **100,** 149–160.

Towfighi, J., Housman, C., Mauger, D., and Vannucci, R. C. (1999). Effect of seizures on cerebral hypoxic-ischemic lesions in immature rats. *Dev. Brain Res.* **113,** 83–95.

Towfighi, J., Housman, C., Brucklacher, R., and Vannucci, R. C. (2004). Neuropathology of seizures in the immature rabbit. *Dev. Brain Res.* **152,** 143–152.

Tremblay, E., Nitecka, L., Berger, M. L., and Ben-Ari, Y. (1984). Maturation of kainic acid seizure-brain damage syndrome in the rat. I. Clinical, electrographic and metabolic observations. *Neuroscience* **13,** 1051–1072.

Trippenbach, T., Richter, D. W., and Acker, H. (1990). Hypoxia and ion activities within the brain stem of newborn rabbits. *J. Appl. Physiol.* **68,** 2494–2503.

Tyzio, R., Ivanov, A., Bernard, C., Holmes, G. L., Ben-Ari, Y., and Khazipov, R. (2003). Membrane potential of CA3 hippocampal pyramidal cells during postnatal development. *J. Neurophysiol.* **90,** 2964–2972.

Vernadakis, A., and Woodbury, D. M. (1962). Electrolyte and amino acid changes in rat brain during maturation. *Am. J. Physiol.* **203,** 748–752.

Vernadakis, A., and Woodbury, D. M. (1965). Cellular and extracellular spaces in developing rat brain. *Archiv. Neurol.* **12,** 284–293.

Világi, I., Tarnawa, I., and Banczerowski-Pelyhe, I. (1991). Changes in seizure activity of the neocortex during the early postnatal development of the rat: An electrophysiological study on slices in Mg^{2+}-free medium. *Epilepsy Res.* **8,** 102–106.

Volpe, J. J. (2000). "Neurology of the Newborn." W. B. Saunders, Philadelphia, PA.

Vorísek, I., and Syková, E. (1997). Ischemia-induced changes in the extracellular space diffusion parameters, K^+, and pH in the developing rat cortex and corpus callosum. *J. Cereb. Blood Flow Metab.* **17,** 191–203.

Warren, R. A., and Jones, E. G. (1997). Maturation of neuronal form and function in a mouse thalamo-cortical circuit. *J. Neurosci.* **17,** 277–295.

Wasterlain, C. G. (1997). Recurrent seizures in the developing brain are harmful. *Epilepsia* **38**, 728–734.

Welsh, F. A., Vannucci, R. C., and Brierley, J. B. (1982). Columnar alterations of NADH fluorescence during hypoxia-ischemia in immature rat brain. *J. Cereb. Blood Flow Metab.* **2**, 221–228.

Wong, M., and Yamada, K. A. (2001). Developmental characteristics of epileptiform activity in immature rat neocortex: A comparison of four *in vitro* seizure models. *Dev. Brain Res.* **128**, 113–120.

Xia, Y., Jiang, C., and Haddad, G. G. (1992). Oxidative and glycolytic pathways in rat (newborn and adult) and turtle brain: Role during anoxia. *Am. J. Physiol.* **262**, R595–603.

Yager, J. Y., Shuaib, A., and Thornhill, J. (1996). The effect of age on susceptibility to brain damage in a model of global hemispheric hypoxia-ischemia. *Dev. Brain Res.* **93**, 143–154.

Yamada, J., Okabe, A., Toyoda, H., Kilb, W., Luhmann, H. J., and Fukuda, A. (2004). Cl$^-$ uptake promoting depolarizing GABA actions in immature rat neocortical neurones is mediated by NKCC1. *J. Physiol.* **557**, 829–841.

Young, R. S., Osbakken, M. D., Briggs, R. W., Yagel, S. K., Rice, D. W., and Goldberg, S. (1985). ^{31}P NMR study of cerebral metabolism during prolonged seizures in the neonatal dog. *Ann. Neurol.* **18**, 14–20.

Zhang, L., Spigelman, I., and Carlen, P. L. (1991). Development of GABA-mediated, chloride-dependent inhibition in CA1 pyramidal neurones of immature rat hippocampal slices. *J. Physiol.* **444**, 25–49.

Zhou, F. M., and Hablitz, J. J. (1996). Postnatal development of membrane properties of layer I neurons in rat neocortex. *J. Neurosci.* **16**, 1131–1139.

Further Reading

Attwell, D., and Laughlin, S. B. (2001). An energy budget for signaling in the grey matter of the brain. *J. Cereb. Blood Flow Metab.* **21**, 1133–1145.

Lowry, O. H., Passonneau, J. V., Hasselberger, F. X., and Schulz, D. W. (1964). Effect of ischemia on known substrates and cofactors of the glycolytic pathway in brain. *J. Biol. Chem.* **239**, 18–30.

Pérez-Pinzón, M. A., Tao, L., and Nicholson, C. (1995). Extracellular potassium, volume fraction, and tortuosity in rat hippocampal CA1, CA3, and cortical slices during ischemia. *J. Neurophysiol.* **74**, 565–573.

7

Thinking about Visual Behavior; Learning about Photoreceptor Function

Kwang-Min Choe and Thomas R. Clandinin*
Department of Neurobiology
Stanford University
Stanford, California 94305

Visual behavioral assays in *Drosophila melanogaster* were initially developed to explore the genetic control of behavior, but have a rich history of providing conceptual openings into diverse questions in cell and developmental biology. Here, we briefly summarize the early efforts to employ three of these behaviors: phototaxis, the UV-visible light choice, and the optomotor response. We then discuss how each of these assays has expanded our understanding of neuronal connection specificity and synaptic function. All of these studies have contributed to the development of

*Current affiliation: Department of Biology, Yonsei University, Seoul, Korea

Current Topics in Developmental Biology, Vol. 69
Copyright 2005, Elsevier Inc. All rights reserved.

0070-2153/05 $35.00
DOI: 10.1016/S0070-2153(05)69007-2

sophisticated tools for manipulating gene expression, assessing cell fate specification, and visualizing neuronal development. With these tools in hand, the field is now poised to return to the original goal of understanding visual behavior using genetic approaches. © 2005, Elsevier Inc.

I. Introduction

The questions surrounding the issue of how complex sets of neuronal connections form and function define much of modern neuroscience. Perhaps unsurprisingly, a diverse set of techniques drawn from biochemistry, electrophysiology, genetics, molecular biology, and computation has been applied to exploring these questions. This chapter focuses on genetic approaches in the fruit fly *Drosophila melanogaster*, particularly those employing visual behavior assays, to understanding the genetic control of behavior as well as a variety of aspects of photoreceptor function and development. In particular, we focus on how such screens have identified critical components involved in synapse formation and physiology and how they have identified genes whose functions would have been difficult to uncover by other methods.

A variety of experimental approaches has provided us with a rich understanding of a number of the key steps in neural development, including neuronal fate specification, long-range axon guidance, and synapse formation (reviewed in Bertrand *et al.*, 2002; Huber *et al.*, 2003; Yamagata *et al.*, 2003). In addition, biochemical, molecular, and genetic approaches have identified many of the components of the synaptic vesicle release machinery, and we now understand in some detail how synaptic vesicle cycling occurs (reviewed in Sudhof, 2004). However, important questions regarding both neural development and synaptic function remain. For example, how do individual neurons make choices about where, when, and with whom to form synaptic connections? How is synaptic vesicle release regulated *in vivo*, and how is it integrated into the basic cellular machinery of a neuron?

The incredibly precise connections between neurons in the fruit fly visual system provide a powerful system with which to dissect the molecular mechanisms underlying these processes. Indeed, the stereotyped nature of these connections has fascinated anatomists and developmental biologists beginning with Ramon y Cajal, and the fly visual behavioral repertoire has inspired generations of neuroscientists interested in identifying genes that control behavior. In this system, it is possible to reliably visualize the connectivity of the system using confocal microscopy, quantify the response of specific synapses using electrophysiological techniques, and visualize the structures of many identical synaptic terminals in a single preparation using electron microscopy. This system also allows gene expression to be manipulated at the level of single cells, in both loss-of-function and gain-of-function experiments. These tools have enabled the emergence of this system as a facile model for defining the molecular mechanisms underlying a number of developmental and physiological processes in the retina and the optic ganglia. In this chapter, we first summarize efforts over the past 45 years to make use of fly visual behavior to gain entry into a variety of fields, and then we discuss both the advantages and disadvantages of this approach in the context of future efforts to understand molecular mechanisms underlying neuronal connection specificity and synaptic function.

A. Picking Your Poison

Classical forward genetics has driven much of the progress made thus far in understanding the molecules involved in fly neurodevelopment. In terms of connection specificity, there have been two broad types of screening assays employed. The first, and most direct, of these is a histological screen that visualizes specific axons and identifies mutations that cause morphological change. Such screens have been conducted in both flies and worms, and their significant advantage is that many of the mutations one finds are, by definition, affecting the processes of interest. Indeed, such approaches have identified many of the key molecular components involved in long-range guidance of axons at the ventral midline, the navigation of axons along specific paths in the periphery, as well as key processes involved in regulating synaptic structure at the neuromuscular junction (reviewed in Ackley and Jin, 2004; Huber *et al.*, 2003). However, such screens have a significant limitation: they demand a facile, high-throughput histological assay. Accordingly, it has proven difficult, though not impossible, to implement such an approach in the context of the central nervous system (CNS), where differences between appropriate and inappropriate targets are measured on the scale of a few microns. Although significant and successful efforts have overcome this limitation in the context of specific neurons in the visual and olfactory systems, the resulting screens remain laborious (Maurel-Zaffran *et al.*, 2001; Reuter *et al.*, 2003).

The second approach, which is the focus of this chapter, relies on assays of neuronal function as indirect measures of neuronal connectivity. The notion is that a high-throughput behavioral screen can identify mutations that disrupt neuronal function, and that a subset of such mutations should affect neuronal connectivity. Assays of neural function are of two main types: electrophysiological measures of synaptic transmission and behavioral measures that reflect the activities of one or more neurons in a circuit. Because electrophysiological assays tend to be unsuited to large-scale forward genetic screens, they have typically been applied in combination with behavioral assays (Hotta and Benzer, 1969; Pak *et al.*, 1969). As we will see, the advantage of such function-driven screens is that they can identify mutations that affect connectivity in contexts in which direct histological screens would be laborious or unfeasible. As such, they have identified mutations that affect target specificity in the CNS as well as mutations that affect many aspects of synapse assembly and function (even though these are often associated with morphological defects that are only visible at the ultrastructural level) (Clandinin *et al.*, 2001; Koh *et al.*, 2004; Lee *et al.*, 2001, 2003; Stowers *et al.*, 2002; Verstreken *et al.*, 2003). However, all such screens suffer from the relative lack of specificity in the primary assay: behaviors can typically be disrupted by mutations acting in many different cells and affecting many different aspects of cellular physiology. As a result, success critically depends on tightly controlled behavioral assays, in combination with cell-type specific somatic mosaics.

B. The Functional Anatomy of the Visual System

The *Drosophila* visual system is a model of developmental precision and behavioral specialization. The adult visual system comprises the retina and four optic ganglia, designated the lamina, the medulla, the lobula, and the lobula plate (Fig. 1A) (reviewed in Meinertzhagen and Hanson, 1993). The retina forms a hexagonal array of some 800 unit eyes, called ommatidia, each of which is composed of eight photoreceptors (abbreviated as PRs or R cells) and eleven accessory cells (reviewed in Wolff and Ready, 1993). The eight R cells fall into three broad types, defined by the morphology and position of their light-sensing organs, the rhabdomeres, and by opsin gene expression (Figs. 1B–1E). R1–6 cells, the outer photoreceptors in each ommatidium, have large rhabdomes, express a single opsin, denoted Rh1, and respond to a broad range of wavelengths centered in the green color range. R1–6 cells are the most sensitive detectors of visible light and are critical to the fly's ability to perceive motion. The two inner photoreceptors, R7 and R8, have smaller rhabdomes and a more complex pattern of opsin gene expression. R7 cells express one of two opsins, Rh3 and Rh4, that are both sensitive to light in the ultraviolet range. R8 cells also express one of two opsins, Rh5 and Rh6, one of which is blue-light sensitive and one of which is sensitive to light in the green range. Opsin gene expression is matched between R7 and R8 such that only two combinations (of four possible) of R7 and R8 opsins are found in a single ommatidium (Fig. 1E) (Cook and Desplan, 2001). Within a single ommatidium, each R7 and R8 cell looks at the same point in space, and thus is perfectly positioned to compare the relative intensities of two different wavelength ranges. As a result of this anatomical arrangement, R7 and R8 are thought to mediate color vision. Finally, the eye is still further patterned: R7s and R8s in the dorsal-most ommatidia define a region called the dorsal rim area (DRA) and have an unusual structure: they express the same opsin, Rh3,

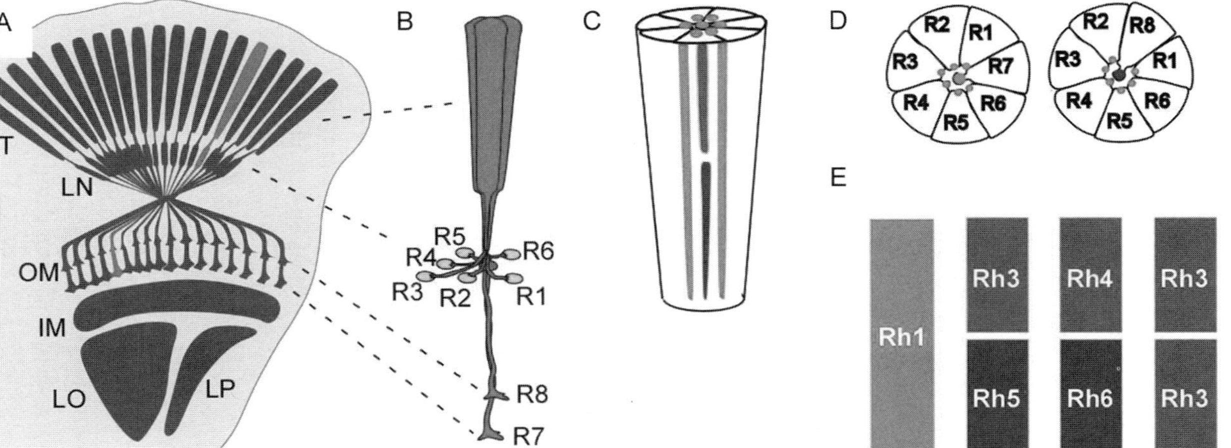

Figure 1 The anatomy of the fly visual system. (A) The visual system comprises the retina (RT) and four morphologically separated ganglia. These four ganglia, designated the lamina (LN), the medulla (OM, outer medulla; IM, inner medulla), the lobula (LO), and the lobula plate (LP), display complex connections with one another (not shown). (B) Photoreceptor axons make characteristic patterns of connections within the lamina and the outer medulla. Each ommatidium (red) contains eight photoreceptor cells, R1–8, and extends a single bundle of axons into the brain in a common fascicle. R1–6 cells defasciculate from this bundle and choose targets arranged in a characteristic relative pattern within the lamina. Axons from R7 and R8 extend through this brain layer and terminate within two distinct layers within the medulla. (C) Photoreceptors are arranged in a characteristic pattern within each ommatidium, with the light-sensing rhabdomes of R1–6 arranged around the outside of each ommatidium, with R7 and R8 tiered on top of one another in the center of the cluster (a side view). (D) A top view of the same ommatidium. Left, cross-sectioned in the top half where R7 cell body is located. Right, cross-sectioned in the bottom half where R8 cell body is located. (E) R1–6 invariably express a single opsin, designated Rh1. R7 and R8 each express one of two opsins, designated Rh3 and Rh4 (for R7) and R5 and Rh6 (for R8). The opsin choice made by these cells is coordinated within each ommatidium such that R7 cells that choose Rh3 are invariably paired with R8 cells that express Rh5, while R7 cells that express Rh4 are paired with R8 cells that express Rh6. This association is altered in the dorsal-most ommatidia in the eye, in which both R7 and R8 express Rh3; these ommatidia are specialized to detect polarized light. (See Color Insert.)

and their rhabdomes are aligned along the same axis of optical polarization (Fig. 1E) (Wernet *et al.*, 2003). As a result, this region of the eye is color-blind, but is dedicated to detecting the e-vector of the polarized light for flight navigation.

Just as in the retina, the pattern of connections that photoreceptors make with their targets in the brain reflects both remarkable precision and behavioral function. The first optic ganglion, the lamina, lies immediately proximal to the retina, separated by a fenestrated membrane. In the lamina, R1–6 axons choose their synaptic partners, lamina target neurons, and form a synaptic unit called the lamina cartridge (Fig. 1B) (reviewed in Meinertzhagen and Hanson, 1993). These synaptic connections are made in a complex pattern that reflects the fact that R1–6 photoreceptors that "see" the same point in space are distributed over the surface of the retina. That is, in order to reconstruct a coherent, topographically accurate map of visual space, axons from different parts of the retina converge onto a common set of targets in the brain. This convergence allows the fly brain to integrate multiple inputs from the same point in space, thereby improving the signal-to-noise sensitivity of the eye and enhancing the ability of the fly to see under dim light conditions. To achieve this goal, R1–6 axons make precise lateral extensions between adjacent columns of target neurons. This pattern of connections has been reconstructed in three dimensions using serial electron microscopy and is both genetically hard-wired and essentially invariant from animal to animal (Meinertzhagen and O'Neil, 1991). As we shall see, this pattern constitutes one of the central systems for the analysis of connection specificity in the visual system.

R7 and R8 axons project directly to the medulla. The connectivity here has not yet been defined precisely, but a large number of neuron types have been identified using Golgi silver stained preparations, and the structure is clearly organized into both columns and layers (Fig. 1A and 1B) (Fischbach and Dittrich, 1989). Each column processes visual information from a single point in space and receives input directly from R7 and R8 as well as indirect input from the corresponding R1–6 cells via the lamina neurons. Each column comprises a series of layers; R7 and R8 each make synaptic connections within a single, specific layer. Spectral information may be computed through interactions between these layers, but direct evidence in favor of this hypothesis is lacking. The ability of R7 and R8 axons to make layer-specific connections is highly reminiscent of the layer-specific connections made by neurons in many parts of the vertebrate CNS (Sanes and Yamagata, 1999) and is also one of the most accessible and closely studied connectivity patterns in the visual system.

Finally, the lobula and lobula plate lie deep within the brain, and their structural organization remains poorly defined. Again, a number of neuron types have been identified using silver-staining techniques, and genetic studies in *Drosophila* as well as electrophysiological studies in larger Diptera have demonstrated that at least some of the interneurons in these areas are involved in higher-order visual processing of motion stimuli (Bausenwein *et al.*, 1986; Fischbach and Dittrich, 1989; Krapp and Hengstenberg, 1996). We know very little about the developmental specification of the connections in this layer; studies of a subset of neurons in the lobula plate indicate that the morphology of these cells is independent of visual input, suggesting that the connectivity patterns seen in these brain regions may be genetically hard-wired (Scott *et al.*, 2003).

C. Visual Behavioral Assays

Three well-established behavioral paradigms in *Drosophila* have been used for three large-scale genetic screens, the phototaxis assay, the UV-visible (UV-Vis) light choice test, and the optomotor response assay. All three paradigms are robust and can be elicited from wild-type flies in the absence of any prior conditioning.

Phototaxis is one of the simplest behaviors known: adult flies simply orient and move toward a light source (Figs. 2A and 2B). All R cells can provide input to the phototaxis response, although R1–6 cells play a dominant role in the response to dim, visible light that reflects their relatively high sensitivity (Harris *et al.*, 1976). Phototaxis assays have been used in a number of genetic screens and have provided critical insights into our understanding of many processes, invertebrate phototransduction predominant among them (reviewed in Smith *et al.*, 1991; Zuker, 1996).

A more complex assay is the UV-Vis light choice test. In this assay, flies are presented with two monochromatic light sources, one an ultraviolet source, the other a dim green light, at the opposite ends of a T-shaped maze (Gerresheim, 1981) (Fig. 2C). Under particular relative light intensities, normal flies prefer to phototax toward the UV light source rather than toward the visible light. This preference is critically dependent on the presence and function of R7 photoreceptors; mutations that eliminate R7, or disrupt its function, cause mutant flies to prefer the green light. This assay has enabled the identification of genes that affect R7 development and connectivity (Banerjee *et al.*, 1987; Harris *et al.*, 1976; Reinke and Zipursky, 1988).

The most complex innate visual behavior that has been used is the optomotor response, a change in fly orientation in response to motion cues in the environment that is thought to reflect a basic component of navigation control. The optomotor response assay takes two forms, in which flies either pursue vertical stripes moving horizontally in front of them or run counter to the movement of horizontal stripes below them (Figs. 2D and 2E). The cause of these different responses is unknown, but input from R1–6

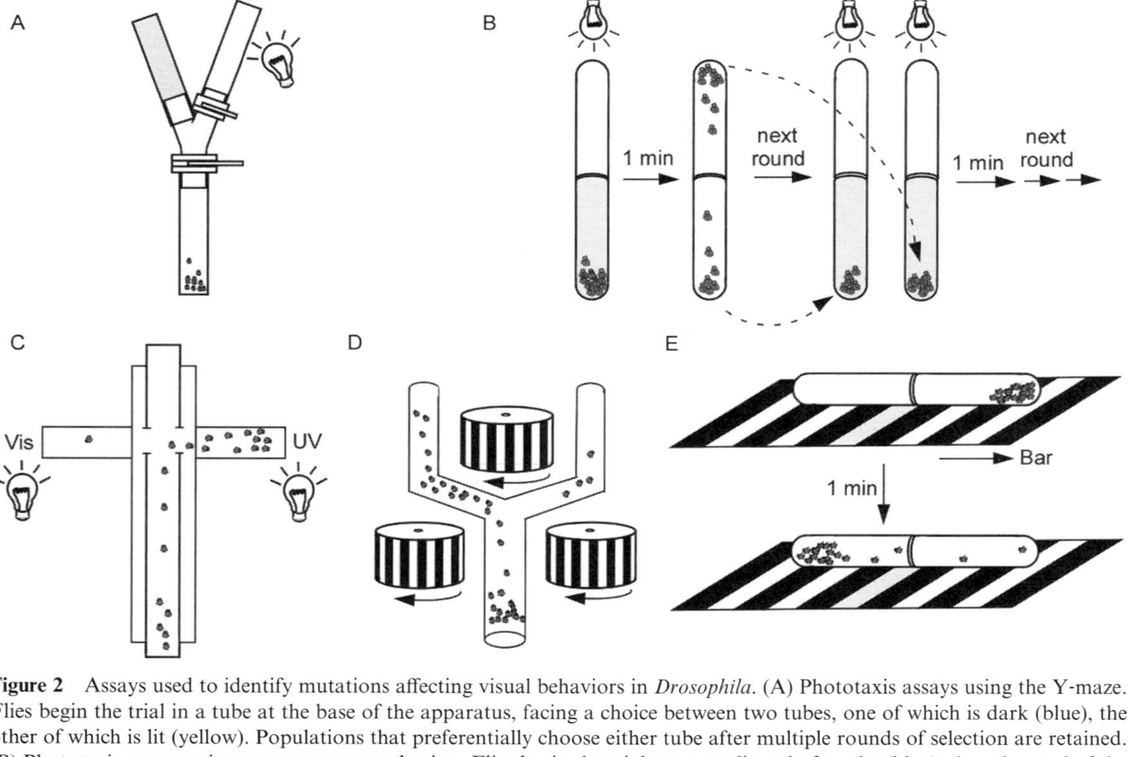

Figure 2 Assays used to identify mutations affecting visual behaviors in *Drosophila*. (A) Phototaxis assays using the Y-maze. Flies begin the trial in a tube at the base of the apparatus, facing a choice between two tubes, one of which is dark (blue), the other of which is lit (yellow). Populations that preferentially choose either tube after multiple rounds of selection are retained. (B) Phototaxis assays using countercurrent selection. Flies begin the trial at one unlit end of a tube (blue); the other end of the tube is lit (yellow). After 1 min, flies that cross the middle of the tube are advanced to a second tube; those that do not remain in the starting tube. After this initial separation, both populations are retested multiple times; in each case, flies that respond

axons is critically important for both. Screens based on the optomotor response have identified mutations that affect phototransduction, optic lobe development, and the formation of the lateral connections of R1–6 axons within the lamina (Clandinin *et al.*, 2001; Fischbach and Heisenberg, 1981; Lee *et al.*, 2001, 2003; Pak, 1975).

Each of these behaviors places different demands on the visual system. For example, as the simplest behavior, phototaxis probes only the ability to detect light and thus makes only limited demands on the underlying neural circuitry. As a result, phototaxis screens provide an efficient means of identifying mutations that cause blindness. The UV-Vis choice test is more complex because it requires a UV-response-specific deficit; entirely blind flies will obviously not prefer green light to UV light and will therefore be excluded. While this exclusion has, to date, simply enriched for mutations that disrupt R7 development and function, this assay also holds the potential of allowing us to decode the neural circuit underlying a simple binary choice. The optomotor response is the most complex behavior, as it requires direction-selective computations to be performed; such computations require integration over both time and space, and likely involve computations occurring at multiple levels in the optic lobe. With this type of assay, one can either focus on cells and connections at the periphery or use the assay to ask questions about deeper brain structures.

Using the right assay is crucial, as each assay focuses on different cells and circuits. Indeed, the three outlined here are unlikely to be the only innate assays possible, and as the field becomes interested in deeper brain regions, the development of new behavioral assays will likely allow us to target different circuits. One might imagine, for example, that an assay based on color discrimination would enable dissection of the neural circuits that compare the inputs from R7 and R8. Focusing on a particular circuit in this way also requires a defined set of genetic tools, including tissue-specific promoters that can be used in mosaic techniques, cell-type specific antibody

---

are advanced and those that do not respond remain in the same tube for the next test. (C) UV-visible light choice test using the T-maze. One end of the T-maze is lit with monochromatic green light (Vis); the other end of the maze is lit with ultraviolet (UV). The relative intensities of each light are adjusted such that flies prefer the UV side of the maze. (D) The optomotor response measured using Y-maze assay. Flies begin the experiment in a tube facing a choice point defined by the coordinated rotation of three striped drums. As flies respond by turning to pursue the rotating stripes, wild-type flies prefer to enter one arm of the apparatus. By constructing an apparatus containing multiple such choice points arranged in a series, flies that fail to orient properly can be identified by their aberrant path through the maze. (E) The optomotor response using a single tube. Flies begin the trial at one end of a long tube placed on top of a moving pattern of stripes. Wild-type flies orient toward the apparent source of motion and run opposite to the flow of the stripes. After 1 min, flies that fail to respond to the bars are retained as candidate mutants. (See Color Insert.)

reagents that can be used to assess connectivity, and high-resolution visualization techniques that allow synaptic connections to be assessed in wild-type animals. Therefore, the future promises both the development of new tools for making mosaics and manipulating cells in other parts of the nervous system and the establishment of new behavioral paradigms.

II. The Early Days

During the 1950s, 1960s, and 1970s, pioneering efforts to genetically dissect visual behaviors using *Drosophila* were initiated by three different groups with very distinct goals. The first of these groups sought to link quantitative genetic techniques with behavioral assays (Hirsch and Boudreau, 1958). The second sought single gene disruptions that altered specific visual (and nonvisual) behaviors (Benzer, 1967). Finally, a third group began to study the insect visual system as a model for specific neural computations (Heisenberg and Götz, 1975). Although their views on the genetics of behavior were quite different, the three groups shared the goal of using genetics to study visual behavior per se, a heretical idea at the time, and all three expended considerable effort in identifying genes involved in their respective behaviors. As we will discuss, the genes identified by each group led to investigations into different aspects of neurobiology, quite independent from the behaviors themselves. As research on these side investigations has matured, the tools that these efforts created have enabled workers to return to the original questions surrounding the genetic basis of visual behavior.

A. Multigenic Influences on Phototaxis

Are behavioral differences among animals in populations under the control of one general genetic factor or many specific factors (Hirsch and Boudreau, 1958)? Using both phototaxis and geotaxis (flies moving against gravity), flies were sorted through an iterative combination of binary choices designed to progressively separate behaviorally different flies from a genetically heterogeneous population (Fig. 2A). Using this approach, both behaviors could be modified by selection and were stably heritable, and thus reflected genetic control (Hirsch and Boudreau, 1958). However, in these experiments, the heritable differences in these experiments were multigenic—multiple genes contributing relatively little to the deviations from an average initial value. As a result, technical limitations thwarted efforts to assign key sources of heritable variation to specific genes, though these efforts were able to identify particular chromosome regions at least for geotaxis (Hirsch and Erlenmeyer-Kimling, 1962; Hirsch and Ksander, 1969; Ricker and Hirsch, 1988). More recently, genomic approaches have identified some key contributors to the geotaxis response, including *cry* (cryptochrome, encoding a photopigment), *Pdf* (encoding a neuropeptide), and *Pen* (encoding a nuclear importin) (Toma *et al.*, 2002). However, it remains unclear precisely how these genes contribute to geotactic behavior.

B. Flies as Atoms of Behavior

An alternate approach to fly behavior, analogous to approaches employed in the dissection of phage and bacterial gene regulation, was to try to disrupt single genes that perturb visual behavior (Benzer, 1967). In this approach, flies were sequentially fractionated through an apparatus based on their phototactic ability using a method named the countercurrent separation, based on a method conceptually similar to methods used in physics to separate molecules from a complex mixture (Fig. 2B). By combining successive rounds of fractionation, non-phototactic mutant lines were isolated from the progeny of mutagenized isogenic wild-type flies (Benzer, 1967). Since the starting population of animals by design was largely isogenic, the behavioral deficits identified were caused by single genetic lesions, rendering the changes far more tractable to further analysis than polygenic traits. Many of the mutants identified in this way disrupted critical components in the phototransduction cascade. Indeed, in this successful approach, a change in animal behavior was linked to a single gene (Hotta and Benzer, 1969).

A more directed version of this approach that pioneered the use of the electroretinogram (ERG) as a screening tool was used to identify additional phototransduction components (Pak *et al.*, 1969; reviewed in Pak, 1979). In this approach, phototaxis assays were used as a primary screen to identify candidate mutations that disrupted vision; once isolated, each candidate was subjected to a secondary screen in which the electrophysiological responses of photoreceptors to flashes of light were measured directly using a field electrode. This ERG records light stimulus-induced changes in electric potential between an electrode placed on the cornea and a reference electrode, and measures the electrical activity generated by populations of photoreceptors. A typical ERG consists of three distinct components: a corneal negative component, which is caused by depolarization of R cell membranes by light; the "on transient," which is a rapid corneal-positive transient observed when the light is first turned on, and the "off transient," which is a rapid, corneal-negative transient that occurs when the light is turned off (reviewed in Pak, 1975). Intriguingly, the on and off transients reflect neuronal activity in the lamina (Goldsmith, 1965; Heisenberg, 1971). As a result, an ERG can localize a visual defect either to the phototransduction cascade itself (blocking the corneal negative current) or to the downstream aspects of the neuronal response (affecting the on and off transients).

The UV-Vis choice test proved to be an effective means of identifying mutations that affect a slightly more complex behavior (Fig. 2C) (Gerresheim, 1981; Harris *et al.*, 1976). These studies opened an interesting, and ultimately highly successful, field of its own, a theme we will return to repeatedly. As we noted, the UV preference of wild-type flies in the choice test is critically dependent upon the function of the UV-sensitive R7 cell. Due to an evolutionary quirk, the specification of R7 cells in the developing retina is critically dependent upon a receptor tyrosine kinase designated sevenless, a gene with no other apparent function during fly development. As a result, many of the initial mutations identifying defects in the choice test turned out to be sevenless alleles (Banerjee *et al.*, 1987; Harris *et al.*, 1976). Since the absence of R7 cells was thus straightforward to detect, a number of groups used this assay as a means of identifying additional genes involved in R cell fate specification, and various extensions of this work led to the elucidation of the RAS/Mitogen-activated protein (MAP) kinase cascade, a critical signal transduction cascade of widespread importance (Nagaraj and Banerjee, 2004). As we shall see, this work also led to the development of a large number of antibody markers, promoter constructs, and genetic strains that allow facile manipulation of gene expression in a variety of R subtypes. These tools have, collectively, proven invaluable in more recent behavioral studies.

C. Computational Approaches to Insect Vision

Insect vision provides an opportunity to unravel the computational capacities of a relatively simple nervous system (Gotz, 1964; Heisenberg and Götz, 1975). Of particular interest are questions related to motion perception. For example, how are motion cues associated with optic flow, the movement vectors associated with stationary objects in the environment that are acquired as the fly moves, used to guide fly navigation? With such questions in mind, the responses of individual wild-type flies to a range of motion stimuli presented under tightly controlled presentation condition were characterized in detail (Heisenberg and Wolf, 1979). Using data obtained first in beetles and later in flies, powerful mathematical models of motion perception were constructed, which made specific predictions regarding the underlying architecture of the circuits involved (Hassenstein and Reichardt, 1956; Poggio and Reichardt, 1973; Reichardt, 1957). As one way of understanding this circuit, an ingenious screen to identify mutations that disrupted the function of this circuit was conducted (Heisenberg and Götz, 1975). This screen used a large maze consisting of a series of binary choice points (Fig. 2D). At each choice, flies were confronted by a set of three rotating stripe patterns, the effect of which was to bias normal flies to turn in a particular direction. After fractionating flies through 16 such choice points, individual flies that consistently made stochastic choices were isolated and characterized. Unfortunately, many of the mutations isolated in this screen overlapped with the mutations identified in the photo-taxis screens: blind flies are also defective in motion vision. However, the discovery of a single mutation lent support to the initial idea behind the screen. This mutation defines a locus, designated *optomotor-blind*, and disrupts a developmentally critical transcription factor; the allele isolated causes a strong and specific defect in the development of the lobula, a key motion-processing center (Heisenberg *et al.*, 1978).

D. Lessons from These Early Screens

None of these efforts identified genes "for" specific behaviors. Rather, each was spectacularly successful in opening new lines of investigation into different, specific processes in neurobiology. The phototaxis screens and the optomotor assay led to elucidation of the phototransduction cascade, and the UV-Vis choice test led to an explosion of studies into the genetic and molecular bases of eye development and fate specification.

A handful of mutants, including *norpA*, *tan*, *nonA*, *rdgA,* and *rdgB,* was isolated independently in several screens (reviewed in Pak, 1975; Fig. 3B). Interestingly, all of these genes act in photoreceptors, even though only those screens that incorporated the ERG specifically demanded mutant foci in the retina. Why were no mutations causing blindness attributable to deficits in the optic ganglia? In hindsight, there is a straightforward explanation that reflects the fact that all of these screens were designed to identify recessive, viable mutations. That is, since the eye is not essential for organismal viability (at least in the laboratory), mutations that disrupt eye-specific functions such as phototransduction are likely to be overrepresented in such a screen, while mutations that disrupt functions common to more neuron types are likely to be homozygous lethal. Consistent with this being a significant part of the explanation, the one exceptional mutation identified in the optomotor screen, *optomotor-blind*, turned out to be an unusual, tissue-specific allele of an otherwise essential gene (Pflugfelder *et al.*, 1992). As we will see, the development of sophisticated somatic mosaic techniques has been critical to bypassing this limitation.

None of the loci involved in contributing to the population differences in phototaxis behavior have been molecularly identified. It is therefore impos-sible to judge whether these loci would differ from the genes identified by the other groups. Given that at least some of loci involved in geotaxis turned out to be genes that are difficult to link directly to the behavioral response, it seems possible that at least some of the loci involved in phototaxis are distinct from those identified in the loss-of-function studies. However, since subtle polymorphisms in genes critical to behavior do cause natural phenotypic variation in some cases, it remains possible that the two sets of genes overlap extensively (Greenspan, 2004). Development of new genomics approaches to analyzing gene expression will hopefully allow us to soon be able to effectively dissect even such highly polygenic traits.

III. New Tools, New Goals

After these initial studies, efforts to study neural development and function in the fly took off, while, ironically, efforts to understand the genetic basis of fly visual behavior fell into a relatively fallow period lasting approximately 20 years from 1980 to 2000. During this time, no fundamentally new behavioral screens were initiated (though both the phototaxis efforts and the UV-Vis choice tests were expanded), and fundamental discoveries in the basic processes of phototransduction and cell fate specification were made. Moreover, the fly visual system came to be viewed as a flexible platform in which to investigate general questions in neurobiology, including pattern formation, cellular morphogenesis, axon guidance, and synaptic function. While these efforts suggested new questions that could be addressed, they also created new tools and experimental approaches that could be applied to the eye. Because it has empowered the more recent approaches to visual behavior, we will briefly summarize some of this work.

Although this section focuses on approaches aimed at identifying mutations affecting neuronal target selection and neuronal physiology, it is important to note that the efforts to understand photoreceptor fate specification and retinal patterning made critical contributions (reviewed in Nagaraj and Banerjee, 2004). In particular, this work defined a large set of tools for manipulating gene expression in subsets of photoreceptors, identified a number of molecular markers specific to different R cell subtypes, and described the sequence of developmental events in the eye in great detail, and it has proven invaluable to efforts to dissect neuronal targeting and synaptic physiology.

Two broadly different approaches have had a significant impact on the identification of new genes with specific functions related to axon guidance and targeting in the visual system. The first approach used histological screens to identify mutants that affect optic lobe structure in an effort to map functional circuits (Heisenberg and Böhl, 1979) or to study neural connectivity (Ebens *et al.*, 1993; Garrity *et al.*, 1996; Martin *et al.*, 1995). The second approach hypothesized that genes expressed specifically in the eye or in the optic lobes might have important functions. Using three different experimental methods, including monoclonal antibodies, subtractive hybridization, and enhancer trapping methods, a number of important genes and reagents were identified (Cheyette *et al.*, 1994; Fujita *et al.*, 1982; Hyde *et al.*, 1990; Pignoni *et al.*, 1997; Shieh *et al.*, 1989; Smith *et al.*, 1990; Van Vactor *et al.*, 1988; Zipursky *et al.*, 1985). Here, we discuss only histological screens aimed at identifying mutations affecting neuronal connectivity.

A. Isolating Mutants That Affect Brain Wiring

Two distinct motivations, one based on the desire to understand behavior itself, and one based on the desire to understand neurodevelopment, have driven work to identify mutations that affect visual system organization. In one approach, the goal is to come to a deeper understanding of behavioral mechanisms through selective lesioning of particular brain regions. The notion here is that by correlating specific behavioral deficits with developmental errors in particular parts of the visual system, one can make inferences regarding the neural computations normally performed by those cells. Using histological analyses of adult brain structure, a number of such mutations were identified (Heisenberg and Böhl, 1979). One such mutant was *small optic lobes* (*sol*), which displayed a 50% reduction in the size of the medulla and the lobula complex; in this mutant, certain classes of medulla neurons were completely missing, but others were left relatively unaffected (Fischbach and Heisenberg, 1981). Similarly, only specific visual behaviors were disrupted. Unfortunately, given the complex nature of the histological phenotype as well as the likely highly indirect nature of the gene involved, a clear correlation between the structural disruption and the behavioral outcome is difficult to establish.

The goal of understanding the molecular bases of brain development, particularly the mechanisms underlying axon guidance and targeting, has driven extensive efforts to identify mutations causing structural defects in the visual system (reviewed in Clandinin and Zipursky, 2002). In this approach, histological efforts are focused on only a small part of the visual system, particularly the connections of photoreceptors and their targets, and mutations were initially identified in homozygous animals. A particular advance was the use of cell-type specific markers for photoreceptors and their axons as a means of identifying and characterizing phenotypes of interest. A second advance was to examine brains during larval development, allowing the identification of genes that are homozygous lethal but whose lethal phases occur during pupal development. Using these approaches, mutations affecting the targeting of photoreceptor axons into the brain during larval development, the global organization of the optic ganglia, and the proliferation of neurons in the medulla have been identified (Ebens *et al.*, 1993;

Garrity *et al.*, 1996; Martin *et al.*, 1995; Poeck *et al.*, 2001; Song *et al.*, 2003; Tayler *et al.*, 2004). Among the largest contributions of this work was the definition of a key signaling pathway comprising a surprising axon guidance receptor, the insulin receptor, that is thought to be coupled to a series of cytoskeletal regulators via the SH2/SH3 adaptor protein Dock.

This approach of identifying mutations affecting R cell axon guidance was greatly improved by the development of a method to generate somatic mosaic animals in which all photoreceptors, but not the rest of the cells in the optic lobes, are made homozygous mutant (Newsome *et al.*, 2000). This approach offers two distinct advantages over past efforts. First, the method requires that mutations identified are, by definition, affecting genes that act in photoreceptor axons for their guidance function. Second, mutations that would be lethal if the whole animal was to be rendered homozygous could now be isolated. Using this approach, a number of mutations that affect the layer-specific targeting of photoreceptor axons to the lamina, as well as the topographic mapping of photoreceptors axons in the medulla, have been identified. These genes include *trio*, a regulator of small GTPases that appears to act downstream of *dock*, *brakeless*, a putative zing finger protein (Rao *et al.*, 2000; Senti *et al.*, 2000), and *flamingo*, a protocadherin (Senti *et al.*, 2003 summarized in Fig. 3A). In addition, by scoring the projections of R cell axons made in adult animals using the same reagents, it became possible to identify genes that affect the layer-specific targeting of photoreceptor axons in the medulla, including the receptor tyrosine phosphatases PTP69D and LAR (Maurel-Zaffran *et al.*, 2001; Newsome *et al.*, 2000). As we describe below, some of these genes were also identified using behavioral approaches.

B. Lessons from These Histological Screens

These histological screens were in large part enabled by the resources, both physical and intellectual, taken from the earlier work focusing on the basic developmental mechanisms involved in photoreceptor fate specification. That is, once the reagents and assays to visualize photoreceptors and their axons were available, it became possible to identify mutations that affected photoreceptor axon guidance. These new studies opened the fly visual system to developmental neurobiological questions, such as topographic mapping and layer-specific target selection, that had previously been addressed in other systems in which forward genetic approaches were less facile. As a result, a number of novel molecules and mechanisms were identified. However, these studies also were limited by their assays: while histological screens for mutations affecting the early stages of photoreceptor axon guidance into the brain that take place during larval development are relatively straightforward to conduct on a large scale, the identification of mutations that specifically disrupt later steps, taking place during pupal development, was laborious. That is, as the interests of the field have shifted to processes taking place during later developmental stages, the intrinsic limitations associated with histological processing of the tissue have begun to impair forward genetic studies based on such assays.

The success of these approaches depended upon a set of assays to sort through and identify only those mutations that were guidance specific. This specificity was achieved in two ways: through the use of extensive tests to identify defects in photoreceptor and target field differentiation and through the use of photoreceptor-specific somatic mosaics. The result was that molecules with clear-cut roles in the growth cone were identified. The necessity of such tools was made clear by the results obtained from the screens that identified *small optic lobe*: in the absence of tools to characterize mutant phenotype, it is difficult to assign the developmental defects associated with this mutation to specific cells. In addition, molecular analysis of the locus revealed a member of the calpain protease family, whose function, if any, during axon guidance and targeting remains inscrutable (Delaney *et al.*, 1991). This requirement for specific mosaics, as well as for significant sets of reagents to assess cell fate, has had a significant influence on the development of new behavioral screens.

IV. Mosaic Screens Using Behavioral Assays

The development of the sophisticated new somatic mosaic approaches for doing histological screens has re-awakened the possibility of doing behavioral screens aimed at identifying genes that are required for photoreceptor function and connectivity. There are two broad goals of this work: to understand synaptic function and to identify mutations that affect connectivity. Two different mosaic approaches have been employed; in the first, somatic mosaic methods were used that generate eyes in which photoreceptors are made homozygous mutant, while the rest of the brain is left largely heterozygous (and thus phenotypically wild-type) (Newsome *et al.*, 2000; Stowers and Schwarz, 1999). Using this approach, two different behavioral paradigms, phototaxis and the optomotor response, were employed to identify mutations that affect photoreceptor function and R1–6 cell connectivity within the lamina (Babcock *et al.*, 2003; Clandinin *et al.*, 2001; Koh *et al.*, 2004; Lee *et al.*, 2001, 2003; Stowers *et al.*, 2002; Verstreken *et al.*, 2003). The second mosaic approach, in which only a subset of R7 cells is made homozygous mutant while the remaining R7 cells are synaptically silent, was used to screen for mutations that affect R7 connectivity using the UV-Vis choice test (Lee *et al.*, 2001; Nern *et al.*, unpublished data).

A. Phototaxis and the Electroretinogram

The phototaxis screen, in combination with the electoretinogram, is a proven method for identifying genes involved in photoreceptor function (Hotta and Benzer, 1969; Pak *et al.*, 1969). As we noted, these studies only identified genes involved in phototransduction, rather than affecting other aspects of photoreceptor (and brain) function, in part, because these early screens were constrained to the study of viable mutations. When this constraint was lifted through the development of somatic mosaic methods in which all photoreceptor cells were made homozygous mutant, large-scale screens were performed to identify mutations that disrupt photoreceptor function (Babcock *et al.*, 2003; Koh *et al.*, 2004; Stowers *et al.*, 2002; Verstreken *et al.*, 2003). Using this approach, new classes of genes have been identified. One important set of mutations appears to disrupt synaptic vesicle release and recovery, and includes two loci, *synaptojanin* and *intersectin* (Koh *et al.*, 2004; Verstreken *et al.*, 2003 summarized in Fig. 3C). In addition, a new angle into mitochondrial regulation was found: mutations in a kinesin interacting protein, designated *milton*, appear to affect the transport of mitochondria down axons into the synaptic terminal (Stowers *et al.*, 2002). As a completely novel biological process not previously considered in another system, this last observation makes clear the point that a forward genetic screen based on a behavioral approach can lead to novel discoveries. Together, these screens also demonstrate that the combination of a somatic mosaic approach with a behavioral assay can enable the isolation of mutations that affect fundamental cell biological processes. As the currently published work represents only a small fraction of the mutations identified in these screens, it seems clear that these efforts are likely to open up a wide range of new directions.

B. The Optomotor Response and the UV-Vis Choice Test

Two additional behavioral screens using assays adapted from those first described 30 years ago have been conducted, with the goal of identifying new mutations that affect photoreceptor connectivity. One of these screens used the optomotor response, a visually evoked behavior known to be dependent upon the function of R1–6 (Heisenberg and Buchner, 1977). In its original application, a multiple Y-maze apparatus was constructed in which fly decisions at each choice point were influenced by the presentation of rotating drums of black and white stripes. In the modified form of the assay, flies in a tube are presented with a moving pattern of black and white stripes; the normal fly response to this presentation is to run in the direction opposite to the flow of the stripes (Fig. 2E). Using such an assay, flies that can see the stripes can be quickly separated from flies that cannot. Mutations that cause behavioral deficits are then re-screened using a variety of histological techniques to visualize R cell axons at a variety of developmental time points. From this screen, mutations that affect R1–6 connections can be selected; a number of these also cause defects in the layer-specific targeting of R7 in the medulla. These genes include two cadherins, N-cadherin and flamingo, as well as the receptor tyrosine phosphatase LAR. N-cadherin and LAR are required in R1–6 axons for their projection toward the lamina targets and in R7 for layer-specific targeting in the medulla (Clandinin *et al.*, 2001; Lee *et al.*, 2001, sumarized in Fig. 3A, B). Flamingo is required in R1–6 axons for projection toward the correct lamina targets and in R8 axons for setting up topography during innervation into the optic lobes (Lee *et al.*, 2003).

A second behavioral assay combining R7-specific mosaics with an R7-dependent behavior, the UV-Vis choice test, has proven to be an efficient means of identifying mutations that affect R7 layer-specific targeting in the medulla. For example, preliminary results from this screen identified multiple mutations affecting both N-cadherin and LAR (Clandinin *et al.*, 2001; Lee *et al.*, 2001). In addition, the design of the screen is amenable to very large-scale efforts and will undoubtedly identify more mutations affecting different aspects of R7 connectivity.

C. Lessons from These New Screens

Together, these screens demonstrate that, with the availability of cell-specific somatic mosaic techniques, behavioral assays can effectively identify mutations affecting a wide range of cellular functions. Chief among their advantages is the non-invasive nature of the primary screen—behavioral assays can be performed on living flies that can be recovered for genetic analysis. This enables genetic screens to be conducted on the F1 generation, bypassing the requirement for the establishment of F2 lines, and thereby greatly increasing the scale and ease of the screen itself. Moreover, by enriching for the mutations of interest by excluding flies with normal photoreceptor function, the subsequent, more laborious electrophysiological or histological techniques are applied to only a small subset of the mutations. In addition, these screens also demonstrate the advantages of using an assay that asks only for mutations that broadly disrupt photoreceptor function. That is, using these approaches, a remarkably diverse collection of biological processes were exposed to further analysis. Indeed, the screens have provided entry points into problems that were previously not thought to be genetically tractable.

These advantages are not without limitations. Foremost among these is the fact that a behavioral screen casts a very wide net for mutations: only a subset of the mutations identified in any of these behavioral assays affects photoreceptor development or function in ways we can assess. That is, many of the mutations originally identified from their behavioral phenotype have yet to be tied to any cellular process. In this sense, behavioral screens are a relatively inefficient means of identifying mutations of interest, and provide only a means of enrichment. In some cases, however, this enrichment has been critical. With this difficulty in mind, it seems likely that while behavioral screens have provided entry points into a number of biological processes, the second-generation screens aimed at understanding those processes will likely be focused specifically on the process of interest, and will be independent of behavior.

These screens also emphasize that the sensitivity and specificity of the behavioral assay used are critical to the success of the approach. In particular, the phototaxis screens were essentially predicated on the identification of blind flies; as a result, most of the mutations identified have severe consequences for photoreceptor function. On the other hand, the optomotor response screen and the UV-Vis screen were both sensitized such that the mutants identified did not have to be so severe as to render the photoreceptors in question completely nonfunctional. Indeed, most of the mutations identified in these screens displayed only mild, if any, deficits in phototaxis. As a result, these screens identified mutations that cause only subtle defects in photoreceptor connectivity but, correspondingly, also identified more mutations of unknown effect on photoreceptor function.

V. Perspectives

A. Using Behavior versus Studying Behavior

Two broad goals have been defined. The first is to describe neuronal circuits at the level of the developmental and molecular mechanisms that determine their connectivity and physiology; the second is to understand how each circuit computes behavioral outputs in response to environmental stimuli. Both of these tasks are daunting in their complexity, even for a relatively simple nervous system like that found in *Drosophila*. In the context of the visual system, however, we have arguably made greater progress toward the first goal. As a result, much of the work on visual behavior in flies has proven to be essentially an enrichment trick in a genetic screen to identify mutations that affect other processes. Indeed, it is striking that much of the behavioral genetics done in *Drosophila* has led to other developmental and physiological questions that have in turn led to experimental goals quite independent from the original behavioral effort. As a result of this success, recent work applying behavioral assays to the visual system has made no pretense of interest in behavior itself; rather, each application has used a relatively simple behavioral assay with the explicit goal of identifying mutations affecting some other cell biological or developmental process. Clearly, this approach has enabled significant progress in a variety of directions, but so far, our understanding of the molecules and circuits underpinning visual behavior has remained primitive in comparison.

B. The Potential Impact of Genome-Based Technologies

This chapter has highlighted some of the ways new technologies have affected our ability to identify and characterize individual loci involved in specific processes relevant to behavior. In the near future it seems likely that genomic information, as well as new technologies for analyzing gene expression and function on a genomic scale, is going to make possible significant new advances. Indeed, behavioral phenotypes are notoriously complex and, in some cases, may be dependent upon the concerted actions of many genes, each making relatively small contributions to the behavioral phenotype. Efforts to select behavioral variation out of a genetically hetero-geneous "wild-type" background have made this strikingly clear (Hirsch and Boudreau, 1958). Until recently, such genetic variation was largely refracto-ry to further analysis. However, recent efforts using techniques to analyze genome-wide patterns of gene expression have begun to identify the con-tributing factors (Toma *et al.*, 2002). As proteomic techniques become more facile, we anticipate that this type of approach will bear new fruit.

C. The Future Returns to the Past

While the application of genetic techniques has so far largely failed to cast light on the fundamental mechanisms of visual behavior, extensive efforts to understand vision from a computational standpoint have been remarkably successful. Using, for example, precise measures of behavioral response, carefully controlled stimulus presentations, and electrophysiological stud-ies, a quantitative model that explains phenomena such as the optomotor response has been extensively tested both in theory and in experimental contexts (for example, see Haag *et al.*, 2004; Poggio and Reichardt, 1973). Indeed, these studies and many others have clearly demonstrated that neu-rons in the fly's brain are capable of sophisticated, behaviorally relevant computations. On the other hand, while there has been significant progress in this direction, these studies still struggle with the complexity of the intact nervous system. For example, it remains difficult to disentangle the compli-cated interactions between neurons in a circuit using physiological studies alone. In addition, in a context in which our detailed understanding of the

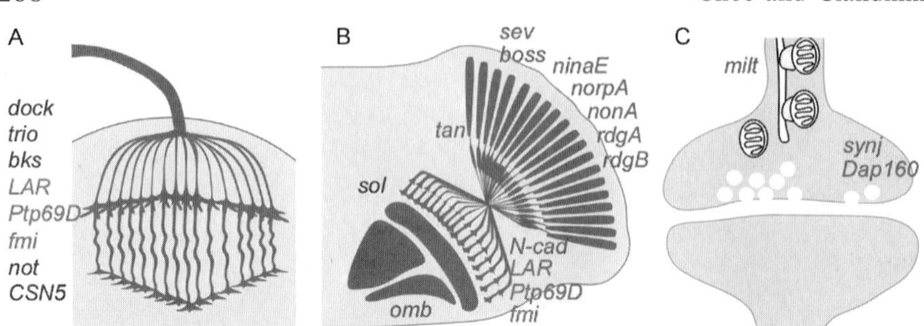

Figure 3 Histological screens and behavioral screens have identified mutations affecting different stages of visual system development. (A) Histological screens for mutations affecting the third instar eye-brain complex have identified loci (labeled in blue) that affect the bundling of R cell axon fascicles as they project into the brain (including *dock, trio, LAR,* and *fmi*), as well as the layer-specific targeting of R1–6 axons to the lamina (including *bks, Ptp69D, not,* and *CSN5*). Purple labels indicate mutants identified from both histological and behavioral screens. (B) Behavioral screens have identified additional loci (labeled red) that affect the pattern of axons seen in the adult brain, as well as the functions and development of the retina. Among these, *fmi, N-cad, LAR,* and *Ptp69D* affect the targeting of R1–6 axons within the lamina as well as the layer-specific connections of R7 and, in some cases, R8 in the medulla. *sev* and *boss* affect R7 fate specification. *ninaE, norpA, nonA, rdgA,* and *rdgB* are involved in controlling photoreceptor function. *tan* is required for neurotransmitter metabolism, and *sol* reduces the overall size of some optic ganglia. (C) Behavioral screens have also identified mutations that affect R1–6 cell synaptic function. Among these, *milt* is required for the proper localization of mitochondria in the synaptic terminal, and *synj* and *Dap160* are involved in synaptic vesicle release and recycling. (See Color Insert.)

circuitry of the nervous system is incomplete, it is difficult to determine precisely how the fly makes use of the computations performed by individual neurons in order to inform its behavioral choices. It is our hope that with the development of synapse-specific, genetically encoded modifiers of neurotransmission, genetically driven lesion studies will begin to cast light on behavior. Indeed, progress in this direction has recently been described in a variety of contexts outside of the visual system, including studies of fly learning and memory, mating behavior, and circadian rhythm control (for example, Grima *et al.,* 2004; Kitamoto, 2001; Manoli and Baker, 2004; McGuire *et al.,* 2003). It seems likely to us that such efforts are in their infancy and, as behavioral assays become more sophisticated, will come to complement the more traditional studies based on electrophysiology and behavioral analysis. As the visual system allows the precise relationship between stimulus and behavioral response to be defined quantitatively, we anticipate significant advances in our understanding of visual circuit function in the years to come.

Acknowledgments

We thank members of the Clandinin lab, including M. Velez, P.-L. Chen, and S. Prakash, for their helpful comments on the manuscript.

References

Ackley, B. D., and Jin, Y. (2004). Genetic analysis of synaptic target recognition and assembly. *Trends Neurosci.* **27,** 540–547.

Babcock, M. C., Stowers, R. S., Leither, J., Goodman, C. S., and Pallanck, L. J. (2003). A genetic screen for synaptic transmission mutants mapping to the right arm of chromosome 3 in. *Genetics* **165,** 171–183.

Banerjee, U., Renfranz, P. J., Pollock, J. A., and Benzer, S. (1987). Molecular characterization and expression of sevenless, a gene involved in neuronal pattern formation in the *Drosophila* eye. *Cell* **49,** 281–291.

Bausenwein, B., Wolf, R., and Heisenberg, M. (1986). Genetic dissection of optomotor behavior in *Drosophila melanogaster.* Studies on wild-type and the mutant optomotor-blind[H31]. *J. Neurogenet.* **3,** 87–109.

Benzer, S. (1967). Behavioral mutants of *Drosophila* isolated by countercurrent distribution. *Proc. Natl. Acad. Sci. USA* **58,** 1112–1119.

Bertrand, N., Castro, D. S., and Guillemot, F. (2002). Proneural genes and the specification of neural cell types. *Nat. Rev. Neurosci.* **3,** 517–530.

Clandinin, T. R., and Zipursky, S. L. (2002). Making connections in the fly visual system. *Neuron* **35,** 827–841.

Clandinin, T. R., Lee, C.-H., Herman, T., Lee, R. C., Yang, A. Y., Ovasapyan, S., and Zipursky, S. L. (2001). *Drosophila* LAR regulates R1-R6 and R7 target specificity in the visual system. *Neuron* **32,** 237–48.

Cook, T., and Desplan, C. (2001). Photoreceptor subtype specification: From flies to humans. *Semin. Cell Dev. Biol.* **12,** 509–18.

Delaney, S. J., Hayward, D. C., Barleben, F., Fischbach, K. F., and Miklos, G. L. (1991). Molecular cloning and analysis of small optic lobes, a structural brain gene of *Drosophila melanogaster. Proc. Natl. Acad. Sci. USA* **88,** 7214–7218.

Ebens, A. J., Garren, H., Cheyette, B. N., and Zipursky, S. L. (1993). The *Drosophila* anachronism locus: A glycoprotein secreted by glia inhibits neuroblast proliferation. *Cell* **74,** 15–27.

Fischbach, K.-F., and Dittrich, A. P. M. (1989). The optic lobe of *Drosophila melanogaster.* I. A Golgi analysis of wild-type structure. *Cell Tissue Res.* **258,** 441–475.

Fischbach, K. F., and Heisenberg, M. (1981). Structural brain mutant of *Drosophila melanogaster* with reduced cell number in the medulla cortex and with normal optomotor yaw response. *Proc. Natl. Acad. Sci. USA* **78,** 1105–1109.

Fujita, S. C., Zipursky, S. L., Benzer, S., Ferrus, A., and Shotwell, S. L. (1982). Monoclonal antibodies against the *Drosophila* nervous system. *Proc. Natl. Acad. Sci. USA* **79,** 7929–7933.

Garrity, P. A., Rao, Y., Salecker, I., McGlade, J., Pawson, T., and Zipursky, S. L. (1996). *Drosophila* photoreceptor axon guidance and targeting requires the dreadlocks SH2/SH3 adapter protein. *Cell* **85,** 639–650.

Gerresheim, F. (1981). Isolation and characterization of mutants with altered phototactic reaction to monochromatic light in *Drosophila* melanogaster. Ph.D. thesis, Munich University, Munich, Germany.

Goldsmith, T. H. (1965). Do flies have a red receptor? *J. Gen. Physiol.* **49**, 265–287.

Gotz, K. G. (1964). Optomotorische Untersuchungen des visuellen systems einiger augenmutaten der fruchtfliege *Drosophila*. *Kybernetik* **2**, 77–92.

Greenspan, R. J. (2004). E pluribus unum, ex uno plura: Quantitative and single-gene perspectives on the study of behavior. *Annu. Rev. Neurosci.* **27**, 79–105.

Grima, B., Chelot, E., Xia, R., and Rouyer, F. (2004). Morning and evening peaks of activity rely on different clock neurons of the *Drosophila* brain. *Nature* **431**, 869–873.

Haag, J., Denk, W., and Borst, A. (2004). Fly motion vision is based on Reichardt detectors regardless of the signal-to-noise ratio. *Proc. Natl. Acad. Sci. USA* **101**, 16333–16338.

Harris, W. A., Stark, W. S., and Walker, J. A. (1976). Genetic dissection of the photoreceptor system in the compound eye of *Drosophila melanogaster*. *J. Physiol.* **256**, 415–439.

Hassenstein, B, and Reichardt, W. (1956). Systemtheoretische Analyse der Zeit, Reihenfolgen- und Vorzeichenauswertung bei der Bewegunsperzeption des Russelkafers Chlorophanus. *Z. Naturforsch.* **11b**, 513–534.

Heisenberg, M. (1971). Separation of receptor and lamina potentials in the electroretinogram of normal and mutant *Drosophila*. *J. Exp. Biol.* **55**, 85–100.

Heisenberg, M., and Bohl, K. (1979). Isolation of anatomical brain mutants of *Drosophila* by histological means. *Z. Naturforsch.* **34**, 143–147.

Heisenberg, M., and Buchner, E. (1977). The role of retinula cell types in the visual behavior of *Drosophila melanogaster*. *J. Comp. Physiol.* **117**, 127–162.

Heisenberg, M., and Gotz, K. G. (1975). The use of mutations for the partial degradation of vision in *Drosophila melanogaster*. *J. Comp. Physiol.* **98**, 217–241.

Heisenberg, M., and Wolf, R. (1979). On the fine structure of yaw torque in visual flight orientation of *Drosophila melanogaster*. *J. Comp. Physiol.* **130**, 113–130.

Heisenberg, M., Wonneberger, R., and Wolf, R. (1978). optomotor-blind[H31]-a *Drosophila* mutant of the lobula plate giant neurons. *J. Comp. Physiol.* **124**, 287–296.

Hirsch, J., and Boudreau, J. C. (1958). Studies in experimental behavior genetics. I. The heritability of phototaxis in a population of *Drosophila melanogaster*. *J. Comp. Physiol. Psychol.* **51**, 647–651.

Hirsch, J., and Erlenmeyer-Kimling, L. (1962). Studies in experimental behavior genetics: IV. Chromosome analyses for geotaxis. *J. Comp. Physiol. Psychol.* **55**, 732–739.

Hirsch, J., and Ksander, G. (1969). Studies in experimental behavior genetics. V. Negative geotaxis and further chromosome analyses in *Drosophila melanogaster*. *J. Comp. Physiol. Psychol.* **67**, 118–122.

Hotta, Y., and Benzer, S. (1969). Abnormal electroretinograms in visual mutants of *Drosophila*. *Nature* **222**, 354–356.

Huber, A. B., Kolodkin, A. L., Ginty, D. D., and Cloutier, J. F. (2003). Signaling at the growth cone: Ligand-receptor complexes and the control of axon growth and guidance. *Annu. Rev. Neurosci.* **26**, 509–563.

Hyde, D. R., Mecklenburg, K. L., Pollock, J. A., Vihtelic, T. S., and Benzer, S. (1990). Twenty *Drosophila* visual system cDNA clones: One is a homolog of human arrestin. *Proc. Natl. Acad. Sci. USA* **87**, 1008–1012.

Kitamoto, T. (2001). Conditional modification of behavior in *Drosophila* by targeted expression of a temperature-sensitive shibire allele in defined neurons. *J. Neurobiol.* **47**, 81–92.

Koh, T. W., Verstreken, P., and Bellen, H. J. (2004). Dap160/intersectin acts as a stabilizing scaffold required for synaptic development and vesicle endocytosis. *Neuron* **43**, 193–205.

Krapp, H. G., and Hengstenberg, R. (1996). Estimation of self-motion by optic flow processing in single visual interneurons. *Nature* **384**, 463–466.

Lee, C.-H., Herman, T., Clandinin, T. R., Lee, R., and Zipursky, S. L. (2001). N-cadherin regulates target specificity in the *Drosophila* visual system. *Neuron* **30**, 437–450.

Lee, R. C., Clandinin, T. R., Lee, C.-H., Chen, P.-L., Meinertzhagen, I. A., and Zipursky, S. L. (2003). The protocadherin Flamingo is required for axon target selection in the *Drosophila* visual system. *Nat. Neurosci.* **6**, 557–563.

Manoli, D. S., and Baker, B. S. (2004). Median bundle neurons coordinate behaviors during *Drosophila* male courtship. *Nature* **430**, 564–569.

Martin, K. A., Poeck, B., Roth, H., Ebens, A. J., Ballard, L. C., and Zipursky, S. L. (1995). Mutations disrupting neuronal connectivity in the *Drosophila* visual system. *Neuron* **14**, 229–240.

Maurel-Zaffran, C., Suzuki, T., Gahmon, G., Treisman, J. E., and Dickson, B. J. (2001). Cell-autonomous and -nonautonomous functions of LAR in R7 photoreceptor axon targeting. *Neuron* **32**, 225–235.

McGuire, S. E., Le, P. T., Osborn, A. J., Matsumoto, K., and Davis, R. L. (2003). Spatiotemporal rescue of memory dysfunction in *Drosophila*. *Science* **302**, 1765–1768.

Meinertzhagen, I. A., and Hanson, T. E. (1993). The development of the optic lobe. *In* "The Development of *Drosophila melanogaster*" (M. Bate and A. Martinez-Arias, Eds.), pp. 1363–1491. Cold Spring Harbor Press, New York.

Meinertzhagen, I. A., and O'Neil, S. D. (1991). Synaptic organization of columnar elements in the lamina of the wild type in *Drosophila melanogaster*. *J. Comp. Neurol.* **305**, 232–263.

Nagaraj, R., and Banerjee, U. (2004). The little R cell that could. *Int. J. Dev. Biol.* **48**, 755–760.

Newsome, T. P., Asling, B., and Dickson, B. J. (2000). Analysis of *Drosophila* photoreceptor axon guidance in eye-specific mosaics. *Development* **127**, 851–860.

Pak, W. L., Grossfield, J., and White, N. V. (1969). Nonphototactic mutants in a study of vision of *Drosophila*. *Nature* **222**, 351–354.

Pak, W. L. (1975). Mutations affecting the vision of *Drosophila melanogaster*. *In* "Handbook of Genetics" (R. C. King, Ed.), Vol. 3, pp. 703–733. Plenum, New York.

Pak, W. L. (1979). Study of photoreceptor function using *Drosophila* mutants. *In* "Neurogenetics, Genetic Approaches to the Nervous System" (X. O. Breakfield, Ed.), pp. 67–99. Elsevier, New York.

Pflugfelder, G. O., Roth, H., Poeck, B., Kerscher, S., Schwarz, H., Jonschker, B., and Heisenberg, M. (1992). The lethal(1)optomotor-blind gene of *Drosophila melanogaster* is a major organizer of optic lobe development: Isolation and characterization of the gene. *Proc. Natl. Acad. Sci. USA* **89**, 1199–1203.

Pignoni, F., Hu, B., and Zipursky, S. L. (1997). Identification of genes required for *Drosophila* eye development using a phenotypic enhancer-trap. *Proc. Natl. Acad. Sci. USA* **94**, 9220–9225.

Poeck, B., Fischer, S., Gunning, D., Zipursky, S. L., and Salecker, I. (2001). Glial cells mediate target layer selection of retinal axons in the developing visual system of *Drosophila*. *Neuron* **29**, 99–113.

Poggio, T., and Reichardt, W. (1973). A theory of the pattern induced orientation of the fly *Musca domestica*. *Kybernetik* **12**, 185–203.

Rao, Y., Pang, P., Ruan, W., Gunning, D., and Zipursky, S. L. (2000). brakeless is required for photoreceptor growth-cone targeting in *Drosophila*. *Proc. Natl. Acad. Sci. USA* **97**, 5966–5971.

Reichardt, W. (1957). Autokorrelations-Auswetung als Funktionsprinzip des Zentralnervensysems. *Z. Naturforschg.* **12b**, 448–457.

Reinke, R., and Zipursky, S. L. (1988). Cell-cell interaction in the *Drosophila* retina: The bride of sevenless gene is required in photoreceptor cell R8 for R7 cell development. *Cell* **55**, 321–30.

Reuter, J. E., Nardine, T. M., Penton, A., Billuart, P., Scott, E. K., Usui, T., Uemura, T., and Luo, L. (2003). A mosaic genetic screen for genes necessary for *Drosophila* mushroom body neuronal morphogenesis. *Development* **130,** 1203–1213.

Ricker, J. P., and Hirsch, J. (1988). Genetic changes occurring over 500 generations in lines of *Drosophila melanogaster* selected divergently for geotaxis. *Behav. Genet.* **18,** 13–25.

Sanes, J. R., and Yamagata, M. (1999). Formation of lamina-specific synaptic connections. *Curr. Opin. Neurobiol.* **9,** 79–87.

Scott, E. K., Reuter, J. E., and Luo, L. (2003). Dendritic development of *Drosophila* high order visual system neurons is independent of sensory experience. *BMC Neurosci.* **4,** 14.

Senti, K., Keleman, K., Eisenhaber, F., and Dickson, B. J. (2000). brakeless is required for lamina targeting of R1-R6 axons in the *Drosophila* visual system. *Development* **127,** 2291–2301.

Senti, K. A., Usui, T., Boucke, K., Greber, U., Uemura, T., and Dickson, B. J. (2003). Flamingo regulates R8 axon-axon and axon-target interactions in the *Drosophila* visual system. *Curr. Biol.* **13,** 828–832.

Shieh, B. H., Stamnes, M. A., Seavello, S., Harris, G. L., and Zuker, C. S. (1989). The ninaA gene required for visual transduction in *Drosophila* encodes a homologue of cyclosporin A-binding protein. *Nature* **338,** 67–70.

Smith, D. P., Shieh, B. H., and Zuker, C. S. (1990). Isolation and structure of an arrestin gene from *Drosophila*. *Proc. Natl. Acad. Sci. USA* **87,** 1003–1007.

Smith, D. P., Stamnes, M. A., and Zuker, C. S. (1991). Signal transduction in the visual system of *Drosophila*. *Annu. Rev. Cell Biol.* **7,** 161–190.

Song, J., Wu, L., Chen, Z., Kohanski, R. A., and Pick, L. (2003). Axons guided by insulin receptor in *Drosophila* visual system. *Science* **300,** 502–505.

Stowers, R. S., and Schwarz, T. L. (1999). A genetic method for generating *Drosophila* eyes composed exclusively of mitotic clones of a single genotype. *Genetics* **152,** 1631–1639.

Stowers, R. S., Megeath, L. J., Gorska-Andrzejak, J., Meinertzhagen, I. A., and Schwarz, T. L. (2002). Axonal transport of mitochondria to synapses depends on milton, a novel *Drosophila* protein. *Neuron* **36,** 1063–1077.

Sudhof, T. C. (2004). The synaptic vesicle cycle. *Annu. Rev. Neurosci.* **27,** 509–547.

Tayler, T. D., Robichaux, M. B., and Garrity, P. A. (2004). Compartmentalization of visual centers in the *Drosophila* brain requires Slit and Robo proteins. *Development* **131,** 5935–5945.

Toma, D. P., White, K. P., Hirsch, J., and Greenspan, R. J. (2002). Identification of genes involved in *Drosophila melanogaster* geotaxis, a complex behavioral trait. *Nat. Genet.* **31,** 349–353.

Van Vactor, D., Jr., Krantz, D. E., Reinke, R., and Zipursky, S. L. (1988). Analysis of mutants in chaoptin, a photoreceptor cell-specific glycoprotein in *Drosophila*, reveals its role in cellular morphogenesis. *Cell* **52,** 281–290.

Verstreken, P., Koh, T. W., Schulze, K. L., Zhai, R. G., Hiesinger, P. R., Zhou, Y., Mehta, S. Q., Cao, Y., Roos, J., and Bellen, H. J. (2003). Synaptojanin is recruited by endophilin to promote synaptic vesicle uncoating. *Neuron* **40,** 733–748.

Wernet, M. F., Labhart, T., Baumann, F., Mazzoni, E. O., Pichaud, F., and Desplan, C. (2003). Homothorax switches function of *Drosophila* photoreceptors from color to polarized light sensors. *Cell* **115,** 267–279.

Wolff, T., and Ready, D. F. (1993). Pattern formation in the Drosophila retina. *In* "The Development of *Drosophila melanogaster*" (M. Bate and A. Martinez-Arias, Eds.), pp. 1277–1325. Cold Spring Harbor Press, New York.

Yamagata, M., Sanes, J. R., and Weiner, J. A. (2003). Synaptic adhesion molecules. *Curr. Opin. Cell Biol.* **15,** 621–632.

Zipursky, S. L., Venkatesh, T. R., and Benzer, S. (1985). From monoclonal antibody to gene for a neuron-specific glycoprotein in *Drosophila*. *Proc. Natl. Acad. Sci. USA* **82,** 1855–1859.
Zuker, C. S. (1996). The biology of vision of *Drosophila*. *Proc. Natl. Acad. Sci. USA* **93,** 571–576.

Further Reading

Cheyette, B. N., Green, P. J., Martin, K., Garren, H., Hartenstein, V., and Zipursky, S. L. (1994). The *Drosophila* sine oculis locus encodes a homeodomain-containing protein required for the development of the entire visual system. *Neuron* **12,** 977–996.

8

Critical Period Mechanisms in Developing Visual Cortex

Takao K. Hensch
Laboratory for Neuronal Circuit Development
RIKEN Brain Science Institute
Saitama 351–0198, Japan

Binocular vision is shaped by experience during a critical period of early postnatal life. Loss of visual acuity following monocular deprivation is mediated by a shift of spiking output from the primary visual cortex. Both synaptic and network explanations have been offered for this heightened brain plasticity. Direct experimental control over its timing, duration, and closure has now been achieved through a consideration of balanced local circuit excitation-inhibition. Notably, canonical models of homosynaptic plasticity at excitatory synapses alone (LTP/LTD) fail to produce predictable manipulations of the critical period *in vivo*. Instead, a late functional maturation of intracortical inhibition is the driving force, with one subtype in particular standing out.

Parvalbumin-positive large basket cells that innervate target cell bodies with synapses containing the α1-subunit of $GABA_A$ receptors appear to be critical. With age, these cells are preferentially enwrapped in peri-neuronal nets of extracellular matrix molecules, whose disruption by chondroitinase treatment reactivates ocular dominance plasticity in adulthood. In fact, critical period plasticity is best viewed as a continuum of local circuit computations ending in structural consolidation of inputs. Monocular deprivation induces an increase of endogenous proteolytic (tPA-plasmin) activity and consequently motility of spines followed by their pruning, then re-growth. These early morphological events faithfully reflect competition

0070-2153/05 $35.00
DOI: 10.1016/S0070-2153(05)69008-4

only during the critical period and lie downstream of excitatory-inhibitory balance on a timescale (of days) consistent with the physiological loss of deprived-eye responses *in vivo*. Ultimately, thalamic afferents retract or expand accordingly to hardwire the rapid functional changes in connectivity.

Competition detected by local inhibitory circuits then implemented at an extracellular locus by proteases represents a novel, cellular understanding of the critical period mechanism. It is hoped that this paradigm shift will lead to novel therapies and training strategies for rehabilitation, recovery from injury, and lifelong learning in adulthood. © 2005, Elsevier Inc.

I. Introduction

For over 40 years the primary visual cortex has stood as the premier model of critical period plasticity (Wiesel and Hubel, 1963). During a brief postnatal period (of weeks to years) proportional to the expected lifespan of the species (Berardi *et al.*, 2000; Daw, 1995), the closure of one eye (but not both) yields a loss of visual acuity. Amblyopia occurs even though there is no damage to the retina or visual thalamus (dorsal lateral geniculate nucleus [dLGN]) and is determined in the neocortex (V1), where the inputs of the two eyes first converge and compete for space (Wiesel and Hubel, 1963). Mouse models are now yielding with greater resolution the molecular, cellular, and structural events underlying experience-dependent circuit refinement. A general understanding of the neural basis for "critical" or "sensitive" windows of brain development is anticipated to inform classrooms and educational policy, drug design, clinical therapy, and strategies for improved learning into adulthood.

Most impressively, only during the critical period can the seemingly innocuous act of covering an eye profoundly alter the physical structure of the brain. Columnar architecture is the fundamental unit of neocortical organization across mammalian species. Morphological clusters of thalamocortical axon terminals serving the right or left eye tile layer 4 of the mature cortex in alternating "ocular dominance" domains (Hubel *et al.*, 1976; Shatz and Stryker, 1978). Monocular occlusion produces an expansion of open eye columns at the expense of deprived-eye afferents, which become reduced in size and complexity (Antonini and Stryker, 1996; Antonini *et al.*, 1999). This physical manifestation of early postnatal experience is preceded by more rapid changes (Trachtenberg and Stryker, 2001; Trachtenberg *et al.*, 2000) of intracortical circuits outside layer 4 that instruct the hardwiring of inputs into an anatomical fingerprint unique to the individual. This chapter considers experience-dependent circuit refinement during the critical period as a cascade of cellular and molecular events linking functional to structural plasticity (Fig. 1).

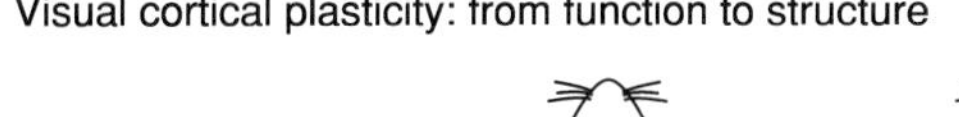

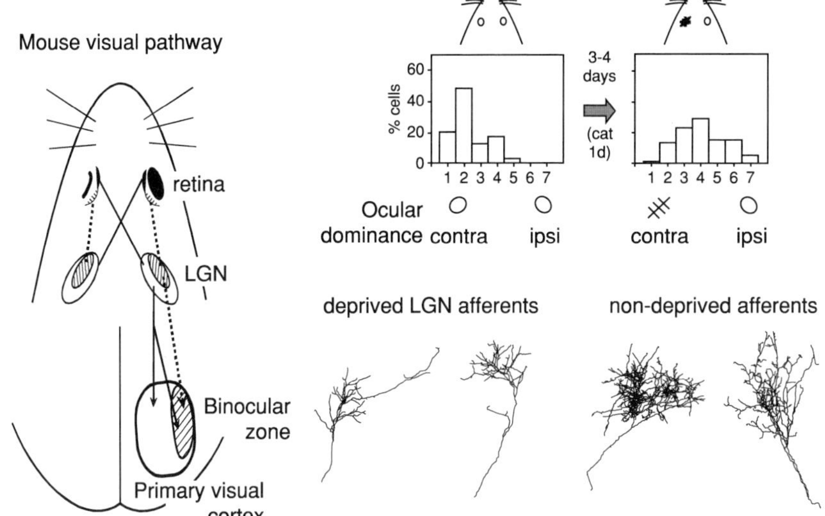

Figure 1 From functional to structural change during the critical period. Visual cortical plasticity begins as a functional shift of spiking response within a few days of monocular deprivation (MD). It first produces a loss of deprived-eye response and gain of open-eye input, as measured by neuronal discharge of single units from mouse visual cortex (Gordon and Stryker, 1996). The ocular dominance of cells rated on a seven-point scale indicates a typical bias toward the contralateral eye (groups 1–3) in the rodent. After 3 or more days of MD, the distribution shifts toward the open, ipsilateral eye (groups >4). Only gradually is this translated into an anatomical shrinkage then expansion of thalamic axons serving the closed or open eye, respectively, in cortical layer 4 (Antonini *et al.*, 1999).

II. Synaptic Mechanisms (LTP/LTD)

What appears essential for vision is the proper communication of output from the primary visual cortex to higher areas. Thus, amblyopia due to monocular deprivation (MD) is faithfully reflected in the relative inability of V1 neurons to fire action potentials through the originally closed eye (ocular dominance; Fig. 1) (Daw, 1995; Prusky and Douglas, 2003). Sensory-evoked field potential amplitudes or the expression of molecular markers (immediate early genes) may remain modifiable beyond the critical period (Pham *et al.*, 2004; Sawtell *et al.*, 2003; Tagawa *et al.*, 2005). However, these largely subthreshold changes in synaptic activity bear little lasting impact on spatial acuity, as perceptual behavior observes a strict critical period that matches single-unit response in V1 (Gordon and Stryker, 1996; Prusky and Douglas, 2003). It is tempting to speculate, nevertheless, that the loss or gain of visual responsiveness represents a homosynaptic long-term depression (LTD) or potentiation (LTP) of synaptic strength somewhere in the visual circuit. Indeed, these *in vitro* models are coregulated by age and sensory experience and are believed to share mechanisms with the hippocampus (Kirkwood *et al.*, 1993, 1995), where a wealth of molecular understanding is already available (Sanes and Lichtman, 1999).

Activation of postsynaptic NMDA-type glutamate receptors is, for example, thought to be a specific mediator of LTP/LTD, and hence experience-dependent plasticity. In particular, it has been proposed that the progressive shortening of NMDA receptor currents by 2A subunit (NR2A) insertion ends the critical period in visual cortex by truncating calcium influx (Nase *et al.*, 1999; Quinlan *et al.*, 1999; Philpot *et al.*, 2001). Surprisingly, mice engineered to maintain prolonged NMDA responses by targeted deletion of NR2A exhibit weaker ocular dominance shifts that are nevertheless restricted to a typical critical period and are delayed normally by dark-rearing from birth (Fagiolini *et al.*, 2003). Postnatal increase of NR2A subunit interactions with specific LTP induction proteins, as in the hippocampus (Liu *et al.*, 2004; Tang *et al.*, 1999), is unnecessary for visual cortical plasticity *in vivo,* since NR2A knockout mice are rescued without re-introducing NR2A itself (Fagiolini *et al.*, 2003).

Synaptic depression is thought to underlie the loss of cortical responsiveness to an eye deprived of vision during the critical period (Frenkel and Bear, 2004). Type 2 metabotropic glutamate receptors (mGluR2) play a fundamental role in visual cortical LTD (Renger *et al.*, 2002). Direct mGluR2 activation by the selective agonist DCG-IV persistently depresses layer 2/3 field potentials in slices of mouse binocular zone, which occludes conventional LTD by low-frequency stimulation (LFS), indicating shared downstream events. In contrast, Schaeffer collateral synapses do not exhibit this chemical LTD, revealing hippocampal area CA1 (naturally devoid of mGluR2) to be an inappropriate model for neocortical plasticity. Antagonists or gene-targeted disruption of mGluR2 prevents LTD induction in visual cortex by electrical LFS to layer 4. However, monocular deprivation remains effective in mice lacking mGluR2 (Renger *et al.*, 2002), and receptor expression levels are unchanged during the critical period in wild-type mice, indicating that experience-dependent plasticity is independent of LTD induction in visual cortex.

Repeated LFS saturates LTD at a weaker level contralateral to an eye deprived for 24 hr when compared to the opposite (ipsilateral) hemisphere (Heynen *et al.*, 2003). This "occlusion" has been viewed as evidence for LTD as a mechanism of ocular dominance plasticity. Multiple, spaced stimuli, however, typically engage protein synthesis and additional molecular machinery in order to convert early synaptic plasticity into longer-lasting forms (Frey *et al.*, 1993). Again, distinct from that found in hippocampal area CA1, this late LTD in neocortex involves the *zif268* immediate early gene, transcription, and translation to saturate LTD in an input- and frequency-specific manner (Atapour *et al.*, unpublished data). Suturing one eye for 24 hr (a time period too short to produce ocular dominance shifts *in vivo*; Gordon and Stryker, 1996) yields a well-known activity-dependent decrease of *zif268* expression, which then impairs LTD saturation contralateral to the deprived eye in a gene dose-dependent manner (Atapour *et al.*, unpublished data). But notably, *zif268* is entirely unnecessary for visual plasticity *in vivo* (Mataga *et al.*, 2001), dissociating it from LTD saturation. Moreover, early depression by a single LFS is not occluded by 1 day of MD (Heynen *et al.*, 2003), and these early forms of LTP/LTD that persist in the presence of protein synthesis inhibitors (Frey *et al.*, 1993) are insufficient to shift ocular dominance *in vivo* (Taha and Stryker, 2002).

Phosphorylation state and membrane trafficking of AMPA receptor subunits are signature events of LTP/LTD at a variety of central synapses that have also been observed after natural sensory experience *in vivo* (Barry and Ziff, 2002; Takahashi *et al.*, 2003). One-day MD produces a constellation of phosphorylation state changes on GluR1 subunits in V1 by protein kinase A (PKA) akin to hippocampal LTD (Heynen *et al.*, 2003). But mimicry need not be causal, since no loss of visual response or acuity occurs until several days of eyelid suture have elapsed (Gordon and Stryker, 1996; Prusky and Douglas, 2003). Phosphorylation of an amino acid residue alone is unlikely to explain the complex functional and structural events that constitute the critical period. Overall, no consistent relationship between the ability to induce homosynaptic plasticity *in vitro* and the capacity for visual plasticity *in vivo* has been found (Bartoletti *et al.*, 2002; Daw, 2004; Hensch, 2003; Fischer *et al.*, 2004; Shimegi *et al.*, 2003). The correlation is not straightforward, as LTP/LTD mechanisms may differ further depending on cortical layer (Daw *et al.*, 2004). Most dramatically, homosynaptic models based on NMDA receptor activation predict that the maturation of inhibition will terminate plasticity (Feldman, 2000; Kirkwood *et al.*, 1995), when in fact quite the opposite is true: GABA function is required to trigger the critical period *in vivo* (Hensch *et al.*, 1998).

III. Network Mechanisms (Excitatory-Inhibitory Balance)

Gross pharmacological perturbations of neuronal activity, such as hyperexcitation (Ramoa *et al.*, 1988; Shaw and Cynader, 1984) or total silencing (Bear *et al.*, 1990; Hata and Stryker, 1994; Reiter and Stryker, 1988; Reiter *et al.*, 1986), not surprisingly disrupt plasticity but fail to inform us about intrinsic network behavior. Even small changes in the relative amounts of excitation and inhibition can dramatically alter information processing

(Hensch and Fagiolini, 2004; Liu, 2004). This exquisite balance is dynamically adjusted by the cortical layer (Desai *et al.*, 2002; Turrigiano and Nelson, 2004), especially because inhibitory connections emerge later than excitation in the pre-critical period for ocular dominance (Del Rio *et al.*, 1994). To dissect the role of local circuit elements, a gentle titration of endogenous neurotransmission through gene-targeted disruption in mice was instrumental.

Fortuitously, GABA is synthesized by glutamic acid decarboxylase made from two distinct genes, GAD65 and GAD67; the former is concentrated in axon terminals and is bound to synaptic vesicles, while the latter is found throughout the cell (Soghomonian and Martin, 1998). Reducing stimulated GABA release by GAD65 deletion (knockout) prevents ocular dominance plasticity until inhibition is acutely restored with diazepam (Hensch *et al.*, 1998). Remarkably, this rescue is possible at any age, indicating that the critical period lies in wait of the proper level of inhibition (Fagiolini and Hensch, 2000). Conversely, critical period onset can be accelerated by prematurely enhancing inhibition by direct infusion of benzodiazepines in immature mice (Fagiolini and Hensch, 2000; Fagiolini *et al.*, 2004; Iwai *et al.*, 2003), as well as transgenic overexpression of brain-derived neurotrophic factor (BDNF) to promote GABAergic maturation (Hanover *et al.*, 1999; Huang *et al.*, 1999).

In the absence of NR2A, the depolarizing action of NMDA currents is prolonged, tipping local circuit equilibrium in favor of excitation (like GAD65 deletion) that disrupts plasticity and is restored by diazepam (Fagiolini *et al.*, 2003). Although it seems counterintuitive from a purely LTP perspective (Feldman, 2000), inhibition is required for plasticity *in vivo* when GABAergic transmission is low or NMDA receptor function is high. Conversely, postsynaptic silencing by either GABA agonist (Hata and Stryker, 1994; Reiter and Stryker, 1988) or NMDA receptor antagonist (Bear *et al.*, 1990) yields a paradoxical loss of open eye input (although LTD is blocked by APV). While maturation of other receptive field properties (e.g., orientation bias) may reflect NR2A signaling pathways more directly (Fagiolini *et al.*, 2003), the yin and yang relationship of excitatory-inhibitory balance is essential for ocular dominance plasticity.

Focus on balanced networks thus offers direct control over the timing of the critical period (Fig. 2), an area in which single-synapse models were wanting. Accelerated plasticity *in vivo* is not predicted by homosynaptic rules, given that diazepam (Wan *et al.*, 2004) or endogenous BDNF blocks LTD induction in the cortex (Jiang *et al.*, 2003). The close interrelationship of GABA, BDNF, and neuronal activity also explains the classic effect of dark-rearing. Raising animals without visual experience from birth naturally reduces GABAergic transmission in the visual cortex (Chen *et al.*, 2001; Morales *et al.*, 2002) and delays the critical period profile into adulthood

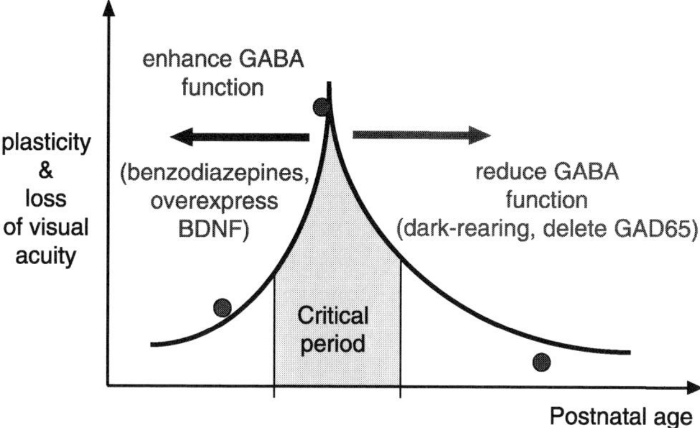

Figure 2 GABAergic control of the critical period. Sensitivity (of spiking response) to monocular deprivation (MD) is restricted to a critical period beginning around 1 week after eye-opening (at P13) and peaking 1 month after birth. Amblyopia as a behavioral consequence is also observed only following MD during the same critical period (red circles). The onset of plasticity can be delayed by directly preventing maturation of GABAergic transmission with gene-targeted deletion of GABA-synthetic enzyme (GAD65) or by dark-rearing from birth (red arrow). Conversely, the critical period may be accelerated by enhancing GABAergic transmission directly with benzodiazepines just after eye-opening or by promoting the rapid maturation of interneurons through excess brain-derived neurotrophic factor (BDNF) expression (blue arrow). See text for references. (See Color Insert.)

(Fagiolini *et al.*, 2003; Iwai *et al.*, 2003; Mower, 1991). Either diazepam infusion (2 days) or BDNF overexpression in complete darkness abolishes the expected delay of plasticity (Gianfranceschi *et al.*, 2003; Iwai *et al.*, 2003). A minimum of 2 days of diazepam treatment at the start of MD that does not need to overlap the deprivation per se is also sufficient to fully activate plasticity in GAD65 knockout mice (Iwai *et al.*, 2003). Thus, tonic signaling through $GABA_A$ receptors rapidly creates a milieu for plasticity within neocortex capable of initiating a critical period for ocular dominance independent of visual experience itself.

IV. Specific GABA Circuits for Plasticity (Large Basket Cells)

Interestingly, not all GABAergic connections are involved in critical period regulation. Several lines of evidence point toward a single class of interneuron with the potential for controlling long-range inhibition and synchrony in visual cortex. Among the large diversity of GABA cells (DeFelipe, 1997; Kawaguchi and Kubota, 1997; Markram *et al.*, 2004), neurochemical markers such as calcium-binding proteins reveal a close correspondence of critical period onset and the emergence of parvalbumin (PV)-positive cells

(Del Rio *et al.*, 1994), both of which are accelerated by BDNF overexpression (Huang *et al.*, 1999). Deletion of a potassium current ($K_v 3.1$) that uniquely regulates the fidelity of fast-spiking behavior (and hence GABA release) specifically from PV-positive interneurons (Erisir *et al.*, 1999; Lien and Jonas, 2003; Rudy and McBain, 2001) slows the rate of ocular dominance plasticity (Matsuda *et al.*, unpublished data).

Although widely heterogeneous, GABA cells in the neocortex are remarkably precise in their connectivity (DeFelipe, 1997; Somogyi *et al.*, 1998). Formed largely through molecular cues then refined by neuronal activity (Chattopadhyaya *et al.*, 2004; Di Cristo *et al.*, 2004), PV-positive contacts include axon-ensheathing Chandelier cells and soma-targeting large basket cells. The latter extend a wide-reaching, horizontal axonal plexus, which can span ocular dominance columns in the cat (Buzas *et al.*, 2001). Immuno-electron microscopy indicates that individual $GABA_A$ receptor α-subunits are trafficked to discrete postsynaptic sites on the pyramidal cell axon, soma, and dendrites. For example, $\alpha 2$-subunits are preferentially enriched at the axon initial segment and basket cell synapses innervated by cholecystokinin (CCK)-positive axon terminals (Klausberger *et al.*, 2002). Importantly, the α-subunits determine benzodiazepine binding through a single amino acid residue in their N terminus (Cherubini and Conti, 2001; Sieghart, 1995). Knock-in of a point mutation at this site renders individual $GABA_A$ receptor subtypes insensitive to benzodiazepines in separate lines of mice (Rudolph *et al.*, 2001).

Weak inhibition within visual cortex early in life (like GAD65 deletion) prevents experience-dependent plasticity (Fagiolini and Hensch, 2000; Iwai *et al.*, 2003). Loss of responsiveness to an eye deprived of vision can be initiated prematurely by enhancing GABA-mediated transmission with zolpidem (Fig. 3), a $GABA_A$ $\alpha 1$, 2, 3-selective ligand (Fagiolini *et al.*, 2004). Systematic use of the mouse knock-in mutation further demonstrates that only one of these subtypes controls the critical period. The $\alpha 1$-containing circuits were found to drive cortical plasticity (Fig. 3), whereas $\alpha 2$-enriched connections separately regulated neuronal firing (Fagiolini *et al.*, 2004). This dissociation carries implications not only for models of brain development, but also for the safe design of benzodiazepines for use in human infants (De Negri *et al.*, 1993).

Indeed, the GABA circuit control of visual cortical plasticity in mice may extend to human brain development. In autopsy samples of visual cortex (Murphy *et al.*, 2005), the maturation of NMDA receptor 2A subunits is complete within the first 9 months. In contrast, GAD65 expression and the $GABA_A$ receptor conversion from $\alpha 3$ to $\alpha 1$ exhibits a slower time constant of several years, consistent with the extended length of the critical period for binocular vision (amblyopia) in humans (Berardi *et al.*, 2000; Daw, 1995). Strikingly, the levels of GAD67 and $\alpha 2$-subunits are constant over the same

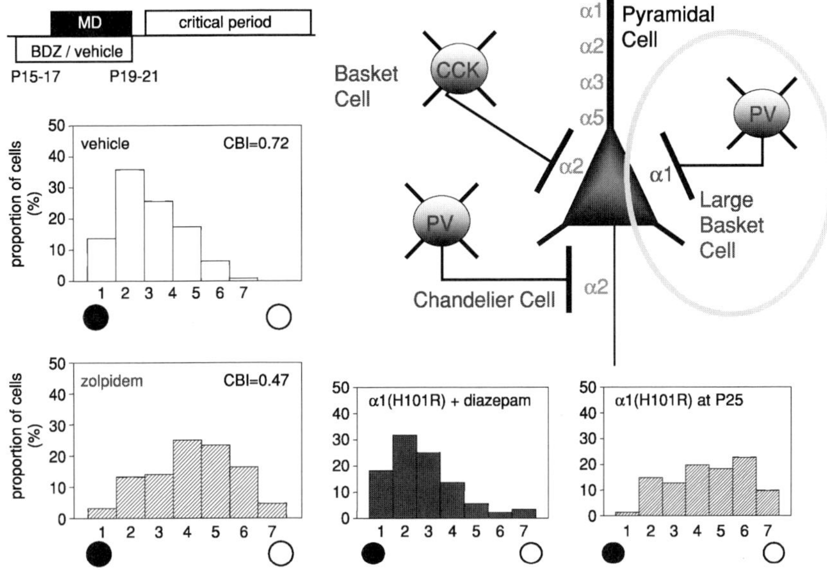

Figure 3 An essential inhibitory subcircuit for critical period plasticity in the visual cortex. Premature ocular dominance plasticity is triggered by the $GABA_A$ receptor $\alpha1$-subunit-selective benzodiazepine agonist zolpidem. Large parvalbumin (PV)-positive basket cells make somatic synapses that utilize $GABA_A$ receptors containing the $\alpha1$-subunit. Knock-in of a point mutation rendering only the $\alpha1$-receptors insensitive to diazepam prevents critical period acceleration by these drugs (red bars). Note that plasticity emerges naturally at the proper time (P25, black bars), since these are still functional GABA receptors. Point mutation of other α-subunits does not interfere with drug-induced premature plasticity. Basket cells extend a wide, horizontal axonal plexus across ocular dominance columns in cats ideally suited for comparing input from the two eyes (Buzas *et al.*, 2001). (See Color Insert.)

early postnatal time period, in agreement with no role in plasticity in animal studies (Fagiolini *et al.*, 2004). Pharmaceutical development of $\alpha2$-selective ligands would avoid the rapid, premature induction of critical period plasticity through $\alpha1$-containing receptors. Moreover, the contribution of kinases to ocular dominance plasticity (traditionally viewed from an LTP perspective) must be re-evaluated with regard to their actions upon $GABA_A$ receptors incorporating the $\alpha1$-subunit (Fischer *et al.*, 2004; Hinkle and Macdonald, 2003).

The $GABA_A$ $\alpha1$-receptors are preferentially sent to receive PV-positive (but not CCK-positive) synapses upon the soma (Klausberger *et al.*, 2002), further implicating these large basket cell circuits. With age, large PV cells are preferentially enwrapped by peri-neuronal nets of extracellular matrix (ECM) molecules and sugars (Härtig *et al.*, 1999). When these are disrupted, peri-somatic inhibition is reduced (Saghatelyan *et al.*, 2001), and MD is once again able to induce ocular dominance shifts even in adulthood (Pizzorusso *et al.*, 2002), perhaps by resetting and tapping its original GABAergic trigger (Fig. 4, left) (Fagiolini and Hensch, 2000). Peri-neuronal nets likely control the extracellular ionic milieu (e.g., potassium/GABA concentration; Härtig *et al.*, 1999) surrounding PV cells to establish their firing efficiency (Erisir *et al.*, 1999; Lien and Jonas, 2003; Rudy and McBain, 2001), or may otherwise sequester molecular regulators of PV cell maturation.

V. From Functional to Structural Rewiring (Extracellular Matrix)

The ECM is increasingly emerging as a major site for critical period plasticity (Berardi *et al.*, 2004). Sensory experience physically rewires the brain in early postnatal life through unknown mechanisms. To convert physiological events (altered input) into structural refinements, connections must

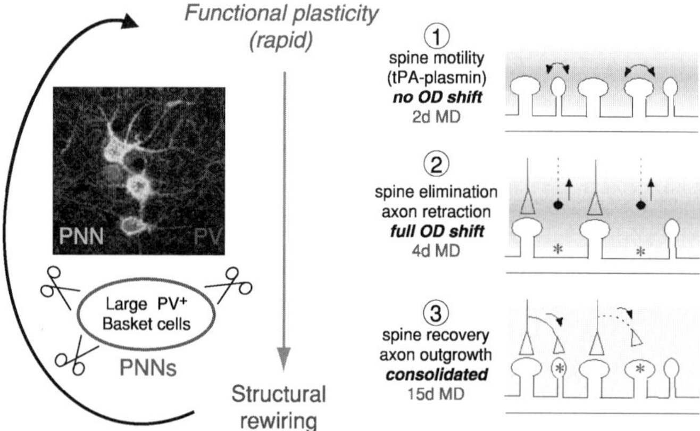

Figure 4 Structural consolidation during the critical period. The structural events that link the functional detection of imbalanced sensory input by GABAergic circuits to anatomical rewiring. A three-step process of increased spine motility (Oray *et al.*, 2004), transient elimination, then regrowth (Mataga *et al.*, 2004) is mediated by a biochemical increase of proteolytic activity (tPA-plasmin; gray background) between 2 and 7 days of monocular deprivation (MD) (Mataga *et al.*, 2002). Spine pruning is the first anatomical correlate of the rapid physiological shifts in ocular dominance (OD) by brief MD. Taking this structural view, plasticity is successfully restored to adult visual cortex only by loosening up the extracellular matrix (ECM) by infusion of chondroitinases (left; Pizzorusso *et al.*, 2002). Interestingly, this treatment (unlike tPA) degrades peri-neuronal net (PNN) structures, which preferentially enwrap the large PV-positive basket cells believed to be the endogenous trigger for the critical period (see text). (See Color Insert.)

ultimately be broken and neuronal wiring rerouted. Proteases are ideally suited to clear the way for growing neurites (Liu *et al.*, 1994). Tissue-type plasminogen activator (tPA) is the major serine protease in the postnatal mammalian brain (Shiosaka and Yoshida, 2000). Originally identified as an immediate early gene upon hippocampal seizures (Qian *et al.*, 1993), tPA activity is gradually upregulated in visual cortex by 2 days of MD during the critical period (gray background, Fig. 4, right), but not in adults or GAD65 knockout mice (Mataga *et al.*, 2002). Conversely, a minimum of 2 days of diazepam treatment is required to rescue plasticity in the absence of GAD65 (Iwai *et al.*, 2003). Functional ocular dominance plasticity is impaired when tPA action is blocked and is rescued by exogenous tPA (but not diazepam) (Mataga *et al.*, 2002; Müller and Griesinger, 1998).

Permissive amounts of tPA may, thus, couple functional changes to structural changes downstream of the excitatory-inhibitory balance that triggers visual cortical plasticity. Second messenger systems recruited in the process (reviewed in Berardi *et al.*, 2003) lie satisfyingly along a molecular cascade linking neuronal activity to tPA release (Fig. 5) (Hensch, 2004), the structural consequences of which have recently been clarified. Occluding an eye of vision during development classically trims that input to the neocortex, while thalamic axons serving the open eye progressively expand (Antonini and Stryker, 1996; Antonini *et al.*, 1999). Yet, this process is far too slow to explain the rapid shift of ocular dominance within days of MD (Gordon and Stryker, 1996; Silver and Stryker, 1999). The most immediate and potent cortical plasticity occurs beyond thalamo-recipient layer 4, for which the structural basis remains obscure (Gordon and Stryker, 1996; Trachtenberg *et al.*, 2000). Morphological plasticity is initiated along the apical dendrites of target pyramidal cells in the cerebral cortex, where spines serve as pleiomorphic sites of excitatory synaptic connections (Yuste and Bonhoeffer, 2001).

Spine shape is highly dynamic when viewed by two-photon laser scanning microscopy in living transgenic mice expressing green fluorescent protein (GFP) in a subset of layer 5 cells. Motility of spines decreases with age in the visual cortex (Grutzendler *et al.*, 2002), but can be transiently elevated by 2-day MD only during the critical period (step 1, Fig. 4, right) (Oray *et al.*, 2004). This occludes the motility triggered by direct protease application to brain slices, indicating that tPA-plasmin may be the endogenous mediator. Increased proteolysis after 2-day MD will degrade ECM and cell adhesion proteins before any ocular dominance shift is detectable. Even along the same apical dendrite (Oray *et al.*, 2004), spines are first set in motion by brief MD only in layers 2, 3, and 5, consistent with early extragranular changes instructing later events in layer 4 (Trachtenberg *et al.*, 2000).

The robust anatomical consequence of 4-day MD in layer 2/3 of visual cortex that corresponds to full, functional loss of responsiveness is spine pruning (Mataga *et al.*, 2004). Protrusions on the apical dendrite of

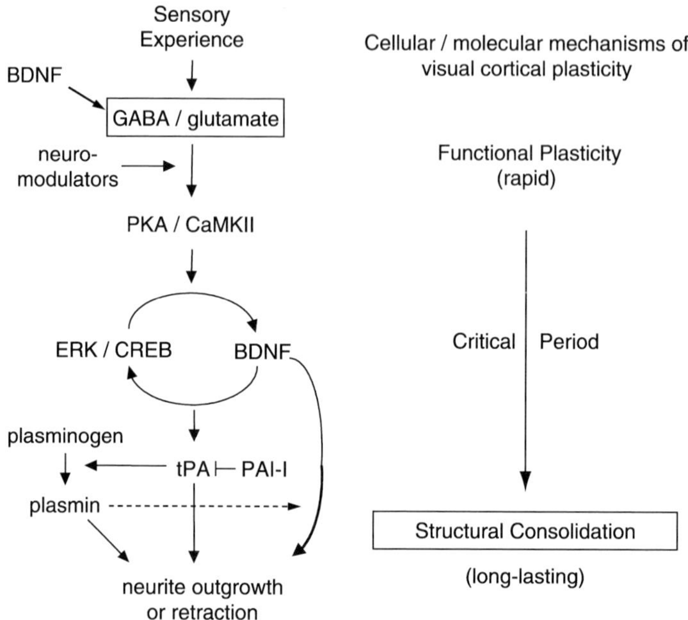

Figure 5 Molecular mechanisms of visual cortical plasticity. Many candidate plasticity factors have been screened in the monocular deprivation paradigm by pharmacology in kittens or gene-targeted disruption in mice (see Berardi *et al.*, 2003). Only a handful of second messenger molecules have been found to play a direct role in plasticity without perturbing global neuronal activity, including protein kinase A (PKA), calcium/calmodulin-dependent protein kinase II (CaMKII), extracellularly regulated kinase (ERK), cyclic AMP response element binding protein (CREB) (Mower *et al.*, 2002; Pham *et al.*, 1999), protein synthesis (Taha and Stryker, 2002), and the plasmin system (tPA-plasmin) regulated by its inhibitors (PAI-1) (Mataga *et al.*, 2002, 2004). Brain-derived neurotrophic factor (BDNF) plays an early role to establish the GABA cells that will later discriminate competing sensory inputs to trigger the critical period (Huang and Reichardt, 2001; Huang *et al.*, 1999). Mature BDNF, produced from the cleavage of pro-BDNF by plasmin (Pang *et al.*, 2004), in turn stimulates expression and release of more tPA (Fiumelli *et al.*, 1999). Both tPA and BDNF can then contribute to the final anatomical rewiring of the cortical circuit by promoting neurite growth through the extracellular matrix (ECM). Plasticity may end when permissive factors are gradually lost or when further growth is actively suppressed by late-emerging inhibitory factors in the ECM (Berardi *et al.*, 2004; Schoop *et al.*, 1997). Gene expression analyses support the view that the critical period offers a unique molecular milieu for plasticity (Ossipow *et al.*, 2004; Prasad *et al.*, 2002), consistent with dendritic spine and axonal rearrangement being limited to this time in life (Antonini and Stryker, 1996; Mataga *et al.*, 2004; Oray *et al.*, 2004).

pyramidal cells increase steadily in number with postnatal age, but are rapidly lost after MD only during the physiological critical period (step 2, Fig. 4, right). Importantly, spine density is not decreased by brief MD in tPA or GAD65 knockout mice, but can be made to decrease pharmacologically with exogenous tPA or diazepam infusion, respectively (Mataga *et al.*, 2004).

Moreover, pruning faithfully reflects competitive interactions between the two eyes, as it fails to occur when both eyes are closed (Majewska and Sur, 2003) or in the adjacent monocular segment receiving input only from the contralateral eye (Mataga *et al.*, 2004). Deletion of the tPA substrate plasminogen mimics the impaired ocular dominance plasticity observed in tPA knockout mice with brief MD. The tPA-plasmin axis may, thus, mediate rapid structural rearrangement underlying experience-dependent plasticity on a timescale (several days) that is more consistent with plasticity *in vivo* than LTP/LTD models (Heynen *et al.*, 2003).

After this postsynaptic pruning, deprived-eye afferents first retract before axonal arbors serving the open eye migrate to spaces cleared away by tPA-plasmin along the dendrite (asterisks, Fig. 4). Ultimately, territory representing the open eye is expanded (step 3, Fig. 4, right). As axons serving the open eye grow, spines emerge to meet them, and spine density largely recovers after prolonged MD (Mataga *et al.*, 2004). Thus, competition detected by an appropriate excitatory-inhibitory balance may gradually be converted into structural changes through a multistep proteolytic action of the secreted tPA-plasmin cascade. This structural model considers an extracellular locus of competition quite distinct from intracellular mechanisms of LTP/LTD. Axons and dendritic spines may be exposed to a permissive growth environment in an activity-dependent manner (Dityatev and Schachner, 2003; Mataga *et al.*, 2002, 2004).

The source and dynamics of tPA-plasmin release in the brain remain unclear due to the lack of specific reagents. Laminar motility of spines (Oray *et al.*, 2004) and their rapid pruning (Mataga *et al.*, 2004) by brief MD could reflect calcium-dependent secretion of proteases (Gualandris *et al.*, 1996; Parmer *et al.*, 1997) from axons of fast-spiking cells themselves, in which PV is a major contributor to presynaptic calcium signals and synaptic integration (Collin *et al.*, 2005). This may explain why spines nearest the soma of layer 2/3 pyramidal cells are most robustly lost (Mataga *et al.*, 2004), as they lie nearest the PV-cell-rich layer (Del Rio *et al.*, 1994). How a competitive outcome arises by uniformly bathing dendrites in proteases also needs to be considered. Cell adhesion and ECM molecules may become insensitive to proteases during high levels of activity (Murase *et al.*, 2002; Tanaka *et al.*, 2000). Less-active synapses would further release fewer endogenous protease inhibitors (Wannier-Morino *et al.*, 2003), tilting the overall balance nearby toward pruning.

VI. Normal Columnar Development

The segregation of columns by normal vision during the critical period is believed to result from the same activity-dependent rules acting upon an initially overlapping continuum of thalamic afferents. This dogma has recently been challenged by the finding that single thalamic arbors may in part be clustered well before the critical period (Katz and Crowley, 2002). If molecular cues were to establish columnar architecture, a substantial genetic similarity of maps should emerge among siblings, for which there is now some evidence (Kaschube *et al.*, 2002). A significant role for sensory experience is nevertheless predicted to individualize these ocular dominance maps during the critical period.

Even the focal deprivation produced by shadows of blood vessels within an individual eye is embossed as an image of the retinal vasculature onto primary visual cortex (Adams and Horton, 2002). In computational models of self-organization, it is the recipient cortical circuits that largely determine the final spacing of columns (Miller *et al.*, 1989; Willshaw and von der Malsburg, 1976). Overlapping inputs segregate into clusters as "neurons that fire together wire together" through a neocortical organization that spreads excitation locally but is limited at a distance by farther-reaching inhibition. On theoretical grounds, homosynaptic rules of excitatory synaptic plasticity alone are insufficient to produce a competitive outcome (Miller, 1996), requiring other complex mechanisms such as sliding thresholds or metaplasticity. Lateral inhibitory interactions provide a straightforward scaffolding with which to discriminate individual sensory inputs.

By adjusting the canonical "Mexican hat" profile of intracortical activation during development (Fig. 6), lateral inhibition in particular can establish narrow or wide columns *in silico* (Miller *et al.*, 1989). These long-standing theoretical predictions have recently been validated *in vivo* through the direct infusion of benzodiazepines during the critical period in kitten visual cortex (Hensch and Stryker, 2004). Such drugs come in three varieties, including agonists such as diazepam (valium), inverse agonists such as the β-carbolines (e.g., DMCM), and antagonists that block the actions of both (Sieghart, 1995). All are known to act on particular $GABA_A$ receptor subtypes with the opposite effect on chloride flux. Enhancing inhibition with benzodiazepine agonists throughout the critical period leads to a 30% increase in column width, while inverse agonists produce column shrinkage (Fig. 6) (Hensch and Stryker, 2004).

Bidirectional control of columnar architecture is unprecedented and simulated in computer models when long-range (rather than local) inhibition is preferentially altered. Interestingly, increased column spacing is also observed with strabismus following exotropic deviation of the eyes during the critical period (Löwel, 1994). Both enhanced lateral inhibition by direct intracortical infusion of diazepam and decorrelation of visual input by artificial squint are conditions that favor the maximal segregation of ocular dominance. Taken together, local imbalances in neuronal activity influence

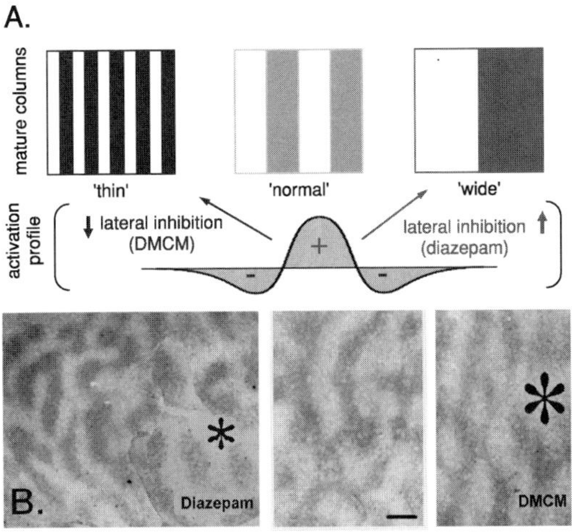

Figure 6 Local circuit control of developing columnar architecture. Activity-dependent models of segregation predict a role for cortical GABAergic circuits in determining final column spacing from an initially overlapping mosaic of afferents (Miller *et al.*, 1989). (A) Neuronal activity from thalamic input serving the right or left eye is spread by local excitatory connections (red cell) within the neocortex but inhibited at farther distances (blue cell). When this "Mexican hat" activation profile is modulated during development by enhancing or reducing horizontal, long-range inhibition preferentially (Hensch and Stryker, 2004), columns emerge that are wider or thinner than normal, respectively. (B) This hypothesis was verified *in vivo* by modulating $GABA_A$ currents with benzodiazepine agonists (diazepam) or inverse agonists (DMCM) throughout the critical period (Hensch and Stryker, 2004). Asterisks, infusion sites; control, middle panel. Scale bar: 1 mm. (See Color Insert.)

column formation during normal development, which cannot be explained solely by genetic instruction.

It is paradoxical to think how inhibition might shape plasticity in the developing brain. Powerful, fast somatic inhibition can edit action potentials that may back-propagate into the dendritic arbor. Spike-timing-dependent models of synaptic plasticity rely upon a precise millisecond time window for such postsynaptic spikes to meet presynaptic input (Bi and Poo, 2001; Song *et al.*, 2000). Sloppy gating by weak inhibition at the soma (DMCM) would reduce competition by allowing spurious coincident activity of overlapping inputs. Contrast enhancement by feed-forward GABA circuits (diazepam) would instead sharpen the edges of emerging columnar borders by suppressing back-propagation of unwanted spikes (Pouille and Scanziani, 2001). This simple circuit relieves the burden of discriminating competitors by homosynaptic mechanisms alone. It entrusts the wide-reaching axons of basket cells receiving input from one eye to inhibit targets of the other eye (Buzas *et al.*, 2001), and thus to sculpt cortical architecture.

VII. Critical Period Reactivation

The critical period, in general, is a time period when the best neural representation of the world is selected from among the many competing inputs that bombard the maturing nervous system. The growth and function of lateral inhibitory circuits offer a rational, cellular substrate that can now be compared and modeled across regions to gain broader insight into brain development and its disorders (Möhler *et al.*, 2004; Rubenstein and Merzenich, 2003).

Critical period closure may reflect sequential locks placed on the molecular pathway as it flows from mature GABA function toward structural consolidation (Fig. 5). In fact, subthreshold, synaptic plasticity is still possible after MD in adulthood (Pham *et al.*, 2004; Sawtell *et al.*, 2003; Tagawa *et al.*, 2005) but has no further impact on spiking output from primary visual cortex or on visuo-spatial acuity (Fagiolini and Hensch, 2000; Gordon and Stryker, 1996; Prusky and Douglas, 2003). To fully reactivate plasticity, it may be necessary to reset the entire cascade from its GABAergic trigger onward (Pizzorusso *et al.*, 2002). Interestingly, cortical lesions or retinal scotomas reconfigure local circuit excitation-inhibition to an immature state (Arckens *et al.*, 2000). This rationalizes the administration of diazepam after acute stroke (to reduce excitotoxicity), which is also used to aid in recovery by triggering plasticity (Lodder *et al.*, 2000).

Conversely, in the somatosensory cortex GABA circuits are formed and reorganized throughout life (Fuchs and Salazar, 1998; Knott *et al.*, 2002; Welker *et al.*, 1989) and are associated with lifelong plasticity (Diamond *et al.*, 1993; Wang *et al.*, 1995). This is also the case in the mammalian olfactory system, where constant neurogenesis is responsible for odor discrimination underlying memory in adulthood (Cecchi *et al.*, 2001; Gheusi *et al.*, 2000). These newly born cells are GABAergic granule cells, whose dual functions include lateral inhibition and synchronization of neuronal activity (Lagier *et al.*, 2004; Yokoi *et al.*, 1995). Would an olfactory critical period emerge in the absence of neurogenesis, or can visual plasticity be maintained at juvenile levels by prolonging cell proliferation in the neocortex? Alternatively, the loss of factors permissive for growth (Mataga *et al.*, 2004) and/or the emergence of active inhibitors of growth (myelin) may terminate structural plasticity (Daw, 1995; Schoop *et al.*, 1997). The true test will be a reliable cure for amblyopia in adulthood.

VIII. Summary

Neuronal circuits in the brain are shaped by experience during "critical periods" of early postnatal life. Surprisingly, it is the functional maturation of local inhibitory connections that triggers this classical activity-dependent development in primary visual cortex. Among the large diversity of interneurons, a late-developing subset employing specific $GABA_A$ receptors and widespread axons drives plasticity *in vivo* before becoming ensheathed by peri-neuronal nets in adulthood. Ultimately, structural consolidation of competing sensory input is mediated by a proteolytic reorganization of the extracellular matrix only during the critical period. Its reactivation and recovery of impaired function (amblyopia) can now be based on realistic circuit models and may generalize across systems.

References

Adams, D. L., and Horton, J. C. (2002). Shadows cast by retinal blood vessels mapped in primary visual cortex. *Science* **298**, 572–576.

Antonini, A., and Stryker, M. P. (1996). Plasticity of geniculocortical afferents following brief or prolonged monocular occlusion in the cat. *J. Comp. Neurol.* **369**, 64–82.

Antonini, A., Fagiolini, M., and Stryker, M. P. (1999). Anatomical correlates of functional plasticity in mouse visual cortex. *J. Neurosci.* **19**, 4388–4406.

Arckens, L., Schweigart, G., Qu, Y., Wouters, G., Pow, D. V., Vandesande, F., Eysel, U. T., and Orban, G. A. (2000). Cooperative changes in GABA, glutamate and activity levels: The missing link in cortical plasticity. *Eur. J. Neurosci.* **12**, 4222–4232.

Barry, M. F., and Ziff, E. B. (2002). Receptor trafficking and the plasticity of excitatory synapses. *Curr. Opin. Neurobiol.* **12**, 279–286.

Bartoletti, A., *et al.* (2002). Heterozygous knock-out mice for brain-derived neurotrophic factor show a pathway-specific impairment of long-term potentiation but normal critical period for monocular deprivation. *J. Neurosci.* **22**, 10072–10077.

Bear, M. F., Kleinschmidt, A., Gu, Q. A., and Singer, W. (1990). Disruption of experience-dependent synaptic modifications in striate cortex by infusion of an NMDA receptor antagonist. *J. Neurosci.* **10**, 909–925.

Berardi, N., Pizzorusso, T., and Maffei, L. (2000). Critical periods during sensory development. *Curr. Opin. Neurobiol.* **10**, 138–145.

Berardi, N., Pizzorusso, T., Ratto, G. M., and Maffei, L. (2003). Molecular basis of plasticity in the visual cortex. *Trends Neurosci.* **26**, 369–378.

Berardi, N., Pizzorusso, T., and Maffei, L. (2004). Extracellular matrix and visual cortical plasticity; freeing the synapse. *Neuron* **44**, 905–908.

Bi, G., and Poo, M. (2001). Synaptic modification by correlated activity: Hebb's postulate revisited. *Annu. Rev. Neurosci.* **24**, 139–166.

Buzas, P., Eysel, U. T., Adorjan, P., and Kisvarday, Z. F. (2001). Axonal topography of cortical basket cells in relation to orientation, direction, and ocular dominance maps. *J. Comp. Neurol.* **437**, 259–285.

Cecchi, G. A., Petreanu, L. T., Alvarez-Buylla, A., and Magnasco, M. O. (2001). Unsupervised learning and adaptation in a model of adult neurogenesis. *J. Comput. Neurosci.* **11**, 175–182.

Chattopadhyaya, B., *et al.* (2004). Experience and activity-dependent maturation of perisomatic GABAergic innervation in primary visual cortex during a postnatal critical period. *J. Neurosci.* **24**, 9598–9611.

Chen, L., Yang, C., and Mower, G. D. (2001). Developmental changes in the expression of GABA(A) receptor subunits (alpha(1), alpha(2), alpha(3)) in the cat visual cortex and the effects of dark rearing. *Mol. Brain Res.* **88**, 135–143.

Cherubini, E., and Conti, F. (2001). Generating diversity at GABAergic synapses. *Trends Neurosci.* **24,** 155–162.

Collin, T., Chat, M., Lucas, M. G., Moreno, H., Racay, P., Schwaller, B., Marty, A., and Llano, I. (2005). Developmental changes in parvalbumin regulate presynaptic Ca^{2+} signaling. *J. Neurosci.* **25,** 96–107.

Daw, N. (1995). Mechanisms of plasticity in the visual cortex. *In* "The Visual Neurosciences" (L. Chalupa and J. S. Werner, Ed.), Vol. 1, pp. 126–145. MIT Press, Cambridge, MA.

Daw, N. (2004). "Visual Development." Plenum, New York.

Daw, N., Rao, Y., Wang, X. F., Fischer, Q., and Yang, Y. (2004). LTP and LTD vary with layer in rodent visual cortex. *Vision Res.* **44,** 3377–3380.

De Negri, M., Baglietto, M. G., and Biancheri, R. (1993). Electrical status epilepticus in childhood: Treatment with short cycles of high dosage benzodiazepine. *Brain Dev.* **15,** 311–312.

DeFelipe, J. (1997). Types of neurons, synaptic connections and chemical characteristics of cells immunoreactive for calbindin-D28K, parvalbumin and calretinin in the neocortex. *J. Chem. Neuroanat.* **14,** 1–19.

Del Rio, J. A., De Lecea, L., Ferrer, I., and Soriano, E. (1994). The development of parvalbumin-immunoreactivity in the neocortex of the mouse. *Dev. Brain Res.* **81,** 247–259.

Desai, N. S., Cudmore, R. H., Nelson, S. B., and Turrigiano, G. G. (2002). Critical periods for experience-dependent synaptic scaling in visual cortex. *Nat. Neurosci.* **5,** 783–789.

Di Cristo, G., Wu, C., Chattopadhyaya, B., Ango, F., Knott, G., Welker, E., Sroboda, K., and Huang, Z. J. (2004). Subcellular domain-restricted GABAergic innervation in primary visual cortex in the absence of sensory and thalamic inputs. *Nat. Neurosci.* **7,** 1184–1186.

Diamond, M. E., Armstrong-James, M., and Ebner, F. F. (1993). Experience-dependent plasticity in adult rat barrel cortex. *Proc. Natl. Acad. Sci. USA* **90,** 2082–2086.

Dityatev, A., and Schachner, M. (2003). Extracellular matrix molecules and synaptic plasticity. *Nat. Rev. Neurosci.* **4,** 456–468.

Erisir, A., Lau, D., Rudy, B., and Leonard, C. S. (1999). Function of specific K(+) channels in sustained high-frequency firing of fast-spiking neocortical interneurons. *J. Neurophysiol.* **82,** 2476–2489.

Fagiolini, M., and Hensch, T. K. (2000). Inhibitory threshold for critical-period activation in primary visual cortex. *Nature* **404,** 183–186.

Fagiolini, M., Katagiri, H., Míyamoto, H., Mori, H., Grant, S. G., Mishina, M., and Hensch, T. K. (2003). Separable features of visual cortical plasticity revealed by N-methyl-D-aspartate receptor 2A signaling. *Proc. Natl. Acad. Sci. USA* **100,** 2854–2859.

Fagiolini, M., Fritschy, J. M., Low, K., Mohler, H., Rudolf, U., and Hensch, T. K. (2004). Specific $GABA_A$ circuits for visual cortical plasticity. *Science* **303,** 1681–1683.

Feldman, D. E. (2000). Inhibition and plasticity. *Nat. Neurosci.* **3,** 303–304.

Fischer, Q. S., Beaver, C. J., Yang, Y., Rao, Y., Jakobsdottir, K. B., Storm, D. R., McKnight, G. S., and Daw, N. W. (2004). Requirement for the $RII\beta$ isoform of PKA, but not calcium-stimulated adenylyl cyclase, in visual cortical plasticity. *J. Neurosci.* **24,** 9049–9058.

Fiumelli, H., Jabaudon, D., Magistretti, P. J., and Martin, J.-L. (1999). BDNF stimulates expression, activity and release of tissue-type plasminogen activator in mouse cortical neurons. *Eur. J. Neurosci.* **11,** 1639–1646.

Frenkel, M. Y., and Bear, M. F. (2004). How monocular deprivation shifts ocular dominance in visual cortex of young mice. *Neuron* **44,** 917–923.

Frey, U., Huang, Y. Y., and Kandel, E. R. (1993). Effects of cAMP simulate a late-stage of LTP in hippocampal CA1 neurons. *Science* **260,** 1661–1664.

Fuchs, J. L., and Salazar, E. (1998). Effects of whisker trimming on GABA(A) receptor binding in the barrel cortex of developing and adult rats. *J. Comp. Neurol.* **395,** 209–216.

Gheusi, G., Cremer, H., McLean, H., Chazal, G., Vincent, J. D., and Lledo, P.-M. (2000). Importance of newly generated neurons in the adult olfactory bulb for odor discrimination. *Proc. Natl. Acad. Sci. USA* **97**, 1823–1828.

Gianfranceschi, L., *et al.* (2003). Visual cortex is rescued from the effects of dark rearing by overexpression of BDNF. *Proc. Natl. Acad. Sci. USA* **100**, 12486–12491.

Gordon, J. A., and Stryker, M. P. (1996). Experience-dependent plasticity of binocular responses in the primary visual cortex of the mouse. *J. Neurosci.* **16**, 3274–3286.

Grutzendler, J., Kasthuri, N., and Gan, W. B. (2002). Long-term dendritic spine stability in the adult cortex. *Nature* **420**, 812–816.

Gualandris, A., Jones, T. E., Strickland, S., and Tsirka, S. E. (1996). Membrane depolarization induces calcium-dependent secretion of tissue plasminogen activator. *J. Neurosci.* **16**, 2220–2225.

Hanover, J. L., Huang, Z. J., Tonegawa, S., and Stryker, M. P. (1999). Brain-derived neurotrophic factor overexpression induces precocious critical period in mouse visual cortex. *J. Neurosci.* **19**, RC40.

Härtig, W., Deroniche, A., Welt, K., Broner, K., Grosche, J., Mader, M., Reichenbach, A., and Bruckner, G. (1999). Cortical neurons immunoreactive for the potassium channel $K_v3.1b$ subunit are predominantly surrounded by perineuronal nets presumed as a buffering system for cations. *Brain Res.* **842**, 15–29.

Hata, Y., and Stryker, M. P. (1994). Control of thalamocortical afferent rearrangement by postsynaptic activity in developing visual cortex. *Science* **265**, 1732–1735.

Hensch, T. K. (2003). Controlling the critical period. *Neurosci. Res.* **47**, 17–22.

Hensch, T. K. (2004). Critical period regulation. *Annu. Rev. Neurosci.* **27**, 549–579.

Hensch, T. K., and Fagiolini, M. (Eds.) (2004). "Excitatory-Inhibitory Balance: Synapses, Circuits, Systems." Kluwer/Plenum, New York.

Hensch, T. K., and Stryker, M. P. (2004). Columnar architecture sculpted by GABA circuits in developing cat visual cortex. *Science* **303**, 1678–1681.

Hensch, T. K., Fagiolini, M., Mataga, N., Stryker, M. P., Baekkeskov, S., and Kash, S. F. (1998). Local GABA circuit control of experience-dependent plasticity in developing visual cortex. *Science* **282**, 1504–1508.

Heynen, A. J., Yoon, B. J., Liu, C. H., Chung, H. J., Hunganir, R. L., and Bear, M. F. (2003). Molecular mechanism for loss of visual cortical responsiveness following brief monocular deprivation. *Nat. Neurosci.* **6**, 854–862.

Hinkle, D. J., and Macdonald, R. L. (2003). Beta subunit phosphorylation selectively increases fast desensitization and prolongs deactivation of $\alpha1\beta1\gamma2L$ and $\alpha1\beta3\gamma2L$ GABA(A) receptor currents. *J. Neurosci.* **23**, 11698–11710.

Huang, E. J., and Reichardt, L. F. (2001). Neurotrophins: Roles in neuronal development and function. *Annu. Rev. Neurosci.* **24**, 677–736.

Huang, Z. J., Kirkwood, A., Pizzorusso, T., Porcïatti, V., Morales, B., Bear, M. F., Mattei, L., and Tonegawa, S. (1999). BDNF regulates the maturation of inhibition and the critical period of plasticity in mouse visual cortex. *Cell* **98**, 739–755.

Hubel, D. H., Wiesel, T. N., and Le Vay, S. (1976). Functional architecture of area 17 in normal and monocularly deprived macaque monkeys. *Cold Spring Harb. Symp. Quant. Biol.* **40**, 581–589.

Iwai, Y., Fagiolini, M., Obata, K., and Hensch, T. K. (2003). Rapid critical period induction by tonic inhibition in visual cortex. *J. Neurosci.* **23**, 6695–6702.

Jiang, B., Akaneya, Y., Hata, Y., and Tsumoto, T. (2003). Long-term depression is not induced by low-frequency stimulation in rat visual cortex *in vivo*: A possible preventing role of endogenous brain-derived neurotrophic factor. *J. Neurosci.* **23**, 3761–3770.

Kaschube, M., Wolf, F., Geisel, T., and Lowel, S. (2002). Genetic influence on quantitative features of neocortical architecture. *J. Neurosci.* **22**, 7206–7217.

Katz, L. C., and Crowley, J. C. (2002). Development of cortical circuits: Lessons from ocular dominance columns. *Nat. Rev. Neurosci.* **3**, 34–42.

Kawaguchi, Y., and Kubota, Y. (1997). GABAergic cell subtypes and their synaptic connections in rat frontal cortex. *Cereb. Cortex* **7**, 476–486.

Kirkwood, A., Dudek, S. M., Gold, J. T., Aizenman, C. D., and Bear, M. F. (1993). Common forms of synaptic plasticity in the hippocampus and neocortex in vitro. *Science* **260**, 1518–1521.

Kirkwood, A., Lee, H. K., and Bear, M. F. (1995). Co-regulation of long-term potentiation and experience-dependent synaptic plasticity in visual cortex by age and experience. *Nature* **375**, 328–331.

Klausberger, T., Roberts, J. D., and Somogyi, P. (2002). Cell type- and input-specific differences in the number and subtypes of synaptic GABA(A) receptors in the hippocampus. *J. Neurosci.* **22**, 2513–2521.

Knott, G. W., Quairiaux, C., Genoud, C., and Welker, E. (2002). Formation of dendritic spines with GABAergic synapses induced by whisker stimulation in adult mice. *Neuron* **34**, 265–273.

Lagier, S., Carleton, A., and Lledo, P. M. (2004). Interplay between local GABAergic interneurons and relay neurons generates gamma oscillations in the rat olfactory bulb. *J. Neurosci.* **24**, 4382–4392.

Lien, C. C., and Jonas, P. (2003). K$_v$3 potassium conductance is necessary and kinetically optimized for high-frequency action potential generation in hippocampal interneurons. *J. Neurosci.* **23**, 2058–2068.

Liu, G. (2004). Local structural balance and functional interaction of excitatory and inhibitory synapses in hippocampal dendrites. *Nat. Neurosci.* **7**, 373–379.

Liu, L., Wong, T. P., Pozza, M. F., Lingenhoel, K., Wang, Y., Sheng, M., Auberson, Y. P., and Wang, Y. T. (2004). Role of NMDA receptor subtypes in governing the direction of hippocampal synaptic plasticity. *Science* **304**, 1021–1024.

Liu, Y., Fields, R. D., Fitzgerald, S., Festoff, B. W., and Nelson, P. G. (1994). Proteolytic activity, synapse elimination, and the Hebb synapse. *J. Neurobiol.* **25**, 325–335.

Lodder, J., Luijckx, G., van Raak, L., and Kessels, F. (2000). Diazepam treatment to increase the cerebral GABAergic activity in acute stroke: A feasibility study in 104 patients. *Cerebrovasc. Dis.* **10**, 437–440.

Löwel, S. (1994). Ocular dominance column development: Strabismus changes the spacing of adjacent columns in cat visual cortex. *J. Neurosci.* **14**, 7451–7468.

Majewska, A., and Sur, M. (2003). Motility of dendritic spines in visual cortex in vivo: Changes during the critical period and effects of visual deprivation. *Proc. Natl. Acad. Sci. USA* **100**, 16024–16029.

Markram, H., Toledo-Rodriguez, M., Wang, Y., Gupta, A., Silberberg, G., and Wu, C. (2004). Interneurons of the neocortical inhibitory system. *Nat. Rev. Neurosci.* **5**, 793–807.

Mataga, N., Fujishima, S., Condie, B. G., and Hensch, T. K. (2001). Experience-dependent plasticity of mouse visual cortex in the absence of the neuronal activity-dependent marker egr1/zif268. *J. Neurosci.* **21**, 9724–9732.

Mataga, N., Mizuguchi, Y., and Hensch, T. K. (2004). Experience-dependent pruning of dendritic spines in visual cortex by tissue plasminogen activator. *Neuron* **44**, 1031–1041.

Mataga, N., Nagai, N., and Hensch, T. K. (2002). Permissive proteolytic activity for visual cortical plasticity. *Proc. Natl. Acad. Sci. USA* **99**, 7717–7721.

Miller, K. D. (1996). Synaptic economics: Competition and cooperation in synaptic plasticity. *Neuron* **17**, 371–374.

Miller, K. D., Keller, J. B., and Stryker, M. P. (1989). Ocular dominance column development: Analysis and simulation. *Science* **245**, 605–615.

Möhler, H., Fritschy, J. M., Crestani, F., Hensch, T., and Rudolph, U. (2004). Specific GABA (A) circuits in brain development and therapy. *Biochem. Pharmacol.* **68,** 1685–1690.

Morales, B., Choi, S. Y., and Kirkwood, A. (2002). Dark rearing alters the development of GABAergic transmission in visual cortex. *J. Neurosci.* **22,** 8084–8090.

Mower, A. F., Liao, D. S., Nestler, E. J., Neve, R. L., and Ramoa, A. S. (2002). cAMP/Ca2+ response element-binding protein function is essential for ocular dominance plasticity. *J. Neurosci.* **22,** 2237–2245.

Mower, G. D. (1991). The effect of dark rearing on the time course of the critical period in cat visual cortex. *Dev. Brain Res.* **58,** 151–158.

Murase, S., Mosser, E., and Schuman, E. M. (2002). Depolarization drives β-catenin into neuronal spines promoting changes in synaptic structure and function. *Neuron* **35,** 91–105.

Murphy, K. M., Beston, B. R., Boley, P. M., and Jones, D. G. (2005). Development of human visual cortex: A balance between excitatory and inhibitory plasticity mechanisms. *Dev. Psychobiol.* **46,** 209–221.

Müller, C. M., and Griesinger, C. B. (1998). Tissue plasminogen activator mediates reverse occlusion plasticity in visual cortex. *Nat. Neurosci.* **1,** 47–53.

Nase, G., Weishaupt, J., Stern, P., Singer, W., and Monyer, H. (1999). Genetic and epigenetic regulation of NMDA receptor expression in the rat visual cortex. *Eur. J. Neurosci.* **11,** 4320–4326.

Oray, S., Majewska, A., and Sur, M. (2004). Dendritic spine dynamics are regulated by monocular deprivation and extracellular matrix degradation. *Neuron* **44,** 1021–1030.

Ossipow, V., Pellissier, F., Schaad, O., and Ballivet, M. (2004). Gene expression analysis of the critical period in the visual cortex. *Mol. Cell Neurosci.* **27,** 70–83.

Pang, P., Teng, H. K., Zaitsev, E., Woo, N. T., Sakata, K., Zhen, S., Teng, K. K., Yung, W. H., Hempstead, B. L., and Lu, B. (2004). Cleavage of ProBDNF by tPA/plasmin is essential for long-term hippocampal plasticity. *Science* **306,** 487–491.

Parmer, R. J., Mahata, M., Mahata, S., Sebald, M. T., O'Conner, D. T., and Miles, L. A. (1997). Tissue plasminogen activator (tPA) is targeted to the regulated secretory pathway. *J. Biol. Chem.* **272,** 1976–1982.

Pham, T. A., Impey, S., Storm, D. R., and Stryker, M. P. (1999). CRE-mediated gene transcription in neocortical neuronal plasticity during the developmental critical period. *Neuron* **22,** 63–72.

Pham, T. A., Graham, S. J., Suzuki, S., Barco, A., Kandel, E. R., Gordon, B., and Lickey, M. E. (2004). A semi-persistent adult ocular dominance plasticity in visual cortex is stabilized by activated CRE. B. *Learn Mem.* **11,** 738–747.

Philpot, B. D., Sekhar, A. K., Shouval, H. Z., and Bear, M. F. (2001). Visual experience and deprivation bidirectionally modify the composition and function of NMDA receptors in visual cortex. *Neuron* **29,** 157–169.

Pizzorusso, T., Medini, P., Berardi, N., Chierzi, S., Faucett, J. W., and Maffei, L. (2002). Reactivation of ocular dominance plasticity in the adult visual cortex. *Science* **298,** 1248–1251.

Pouille, F., and Scanziani, M. (2001). Enforcement of temporal fidelity in pyramidal cells by somatic feed-forward inhibition. *Science* **293,** 1159–1163.

Prasad, S. S., *et al.* (2002). Gene expression patterns during enhanced periods of visual cortex plasticity. *Neuroscience* **111,** 35–45.

Prusky, G. T., and Douglas, R. M. (2003). Developmental plasticity of mouse visual acuity. *Eur. J. Neurosci.* **17,** 167–173.

Qian, Z., Gilbert, M. E., Colicos, M. A., Kandel, E. R., and Kuhl, D. (1993). Tissue-plasminogen activator is induced as an immediate-early gene during seizure, kindling and long-term potentiation. *Nature* **361,** 453–457.

Quinlan, E. M., Philpot, B. D., Huganir, R. L., and Bear, M. F. (1999). Rapid, experience-dependent expression of synaptic NMDA receptors in visual cortex *in vivo*. *Nat. Neurosci.* **2**, 352–357.

Ramoa, A. S., Paradiso, M. A., and Freeman, R. D. (1988). Blockade of intracortical inhibition in kitten striate cortex: Effects on receptive field properties and associated loss of ocular dominance plasticity. *Exp. Brain Res.* **73**, 285–296.

Reiter, H. O., and Stryker, M. P. (1988). Neural plasticity without postsynaptic action potentials: Less-active inputs become dominant when kitten visual cortical cells are pharmacologically inhibited. *Proc. Natl. Acad. Sci. USA* **85**, 3623–3627.

Reiter, H. O., Waitzman, D. M., and Stryker, M. P. (1986). Cortical activity blockade prevents ocular dominance plasticity in the kitten visual cortex. *Exp. Brain Res.* **65**, 182–188.

Renger, J. J., Hartman, K. N., Tsuchimoto, Y., Yokoi, M., Nakanishi, S., and Hensch, T. K. (2002). Experience-dependent plasticity without long-term depression by type 2 metabotropic glutamate receptors in developing visual cortex. *Proc. Natl. Acad. Sci. USA* **99**, 1041–1046.

Rubenstein, J. L., and Merzenich, M. M. (2003). Model of autism: Increased ratio of excitation/inhibition in key neural systems. *Genes Brain Behav.* **2**, 255–267.

Rudolph, U., Crestani, F., and Möhler, H. (2001). GABA(A) receptor subtypes: Dissecting their pharmacological functions. *Trends Pharmacol. Sci.* **22**, 188–194.

Rudy, B., and McBain, C. J. (2001). K_v3 channels: Voltage-gated K+ channels designed for high-frequency repetitive firing. *Trends Neurosci.* **24**, 517–526.

Saghatelyan, A. K., Dityater, A., Schmidt, S., Schuster, T., Bartsch, U., and Schachn, M. (2001). Reduced perisomatic inhibition, increased excitatory transmission, and impaired long-term potentiation in mice deficient for the extra-cellular matrix glycoprotein tenascin-R. *Mol. Cell Neurosci.* **17**, 226–240.

Sanes, J. R., and Lichtman, J. W. (1999). Can molecules explain long-term potentiation? *Nat. Neurosci.* **2**, 597–604.

Sawtell, N. B., Frenkel, M. Y., Philpot, B. D., Nakazawa, K., Tonegawa, S., and Bear, M. F. (2003). NMDA receptor-dependent ocular dominance plasticity in adult visual cortex. *Neuron* **38**, 977–985. Erratum *Neuron* **39**, 727.

Schoop, V. M., Gardziella, S., and Muller, C. M. (1997). Critical period-dependent reduction of the permissiveness of cat visual cortex tissue for neuronal adhesion and neurite growth. *Eur. J. Neurosci.* **9**, 1911–1922.

Shatz, C. J., and Stryker, M. P. (1978). Ocular dominance in layer IV of the cat's visual cortex and the effects of monocular deprivation. *J. Physiol. (Lond.)* **281**, 267–283.

Shaw, C., and Cynader, M. (1984). Disruption of cortical activity prevents ocular dominance changes in monocularly deprived kittens. *Nature* **308**, 731–734.

Shimegi, S., Fischer, Q. S., Yang, Y., Sato, H., and Daw, N. W. (2003). Blockade of cyclic AMP-dependent protein kinase does not prevent the reverse ocular dominance shift in kitten visual cortex. *J. Neurophysiol.* **90**, 4027–4032.

Shiosaka, S., and Yoshida, S. (2000). Synaptic microenvironments-structural plasticity, adhesion molecules, proteases and their inhibitors. *Neurosci. Res.* **37**, 85–89.

Sieghart, W. (1995). Structure and pharmacology of γ-aminobutyric acid$_A$ receptor subtypes. *Pharmacol. Rev.* **47**, 181–234.

Silver, M. A., and Stryker, M. P. (1999). Synaptic density in geniculocortical afferents remains constant after monocular deprivation in the cat. *J. Neurosci.* **19**, 10829–10842.

Soghomonian, J. J., and Martin, D. L. (1998). Two isoforms of glutamate decarboxylase: Why? *Trends Pharmacol.* **19**, 500–505.

Somogyi, P., Tamas, G., Lujan, R., and Buhl, E. H. (1998). Salient features of synaptic organisation in the cerebral cortex. *Brain Res. Rev.* **26**, 113–135.

Song, S., Miller, K. D., and Abbott, L. F. (2000). Competitive Hebbian learning through spike-timing-dependent synaptic plasticity. *Nat. Neurosci.* **3**, 919–926.

Tagawa, Y., Kanold, P. O., Majdan, M., and Shatz, C. J. (2005). Multiple periods of functional ocular dominance plasticity in mouse visual cortex. *Nat. Neurosci.* **8**, 380–388.

Taha, S., and Stryker, M. P. (2002). Rapid ocular dominance plasticity requires cortical but not geniculate protein synthesis. *Neuron* **34**, 425–436.

Takahashi, T., Svoboda, K., and Malinow, R. (2003). Experience strengthening transmission by driving AMPA receptors into synapses. *Science* **299**, 1585–1588.

Tanaka, H., Shan, W., Phillips, G. R., Arndt, K., Bozdagi, O., Shapiro, L., Huntley, G. W., Benson, D. L., and Colman, D. R. (2000). Molecular modification of N-cadherin in response to synaptic activity. *Neuron* **25**, 93–107.

Tang, Y. P., Shimizu, E., Dube, G. R., Rampon, C., Kerchner, G. A., Zhuo, M., Liu, G., and Tsien, J. Z. (1999). Genetic enhancement of learning and memory in mice. *Nature* **401**, 63–69.

Trachtenberg, J. T., and Stryker, M. P. (2001). Rapid anatomical plasticity of horizontal connections in the developing visual cortex. *J. Neurosci.* **21**, 3476–3482.

Trachtenberg, J. T., Trepel, C., and Stryker, M. P. (2000). Rapid extragranular plasticity in the absence of thalamocortical plasticity in the developing primary visual cortex. *Science* **287**, 2029–2032.

Turrigiano, G. G., and Nelson, S. B. (2004). Homeostatic plasticity in the developing nervous system. *Nat. Rev. Neurosci.* **5**, 97–107.

Wan, H., Warburton, E. C., Zhu, X. O., Koder, T. J., Park, Y., Aggleton, J. P., Cho, K., Bashir, Z. I., and Brown, M. W. (2004). Benzodiazepine impairment of perirhinal cortical plasticity and recognition memory. *Eur. J. Neurosci.* **20**, 2214–2224.

Wang, X., Merzenich, M. M., Sameshima, K., and Jenkins, W. M. (1995). Remodelling of hand representation in adult cortex determined by timing of tactile stimulation. *Nature* **378**, 71–75.

Wannier-Morino, P., Rager, G., Sonderegger, P., and Grabs, D. (2003). Expression of neuroserpin in the visual cortex of the mouse during the developmental critical period. *Eur., J. Neurosci.* **17**, 1853–1860.

Welker, E., Soriano, E., and Van der Loos, H. (1989). Plasticity in the barrel cortex of the adult mouse: Effects of peripheral deprivation on GAD-immunoreactivity. *Exp. Brain Res.* **74**, 441–452.

Wiesel, T. N., and Hubel, D. H. (1963). Single-cell responses in striate cortex of kittens deprived of vision in one eye. *J. Neurophysiol.* **26**, 1003–1017.

Willshaw, D. J., and von der Malsburg, C. (1976). How patterned neural connections can be set up by self-organization. *Proc., R. Soc. Lond. B Biol. Sci.* **194**, 431–445.

Yokoi, M., Mori, K., and Nakanishi, S. (1995). Refinement of odor molecule tuning by dendro-dendritic synaptic inhibition in the olfactory bulb. *Proc. Natl. Acad. Sci. USA* **92**, 3371–3375.

Yuste, R., and Bonhoeffer, T. (2001). Morphological changes in dendritic spines associated with long-term synaptic plasticity. *Annu. Rev. Neurosci.* **24**, 1071–1089.

Further Reading

Tyler, W. J., and Pozzo-Miller, L. (2004). Miniature synaptic transmission and BDNF modulates dendritic spine growth and form in rat CA1 neurons. *J. Physiol. (Lond.)* **553.2**, 497–509.

9

Brawn for Brains: The Role of MEF2 Proteins in the Developing Nervous System

Aryaman K. Shalizi and Azad Bonni[†]*
*Biological and Biomedical Sciences Program
[†]Department of Pathology
Harvard Medical School
Boston, Massachusetts 02115

The myocyte enhancer factor 2 (MEF2) transcription factors were originally identified, as their family name implies, on the basis of their role in muscle differentiation. Expression of the four MEF2 proteins, however, is not restricted to contractile tissue. While it has been known for more than a decade that MEF2s are abundantly expressed in neurons, their contributions to the development and function of the nervous system are only now being elucidated. Interestingly, the emerging mechanisms regulating MEF2 in neurons have significant parallels with the regulatory mechanisms in muscle, despite the quite distinct identities of these two electrically excitable tissues. The goal of this chapter is to provide an introduction to those regulatory mechanisms and their consequences for brain development. As such, we first provide an overview of MEF2 itself and its expression within the central nervous system. The second part of this chapter describes the signaling molecules that regulate MEF2 transcriptional activity and their contributions to MEF2 function. The third part of this chapter discusses the role of MEF2 proteins in the developing nervous system and compares the analogous functions of this protein family in muscle and brain. © 2005, Elsevier Inc.

Current Topics in Developmental Biology, Vol. 69
Copyright 2005, Elsevier Inc. All rights reserved.

0070-2153/05 \$35.00
DOI: 10.1016/S0070-2153(05)69009-6

I. Structure and Expression of the MEF2 proteins

The four myocyte enhancer factor 2 (MEF2) proteins (MEF2A–D) belong
to the minichromosome maintenance 1-agamous-deficiens-serum response
factor (MADS) box family of transcription factors, which have important
roles in proliferation and differentiation in organisms as diverse as plants,
fungi, and metazoans (Theissen *et al.*, 1996). Transcripts of MEF2 are
present in a variety of tissues, but the proteins are most abundant in muscle,
brain, and lymphocytes (Dodou *et al.*, 1995; Martin *et al.*, 1994; Yu *et al.*,
1992). The MEF2 family is characterized by a conserved DNA-binding
and dimerization domain, the MADS box, that targets proteins to A/T-
rich sequences in gene regulatory regions (West *et al.*, 1997). The four
MEF2 proteins also share a highly conserved "MEF2 domain," which
mediates homo- and heterodimerization of the various MEF2 proteins
and provides a surface for protein–protein interactions (Molkentin *et al.*,
1996a; Yu, 1996). The C terminus of MEF2 comprises a transactivation
domain (TAD), which diverges between the four MEF2 proteins (Black and
Olson, 1998).

MEF2 shows a high degree of conservation across all metazoan lineages
examined. Gene duplication events have led to the evolution of four distinct
MEF2 genes in the vertebrate lineage from a single ancestral MEF2 gene
present in invertebrates (Breitbart *et al.*, 1993; Dichoso *et al.*, 2000; Lilly
et al., 1994; McDermott *et al.*, 1993; Nguyen *et al.*, 1994; Rescan, 2001;
Spring *et al.*, 2002; Yu *et al.*, 1992). The MADS and MEF2 domains are
highly conserved across species, as are multiple sites for posttranslational
modifications in the transactivation domain (Black and Olson, 1998). An
additional layer of regulation of vertebrate MEF2 genes arises from alternative splicing, often in a tissue-specific fashion, generating a variety of distinct
isoforms (Breitbart *et al.*, 1993; Hobson *et al.*, 1995; Martin *et al.*, 1994;
McDermott *et al.*, 1993; Morisaki *et al.*, 1997; Yu *et al.*, 1992; Zhu and
Gulick, 2004).

MEF2 expression occurs in neurons of *Caenorhabditis elegans* and
Kenyon cells of *Drosophila melanogaster* mushroom bodies, but a direct
contribution of MEF2 to neurogenesis in these powerful genetic model
organisms has yet to be established (Dichoso *et al.*, 2000; Schulz *et al.*,
1996). In mammals, expression of MEF2 is observed in the neural crest
beginning at embryonic day (E) 8.5, and in the developing brain from
E12.5 onward (Edmondson *et al.*, 1994). All four MEF2 proteins are
expressed in the cerebral cortex and olfactory bulb, though the precise
cellular distribution has not been thoroughly characterized for each.
MEF2D transcripts are found throughout the developing central nervous
system (CNS) through adulthood (Lyons *et al.*, 1995). In contrast, the other

MEF2 proteins show more restricted patterns of expression. In addition to cortex and olfactory bulb, MEF2A transcripts are found in the hippocampus, thalamus, and internal granular layer of the cerebellum (Lyons *et al.*, 1995). In the cerebellum, MEF2A protein levels are highly correlated with expression of markers of granule neuron differentiation, such as the gamma aminobutyric acid (GABA) receptor α-6 subunit (Lin *et al.*, 1996). The pattern of MEF2B expression largely follows that of MEF2A developmentally, but it is undetectable outside the olfactory bulb, cortex, and dentate gyrus by adulthood (Lyons *et al.*, 1995).

The expression pattern of MEF2C in the CNS is the most extensively characterized of the four MEF2 proteins. Indeed, MEF2C was initially cloned on the basis of its enrichment in the brain (Allen *et al.*, 2002; Leifer *et al.*, 1993; McDermott *et al.*, 1993). Cortical expression of MEF2C protein is restricted to a subset of cortical neurons in layers II, IV, and VI. Interestingly, an alternative exon of MEF2C that extends the transactivation domain is expressed in the adult, but not in the developing cerebral cortex (Allen *et al.*, 2002; Leifer *et al.*, 1994; McDermott *et al.*, 1993). This alternative exon encodes a conserved phosphorylation site also present in MEF2A and -D that negatively regulates MEF2 transcriptional activity (Gong *et al.*, 2003; Zhu and Gulick, 2004). The reason for this developmental switch in MEF2C splicing has yet to be established, but it appears to render MEF2C sensitive to oxidative stress (Gong *et al.*, 2003).

II. MEF2 and the Regulation of Transcription

Eukaryotic gene expression is controlled by proteins that recruit the basal transcription machinery or alter chromatin structure, either through direct posttranslational modifications of histone proteins or through energy-dependent translocation of nucleosomes (Featherstone, 2002). As with many DNA-binding transcription factors, MEF2 alters target gene expression by recruiting specific chromatin-modifying activities to promoter regions (McKinsey *et al.*, 2002b). Interestingly, MEF2s can be thought of as bifunctional regulators of transcription: in the absence of transactivating stimuli, target genes are maintained in a repressed state through interaction with class IIa histone deacetylases (HDACs) and other corepressor molecules (McKinsey *et al.*, 2002a). In the presence of transactivating stimuli, the association with class-II HDACs is disrupted, and MEF2 recruits histone acetyltransferases (HATs) and other co-activators (McKinsey *et al.*, 2001a, 2002a). Furthermore, appropriate target gene regulation by MEF2 involves the direct posttranslational modification of MEF2 by kinases, phosphatases, and other enzymes (McKinsey *et al.*, 2002a). These protein–protein interactions and posttranslational modifications are discussed below, and are outlined in Fig. 1.

A. MEF2 and Transcriptional Repressors: Class IIa HDACs

HDACs are enzymes that regulate chromatin structure by catalyzing the removal of acetyl groups from the -amine group of lysines in target proteins, principally histones. There are three known classes of HDAC, which comprise distinct gene families (de Ruijter *et al.*, 2003): class I HDACs are ubiquitously expressed homologues of yeast rpd3; class II HDACs are tissue-restricted proteins related to yeast hda1; and class III HDACs, the sirtuins, are ubiquitously expressed homologues of yeast sir2 (Blander and Guarente, 2004; de Ruijter *et al.*, 2003). Both class I and II HDACs promote histone deacetylation by similar catalytic mechanisms, while the sirtuins require reduced nicotinamide adenine dinucleotide (NADH) for their catalytic activity (Blander and Guarente, 2004; de Ruijter *et al.*, 2003). With relatively few exceptions, these proteins have the net effect of repressing gene expression, by catalyzing the formation of heterochromatin (Czermin and Imhof, 2003; Kuo and Allis, 1998).

The class IIa HDACs 4, 5, 7, and 9 are defined by a bipartite structure (Bertos *et al.*, 2001; Verdin *et al.*, 2003). The C-terminal domain harbors the

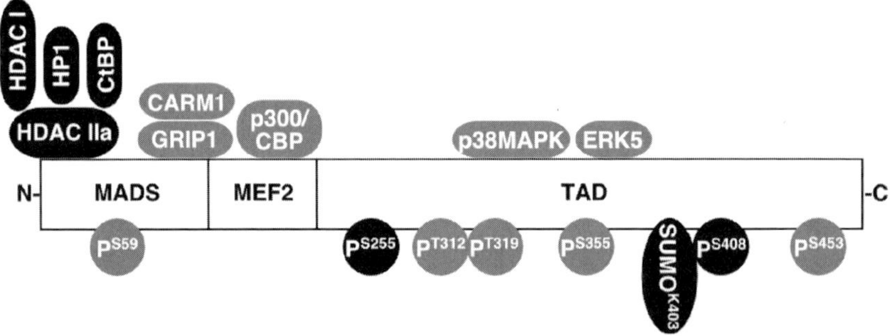

Figure 1   MEF2 structure, regulatory modifications, and physical interactions with corepressors and coactivators. MEF2 proteins are comprised of three domains: the N-terminal MADS domain, the MEF2 domain, and the C-terminal transactivation domain (TAD). Posttranslational modifications that regulate MEF2 function are shown below the box diagram of MEF2. P indicates phosphorylation; SUMO indicates sumoylation. Superscript identifies the modified residue in human MEF2A, but all modifications shown occur on residues conserved in MEF2C and D. Modifications that enhance or repress MEF2-dependent transcription are shown in grey or black, respectively. Transcriptional coactivators and corepressors that have demonstrated physical interactions with MEF2 are shown above the box diagram of MEF2. Coactivators and corepressors are shown in grey or black, respectively. Interacting proteins are positioned relative to the MEF2 domains they interact with. Abbreviations are as in the text.

deacetylase activity and is structurally similar to the deacetylase domain of class I HDACs (Lemercier *et al.*, 2000; Miska *et al.*, 1999; Sparrow *et al.*, 1999; Wang *et al.*, 1999). Both classes are sensitive to inhibition by the pharmacologic agents trichostatin A, valproic acid, and sodium butyrate (de Ruijter *et al.*, 2003). The N-terminal domain of class IIa HDACs binds specifically to the MADS and MEF2 domains of MEF2 proteins and has a repressive function independent of the C-terminal deacetlyase activity (Bertos *et al.*, 2001; Verdin *et al.*, 2003). Indeed, the deacetylase activity of class IIa HDACs is dispensable for transcriptional repression of MEF2 target genes, and the MEF2-interacting transcriptional repressor (MITR), a splice variant of HDAC9, lacks a functioning deacetylase activity altogether (Sparrow *et al.*, 1999; Zhang *et al.*, 2002). Instead, the N-terminal domain of class IIa HDACs recruits additional transcriptional repressors, including class I HDACs, C-terminal (CtBP), and the histone methyltransferase binding protein heterochromatin protein 1 (HP1) (Chan *et al.*, 2003; Dressel *et al.*, 2001; Zhang *et al.*, 2001a, 2002). In contrast, the class IIb HDACs HDAC6 and HDAC10 lack this N-terminal domain (Verdin *et al.*, 2003).

Finally, it has been reported recently that class IIa HDACs promote the sumoylation of the transactivation domain of MEF2D (Gregoire and Yang, 2005). The small ubiquitin-like modifier (SUMO) represses the transcriptional activity of target proteins, in part through its ability to recruit the class I HDACs HDAC1 and HDAC2 (Yang and Sharrocks, 2004). The ability of HDACs 4, 5, 7, and 9 to promote MEF2D sumoylation depends on the N-terminal repressor domain and is independent of the deacetlyase domain, in line with the observations described above (Gregoire and Yang, 2005). Kinase pathways that activate MEF2D inhibit sumoylation, although the precise regulatory mechanism for this modification remains to be elucidated (Gregoire and Yang, 2005).

Dynamic, signal-responsive subcellular localization of class IIa HDACs has emerged as the major regulatory mechanism of transcriptional repression for this class of proteins. Class IIa HDACs contain both nuclear localization signals (NLS) and nuclear export signals (NES), and these are necessary for appropriate subcellular targeting (Borghi *et al.*, 2001; Wang and Yang, 2001). Activation of calmodulin (CaM)-dependent protein kinases (CaMKs) results in the phosphorylation of distinct conserved serine and threonine residues in the N terminus of class IIa HDACs. These phosphorylation events disrupt the class IIa HDAC-MEF2 interaction by creating binding sites for 14-3-3 proteins. Association of class IIa HDACs with 14-3-3 leads to nuclear export of the HDACs and liberates MEF2 to interact with transcriptional activators (Choi *et al.*, 2001; Grozinger and Schreiber, 2000; Kao *et al.*, 2001; Li *et al.*, 2004; McKinsey *et al.*, 2001b; Miska *et al.*, 2001; Wang and Yang, 2001; Wang *et al.*, 2000; Zhao *et al.*, 2001). Alternatively, HDACs 4, 5, and 9 can interact directly with CaM in response to calcium signals. The interaction with CaM involves the MEF2-binding portion of the HDAC's N terminus, and thus enhances MEF2-dependent transcription by disrupting HDAC-MEF2 binding (Berger *et al.*, 2003; Youn *et al.*, 2000b).

In the context of muscle differentiation, evidence is emerging that specific class IIa HDACs regulate distinct aspects of the myogenic program. For example, HDAC9 is a potent repressor of MEF2-dependent transcription, but it has only modest inhibitory effects on muscle differentiation (Zhang *et al.*, 2001b). In contrast, HDAC9 is an essential regulator of the transcriptional response of differentiated muscle to motor neuron innervation (Mejat *et al.*, 2005). HDACs 4, 5, and 7 all exhibit some form of nuclear-to-cytoplasmic shuttling during execution of the myogenic program, and inhibition of this nuclear export stifles differentiation (Borghi *et al.*, 2001; Dressel *et al.*, 2001; Kao *et al.*, 2001; Lu *et al.*, 2000a,b; McKinsey *et al.*, 2000a,b, 2001b; Miska *et al.*, 2001; Wang and Yang, 2001; Youn *et al.*, 2000b). Interestingly, HDAC4 returns to the nucleus following myotube fusion, suggesting an additional role in maintenance of the muscle phenotype (Miska *et al.*, 2001).

B. MEF2 and Transcriptional Activators: HATs, GRIPs, and Other Transactivators

Interactions between MEF2 and transcriptional coactivators have been studied substantially less than those between MEF2 and the class IIa HDACs. However, it is clear that MEF2 function depends on its ability to recruit such factors for appropriate target gene transcription. Although the transactivation domain of MEF2 is dispensable for the interaction with most of the known MEF2 coactivators, a truncated form of MEF2 containing only the MADS and MEF2 domains acts as a dominant-negative repressor of transcription, even in the presence of activating stimuli (Ornatsky *et al.*, 1997). This suggests that despite the ability of coactivating proteins to bind the MADS and MEF2 domain alone, gene activation requires more than DNA binding and dimerization of MEF2.

The related HATs cyclic adenosine monophosphate (cAMP) response element binding protein (CREB)-binding protein (CBP) and p300 interact directly with the MEF2 domain of MEF2 proteins, a surface that overlaps significantly with the binding site for class IIa HDACs (De Luca *et al.*, 2003; Sartorelli *et al.*, 1997). Displacement of HDACs and recruitment of HAT activity following myogenic stimuli have been shown to enhance lysine acetylation in chromatin at MEF2-responsive genes, a common correlate of transcriptional activation (Zhang *et al.*, 2002). In T-lymphocytes, MEF2 is required for apoptosis upon activation of the T-cell receptor (McKinsey *et al.*, 2002a). This process requires dissociation of the calcium-sensing repressor Cabin1, which competes with p300 for MEF2 binding, from the MADS/MEF2 domain (Youn and Liu, 2000). Disruption of Cabin1 binding to MEF2 by calcium signaling leads to transcription of the proapoptotic *nur77* gene in a p300- and MEF2-dependent manner (Youn and Liu, 2000; Youn *et al.*, 1999, 2000a). In addition to HATs, MEF2 interacts indirectly with the coactivator-associated arginine methyltransferase-1 (CARM1). The enzymatic activity of CARM1, which promotes histone arginine methylation, is required for myoblast differentiation (Chen *et al.*, 2002).

The interaction of MEF2 and CARM1 is dependent upon a nuclear receptor coactivator, glucocorticoid receptor interacting protein-1 (GRIP-1). GRIP-1 is a member of the steroid receptor coactivator (SRC)/p160 family of proteins that facilitate chromatin remodeling through the recruitment of histone acetyl- and methyltransferases (Xu and Li, 2003). GRIP-1 targets MEF2 to dot-like subnuclear structures upon differentiation and enhances MEF2-dependent transcription (Chen *et al.*, 2001). Conversely, stimuli that block muscle differentiation, such as transforming growth factor (TGF)-β activation of SMA- and MAD-homolog 3(SMAD3) or the activity of cyclin-dependent kinases, disrupt the association of MEF2 and GRIP-1 and prevent the GRIP-1-dependent subnuclear targeting of MEF2 (Lazaro *et al.*, 2002; Liu *et al.*, 2004). The precise nature of the subnuclear dots, and their role in MEF2 function, remains unclear.

An alternative coactivator of MEF2-dependent transcription is the peroxisome proliferators-activated receptor γ (PPARγ) coactivator 1-α (PGC-1α). PGC-1α is a master regulator of mitochondrial biogenesis and energy homeostasis (Puigserver and Spiegelman, 2003). PGC-1α promotes transcription through its ability to recruit chromatin-modifying and RNA-processing complexes (Puigserver and Spiegelman, 2003). In addition to serving as a MEF2A coactivator, expression of the PGC-1α gene is stimulated by calcium signaling and MEF2-dependent transcription (Czubryt *et al.*, 2003). PGC-1α thus positively regulates its own production by activating MEF2-dependent transcription (Czubryt *et al.*, 2003; Handschin *et al.*, 2003; Wu *et al.*, 2002). PGC-1α is not required for muscle formation, but instead promotes fast-to-slow fiber type switching and enhances the expression of genes required for oxidative metabolism (Lin *et al.*, 2002).

The association between coactivators such as GRIP-1 and PGC-1α and distinct MEF2 isoforms may provide some insight into the unique functions of these seemingly interchangeable transcription factors. For example, disruption of the association between MEF2C and GRIP-1 is sufficient to block muscle differentiation (Lazaro *et al.*, 2002). In contrast, the interaction with PGC-1α may be essential for MEF2A function, as MEF2A knockout mice show a profound disruption of mitochondrial organization and gene expression (Naya *et al.*, 2002). These findings underscore the importance of direct chromatin-modifying activities in the regulation of MEF2 function in a variety of cell types. The contribution of the coactivators discussed above to MEF2 function in the nervous system remains to be elucidated.

C. Regulation of MEF2 Activity by p38MAP Kinase and ERK5

There are numerous conserved phosphorylation sites in MEF2 that are essential for target gene activation. The MADS and MEF2 domains, despite providing interaction surfaces for most of the regulators of MEF2-dependent transcription, are relatively devoid of sites for posttranslational modifications. There is a single phosphorylated serine residue in the MADS domain that enhances the affinity of MEF2 for target DNA sequences (Molkentin *et al.*, 1996c). To date, at least eight sites of phosphorylation have been identified in the transactivation domain of MEF2A through a combination of mutational and biochemical analyses (Cox *et al.*, 2003; Gong *et al.*, 2003; Kato *et al.*, 1997; Ornatsky *et al.*, 1999; Yang *et al.*, 1998, 1999; Zhao *et al.*, 1999). The majority of these sites are proline-directed serine (SP) or threonine (TP) residues, which are preferred substrates of the mitogen-activated protein kinase (MAPK) and cyclin-dependent kinase (Cdk) families of protein kinases. A pair of TP sites is conserved in all four MEF2 genes and is essential for transcriptional activity (Molkentin *et al.*, 1996b). Additional SP sites are conserved between MEF2A, C, and D and have been found to regulate protein stability and repress transcriptional activity (Cox *et al.*, 2003; Gong *et al.*, 2003).

A variety of signal transduction cascades promote the phosphorylation of MEF2 proteins. In response to osmotic stress or myogenic or inflammatory stimuli, MEF2 is phosphorylated by p38MAPK (Han *et al.*, 1997; Marinissen *et al.*, 1999; Zetser *et al.*, 1999). Following serum treatment, receptor tyrosine kinase activation, G-protein-coupled receptor activation, or oxidative stress, MEF2 is phosphorylated by extracellular signal-regulated kinase 5 (ERK5) (Fukuhara *et al.*, 2000; Kato *et al.*, 1997; Marinissen *et al.*, 1999; Suzaki *et al.*, 2002). Some specificity of the MEF2 transcriptional response may be a byproduct of the activating stimuli, although a systematic analysis of the gene targets for these stimuli has yet to be performed. However, both ERK5 and p38MAPK phosphorylate similar residues of MEF2 (Han *et al.*, 1997; Kato *et al.*, 1997).

The p38MAPK is a major regulator of MEF2 phosphorylation and MEF2-dependent gene expression. There are several isoforms of p38MAPK, and only a subset phosphorylate the MEF2 transactivation domain efficiently (Marinissen *et al.*, 1999; Yang *et al.*, 1999). Furthermore, p38MAPK requires a docking site to interact with MEF2. The "D domain" is sufficient to target p38MAPK to heterologous substrates and is present only in the transactivation domains of MEF2A and MEF2C (Ornatsky *et al.*, 1999;

Yang *et al.*, 1999; Zhao *et al.*, 1999). Activity of p38MAPK is necessary for muscle formation, and mutation of the primary phosphorylation sites in the transactivation domain of MEF2A or C is sufficient to block myogenesis (Kolodziejczyk *et al.*, 1999; Penn *et al.*, 2004; Puri *et al.*, 2000; Zetser *et al.*, 1999). p38MAPK also mediates the pathological effects of MEF2 in cardiac hypertrophy and some forms of myotonia, underscoring the importance of posttranslational modification for normal MEF2 function (Kolodziejczyk *et al.*, 1999; Wu and Olson, 2002).

ERK5, also known as big map kinase 1 (BMK1) is an unusual member of the ERK family. In addition to the N-terminal kinase domain, ERK5 has a large C-terminal transactivation domain, which is necessary for interaction with MEF2 and induction of target genes such as *nur77* (Kasler *et al.*, 2000; Zhou *et al.*, 1995). ERK5 is an essential developmental gene; its function is required for muscle differentiation, and gene disruption closely phenocopies the cardiac developmental defect observed in MEF2C knockouts (Dinev *et al.*, 2001; Regan *et al.*, 2002). MEF2 proteins recruit ERK5 in a manner similar to p38MAPK, through an interaction motif located in the transactivation domain, as well as through an interaction with the MADS/MEF2 domain (Barsyte-Lovejoy *et al.*, 2004; Yang *et al.*, 1998). However, ERK5 has a broader substrate range and can efficiently phosphorylate MEF2D in addition to MEF2A and C, but does not phosphorylate MEF2B (Kato *et al.*, 1997, 2000; Yang *et al.*, 1998).

The ability of activating kinases such as p38MAPK and ERK5 to discriminate between different MEF2 proteins may be an important means of regulating MEF2-dependent transcription. For example, although MEF2 DNA-binding activity can be detected in a variety of cell lines, it is transcriptionally inert in non-muscle cells (Dodou *et al.*, 1995; Ornatsky and McDermott, 1996). Furthermore, MEF2–DNA complexes in muscle cell lines consist almost exclusively of MEF2A homodimers, while MEF2–DNA complexes in non-muscle cells are predominantly MEF2A–MEF2D heterodimers (Dodou *et al.*, 1995; Ornatsky and McDermott, 1996). The inability of MEF2 to robustly activate transcription in certain cellular contexts may thus arise from discrimination between the transactivation domains present in DNA-bound MEF2 dimers by upstream kinases (Zhao *et al.*, 1999).

D. Regulation of MEF2 Activity by Calcineurin

Protein phosphatase 2B (PP2B), or calcineurin, is a serine-threonine phosphatase that is activated by calcium signals (Aramburu *et al.*, 2000, 2004; Crabtree, 2001; Hemenway and Heitman, 1999). In the absence of calcium signaling, calcineurin exists as an inactive dimer, composed of the regulatory

B chain and the catalytic A chain (Aramburu *et al.*, 2000, 2004). Upon calcium influx, the EF-hand protein calmodulin binds to the A chain and disrupts the association with the B chain, allowing the calmodulin–A chain complex to dephosphorylate target proteins (Aramburu *et al.*, 2000, 2004).

In recent years, calcineurin has emerged as a key regulator of muscle fiber phenotype (Olson and Williams, 2000). In response to distinct patterns of sustained excitatory activity, muscle fibers will switch from a primarily glycolytic, or "fast-twitch," phenotype to a primarily oxidative, or "slow-twitch," phenotype (Bigard *et al.*, 2000; Chin *et al.*, 1998; Hughes, 1998; Serrano *et al.*, 2001). This calcineurin-regulated transition involves the *de novo* expression of slow-twitch-specific metabolic and cytoskeletal genes (Olson and Williams, 2000). The slow-twitch gene expression program requires the nuclear factor of activated T-cells (NFAT) and MEF2-dependent transcription (Delling *et al.*, 2000; Dunn *et al.*, 2001; Swoap *et al.*, 2000; Wu *et al.*, 2000). Calcineurin dephosphorylates MEF2 directly, but the target sites remain undefined (Mao and Wiedmann, 1999; Wu *et al.*, 2000). Interestingly, calcineurin promotes MEF2-dependent transcription only from a subset of MEF2 enhancer elements found primarily upstream of genes associated with the slow-twitch phenotype that harbor a small but consistent change in the MEF2-binding sequence (Wu *et al.*, 2000). Calcium signaling in neurons appears to modulate the affinity of MEF2 for its consensus DNA-binding sequence through dephosphorylation, but the functional consequences of this are unclear (Linseman *et al.*, 2003b; Mao and Wiedmann, 1999).

The ability of calcineurin to activate a subset of MEF2 enhancer elements and modulate the consensus-DNA-binding affinity of MEF2 suggests that calcineurin alters MEF2 function through changes in DNA-binding specificity. However, activation of MEF2 by PGC-1α, which regulates fiber type switching through potentiation of MEF2-dependent transcription, is also enhanced by calcineurin (Handschin *et al.*, 2003; Lin *et al.*, 2002). Calcineurin may thus regulate the association of MEF2 not only with specific enhancer elements, but also with transcriptional coactivators and corepressors to generate transcriptional outcomes tuned to specific patterns of calcium flux.

III. MEF2 in the CNS: A Multifunctional Regulator of Survival ... and More?

As discussed above, the four MEF2 proteins are expressed with distinctive spatial and temporal patterns throughout the developing and adult CNS. Furthermore, there are indications that MEF2 proteins regulate acquisition of the neuronal phenotype (Okamoto *et al.*, 2000), although this remains to be established in greater detail. To date, the preponderance of work on MEF2 function in the CNS has focused on the role of these transcription factors in controlling neuronal survival and apoptosis in response to a variety of extracellular stimuli. Our current understanding of the signaling pathways that regulate MEF2-dependent survival in the nervous system is discussed in the following sections, and is outlined in Fig. 2.

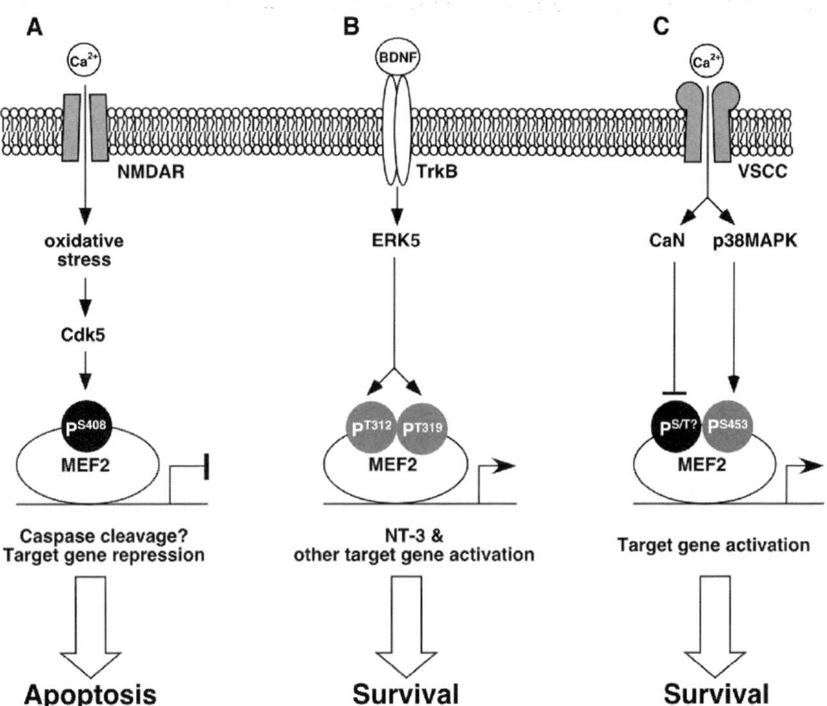

Figure 2   MEF2-dependent signaling pathways that regulate neuronal survival and apoptosis. (A) Excitotoxic stimuli promote apoptosis through Cdk5-dependent phosphorylation of MEF2. Phosphorylation by Cdk5 blocks MEF2-dependent transcription and may regulate cleavage by caspases in response to excitotoxic insults and/or oxidative stress. (B) BDNF promotes survival through ERK5-dependent phosphorylation of MEF2. Activation of MEF2 by ERK5 in newly generated neurons leads to NT-3 production, which promotes survival in response to BDNF stimulation. (C) MEF2 promotes activity-dependent survival. Calcium influx activates calcineurin and p38MAPK, both of which stimulate MEF2-dependent transcription. Phosphorylation of MEF2 by p38MAPK promotes survival. In all cases, activating and repressing phosphorylation events are shown in grey and black, respectively. Numbering is relative to the human MEF2A. NMDAR, N-methyl-D-aspartate receptor; VSCC, voltage-sensitive calcium channel. Other abbreviations are as in the text.

A. Use It or Lose It: MEF2 as a Calcium-Sensitive Survival Factor

Intracellular calcium levels are tightly regulated in neurons, because of calcium's potent activity as a second messenger. Small alterations of intracellular calcium content can have profound effects on neuronal function (Ghosh and Greenberg, 1995). The specific route of calcium entry to the neuronal cytoplasm from ligand-gated excitatory channels, voltage-sensitive calcium channels, or release from stores in the endoplasmic reticulum leads to distinct functional outcomes (Gallin and Greenberg, 1995).

In response to sustained neuronal activity, calcium influx through extra-synaptic voltage-sensitive calcium channels (VSCCs) promotes neuronal survival in part through the activation of nuclear transcription factors, including MEF2 (Bito and Takemoto-Kimura, 2003; West *et al.*, 2001). The transcriptional activity of MEF2 is enhanced by calcium influx, in a manner that requires both MEF2 DNA binding and phosphorylation of the transactivation domain (Li *et al.*, 2001; Linseman *et al.*, 2003a; Mao and Wiedmann, 1999; Mao *et al.*, 1999). Mutations of MEF2 that render it transcriptionally inactive block activity-dependent survival in both cortical and cerebellar granule neurons, while constitutively active MEF2, in which the MADS/MEF2 domains are fused to the strong transcriptional activator VP16, can promote survival in the absence of depolarization (Li *et al.*, 2001; Linseman *et al.*, 2003a; Mao *et al.*, 1999).

RNAi-mediated knockdown experiments in cerebellar granule neurons indicate that endogenous MEF2A function is necessary for neuronal survival in response to neuronal activity (Gaudilliere *et al.*, 2002). Furthermore, since MEF2D is abundantly expressed in cerebellar granule neurons, it suggests that MEF2D cannot compensate for the activity-dependent survival function of MEF2A (Gaudilliere *et al.*, 2002). Since p38MAPK phosphorylates MEF2A but not MEF2D (Han *et al.*, 1997; Yang *et al.*, 1999; Zhao *et al.*, 1999), this is consistent with the requirement for p38MAPK activity for MEF2-mediated survival in response to calcium influx following depolarization (Mao *et al.*, 1999). In hyperpolarized neurons, p38MAPK is inactive, and it is stimulated upon depolarization (Mao *et al.*, 1999). Activation of p38MAPK is necessary for MEF2-mediated neuronal survival in both primary neurons and neuronally differentiated P19 cells (Mao *et al.*, 1999; Okamoto *et al.*, 2000). These findings indicate some mechanistic similarities between muscle differentiation and the regulation of neuronal survival by VSCCs.

Further evidence of functional analogies between muscle differentiation and activity-dependent neuronal survival comes from the study of class IIa HDACs in neurons. In cultured hippocampal neurons, spontaneous and synaptic activity promotes the nuclear exclusion of HDAC4 and HDAC5, respectively (Chawla *et al.*, 2003). The cytoplasmic shuttling of HDAC5, and to a lesser extent, HDAC4, in neurons is sensitive to CaMK inhibition, as seen in muscle (Chawla *et al.*, 2003; Linseman *et al.*, 2003a). Nuclear import of HDAC5 upon CaMK inhibition results in reduced MEF2-dependent transcription and increased apoptosis of depolarized neurons (Linseman *et al.*, 2003a). Whether other class IIa HDACs have a role in activity-dependent survival, or whether they are involved in distinct aspects of neuronal maturation, is open to further study. Given the novel finding that class IIa HDACs promote MEF2 sumoylation (Gregoire and Yang, 2005), it will be of great interest to determine whether calcium signaling regulates MEF2 sumoylation in neurons, and what the functional consequences of this modification are for neuronal biology.

In addition to the phosphorylation of MEF2 at distinct sites, the dephosphorylation of MEF2 in response to membrane depolarization also regulates MEF2 function. Activity withdrawal and consequent inactivation of voltage-gated calcium channels leads to the hyperphosphorylation of MEF2A and MEF2D (Li *et al.*, 2001; Linseman *et al.*, 2003b; Mao and Wiedmann, 1999). This hyperphosphorylation reduces the transcriptional activity of MEF2 (Butts *et al.*, 2003; Li *et al.*, 2001; Linseman *et al.*, 2003b). The sites in MEF2 that are phosphorylated by activity withdrawal are unknown, but they are likely to be distinct from those involved in promoting MEF2-dependent transcription. Hyperpolarization-induced MEF2 phosphorylation is sensitive to the activity of calcineurin, as pharmacologic inhibition of calcineurin in depolarized neurons leads to hyperphosphorylation identical to that seen with activity withdrawal (Mao and Wiedmann, 1999). Pharmacologic inhibition of calcineurin activity also reduces MEF2-dependent transcription in depolarized cerebellar granule neurons, and causes a modest reduction in MEF2 DNA-binding activity (Mao and Wiedmann, 1999). Whether the reduction in MEF2-dependent transcription observed with calcineurin inhibition occurs on a subset of MEF2 target genes, similar to calcineurin-dependent regulation of MEF2 in fiber type switching, or on all MEF2-responsive genes remains to be established. It is worth noting, however, that the immunosuppressants cyclosporine A and FK506, which inhibit calcineurin's phosphatase activity, have little effect on MEF2-dependent neuronal survival (Mao and Wiedmann, 1999).

In addition to altering the DNA-binding affinity of MEF2, the hyperphosphorylation observed following activity withdrawal may also control the stability of MEF2 protein. Activity withdrawal leads to apoptosis in the absence of other neurotrophic stimuli, a process executed by the activation of caspases, proteases that degrade target proteins after aspartate residues (Polster and Fiskum, 2004; Rubin *et al.*, 1994; Yuan and Yankner, 2000). Hyperphosphorylated MEF2 is a substrate for caspase cleavage in neurons (Li *et al.*, 2001). The caspase-mediated cleavage of MEF2 generates two fragments, one containing the transactivation domain and one containing the MADS and MEF2 domain, thus decoupling DNA binding and transcriptional activation functions of MEF2 (Li *et al.*, 2001). Overexpression of the N-terminal DNA-binding domain of MEF2 in neurons promotes apoptosis, even in the presence of constitutively active MEF2–VP16 fusion proteins (Li *et al.*, 2001), a phenomenon similar to the dominant-negative effects of the MEF2 DNA-binding domain on myocyte differentiation.

In the presence of trophic factors such as insulin-like growth factor 1 (IGF-1), MEF2 is still hyperphosphorylated upon activity withdrawal, but it is no longer degraded (Butts *et al.*, 2003). The hyperphosphorylation of MEF2 in the presence of IGF-1 nonetheless reduces its transcriptional activity (Butts *et al.*, 2003), suggesting that the fate of MEF2—active, repressed, or degraded—is precisely tied to the cellular environment.

B. Too Much of a Good Thing: Negative Regulation of MEF2 by Excitotoxicity

Physiological levels of excitatory neuronal activity, and the accompanying calcium influx, are essential for neuronal survival. In contrast, pathological levels of excitatory activation, as observed in ischemic events, can cause neuronal death through deregulation of calcium signaling and production of reactive oxygen species (Choi, 1994; Zipfel *et al.*, 2000). The excitotoxic death caused by excessive exposure to excitatory neurotransmitters such as glutamate or reactive oxygen species does not result from random derangement of cellular processes, but involves competition between repair and recovery programs as well as apoptotic programs, both of which may prove to be targets for therapeutic intervention (Choi, 1998; Dugan and Choi, 1994).

MEF2 itself may be a target of both survival-promoting and survival-inhibiting pathways in response to excitotoxic stress. In PC12 cells, a commonly used model cell for neurons, oxidative stress leads to the activation of ERK5 (Suzaki *et al.*, 2002). Oxidative stress enhanced MEF2C DNA binding in an ERK5-sensitive manner, and pharmacologic inhibitors that block ERK5 activity enhanced apoptosis following oxidative stress (Suzaki *et al.*, 2002). These results suggest that MEF2C activation by ERK5 may serve a neuroprotective role in response to oxidative stress, though the direct contribution of MEF2C to survival in response to oxidative stress was not examined.

In contrast, oxidative stress appears to inhibit the pro-survival function of MEF2 in cortical neurons. Cdk5 is a cyclin-dependent kinase with activity restricted to postmitotic neurons that has been implicated in a variety of neuropathological processes. In response to oxidative stress in cortical neurons, Cdk5 has been reported to phosphorylate MEF2 on a distinct site in the transactivation domain (Gong *et al.*, 2003). The site of Cdk5 phosphorylation in MEF2 is conserved between the A, C, and D proteins but is not found in MEF2B (Gong *et al.*, 2003). Phosphorylation on this site following oxidative stress results in a Cdk5-dependent inhibition of MEF2-dependent transcription (Gong *et al.*, 2003). Mutation of the phosphorylation site to alanine makes MEF2 resistant to Cdk5 activation and protects cortical neurons against apoptosis caused by excitotoxicity or oxidative stress (Gong *et al.*, 2003). Whether the Cdk5 phosphorylation site in MEF2 is regulated by additional stimuli or is a target for other proline-directed serine/threonine kinases remains to be determined.

The mechanism by which excitotoxic stimuli such as Cdk5 activation inhibit MEF2 function is unclear at present, but two possibilities exist. One possibility is the direct degradation of MEF2 by the apoptotic machinery. This model is supported by the observation that N-methyl-D-aspartate (NMDA) treatment of cortical neurons, a potent apoptosis-inducing stimulus, led to caspase-dependent cleavage of MEF2A, C, and D (Okamoto *et al.*, 2002). Similar to findings with activity withdrawal, overexpression of MEF2 proteins corresponding to the caspase-cleaved fragments of MEF2 blocked MEF2–VP16-dependent transcriptional activation (Okamoto *et al.*, 2002). Furthermore, constitutively active MEF2–VP16 was neuroprotective against NMDA-stimulated apoptosis, but this effect was abrogated by co-expression of caspase-cleaved fragments of MEF2 (Okamoto *et al.*, 2002). However, a direct link from MEF2 phosphorylation generally, much less the activation of a specific kinase, to caspase-mediated degradation of MEF2 has not been established.

Alternatively, phosphorylation of the Cdk5 site in MEF2 may regulate the recruitment of negative regulators of transcription. This possibility is supported by the recent observation in COS-7 and C2C12 cells that the Cdk5 site lies within a repressor domain encoded by an alternative exon of MEF2C, the "γ-domain," which is expressed in only a subset of MEF2C transcripts from muscle, brain, and spleen (Zhu and Gulick, 2004). Phosphorylation of this site occurs in a variety of cell types, suggesting that Cdk5 may not be the sole, or even the primary, regulatory kinase (Zhu and Gulick, 2004). Furthermore, this exon defines an independent repressive domain that is phosphorylation-dependent: replacement of the serine with alanine leads to profound induction of MEF2-dependent transcription, despite having no effect on MEF2 stability or DNA-binding activity (Zhu and Gulick, 2004). Interestingly, substantial portions of the γ-domain beyond the phosphorylation site are conserved in both MEF2A and MEF2D, and have similar repressive effects on the transcriptional activities of these proteins (Zhu and Gulick, 2004). This suggests that phosphorylation of the Cdk5 site may influence the interaction of MEF2 with transcriptional regulators, most likely through recruitment of corepressors distinct from the MADS/MEF2-binding class IIa HDACs. It is worth noting that the Cdk5 site in MEF2D is adjacent to the recently identified sumoylation site (Gregoire and Yang, 2005). Because of its profound effects on MEF2 function in diverse cell types, understanding the role of this novel phosphorylation site in controlling MEF2 in the nervous system may in fact provide insight into the regulation of MEF2 in a variety of tissues.

C. Good Neighbors: MEF2 as a Neurotrophin-Sensitive Survival Factor

The neurotrophic hypothesis states that neurons compete for limiting amounts of target-derived trophic factors: those that receive sufficient levels of neurotrophin survive, while those that do not are lost through apoptosis (Kaplan and Miller, 1997, 2000; Levi-Montalcini, 1987; Miller and Kaplan, 2001). The four vertebrate neurotrophins, nerve growth factor (NGF), brain-derived neurotrophic factor (BDNF), neurotrophin-3 (NT-3), and neurotrophin-4/5 (NT-4/5), regulate survival, and a variety of other cellular functions, through the activation of specific receptor tyrosine kinases, the tropomyopsin-related kinases (Trk) A, B, and C (Kaplan and Miller, 1997, 2000; Miller and Kaplan, 2001). Following receptor ligation, multiple down-stream signaling pathways that regulate neuronal survival in part through the activation of nuclear transcription factors are activated (Kaplan and Miller, 1997, 2000; Miller and Kaplan, 2001). MEF2 has emerged in recent years as a key mediator of the pro-survival transcriptional response to neurotrophin stimulation.

Activation of MEF2 by neurotrophic stimuli appears to be mediated exclusively by ERK5. The activation of ERK5 in cortical and granule neurons is dependent on the upstream kinase MEK5 following neurotrophin stimulation (Cavanaugh et al., 2001; Shalizi et al., 2003). In marked contrast to the closely related kinases ERK1/2, ERK5 activation appears to be specific to neurotrophins, as ERK5 is not activated by either cyclic adenosine monophosphate (cAMP) or neuronal activity (Cavanaugh et al., 2001). Activation of ERK5 by neurotrophins stimulates MEF2-dependent transcription in cortical neurons and PC12 cells, suggesting that MEF2 activation by ERK5 is a consequence of neurotrophin signaling in a variety of cell types (Cavanaugh et al., 2001). Interestingly, ERK5, but not ERK1/2, is activated by endocytosed Trk receptors involved in the retrograde signaling response to neurotrophins (Watson et al., 2001). Although a direct role for MEF2 in the retrograde signaling has not been demonstrated, the activation of ERK5 raises the possibility that MEF2 activation may confer spatial specificity to the neurotrophin response in neurons.

One of the interesting features of neurotrophins is the pleiotropic nature of their effects on target cells. At distinct developmental time points, the same neurotrophin may regulate different processes for the same cell population (Huang and Reichardt, 2001). The appropriate response to a particular neurotrophin thus requires the regulated activation of distinct intracellular signaling pathways. Interestingly, the activation of ERK5 by neurotrophins appears to be a temporally and spatially regulated signaling phenomenon (Cavanaugh, 2004; Heerssen and Segal, 2002).

ERK5 is abundantly expressed in the developing brain, but declines to undetectable levels by adulthood (Liu *et al.*, 2003). BDNF induces the activity of ERK5 in newly generated cerebellar granule neurons but fails to activate ERK5 in mature cerebellar granule neurons (Shalizi *et al.*, 2003). The ability of BDNF to promote granule neuron survival correlates with its ability to activate ERK5, and inhibition of ERK5 by dominant-negative MEK5 blocks BDNF-induced survival of newly generated granule neurons (Shalizi *et al.*, 2003). In a similar fashion, dominant-negative ERK5 blocks BDNF-dependent survival in E17, but not postnatal day zero (P0), cerebral cortical neurons, suggesting that ERK5 is required for the trophic response to BDNF in young cortical neurons (Liu *et al.*, 2003). In both cerebral cortical and cerebellar granule neurons, the ability of neurotrophins to promote survival in newly generated cells requires MEF2 activation. Dominant-negative forms of MEF2, or RNAi-mediated knockdown of MEF2, block BDNF-dependent survival in newly generated cerebellar granule neurons (Shalizi *et al.*, 2003). Constitutively active MEF2 can protect E17 cortical neurons from BDNF withdrawal or the apoptotic effects of dominant-negative ERK5 (Liu *et al.*, 2003).

At least one MEF2 target gene that is necessary for BDNF-mediated survival of cerebellar granule neurons has been found. BDNF induces the expression of the neurotrophin NT-3 in newly generated, but not mature, cerebellar granule neurons (Shalizi *et al.*, 2003). The BDNF-induced transcription requires the coordinate activation of both the ERK1/2 and the ERK5 signaling pathways, which in turn stimulate CREB and MEF2, respectively (Shalizi *et al.*, 2003). Because in mature granule neurons BDNF does not activate the ERK5-MEF2 signaling pathway, BDNF fails to stimulate NT-3 transcription in these neurons (Shalizi *et al.*, 2003). Depletion of NT-3 with neutralizing antibodies, or by gene knockout, blocks the survival-promoting effects of BDNF (Shalizi *et al.*, 2003). These results indicate that NT-3 is a key mediator of BDNF-induced MEF2-dependent neuronal survival in the developing cerebellum. Consistent with this conclusion, the BDNF- and CNS-specific NT-3 knockout mice phenocopy each other in regards to granule neuron apoptosis (Bates *et al.*, 1999; Schwartz *et al.*, 1997). However, additional targets of ERK5/MEF2 signaling must be required for the full survival response to BDNF, since NT-3 alone is not sufficient to promote survival (Shalizi *et al.*, 2003). Together, studies in this area suggest that MEF2 is a key mediator of the response to neurotrophic factors in the developing CNS.

D. Does MEF2 Make Neurons?

Prevention of apoptosis in response to trophic or calcium signaling is the most clearly established function for MEF2 proteins in the nervous system, but MEF2s may play additional roles in neuronal maturation. Evidence suggesting a role in neuronal maturation comes from studies of the P19 cell line, a pluripotent embryonal carcinoma that differentiates into neuron-like cells upon exposure to retinoic acid. When aggregated in the presence of retinoic acid, P19 cells upregulate MEF2C and D proteins. Inhibition of MEF2-dependent transcription in retinoic acid-differentiated P19 cells by a dominant-negative MEF2 reduces expression of the neuron-specific microtubule-binding protein MAP2, as well as the NR-1 subunit of the NMDA receptor. Furthermore, P19 cells transfected with MEF2C express neuronal markers such as NeuN, neurofilament, and mammalian achaete-scute homolog 1 (MASH1) in the absence of retinoic acid (Krainc et al., 1998; Okamoto et al., 2000; Skerjanc and Wilton, 2000). Despite these tantalizing findings, evidence of involvement of MEF2 proteins in the differentiation of primary neurons remains scant. Unfortunately, the two reported MEF2 knockout mice are uninformative about the contribution of MEF2 proteins to neurogenic processes: MEF2C nullizygous mice are embryonic lethal by E8.5 due to malformation of the right ventricle, while MEF2A knockout mice show no overt neurological defects due to compensation by the closely related MEF2D protein (Lin et al., 1997; Naya et al., 2002). Development of anatomically and temporally restricted conditional knock-out alleles of the MEF2 proteins may prove essential for the study of the contribution of these factors to neuronal development.

IV. Perspectives

The MEF2 proteins are emerging as key players in the development of the CNS, and the variety of neuronal functions regulated by MEF2 may eventually rival or even surpass the diverse roles of the transcription factor CREB in the nervous system, which have taken more than a decade to characterize (Shaywitz and Greenberg, 1999). Beyond its established role in promoting neuronal survival, both in response to activity-dependent depolarization and in neurotrophin stimulation, there is suggestive evidence that MEF2s are involved in the acquisition of the neuronal phenotype (Mao et al., 1999; Okamoto et al., 2000; Shalizi et al., 2003). It is likely that as in muscle, the set of MEF2 -responsive genes in neurons will be dynamically responsive to developmental stage and extracellular stimuli. The set of target genes regulated by MEF2 in neurons remains to be defined, and will benefit from both candidate-gene studies and systematic approaches combining chromatin immunoprecipitation (ChIP) and promoter microarrays.

While there has been a tendency to consider MEF2A, C, and D as interchangeable, and to ignore MEF2B altogether, these proteins clearly have overlapping but nonredundant functions in muscle development (Lin *et al.*, 1997; Naya *et al.*, 2002). Due to the lethal cardiac and muscle phenotypes of the MEF2 knockouts described to date (Lin *et al.*, 1997; Naya *et al.*, 2002), characterizing the nonredundant functions of the different MEF2 proteins in neurons will require the generation of CNS-specific knockout mice. The generation of such mice will hopefully reveal additional roles for MEF2 in CNS development and function.

Interestingly, MEF2A, C, and D are upregulated in the visual cortex during retinotopic reorganization, suggesting a role in neuronal plasticity (Leysen *et al.*, 2004). MEF2 activity in neurons is strongly influenced by electrical activity and calcium-signaling pathways. Several of the calcium-responsive signaling proteins that regulate MEF2 function, particularly CaMKs and calcineurin, have been implicated in learning and memory (Mellstrom and Naranjo, 2001). Based on these observations, it is tempting to speculate that MEF2s might participate in the regulation of synaptic function and plasticity. It will be of particular interest to determine whether survival and potential differentiation functions of MEF2s in neurons are governed by distinct sets of posttranslational modifications or through associations with particular subsets of coactivating or corepressing molecular partners, in a manner that parallels the regulation of differentiation and fiber type switching in skeletal muscle.

Acknowledgements

This work was supported by NIH grant R01_NS41021 (A. B.), and an Albert J. Ryan Foundation fellowship (A. S.). We thank Esther Becker for a critical reading of the manuscript.

References

Allen, M. P., Xu, M., Linseman, D. A., Pawlowski, J. E., Bokoch, G. M., Heidenreich, K. A., and Wierman, M. E. (2002). Adhesion-related kinase repression of gonadotropin-releasing hormone gene expression requires Rac activation of the extracellular signal-regulated kinase pathway. *J. Biol. Chem.* **277,** 38133–38140.

Aramburu, J., Rao, A., and Klee, C. B. (2000). Calcineurin: From structure to function. *Curr. Top Cell Regul.* **36,** 237–295.

Aramburu, J., Heitman, J., and Crabtree, G. R. (2004). Calcineurin: A central controller of signalling in eukaryotes. *EMBO Rep.* **5,** 343–348.

Barsyte-Lovejoy, D., Galanis, A., Clancy, A., and Sharrocks, A. D. (2004). ERK5 is targeted to myocyte enhancer factor 2A (MEF2A) through a MAPK docking motif. *Biochem. J.* **381,** 693–699.

Bates, B., Rios, M., Trumpp, A., Chen, C., Fan, G., Bishop, J. M., and Jaenisch, R. (1999). Neurotrophin-3 is required for proper cerebellar development. *Nat. Neurosci.* **2,** 115–117.

Berger, I., Bieniossek, C., Schaffitzel, C., Hassler, M., Santelli, E., and Richmond, T. J. (2003). Direct interaction of Ca2+/calmodulin inhibits histone deacetylase 5 repressor core binding to myocyte enhancer factor 2. *J. Biol. Chem.* **278,** 17625–17635.

Bertos, N. R., Wang, A. H., and Yang, X. J. (2001). Class II histone deacetylases: Structure, function, and regulation. *Biochem. Cell Biol.* **79,** 243–252.

Bigard, X., Sanchez, H., Zoll, J., Mateo, P., Rousseau, V., Veksler, V., and Ventura-Clapier, R. (2000). Calcineurin Co-regulates contractile and metabolic components of slow muscle phenotype. *J. Biol. Chem.* **275,** 19653–19660.

Bito, H., and Takemoto-Kimura, S. (2003). Ca(2+)/CREB/CBP-dependent gene regulation: A shared mechanism critical in long-term synaptic plasticity and neuronal survival. *Cell Calcium* **34,** 425–430.

Black, B. L., and Olson, E. N. (1998). Transcriptional control of muscle development by myocyte enhancer factor-2 (MEF2) proteins. *Annu. Rev. Cell Dev. Biol.* **14,** 167–196.

Blander, G., and Guarente, L. (2004). The Sir2 family of protein deacetylases. *Annu. Rev. Biochem.* **73,** 417–435.

Borghi, S., Molinari, S., Razzini, G., Parise, F., Battini, R., and Ferrari, S. (2001). The nuclear localization domain of the MEF2 family of transcription factors shows member-specific features and mediates the nuclear import of histone deacetylase 4. *J. Cell Sci.* **114,** 4477–4483.

Breitbart, R. E., Liang, C. S., Smoot, L. B., Laheru, D. A., Mahdavi, V., and Nadal-Ginard, B. (1993). A fourth human MEF2 transcription factor, hMEF2D, is an early marker of the myogenic lineage. *Development* **118,** 1095–1106.

Butts, B. D., Linseman, D. A., Le, S. S., Laessig, T. A., and Heidenreich, K. A. (2003). Insulin-like growth factor-I suppresses degradation of the pro-survival transcription factor myocyte enhancer factor 2D (MEF2D) during neuronal apoptosis. *Horm. Metab. Res.* **35,** 763–770.

Cavanaugh, J. E. (2004). Role of extracellular signal regulated kinase 5 in neuronal survival. *Eur. J. Biochem.* **271,** 2056–2059.

Cavanaugh, J. E., Ham, J., Hetman, M., Poser, S., Yan, C., and Xia, Z. (2001). Differential regulation of mitogen-activated protein kinases ERK1/2 and ERK5 by neurotrophins, neuronal activity, and cAMP in neurons. *J. Neurosci.* **21,** 434–443.

Chan, J. K., Sun, L., Yang, X. J., Zhu, G., and Wu, Z. (2003). Functional characterization of an amino-terminal region of HDAC4 that possesses MEF2 binding and transcriptional repressive activity. *J. Biol. Chem.* **278,** 23515–23521.

Chawla, S., Vanhoutte, P., Arnold, F. J., Huang, C. L., and Bading, H. (2003). Neuronal activity-dependent nucleocytoplasmic shuttling of HDAC4 and HDAC5. *J. Neurochem.* **85,** 151–159.

Chen, S. L., Wang, S. C., Hosking, B., and Muscat, G. E. (2001). Subcellular localization of the steroid receptor coactivators (SRCs) and MEF2 in muscle and rhabdomyosarcoma cells. *Mol. Endocrinol.* **15,** 783–796.

Chen, S. L., Loffler, K. A., Chen, D., Stallcup, M. R., and Muscat, G. E. (2002). The coactivator-associated arginine methyltransferase is necessary for muscle differentiation: CARM1 coactivates myocyte enhancer factor-2. *J. Biol. Chem.* **277,** 4324–4333.

Chin, E. R., Olson, E. N., Richardson, J. A., Yang, Q., Humphries, C., Shelton, J. M., Wu, H., Zhu, W., Bassel-Duby, R., and Williams, R. S. (1998). A calcineurin-dependent transcriptional pathway controls skeletal muscle fiber type. *Genes Dev.* **12,** 2499–2509.

Choi, D. (1998). Antagonizing excitotoxicity: A therapeutic strategy for stroke? *Mt. Sinai J. Med.* **65,** 133–138.

Choi, D. W. (1994). Calcium and excitotoxic neuronal injury. *Ann. N Y Acad. Sci.* **747,** 162–171.

Choi, S. J., Park, S. Y., and Han, T. H. (2001). 14–3-3tau associates with and activates the MEF2D transcription factor during muscle cell differentiation. *Nucleic Acids Res.* **29,** 2836–2842.

Cox, D. M., Du, M., Marback, M., Yang, E. C., Chan, J., Siu, K. W., and McDermott, J. C. (2003). Phosphorylation motifs regulating the stability and function of myocyte enhancer factor 2A. *J. Biol. Chem.* **278,** 15297–15303.

Crabtree, G. R. (2001). Calcium, calcineurin, and the control of transcription. *J. Biol. Chem.* **276,** 2313–2316.

Czermin, B., and Imhof, A. (2003). The sounds of silence—Histone deacetylation meets histone methylation. *Genetica* **117,** 159–164.

Czubryt, M. P., McAnally, J., Fishman, G. I., and Olson, E. N. (2003). Regulation of peroxisome proliferator-activated receptor gamma coactivator 1 alpha (PGC-1 alpha) and mitochondrial function by MEF2 and HDAC5. *Proc. Natl. Acad. Sci. USA* **100,** 1711–1716.

De Luca, A., Severino, A., De Paolis, P., Cottone, G., De Luca, L., De Falco, M., Porcellini, A., Volpe, M., and Condorelli, G. (2003). p300/cAMP-response-element-binding-protein ('CREB')-binding protein (CBP) modulates co-operation between myocyte enhancer factor 2A (MEF2A) and thyroid hormone receptor-retinoid X receptor. *Biochem. J.* **369,** 477–484.

de Ruijter, A. J., van Gennip, A. H., Caron, H. N., Kemp, S., and van Kuilenburg, A. B. (2003). Histone deacetylases (HDACs): Characterization of the classical HDAC family. *Biochem. J.* **370,** 737–749.

Delling, U., Tureckova, J., Lim, H. W., De Windt, L. J., Rotwein, P., and Molkentin, J. D. (2000). A calcineurin-NFATc3-dependent pathway regulates skeletal muscle differentiation and slow myosin heavy-chain expression. *Mol. Cell. Biol.* **20,** 6600–6611.

Dichoso, D., Brodigan, T., Chwoe, K. Y., Lee, J. S., Llacer, R., Park, M., Corsi, A. K., Kostas, S. A., Fire, A., Ahnn, J., and Krause, M. (2000). The MADS-Box factor CeMEF2 is not essential for Caenorhabditis elegans myogenesis and development. *Dev. Biol.* **223,** 431–440.

Dinev, D., Jordan, B. W., Neufeld, B., Lee, J. D., Lindemann, D., Rapp, U. R., and Ludwig, S. (2001). Extracellular signal regulated kinase 5 (ERK5) is required for the differentiation of muscle cells. *EMBO Rep.* **2,** 829–834.

Dodou, E., Sparrow, D. B., Mohun, T., and Treisman, R. (1995). MEF2 proteins, including MEF2A, are expressed in both muscle and non-muscle cells. *Nucleic Acids Res.* **23,** 4267–4274.

Dressel, U., Bailey, P. J., Wang, S. C., Downes, M., Evans, R. M., and Muscat, G. E. (2001). A dynamic role for HDAC7 in MEF2-mediated muscle differentiation. *J. Biol. Chem.* **276,** 17007–17013.

Dugan, L. L., and Choi, D. W. (1994). Excitotoxicity, free radicals, and cell membrane changes. *Ann. Neurol.* **35**(Suppl.), S17–S21.

Dunn, S. E., Simard, A. R., Bassel-Duby, R., Williams, R. S., and Michel, R. N. (2001). Nerve activity-dependent modulation of calcineurin signaling in adult fast and slow skeletal muscle fibers. *J. Biol. Chem.* **276,** 45243–45254.

Edmondson, D. G., Lyons, G. E., Martin, J. F., and Olson, E. N. (1994). Mef2 gene expression marks the cardiac and skeletal muscle lineages during mouse embryogenesis. *Development* **120,** 1251–1263.

Featherstone, M. (2002). Coactivators in transcription initiation: Here are your orders. *Curr. Opin. Genet. Dev.* **12,** 149–155.

Fukuhara, S., Marinissen, M. J., Chiariello, M., and Gutkind, J. S. (2000). Signaling from G protein-coupled receptors to ERK5/Big MAPK 1 involves Galpha q and Galpha 12/13 families of heterotrimeric G proteins. Evidence for the existence of a novel Ras AND Rho-independent pathway. *J. Biol. Chem.* **275,** 21730–21736.

Gallin, W. J., and Greenberg, M. E. (1995). Calcium regulation of gene expression in neurons: The mode of entry matters. *Curr. Opin. Neurobiol.* **5,** 367–374.

Gaudilliere, B., Shi, Y., and Bonni, A. (2002). RNA interference reveals a requirement for myocyte enhancer factor 2A in activity-dependent neuronal survival. *J. Biol. Chem.* **277,** 46442–46446.

Ghosh, A., and Greenberg, M. E. (1995). Calcium signaling in neurons: Molecular mechanisms and cellular consequences. *Science* **268,** 239–247.

Gong, X., Tang, X., Wiedmann, M., Wang, X., Peng, J., Zheng, D., Blair, L. A., Marshall, J., and Mao, Z. (2003). Cdk5-mediated inhibition of the protective effects of transcription factor MEF2 in neurotoxicity-induced apoptosis. *Neuron* **38,** 33–46.

Gregoire, S., and Yang, X. J. (2005). Association with class IIa histone deacetylases upregulates the sumoylation of MEF2 transcription factors. *Mol. Cell. Biol.* **25,** 2273–2287.

Grozinger, C. M., and Schreiber, S. L. (2000). Regulation of histone deacetylase 4 and 5 and transcriptional activity by 14–3-3-dependent cellular localization. *Proc. Natl. Acad. Sci. USA* **97,** 7835–7840.

Han, J., Jiang, Y., Li, Z., Kravchenko, V. V., and Ulevitch, R. J. (1997). Activation of the transcription factor MEF2C by the MAP kinase p38 in inflammation. *Nature* **386,** 296–299.

Handschin, C., Rhee, J., Lin, J., Tarr, P. T., and Spiegelman, B. M. (2003). An autoregulatory loop controls peroxisome proliferator-activated receptor gamma coactivator 1alpha expression in muscle. *Proc. Natl. Acad. Sci. USA* **100,** 7111–7116.

Heerssen, H. M., and Segal, R. A. (2002). Location, location, location: A spatial view of neurotrophin signal transduction. *Trends Neurosci.* **25,** 160–165.

Hemenway, C. S., and Heitman, J. (1999). Calcineurin. Structure, function, and inhibition. *Cell Biochem. Biophys.* **30,** 115–151.

Hobson, G. M., Krahe, R., Garcia, E., Siciliano, M. J., and Funanage, V. L. (1995). Regional chromosomal assignments for four members of the MADS domain transcription enhancer factor 2 (MEF2) gene family to human chromosomes 15q26, 19p12, 5q14, and 1q12-q23. *Genomics* **29,** 704–711.

Huang, E. J., and Reichardt, L. F. (2001). Neurotrophins: Roles in neuronal development and function. *Annu. Rev. Neurosci.* **24,** 677–736.

Hughes, S. M. (1998). Muscle development: Electrical control of gene expression. *Curr. Biol.* **8,** R892–R894.

Kao, H. Y., Verdel, A., Tsai, C. C., Simon, C., Juguilon, H., and Khochbin, S. (2001). Mechanism for nucleocytoplasmic shuttling of histone deacetylase 7. *J. Biol. Chem.* **276,** 47496–47507.

Kaplan, D. R., and Miller, F. D. (1997). Signal transduction by the neurotrophin receptors. *Curr. Opin. Cell Biol.* **9,** 213–221.

Kaplan, D. R., and Miller, F. D. (2000). Neurotrophin signal transduction in the nervous system. *Curr. Opin. Neurobiol.* **10,** 381–391.

Kasler, H. G., Victoria, J., Duramad, O., and Winoto, A. (2000). ERK5 is a novel type of mitogen-activated protein kinase containing a transcriptional activation domain. *Mol. Cell. Biol.* **20,** 8382–8389.

Kato, Y., Kravchenko, V. V., Tapping, R. I., Han, J., Ulevitch, R. J., and Lee, J. D. (1997). BMK1/ERK5 regulates serum-induced early gene expression through transcription factor MEF2C. *EMBO J.* **16,** 7054–7066.

Kato, Y., Zhao, M., Morikawa, A., Sugiyama, T., Chakravortty, D., Koide, N., Yoshida, T., Tapping, R. I., Yang, Y., Yokochi, T., and Lee, J. D. (2000). Big mitogen-activated kinase regulates multiple members of the MEF2 protein family. *J. Biol. Chem.* **275,** 18534–18540.

Kolodziejczyk, S. M., Wang, L., Balazsi, K., De Repentigny, Y., Kothary, R., and Megeney, L. A. (1999). MEF2 is upregulated during cardiac hypertrophy and is required for normal post-natal growth of the myocardium. *Curr. Biol.* **9,** 1203–1206.

Krainc, D., Bai, G., Okamoto, S., Carles, M., Kusiak, J. W., Brent, R. N., and Lipton, S. A. (1998). Synergistic activation of the N-methyl-D-aspartate receptor subunit 1 promoter by myocyte enhancer factor 2C and Sp1. *J. Biol. Chem.* **273,** 26218–26224.

Kuo, M. H., and Allis, C. D. (1998). Roles of histone acetyltransferases and deacetylases in gene regulation. *Bioessays* **20,** 615–626.

Lazaro, J. B., Bailey, P. J., and Lassar, A. B. (2002). Cyclin D-cdk4 activity modulates the subnuclear localization and interaction of MEF2 with SRC-family coactivators during skeletal muscle differentiation. *Genes Dev.* **16,** 1792–1805.

Leifer, D., Krainc, D., Yu, Y. T., McDermott, J., Breitbart, R. E., Heng, J., Neve, R. L., Kosofsky, B., Nadal-Ginard, B., and Lipton, S. A. (1993). MEF2C, a MADS/MEF2-family transcription factor expressed in a laminar distribution in cerebral cortex. *Proc. Natl. Acad. Sci. USA* **90,** 1546–1550.

Leifer, D., Golden, J., and Kowall, N. W. (1994). Myocyte-specific enhancer binding factor 2C expression in human brain development. *Neuroscience* **63,** 1067–1079.

Lemercier, C., Verdel, A., Galloo, B., Curtet, S., Brocard, M. P., and Khochbin, S. (2000). mHDA1/HDAC5 histone deacetylase interacts with and represses MEF2A transcriptional activity. *J. Biol. Chem.* **275,** 15594–15599.

Levi-Montalcini, R. (1987). The nerve growth factor 35 years later. *Science* **237,** 1154–1162.

Leysen, I., Van der Gucht, E., Eysel, U. T., Huybrechts, R., Vandesande, F., and Arckens, L. (2004). Time-dependent changes in the expression of the MEF2 transcription factor family during topographic map reorganization in mammalian visual cortex. *Eur. J. Neurosci.* **20,** 769–780.

Li, M., Linseman, D. A., Allen, M. P., Meintzer, M. K., Wang, X., Laessig, T., Wierman, M. E., and Heidenreich, K. A. (2001). Myocyte enhancer factor 2A and 2D undergo phosphorylation and caspase-mediated degradation during apoptosis of rat cerebellar granule neurons. *J. Neurosci.* **21,** 6544–6552.

Li, X., Song, S., Liu, Y., Ko, S. H., and Kao, H. Y. (2004). Phosphorylation of the histone deacetylase 7 modulates its stability and association with 14-3-3 proteins. *J. Biol. Chem.* **279,** 34201–34208.

Lilly, B., Galewsky, S., Firulli, A. B., Schulz, R. A., and Olson, E. N. (1994). D-MEF2: A MADS box transcription factor expressed in differentiating mesoderm and muscle cell lineages during Drosophila embryogenesis. *Proc. Natl. Acad. Sci. USA* **91,** 5662–5666.

Lin, J., Wu, H., Tarr, P. T., Zhang, C. Y., Wu, Z., Boss, O., Michael, L. F., Puigserver, P., Isotani, E., Olson, E. N., Lowell, B. B., Bassel-Duby, R., and Spiegelman, B. M. (2002). Transcriptional co-activator PGC-1 alpha drives the formation of slow-twitch muscle fibres. *Nature* **418,** 797–801.

Lin, Q., Schwarz, J., Bucana, C., and Olson, E. N. (1997). Control of mouse cardiac morphogenesis and myogenesis by transcription factor MEF2C. *Science* **276,** 1404–1407.

Lin, X., Shah, S., and Bulleit, R. F. (1996). The expression of MEF2 genes is implicated in CNS neuronal differentiation. *Brain Res. Mol. Brain Res.* **42,** 307–316.

Linseman, D. A., Bartley, C. M., Le, S. S., Laessig, T. A., Bouchard, R. J., Meintzer, M. K., Li, M., and Heidenreich, K. A. (2003a). Inactivation of the myocyte enhancer factor-2 repressor histone deacetylase-5 by endogenous Ca(2+)/calmodulin-dependent kinase II promotes depolarization-mediated cerebellar granule neuron survival. *J. Biol. Chem.* **278,** 41472–41481.

Linseman, D. A., Cornejo, B. J., Le, S. S., Meintzer, M. K., Laessig, T. A., Bouchard, R. J., and Heidenreich, K. A. (2003b). A myocyte enhancer factor 2D (MEF2D) kinase activated during neuronal apoptosis is a novel target inhibited by lithium. *J. Neurochem.* **85,** 1488–1499.

Liu, D., Kang, J. S., and Derynck, R. (2004). TGF-beta-activated Smad3 represses MEF2-dependent transcription in myogenic differentiation. *EMBO J.* **23,** 1557–1566.

Liu, L., Cavanaugh, J. E., Wang, Y., Sakagami, H., Mao, Z., and Xia, Z. (2003). ERK5 activation of MEF2-mediated gene expression plays a critical role in BDNF-promoted survival of developing but not mature cortical neurons. *Proc. Natl. Acad. Sci. USA* **100,** 8532–8537.

Lu, J., McKinsey, T. A., Nicol, R. L., and Olson, E. N. (2000a). Signal-dependent activation of the MEF2 transcription factor by dissociation from histone deacetylases. *Proc. Natl. Acad. Sci. USA* **97,** 4070–4075.

Lu, J., McKinsey, T. A., Zhang, C. L., and Olson, E. N. (2000b). Regulation of skeletal myogenesis by association of the MEF2 transcription factor with class II histone deacetylases. *Mol. Cell.* **6,** 233–244.

Lyons, G. E., Micales, B. K., Schwarz, J., Martin, J. F., and Olson, E. N. (1995). Expression of mef2 genes in the mouse central nervous system suggests a role in neuronal maturation. *J. Neurosci.* **15,** 5727–5738.

Mao, Z., and Wiedmann, M. (1999). Calcineurin enhances MEF2 DNA binding activity in calcium-dependent survival of cerebellar granule neurons. *J. Biol. Chem.* **274,** 31102–31107.

Mao, Z., Bonni, A., Xia, F., Nadal-Vicens, M., and Greenberg, M. E. (1999). Neuronal activity-dependent cell survival mediated by transcription factor MEF2. *Science* **286,** 785–790.

Marinissen, M. J., Chiariello, M., Pallante, M., and Gutkind, J. S. (1999). A network of mitogen-activated protein kinases links G protein-coupled receptors to the c-jun promoter: A role for c-Jun NH2-terminal kinase, p38s, and extracellular signal-regulated kinase 5. *Mol. Cell. Biol.* **19,** 4289–4301.

Martin, J. F., Miano, J. M., Hustad, C. M., Copeland, N. G., Jenkins, N. A., and Olson, E. N. (1994). A Mef2 gene that generates a muscle-specific isoform via alternative mRNA splicing. *Mol. Cell. Biol.* **14,** 1647–1656.

McDermott, J. C., Cardoso, M. C., Yu, Y. T., Andres, V., Leifer, D., Krainc, D., Lipton, S. A., and Nadal-Ginard, B. (1993). hMEF2C gene encodes skeletal muscle- and brain-specific transcription factors. *Mol. Cell. Biol.* **13,** 2564–2577.

McKinsey, T. A., Zhang, C. L., Lu, J., and Olson, E. N. (2000a). Signal-dependent nuclear export of a histone deacetylase regulates muscle differentiation. *Nature* **408,** 106–111.

McKinsey, T. A., Zhang, C. L., and Olson, E. N. (2000b). Activation of the myocyte enhancer factor-2 transcription factor by calcium/calmodulin-dependent protein kinase-stimulated binding of 14-3-3 to histone deacetylase 5. *Proc. Natl. Acad. Sci. USA* **97,** 14400–14405.

McKinsey, T. A., Zhang, C. L., and Olson, E. N. (2001a). Control of muscle development by dueling HATs and HDACs. *Curr. Opin. Genet. Dev.* **11,** 497–504.

McKinsey, T. A., Zhang, C. L., and Olson, E. N. (2001b). Identification of a signal-responsive nuclear export sequence in class II histone deacetylases. *Mol. Cell. Biol.* **21,** 6312–6321.

McKinsey, T. A., Zhang, C. L., and Olson, E. N. (2002a). MEF2: A calcium-dependent regulator of cell division, differentiation and death. *Trends Biochem. Sci.* **27,** 40–47.

McKinsey, T. A., Zhang, C. L., and Olson, E. N. (2002b). Signaling chromatin to make muscle. *Curr. Opin. Cell Biol.* **14,** 763–772.

Mejat, A., Ramond, F., Bassel-Duby, R., Khochbin, S., Olson, E. N., and Schaeffer, L. (2005). Histone deacetylase 9 couples neuronal activity to muscle chromatin acetylation and gene expression. *Nat. Neurosci.* **8,** 313–321.

Mellstrom, B., and Naranjo, J. R. (2001). Mechanisms of Ca(2+)-dependent transcription. *Curr. Opin. Neurobiol.* **11,** 312–319.

Miller, F. D., and Kaplan, D. R. (2001). Neurotrophin signalling pathways regulating neuronal apoptosis. *Cell Mol. Life Sci.* **58,** 1045–1053.

Miska, E. A., Karlsson, C., Langley, E., Nielsen, S. J., Pines, J., and Kouzarides, T. (1999). HDAC4 deacetylase associates with and represses the MEF2 transcription factor. *EMBO J.* **18,** 5099–5107.

Miska, E. A., Langley, E., Wolf, D., Karlsson, C., Pines, J., and Kouzarides, T. (2001). Differential localization of HDAC4 orchestrates muscle differentiation. *Nucleic Acids Res.* **29,** 3439–3447.

Molkentin, J. D., Black, B. L., Martin, J. F., and Olson, E. N. (1996a). Mutational analysis of the DNA binding, dimerization, and transcriptional activation domains of MEF2C. *Mol. Cell. Biol.* **16,** 2627–2636.

Molkentin, J. D., Firulli, A. B., Black, B. L., Martin, J. F., Hustad, C. M., Copeland, N., Jenkins, N., Lyons, G., and Olson, E. N. (1996b). MEF2B is a potent transactivator expressed in early myogenic lineages. *Mol. Cell. Biol.* **16,** 3814–3824.

Molkentin, J. D., Li, L., and Olson, E. N. (1996c). Phosphorylation of the MADS-Box transcription factor MEF2C enhances its DNA binding activity. *J. Biol. Chem.* **271,** 17199–17204.

Morisaki, T., Sermsuvitayawong, K., Byun, S. H., Matsuda, Y., Hidaka, K., Morisaki, H., and Mukai, T. (1997). Mouse Mef2b gene: Unique member of MEF2 gene family. *J. Biochem. (Tokyo)* **122,** 939–946.

Naya, F. J., Black, B. L., Wu, H., Bassel-Duby, R., Richardson, J. A., Hill, J. A., and Olson, E. N. (2002). Mitochondrial deficiency and cardiac sudden death in mice lacking the MEF2A transcription factor. *Nat. Med.* **8,** 1303–1309.

Nguyen, H. T., Bodmer, R., Abmayr, S. M., McDermott, J. C., and Spoerel, N. A. (1994). D-mef2: A Drosophila mesoderm-specific MADS box-containing gene with a biphasic expression profile during embryogenesis. *Proc. Natl. Acad. Sci. USA* **91,** 7520–7524.

Okamoto, S., Krainc, D., Sherman, K., and Lipton, S. A. (2000). Antiapoptotic role of the p38 mitogen-activated protein kinase-myocyte enhancer factor 2 transcription factor pathway during neuronal differentiation. *Proc. Natl. Acad. Sci. USA* **97,** 7561–7566.

Okamoto, S., Li, Z., Ju, C., Scholzke, M. N., Mathews, E., Cui, J., Salvesen, G. S., Bossy-Wetzel, E., and Lipton, S. A. (2002). Dominant-interfering forms of MEF2 generated by caspase cleavage contribute to NMDA-induced neuronal apoptosis. *Proc. Natl. Acad. Sci. USA* **99,** 3974–3979.

Olson, E. N., and Williams, R. S. (2000). Remodeling muscles with calcineurin. *Bioessays* **22,** 510–519.

Ornatsky, O. I., Andreucci, J. J., and McDermott, J. C. (1997). A dominant-negative form of transcription factor MEF2 inhibits myogenesis. *J. Biol. Chem.* **272,** 33271–33278.

Ornatsky, O. I., Cox, D. M., Tangirala, P., Andreucci, J. J., Quinn, Z. A., Wrana, J. L., Prywes, R., Yu, Y. T., and McDermott, J. C. (1999). Post-translational control of the MEF2A transcriptional regulatory protein. *Nucleic Acids Res.* **27,** 2646–26454.

Ornatsky, O. I., and McDermott, J. C. (1996). MEF2 protein expression, DNA binding specificity and complex composition, and transcriptional activity in muscle and non-muscle cells. *J. Biol. Chem.* **271,** 24927–24933.

Penn, B. H., Bergstrom, D. A., Dilworth, F. J., Bengal, E., and Tapscott, S. J. (2004). A MyoD-generated feed-forward circuit temporally patterns gene expression during skeletal muscle differentiation. *Genes Dev.* **18,** 2348–2353.

Polster, B. M., and Fiskum, G. (2004). Mitochondrial mechanisms of neural cell apoptosis. *J. Neurochem.* **90,** 1281–1289.

Puigserver, P., and Spiegelman, B. M. (2003). Peroxisome proliferator-activated receptor-gamma coactivator 1 alpha (PGC-1 alpha): Transcriptional coactivator and metabolic regulator. *Endocr. Rev.* **24,** 78–90.

Puri, P. L., Wu, Z., Zhang, P., Wood, L. D., Bhakta, K. S., Han, J., Feramisco, J. R., Karin, M., and Wang, J. Y. (2000). Induction of terminal differentiation by constitutive activation of p38 MAP kinase in human rhabdomyosarcoma cells. *Genes Dev.* **14,** 574–584.

Regan, C. P., Li, W., Boucher, D. M., Spatz, S., Su, M. S., and Kuida, K. (2002). Erk5 null mice display multiple extraembryonic vascular and embryonic cardiovascular defects. *Proc. Natl. Acad. Sci. USA* **99,** 9248–9253.

Rescan, P. Y. (2001). Regulation and functions of myogenic regulatory factors in lower vertebrates. *Comp. Biochem. Physiol. B Biochem. Mol. Biol.* **130,** 1–12.

Rubin, L. L., Gatchalian, C. L., Rimon, G., and Brooks, S. F. (1994). The molecular mechanisms of neuronal apoptosis. *Curr. Opin. Neurobiol.* **4,** 696–702.

Sartorelli, V., Huang, J., Hamamori, Y., and Kedes, L. (1997). Molecular mechanisms of myogenic coactivation by p300: Direct interaction with the activation domain of MyoD and with the MADS box of MEF2C. *Mol. Cell. Biol.* **17,** 1010–1026.

Schulz, R. A., Chromey, C., Lu, M. F., Zhao, B., and Olson, E. N. (1996). Expression of the D-MEF2 transcription in the Drosophila brain suggests a role in neuronal cell differentiation. *Oncogene* **12,** 1827–1831.

Schwartz, P. M., Borghesani, P. R., Levy, R. L., Pomeroy, S. L., and Segal, R. A. (1997). Abnormal cerebellar development and foliation in BDNF-/- mice reveals a role for neurotrophins in CNS patterning. *Neuron* **19,** 269–281.

Serrano, A. L., Murgia, M., Pallafacchina, G., Calabria, E., Coniglio, P., Lomo, T., and Schiaffino, S. (2001). Calcineurin controls nerve activity-dependent specification of slow skeletal muscle fibers but not muscle growth. *Proc. Natl. Acad. Sci. USA* **98,** 13108–13113.

Shalizi, A., Lehtinen, M., Gaudilliere, B., Donovan, N., Han, J., Konishi, Y., and Bonni, A. (2003). Characterization of a neurotrophin signaling mechanism that mediates neuron survival in a temporally specific pattern. *J. Neurosci.* **23,** 7326–7336.

Shaywitz, A. J., and Greenberg, M. E. (1999). CREB: A stimulus-induced transcription factor activated by a diverse array of extracellular signals. *Annu. Rev. Biochem.* **68,** 821–861.

Skerjanc, I. S., and Wilton, S. (2000). Myocyte enhancer factor 2C upregulates MASH-1 expression and induces neurogenesis in P19 cells. *FEBS Lett.* **472,** 53–56.

Sparrow, D. B., Miska, E. A., Langley, E., Reynaud-Deonauth, S., Kotecha, S., Towers, N., Spohr, G., Kouzarides, T., and Mohun, T. J. (1999). MEF-2 function is modified by a novel co-repressor, MITR. *EMBO J.* **18,** 5085–5098.

Spring, J., Yanze, N., Josch, C., Middel, A. M., Winninger, B., and Schmid, V. (2002). Conservation of Brachyury, Mef2, and Snail in the myogenic lineage of jellyfish: A connection to the mesoderm of bilateria. *Dev. Biol.* **244,** 372–384.

Suzaki, Y., Yoshizumi, M., Kagami, S., Koyama, A. H., Taketani, Y., Houchi, H., Tsuchiya, K., Takeda, E., and Tamaki, T. (2002). Hydrogen peroxide stimulates c-Src-mediated big mitogen-activated protein kinase 1 (BMK1) and the MEF2C signaling pathway in PC12 cells: Potential role in cell survival following oxidative insults. *J. Biol. Chem.* **277,** 9614–9621.

Swoap, S. J., Hunter, R. B., Stevenson, E. J., Felton, H. M., Kansagra, N. V., Lang, J. M., Esser, K. A., and Kandarian, S. C. (2000). The calcineurin-NFAT pathway and muscle fiber-type gene expression. *Am J. Physiol. Cell Physiol.* **279,** C915–C924.

Theissen, G., Kim, J. T., and Saedler, H. (1996). Classification and phylogeny of the MADS-box multigene family suggest defined roles of MADS-box gene subfamilies in the morphological evolution of eukaryotes. *J. Mol. Evol.* **43,** 484–516.

Verdin, E., Dequiedt, F., and Kasler, H. G. (2003). Class II histone deacetylases: Versatile regulators. *Trends Genet* **19,** 286–293.

Wang, A. H., and Yang, X. J. (2001). Histone deacetylase 4 possesses intrinsic nuclear import and export signals. *Mol. Cell. Biol.* **21,** 5992–6005.

Wang, A. H., Bertos, N. R., Vezmar, M., Pelletier, N., Crosato, M., Heng, H. H., Th'ng, J., Han, J., and Yang, X. J. (1999). HDAC4, a human histone deacetylase related to yeast HDA1, is a transcriptional corepressor. *Mol. Cell. Biol.* **19,** 7816–7827.

Wang, A. H., Kruhlak, M. J., Wu, J., Bertos, N. R., Vezmar, M., Posner, B. I., Bazett-Jones, D. P., and Yang, X. J. (2000). Regulation of histone deacetylase 4 by binding of 14-3-3 proteins. *Mol. Cell. Biol.* **20,** 6904–6912.

Watson, F. L., Heerssen, H. M., Bhattacharyya, A., Klesse, L., Lin, M. Z., and Segal, R. A. (2001). Neurotrophins use the Erk5 pathway to mediate a retrograde survival response. *Nat. Neurosci.* **4,** 981–988.

West, A. E., Chen, W. G., Dalva, M. B., Dolmetsch, R. E., Kornhauser, J. M., Shaywitz, A. J., Takasu, M. A., Tao, X., and Greenberg, M. E. (2001). Calcium regulation of neuronal gene expression. *Proc. Natl. Acad. Sci. USA* **98,** 11024–11031.

West, A. G., Shore, P., and Sharrocks, A. D. (1997). DNA binding by MADS-box transcription factors: A molecular mechanism for differential DNA bending. *Mol. Cell. Biol.* **17,** 2876–2887.

Wu, H., and Olson, E. N. (2002). Activation of the MEF2 transcription factor in skeletal muscles from myotonic mice. *J. Clin Invest* **109,** 1327–1333.

Wu, H., Naya, F. J., McKinsey, T. A., Mercer, B., Shelton, J. M., Chin, E. R., Simard, A. R., Michel, R. N., Bassel-Duby, R., Olson, E. N., and Williams, R. S. (2000). MEF2 responds to multiple calcium-regulated signals in the control of skeletal muscle fiber type. *EMBO J.* **19,** 1963–1973.

Wu, H., Kanatous, S. B., Thurmond, F. A., Gallardo, T., Isotani, E., Bassel-Duby, R., and Williams, R. S. (2002). Regulation of mitochondrial biogenesis in skeletal muscle by CaMK. *Science* **296,** 349–352.

Xu, J., and Li, Q. (2003). Review of the *in vivo* functions of the p160 steroid receptor coactivator family. *Mol. Endocrinol.* **17,** 1681–1692.

Yang, C. C., Ornatsky, O. I., McDermott, J. C., Cruz, T. F., and Prody, C. A. (1998). Interaction of myocyte enhancer factor 2 (MEF2) with a mitogen-activated protein kinase, ERK5/BMK1. *Nucleic Acids Res.* **26,** 4771–4777.

Yang, S. H., and Sharrocks, A. D. (2004). SUMO promotes HDAC-mediated transcriptional repression. *Mol. Cell* **13,** 611–617.

Yang, S. H., Galanis, A., and Sharrocks, A. D. (1999). Targeting of p38 mitogen-activated protein kinases to MEF2 transcription factors. *Mol. Cell. Biol.* **19,** 4028–4038.

Youn, H. D., and Liu, J. O. (2000). Cabin1 represses MEF2-dependent Nur77 expression and T cell apoptosis by controlling association of histone deacetylases and acetylases with MEF2. *Immunity* **13,** 85–94.

Youn, H. D., Sun, L., Prywes, R., and Liu, J. O. (1999). Apoptosis of T cells mediated by Ca2+-induced release of the transcription factor MEF2. *Science* **286,** 790–793.

Youn, H. D., Chatila, T. A., and Liu, J. O. (2000a). Integration of calcineurin and MEF2 signals by the coactivator p300 during T-cell apoptosis. *EMBO J.* **19,** 4323–4331.

Youn, H. D., Grozinger, C. M., and Liu, J. O. (2000b). Calcium regulates transcriptional repression of myocyte enhancer factor 2 by histone deacetylase 4. *J. Biol. Chem.* **275,** 22563–22567.

Yu, Y. T. (1996). Distinct domains of myocyte enhancer binding factor-2A determining nuclear localization and cell type-specific transcriptional activity. *J. Biol. Chem.* **271,** 24675–24683.

Yu, Y. T., Breitbart, R. E., Smoot, L. B., Lee, Y., Mahdavi, V., and Nadal-Ginard, B. (1992). Human myocyte-specific enhancer factor 2 comprises a group of tissue-restricted MADS box transcription factors. *Genes Dev.* **6,** 1783–1798.

Yuan, J., and Yankner, B. A. (2000). Apoptosis in the nervous system. *Nature* **407,** 802–809.

Zetser, A., Gredinger, E., and Bengal, E. (1999). p38 mitogen-activated protein kinase pathway promotes skeletal muscle differentiation. Participation of the Mef2c transcription factor. *J. Biol. Chem.* **274,** 5193–5200.

Zhang, C. L., McKinsey, T. A., Lu, J. R., and Olson, E. N. (2001a). Association of COOH-terminal-binding protein (CtBP) and MEF2-interacting transcription repressor (MITR) contributes to transcriptional repression of the MEF2 transcription factor. *J. Biol. Chem.* **276,** 35–39.

Zhang, C. L., McKinsey, T. A., and Olson, E. N. (2001b). The transcriptional corepressor MITR is a signal-responsive inhibitor of myogenesis. *Proc. Natl. Acad. Sci. USA* **98,** 7354–7359.

Zhang, C. L., McKinsey, T. A., and Olson, E. N. (2002). Association of class II histone deacetylases with heterochromatin protein 1: Potential role for histone methylation in control of muscle differentiation. *Mol. Cell. Biol.* **22,** 7302–7312.

Zhao, M., New, L., Kravchenko, V. V., Kato, Y., Gram, H., di Padova, F., Olson, E. N., Ulevitch, R. J., and Han, J. (1999). Regulation of the MEF2 family of transcription factors by p38. *Mol. Cell. Biol.* **19,** 21–30.

Zhao, X., Ito, A., Kane, C. D., Liao, T. S., Bolger, T. A., Lemrow, S. M., Means, A. R., and Yao, T. P. (2001). The modular nature of histone deacetylase HDAC4 confers phosphorylation-dependent intracellular trafficking. *J. Biol. Chem.* **276,** 35042–35048.

Zhou, G., Bao, Z. Q., and Dixon, J. E. (1995). Components of a new human protein kinase signal transduction pathway. *J. Biol. Chem.* **270,** 12665–13669.

Zhu, B., and Gulick, T. (2004). Phosphorylation and alternative pre-mRNA splicing converge to regulate myocyte enhancer factor 2C activity. *Mol. Cell. Biol.* **24,** 8264–8275.

Zipfel, G. J., Babcock, D. J., Lee, J. M., and Choi, D. W. (2000). Neuronal apoptosis after CNS injury: The roles of glutamate and calcium. *J. Neurotrauma* **17,** 857–869.

10

Mechanisms of Axon Guidance in the Developing Nervous System

Céline Plachez and Linda J. Richards[†]*
*Department of Anatomy and Neurobiology, University of Maryland
School of Medicine, Baltimore, Maryland 21201
[†]University of Queensland, School of Biomedical Sciences and The Queensland
Brain Institute, St. Lucia, Queensland 4072, Australia

Current Topics in Developmental Biology, Vol. 69
Copyright 2005, Elsevier Inc. All rights reserved.

0070-2153/05 $35.00
DOI: 10.1016/S0070-2153(05)69010-2

The human brain assembles an incredible network of over a billion neurons. Understanding how these connections form during development in order for the brain to function properly is a fundamental question in biology. Much of this wiring takes place during embryonic development. Neurons are generated in the ventricular zone, migrate out, and begin to differentiate. However, neurons are often born in locations some distance from the target cells with which they will ultimately form connections. To form connections, neurons project long axons tipped with a specialized sensing device called a growth cone. The growing axons interact directly with molecules within the environment through which they grow. In order to find their targets, axonal growth cones use guidance molecules that can either attract or repel them. Understanding what these guidance cues are, where they are expressed, and how the growth cone is able to transduce their signal in a directionally specific manner is essential to understanding how the functional brain is constructed. In this chapter, we review what is known about the mechanisms involved in axonal guidance. We discuss how the growth cone is able to sense and respond to its environment and how it is guided by pioneering cells and axons. As examples, we discuss current models for the development of the spinal cord, the cerebral cortex, and the visual and olfactory systems. © 2005, Elsevier Inc.

I. Introduction

The establishment of correct neuronal connections is crucial for proper functioning of the nervous system. In the human brain, there are more than a billion neurons, of many different types, that assemble into highly complex neuronal networks with over 10^9 connections. To understand the brain it is essential to understand how this network of neuronal connections is achieved with such high precision as the nervous system develops. Santiago Ramón y Cajal studied the developing mammalian brain and proposed that each axon is attracted to its target cells by diffusible molecules secreted from it. Direct evidence supporting this guidance mechanism has only started to emerge during the past decade, with an increasing number of families of guidance molecules being identified. It is well established that developing axons are guided to their targets by a variety of environmental cues, including long-range diffusible and short-range surface-bound molecules that can either attract or repel the axon. The presence of these guidance cues in a temporal and spatial pattern enables the axon to navigate through the complex environment of the developing embryo to reach its correct target and to stop growing once it arrives. This chapter describes our current understanding of the mechanisms involved in the formation of neuronal connections.

II. The Growth Cone

In order to connect with targets, neurons extend processes called axons. At the tip of the extending axon is a specialized sensing device called a growth cone. In 1880, Santiago Ramón y Cajal observed and named the growth cone (*"cono de crecimiento"*) as the motile structure at the leading edge of extending axons. During development of the nervous system, growth cones navigate along specific pathways, recognize their targets, and then form synaptic connections by elaborating terminal arbors. En route to their target, growth cones show stereotyped behavior by following appropriate pathways and by interacting with intermediate targets. These characteristic behaviors of axons are regulated by a combination of various types of guidance cues, including attractive and repulsive cues. How growth cones respond to guidance cues is not fixed and changes dynamically during navigation. Growth cones derive their directional information from a variety of extracellular guidance cues. These cues can take the form of both secreted and membrane-bound attractive and repulsive factors. To date, a large number of these guidance cues and their receptors have been identified (Mueller, 1999; Tessier-Lavigne and Goodman, 1996). There is also an extensive collection of actin-associated proteins that regulate the structure and dynamics of the actin cytoskeleton that makes up the motile apparatus of the growth cone (Stossel, 1993).

The growth cone has two major domains: the central domain (C domain), rich in microtubules and membranous organelles, and the peripheral domain, rich in actin filaments (Fig. 1). The peripheral domain consists of lamellipodia surrounding a central domain and filopodia extending from the outer edge of the peripheral domain. Filopodial and lamellipodial projections are sent out by the growth cone to sense its environment; they contain dense and highly polymerized actin microfilaments. Both lamellipodia and filopodia are involved in growth cone motility and undergo continuous cycles of expansion and contraction (Bray and Chapman, 1985; Goldberg and Burmeister, 1986). The cytoskeleton of the growth cone is a combination of dynamic cytoplasmic filamentous structures, including microtubules and microfilaments (Forscher and Smith, 1988; Letourneau *et al.*, 1986; Luduena and Wessells, 1973; Yamada *et al.*, 1970, 1971).

The microtubules are involved in axon extension; they extend from the neurite shaft and splay into the central part of the growth cone. Microtubules are polar structures with one end termed the "plus" end and the other end termed the "minus" end. They are composed of heterodimers of α- and β-tubulin that self-assemble to form polymers. Microtubules form dense arrays within the neurite, while they are less stable within the growth cone (Ahmad *et al.*, 1993). As microtubules enter the growth cone from the neurite shaft, the bundled microtubules separate from each other and extend

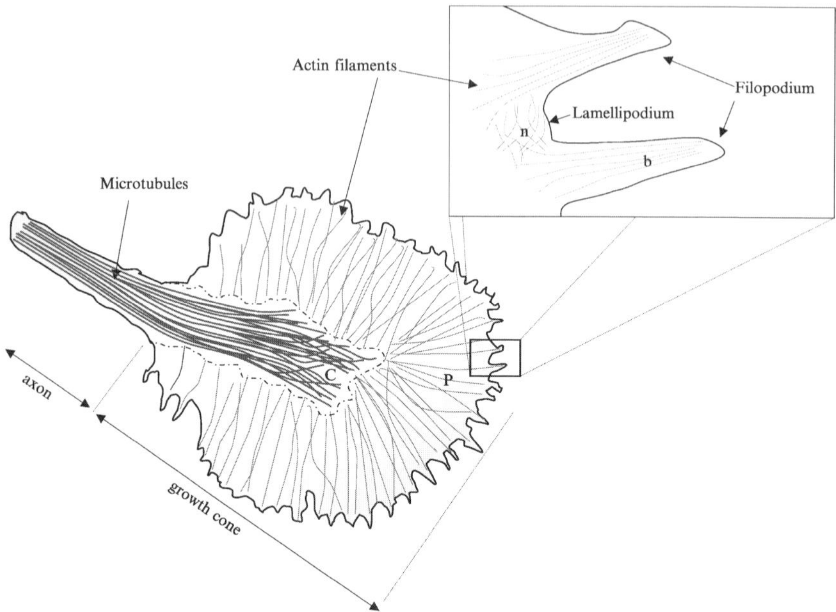

Figure 1 Schematic of the growth cone. The growing tip of the axon consists of a central domain **C** and a peripheral domain **P** with filopodium and lamellopodium. The C domain contains a dense network of microtubules, but some microtubules can also be found in the base of filipodia. Actin filaments are predominant in the P domain. Filopodium move like fingers exploring their environment. These movements are based on actin polymerization and depolymerization. Filopodial actin is organized into bundles (**b**) whereas actin filaments in lamellopodia and in the C region form an intricate network (**n**) (see inset). (See Color Insert.)

into the organelle-rich central domain of the growth cone in the form of individual microtubules (Tanaka and Kirschner, 1991). Growth cones also contain a large pool of soluble tubulin capable of assembling into microtubules (Letourneau and Ressler, 1984). The majority of growth cone microtubules are restricted to the C domain, but individual microtubules can also extend deep into the actin-rich peripheral domain (Gordon-Weeks, 1991; Tanaka and Kirschner, 1991).

The actin cytoskeleton is composed of two distinct microfilament subpopulations: one forms a meshwork of relatively short randomly oriented filaments in the lamellipodia, and the second is composed of parallel bundles of up to a dozen filaments that radiate from the leading edge of the lamellipodia and extend into the filopodium (Forscher and Smith, 1988; Letourneau and Ressler, 1984; Lewis and Bridgman, 1992; Luduena and Wessells, 1973; Yamada *et al.*, 1970, 1971). The actin contributes to the structure of the growth cone as well as the motility. It exits in both an unassembled (often referred to as globular actin [G-actin] or filamentous [F-actin] form). The F-actin is polymerized from the monomeric G-actin subunits. Actin polymerization in growth cones occurs predominantly at the leading edge and is vital to growth cone motility and hence guidance.

A. Cytoskeletal Rearrangements and Growth Cone Turning

Recent studies indicate that the actin and microtubule cytoskeletons are a final common target of many signaling cascades influencing the developing neuron. Regulation of polymer dynamics and transport are crucial for proper growth cone motility. Three stages of axon outgrowth have been termed protrusion, engorgement, and consolidation (Dent and Gertler, 2003; Goldberg and Burmeister, 1986). Protrusion of the growth cone occurs by the rapid extension of filopodia and thin lamellar extensions while the plasma membrane advances forward. These extensions are primarily composed of bundled and mesh-like F-actin networks. In engorgement, the organelles and cytoplasm move forward and, finally, in consolidation, a new axon section is left behind (Bamburg, 2003; Goldberg and Burmeister, 1989). Within the cytoplasm are neurofilaments, which provide structure, and microtubules, which provide a mechanism for rapid vesicle transport (Rivas and Hatten, 1995; Schnapp *et al.*, 1986). Vesicle transport to the growth cone is vital to maintain and re-organize the membrane as it changes. Consolidation occurs when the majority of F-actin depolymerizes at the proximal part of the growth cone. These extension processes also occur during the formation of collateral branches from the main axon shaft (Dent *et al.*, 1999).

Axons are guided in new directions by the reorientation of their growth cones as well as by extension of collateral branches (O'Leary *et al.*, 1990). Axon branch formation occurs by controlled branch extension, retraction (including pruning), and stabilization. Similar to the primary growth cone, collaterals extending from the axon shaft are tipped by growth cones. In axon collateral initiation, extension and navigation require the controlled and coordinated assembly and disassembly of the neuronal cytoskeleton. At axon branch sites, the microtubules become unbundled and undergo local fragmentation within the axon shaft. This is accompanied by focal accumulation of F-actin. Similar to growth cone steering, axon branching is impaired by selective inhibition of microtubule or actin dynamics (Dent and Kalil, 2001; Rodriguez *et al.*, 2003). Cortical axon branching occurs *in vitro* through changes in growth cone morphologies and behaviors (Szebenyi *et al.*, 1998). Using a novel approach of visualizing simultaneous changes in both microtubules and actin filaments during different stages of axon branching in living cortical neurons, Dent and Kalil (2001) observed the cytoskeletal reorganization underlying cortical axon branching. Branching from the growth cone and the axon shaft is always preceded by splaying apart of the looped or bundled microtubules, accompanied by localized accumulation of F-actin. Dynamic microtubules colocalize with F-actin in transition regions of growth cones and axon branch points, consistent with observations in fixed growth cones (Bridgman and Dailey, 1989; Challacombe *et al.*, 1996; Challacombe *et al.*, 1997; Rochlin *et al.*, 1999; Tanaka *et al.*, 1995; Williamson *et al.*, 1996), whereas F-actin is excluded from regions of stable microtubules. Recent evidence shows that several guidance molecules can also influence branching behavior (Dent *et al.*, 2004; Kornack and Giger, 2005). It has been shown that Netrin and Slit promote branching (Dent *et al.*, 2004; Wang *et al.*, 1999b) and Sema3A decreases branching (Dent *et al.*, 2004).

As the axon extends through the complex extracellular environment of the nervous system, its growth cone senses and responds to a variety of molecular guidance cues. For example, it can change from forward elongation to pauses or retraction, or it can change its direction. Growth cone turning is central to axonal navigation and is responsible for changing the direction of neurite elongation (Oakley and Tosney, 1991; Sabry *et al.*, 1991; Sretavan and Reichardt, 1993) (see Section VIII.C for pipette assay). A key event in growth cone turning may be the local realignment and advancement of microtubules (Lin *et al.*, 1994; Mitchison and Kirschner, 1988; Sabry *et al.*, 1991; Tanaka and Sabry, 1995) to establish the dominant side of a turning growth cone. When a growth cone contacts a positive guidance cue, filopodia become stabilized in the direction of neurite elongation, and microtubules are locally reoriented and advance toward the contact site (Bentley and O'Connor, 1994; Lin and Forscher, 1993; O'Connor and Bentley, 1993; Sabry *et al.*, 1991). In a similar manner, actin filaments and microtubules of growth cones may interact to accomplish turning away from a negative cue or stopping the extension of the axon altogether in a process called growth cone "collapse" (Fig. 2). Growth cone collapse involves the loss of lamellipodia and filopodia in response to a negative guidance cue.

In order to understand the mechanisms by which guidance cues direct axon growth it is important to determine how guidance cues affect F-actin dynamics and organization. Indeed, the rate of filopodial tip extension is determined by both the rate of F-actin polymerization at the filopodial tip and the retrograde displacement of polymerization filaments toward the base of the filopodium (Mallavarapu and Mitchison, 1999). The extension cycle of the lamellipodial edge is determined by both F-actin polymerization and retrograde filament transport (Lin and Forscher, 1995). Disruption of this cytoskeletal organization with drugs such as cytochalasin, nocodazole, and taxol results in alterations in growth cone morphology and motility (Forscher and Smith, 1988; Letourneau and Ressler, 1984; Letourneau *et al.*,

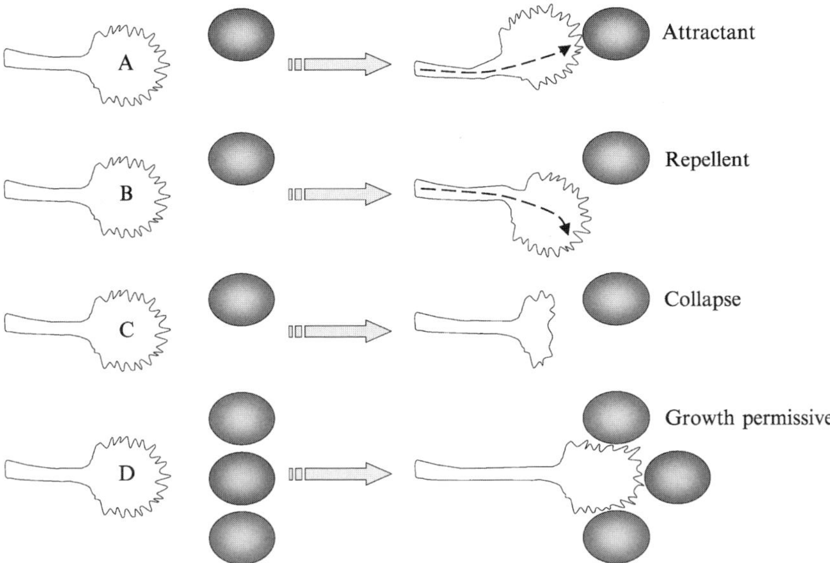

Figure 2 Growth cone behavior. The growth cone is growing toward an attractive molecule (A) or is turning away from a repulsive molecule (B). Some molecules can produce a total collapse of the growth cone (C). Some substrates are growth permissive; the growth cone grows into it with repellent cues on either side (D). (See Color Insert.)

1986; Marsh and Letourneau, 1984; Neely and Gesemann, 1994; Yamada *et al.*, 1970, 1971). Disruption of either the microtubule system, by treatment with nocodozole, or the actin system, by treatment with cytochalasin, halted neuronal migration.

In vivo, growth cones are simultaneously exposed to a number of different guidance cues, both negative and positive. Repulsive guidance cues can either collapse the whole growth cone to arrest neurite outgrowth or cause asymmetric collapse leading to growth cone turning. How signals from repulsive cues are translated by growth cones into a morphological change through rearranging the cytoskeleton is unclear. Many studies examining axon repulsion have focused on the collapse response to semaphorin (Sema) 3A. Sema3A is a member of the class 3 semaphorin family that repels the axons of sensory and sympathetic neurons (Luo *et al.*, 1993). Sema3A-induced growth cone collapse is a process whereby lamellar protrusion and filopodial motility are paralyzed before the retraction of all growth cone specializations (Fan *et al.*, 1993). Fan *et al.* (1993) showed that growth cone collapse induced by Sema3A correlates with a 50% net loss of F-actin in the growth cone, suggesting that growth cone collapse could be mediated by the depolymerization of the F-actin. The morphological response to

Sema3A is similar to that caused by cytochalasin B, a plant alkaloid that inhibits actin polymerization, paralyzes normal growth cone motility, and inhibits axonal extension (Letourneau *et al.*, 1987). Collapsin response mediator proteins (CRMPs) and rac1 have been implicated in the Sema3A signaling pathway. Treatment of neurons with either anti-CRMP-62 antibodies (Goshima *et al.*, 1995) or dominant-negative rac1 (Jin and Strittmatter,1997; Kuhn *et al.*, 1999) prevents growth cone collapse by Sema3A. Stimulation of endocytosis by Sema3A also correlates with growth cone collapse. Sema3A stimulates endocytosis by the focal and coordinated rearrangement of receptor and cytoskeletal elements (Fournier *et al.*, 2000). To examine the cytoskeletal events contributing to growth cone collapse, Zhou and Cohan (2001) used three factors able to induce the collapse of extending *Helisoma* growth cones: serotonin, myosin light chain kinase inhibitor, and phorbol ester. They found that all three factors induced the collapse of extending growth cones and caused actin bundle loss in poly-L-lysine-attached growth cones without loss of actin meshwork. In addition, actin bundle loss correlated with specific filamentous actin redistribution away from the leading edge, characteristic of repulsive factors. Using time-lapse studies of extending growth cones, they showed that actin bundle loss paralleled collapse. Taken together, these results suggest that F-actin reorganization through actin bundles is the cytoskeletal mechanism underlying growth cone collapse. Growth cone collapse, or at least the initiation of collapse, may be mainly a process involving F-actin reorganization resulting in decreased actin assembly at the leading edge rather than direct inhibition of actin polymerization (Zhou and Cohan, 2001). Growth cone collapse is mediated by several signaling pathways, including the Rho-GTPases (Liu and Strittmatter, 2001) ADF (actin depolymerizing factor)/cofilin (Carlier *et al.*, 1997), LIMK (LIM kinases) (Aizawa *et al.*, 2001), and GSK-3β, FYN, and cdk5 (Eickholt *et al.*, 2002; Sasaki *et al.*, 2002).

Positive guidance cues promote and polarize protrusive activity in the direction of growth cone migration. Gundersen and Barrett (1979) were the first to observe the turning of neuronal growth cones towards sources of nerve growth factor (NGF). Since then, additional members of the neurotrophin family have been shown to act as chemoattractants (Paves and Saarma, 1997; Song *et al.*, 1997, 1998). The Arp2/3 complex nucleates the formation of new actin branches and thus initiates the step that causes forward extension of actin meshwork and subsequent extension of the membrane leading edge (Mullins *et al.*, 1998).

Ena/VASP (enabled/vasodilator-stimulated phosphoprotein) proteins play important roles in axon outgrowth and guidance. Ena/VASP activity regulates the assembly and geometry of actin networks within fibroblast lamellipodia. Ena/VASP proteins are found in the leading edge of lamellipodia (Nakagawa *et al.*, 2001) and are concentrated in the F-actin-rich region. They function by binding to the barbed ends of filaments and competing with capping proteins, allowing for longer filament extension (Bear *et al.*, 2002). A more recent study suggests that Ena/VASP proteins play a pivotal role in the formation and elongation of filopodia along the neurite shaft and at the growth cone (Lebrand *et al.*, 2004). Netrin-1-induced filopodia formation is dependent upon Ena/VASP function and is directly correlated with Ena/VASP phosphorylation at a regulatory protein kinase A (PKA) site. Ena/VASP function is required for filopodial formation from the growth cone in response to global PKA activation. Ena/VASP proteins likely control filopodial dynamics in neurons by remodeling the actin network in response to guidance cues (Lebrand *et al.*, 2004). Krause and colleagues (2004) identified Lamellipodin as a novel Ena/VASP-binding protein. Both proteins co-localize at the tips of lamellipodia and filopodia. Lamellipodin overexpression increases lamellipodial protrusion velocity, an effect observed when Ena/VASP proteins are overexpressed or artificially targeted to the plasma membrane. Conversely, knockdown of Lamellipodin expression impairs lamellipodia formation, reduces in velocity of residual lamellipodial protrusion, and decreases in F-actin content. Lamellipodin may act as a key convergence point linking polarized phospholipid signals and small GTPases with Ena/VASP proteins to regulate the actin cytoskeleton (Krause *et al.*, 2003, 2004). A recent study also showed that RIAM, an Ena/VASP and Profilin ligand, interacts with Rap1-GTP and mediates Rap1-induced adhesion. RIAM links Rap1 to integrin activation and plays a role in regulating actin dynamics (Lafuente *et al.*, 2004). More recently, a novel protein with high similarity to the *Caenorhabditis elegans* MIG-10 protein, called PREL1 (proline-rich Ena/VASP homology domain 1 [EVH1] ligand) has been identified. PREL1 directly binds to Ena/VASP proteins and co-localizes with them at lamellipodia tips and at focal adhesions in response to Ras activation. PREL1 provides a link from Ras signaling to the actin cytoskeleton via Ena/VASP proteins (Jenzora *et al.*, 2005). The role of Ena/VASP proteins in axon guidance will be discussed further in Section IV.G.1.

Another important component of the neuronal cytoskeleton is the microtubules. Challacombe *et al.* (1996) have shown that dynamic microtubule ends are rearranged in growth cone repulsion to avoid an inhibitory guidance cue. The importance of microtubule dynamics in axonal growth and guidance has been demonstrated by the pharmacological inhibition of dynamics without affecting microtubule assembly. Low concentrations of drugs such as nocodazole, vinblastine, and taxol not only reduce axonal elongation but also prevent growth cone turning (Challacombe *et al.*, 1997; Rochlin *et al.*, 1996; Williamson *et al.*, 1996).

Studies have shown that stathmin and SCG10 are potent microtubule destabilizing factors. While stathmin is expressed in a variety of cell types and shows a cytosolic distribution, SCG10 is neuron specific and membrane associated. SCG10 accumulates in the central domain of the growth cone, a region that also contains highly dynamic microtubules. SCG10 appears to be an important factor for the dynamic assembly and disassembly of growth cone microtubules during axonal elongation. Phosphorylation negatively regulates the microtubule destabilizing activity of SCG10 and stathmin, suggesting that these proteins may link extracellular signals to the rearrangement of the neuronal cytoskeleton (Grenningloh *et al.*, 2004).

Zhou *et al.* (2002) report that local and specific disruption of actin bundles from the growth cone peripheral domain induced repulsive growth cone turning. Meanwhile, dynamic microtubules within the peripheral domain were oriented in areas where actin bundles remained and were lost from areas where actin bundles disappeared. This resulted in directional microtubule extension, leading to axon bending and growth cone turning. In addition, this local actin bundle loss coincided with localized growth cone collapse, as well as asymmetrical lamellipodial protrusion. Regional actin bundle reorganization can steer the growth cone by coordinating actin reorganization with microtubule dynamics. This suggests that actin bundles are potential targets of signaling pathways downstream of guidance cues, providing a mechanism for coupling changes in leading edge actin with microtubules at the central domain during turning (Zhou *et al.*, 2002).

III. The Role of Pioneering Axons and Glial Guidepost Cells in Axonal Guidance

A. Pioneering Axons

Pioneering axons are the first axons to extend along a given trajectory in the brain (Fig. 3). These axons are tipped with growth cones that respond to repellents and attractants within their environment. Here we discuss the role that pioneering axons play in guiding the formation of large axonal tracts within the brain. Once the pioneer axons have established a correct path of growth, later arriving axons, called follower axons, fasciculate with the pioneers to find their target. Often a series of pioneering axons growing over short distances may be used by the main bundle to find their targets. Ablation experiments in both vertebrates and invertebrates have established the importance of pioneering axons as a mechanism for wiring the nervous system.

1. Pioneering Axons in Invertebrates

Invertebrate organisms such as flies, worms, and grasshoppers have played a prominent role in elucidating the mechanisms that are involved in axonal guidance. The advantages of these organisms include their simple nervous

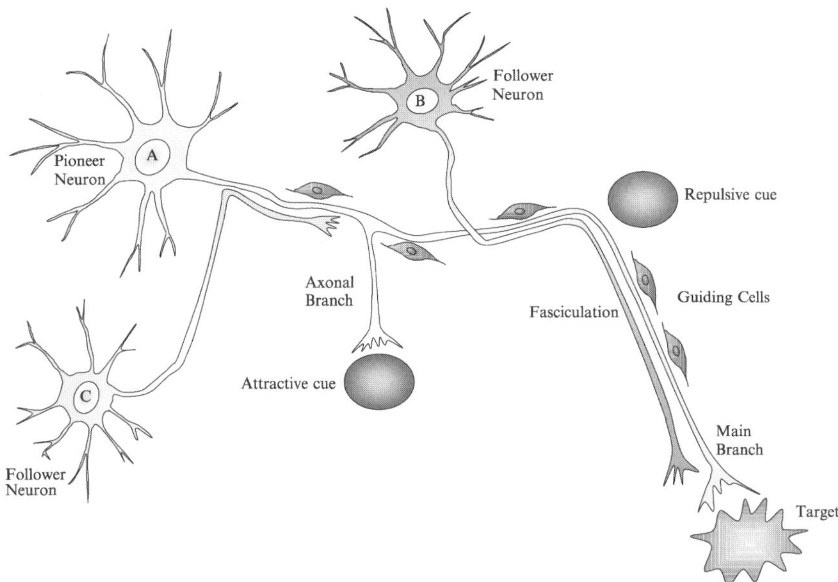

Figure 3   The pioneers. Axon of the pioneer neuron (A) grows along the pathway formed by the presence of guidepost cells as well attractive and repulsive molecules. An axonal branch can form and grow toward an attractive molecule. This axonal branch may become a stable axon, resulting in the retraction of the main branch. As pioneer neurons establish their route, follower neurons (B and C) extend their neurites. Axons of the follower neurons fasciculate with the pioneer to project to the appropriate target. (See Color Insert.)

system architecture, for example, each hemisegment of the *Drosophila melanogaster* ventral nerve cord contains about 300 neurons, and the *C. elegans* nervous system has a total of 302 neurons. It is also possible to identify and follow individual neurons throughout development. Furthermore, the *Drosophila* and *C. elegans* systems allow the use of classical genetic analyses. Mutagenesis screens in both organisms have revealed many essential genes that are necessary for wiring the nervous system. In invertebrates as well as in vertebrates, axons must travel large distances to reach their final targets. Studies examining early neural development in crustacean embryos have shown the important role of pre-existing axons in guiding subsequent axon outgrowth (Whitington, 1993). Once pioneer neurons have established a route for axonal growth, axons that develop later frequently choose to selectively bundle or fasciculate with individual tracts to reach their target area.

Since axons are often pathfinding over long distances, they use guidepost cells such as pioneering axons or glial structures (also called intermediate targets) along the way to help them locate their final target. The first observation of guidepost cells came from an analysis of axon outgrowth in the grasshopper limb bud (Bate, 1976). The Ti1 neuron navigates its way to the central nervous system (CNS) through key landmarks that are provided by neuronal somata. Ablation of these cells results in the failure of the neuron to efficiently complete its pathfinding. Guidepost cells in the limb bud of grasshopper embryos direct pioneer axons to the correct pathway (Bentley and Caudy, 1983; Keshishian and Bentley, 1983).

If particular pioneering axons are removed, the follower neurons stall and fail to extend (Bastiani *et al.*, 1986; du Lac *et al.*, 1986; Hidalgo and Brand, 1997; Raper *et al.*, 1984). Studies in grasshoppers (Bastiani *et al.*, 1984; Goodman *et al.*, 1984; Raper *et al.*, 1983a,b, 1984) demonstrate that follower axons can recognize molecules on the surface of the pioneering axon. The growth cones of two follower axons, the G and C cells, in the developing grasshopper neuropil follow four pioneering neurons (the A1, A2, P1, and P2 cells) to their targets. The G neuron extends anteriorly in the bundle and the C neuron extends posteriorly in the same bundle, but only after several other axons have joined the bundle. These results suggest that the growth cones of the G and C neurons can recognize and extend upon the four pioneering axons. This has been described as the labeled pathways hypothesis (Goodman *et al.*, 1984), which proposes that axon tracts have different molecular labels on their cell surface, labels that follower axons can specifically recognize to extend upon.

Some molecules are beginning to be identified for pioneering axon-mediated guidance. AcP neurons pioneer the anterior commissure of each grasshopper segment and extend their axons toward the midline. A cell surface glycoprotein, Lazarillo, is expressed during grasshopper embryogenesis on the surface of a subset of CNS neurons and by a group of neuroblasts, the precursors of neurons and glia (Sanchez *et al.*, 1995, 2000). When Lazarillo function was blocked, AcP neurons failed to grow into the midline; they either stopped growing or turned anteriorly. These results suggest that Lazarillo could play a role in the guidance of the AcP pioneering neurons (Sanchez *et al.*, 1995, 2000).

Goodman and Jacobs characterized neurons that pioneer the major CNS axon tracts in *Drosophila* (Jacobs and Goodman, 1989). In the embryonic *Drosophila* CNS, RP motor axons make stereotypic pathway choices involving distinct cellular contacts: (i) extension across the midline via contact with the axon and cell body of the homologous contralateral RP motoneuron, (ii) extension down the contralateral longitudinal connective through contact with connective axons and longitudinal glia, and (iii) growth into the intersegmental nerve (ISN) through contact with ISN axons and the segmental boundary glial cell (SBC). Removal of the longitudinal glia or the SBC did not adversely affect pathfinding. This suggests that the motor axons either utilized alternative axonal substrates or could still make filopodial contact with the cues in the next pathway. In contrast, RP motor axons did require contact with the axon and soma of their contralateral RP homologue. Absence of this neuronal substrate frequently impeded RP axon outgrowth, suggesting that the next cues were beyond filopodial reach. Together, these direct ablations of putative guidepost cells in the CNS demonstrated a susceptibility by RP axons to the absence of specific cellular contacts (Whitington *et al.*, 2004).

In the nematode *C. elegans*, the gene unc-6 is required to guide pioneer axons and mesoblasts in dorsal or ventral directions of the body wall (Hedgecock *et al.*, 1990). UNC-6 is required to direct the early circumferential extension of pioneer axons in *C. elegans*, as well as the circumferential migrations of mesoblast cells (which migrate along the basement membrane surfaces facing the central pseudocoelomic cavity). Nerves form as additional axons enter the tract and fasciculate with the pioneering axon and with each other. These followers enter the tract in a set order and at stereotyped positions. Key to the role that UNC-6 plays in this process is its dynamic expression pattern (Wadsworth, 2002; Wadsworth *et al.*, 1996). The unc-129 gene, like the unc-6 gene, is required to guide pioneer motor axons along the dorsoventral axis of *C. elegans*. UNC-129 mediates expression of dorsoventral polarity information required for axon guidance and guided cell migration (Colavita *et al.*, 1998).

In *C. elegans*, synaptic guidepost cells in the vulval epithelium initiate the formation of synapses between the HSNL (hermaphrodite-specific neurons left) and HSNR (hermaphrodite-specific neurons right) and either vulval muscle cells or ventral type C (VC) neurons (Shen and Bargmann, 2003). In the absence of the guidepost cells, synaptic vesicle markers in the HSNL fail to accumulate at the normal synaptic locations, and instead form ectopic aggregates in anterior locations. *Syg-1* encodes a transmembrane immunoglobulin superfamily protein acting in the presynaptic HSN axon. Analysis of the *syg-1* mutant shows that HSNL synapses have defects similar to those observed after guidepost cell ablation, suggesting a role for SYG-1 in guidepost signaling (Shen and Bargmann, 2003). Another protein, SYG-2, binds heterophilically to SYG-1 and is expressed by guidepost cells and vulval epithelial cells. *Syg-2* mutants lack synapses at the normal location and instead form synapses onto inappropriate target cells at ectopic locations (Shen *et al.*, 2004). Molecular interactions between SYG-1 and SYG-2 on neuron and guidepost cells, respectively, contribute to synaptic specificity but also serve to suppress the formation of inappropriate synapses.

2. Pioneering Axons in Vertebrates

Pioneering axons have also been described in zebrafish. The first axons to navigate the neuroepithelium of the zebrafish brain emerge from the ventrocaudal cluster (vcc) at approximately 16 hr postfertilization (hpf) (Chitnis and Kuwada, 1990; Ross *et al.*, 1992). These axons grow caudally to pioneer the medial longitudinal fasciculus (MLF), which is part of the larger ventral longitudinal tract (VLT) (Chitnis and Kuwada, 1990; Ross *et al.*, 1992; Wilson *et al.*, 1990), the major longitudinal tract that connects the midbrain with the hindbrain. By 18 hpf, the first axons in the forebrain emerge from the ventrorostral cluster (vrc) and grow caudally to pioneer the tract of the postoptic commissure (TPOC) (Chitnis and Kuwada, 1990; Ross *et al.*, 1992; reviewed in Hjorth and Key, 2002). Nural and Mastick (2004) also studied the postoptic commissure (POC) and characterized a system of early neurons that establish the first two major longitudinal tracts in the embryonic mouse forebrain. Each of the early axon populations first grows independently, pioneering a short segment of new tract. However, each axon population soon merges with other axons to form one of only two shared longitudinal tracts, both descending: the TPOC, and, in parallel, the stria medullaris. Thus, the forebrain longitudinal tracts are pioneered by a relay of axons, with distinct axon populations pioneering successive segments of these pathways. They identified that the transcription factor Pax6 is critical for tract formation. In Pax6 mutants, both the TPOC and the stria medullaris failed to form due to pathfinding errors of the early pioneering axons. Their results show that Pax6 could regulate longitudinal tract formation by guiding a relay of pioneer longitudinal axons in the embryonic mouse forebrain (Nural and Mastick, 2004).

Fraser and colleagues (2003) studied the POC using *in vivo* microscopy of embryonic zebrafish expressing green fluorescent protein (GFP) in the vrcs of cells in the embryonic forebrain. Their data showed that the growth of the leader pioneering axons slows down at the midline, but not the follower axons. When the leading pioneer axon is ablated, the follower axons change their midline kinetics and behave as leaders. Similarly, once the leader axons have crossed the midline they change their midline kinetics when they encounter the leading axon from the contralateral side (Bak and Fraser, 2003). These data suggest a simple model in which the level of growth cone exposure to midline cues and the presence of other axons as a substrate shape the midline kinetics of commissural axons.

In the neocortex, subplate neurons have been shown to serve as pioneering axons for thalamocortical and corticothalamic axons (De Carlos and O'Leary, 1992; Ghosh and Shatz, 1992, 1993; Ghosh *et al.*, 1990; McConnell *et al.*, 1989). When subplate neurons are ablated both corticothalamic and thalamocortical targeting are disrupted demonstrating a direct requirement for pioneering subplate neurons in forming connections between the thalamus and cortex (Ghosh and Shatz, 1992; Ghosh *et al.*, 1990; McConnell *et al.*, 1994).

In the medial cortical projection the first axons to cross the rostral cortical midline (rostral to the hippocampal commissure) are derived from neurons in the cingulate cortex (Koester and O'Leary, 1994; Ozaki and Wahlsten,

1998; Rash and Richards, 2001). These axons begin to cross the midline at embryonic day (E) 17 in the rat and E15.5 in the mouse. Neurons in the cingulate cortex project to three different regions: across the midline into the contralateral cortex, into the fornix, and ventrally into the medial septum and the diagonal band of Broca. The cingulate axons cross the midline first, followed by the neocortical axons, which grow within the tract of the cingulate pioneering axons, possibly fasciculating with the cingulate axons (Rash and Richards, 2001).

The optic chiasm also contains a subset of early generated neurons, which have been shown to be involved in retinal axon guidance (Easter *et al.*, 1993; Sretavan *et al.*, 1994, 1995). Furthermore, Cajal-Retzius cells in the hippocampus are suggested to play a role in the guidance of entorhinohippocampal axons (Del Rio *et al.*, 1997; Soriano *et al.*, 1994). In fact, two groups of pioneer neurons, Cajal-Retzius cells and GABAergic neurons, form layer-specific scaffolds that overlap with distinct hippocampal afferents at embryonic and early postnatal stages. Before the dendrites of pyramidal neurons develop, these pioneer neurons act as synaptic targets for hippocampal afferents. These findings indicate that distinct pioneer neurons are involved in the guidance and targeting of different hippocampal afferents (Super *et al.*, 1998).

Recent evidence also suggests the presence of guidepost cells in the developing mouse olfactory system. Mitral cell axons, the major efferents of the olfactory bulb, caudally elongate in a very narrow part of the lateral telencephalon and make a stereotyped turn toward the amygdala (Brunjes and Frazier, 1986; Schwob and Price, 1984; Shipley *et al.*, 1995). The axons collectively form a fiber bundle called the lateral olfactory tract (LOT). Organotypic co-culture of olfactory bulb with various regions of the mouse telencephalon showed that mitral cell axons are guided by biochemical cues that are strictly localized to the telencephalon (Sugisaki *et al.*, 1996). Indeed, intrinsic cells in the telencephalon play a directional role in the guidance of mitral cell axons (Sugisaki *et al.*, 1996). Mitral cell axons selectively grew along the LOT cells *in vivo* and in co-culture. Ablation of LOT cells in organotypic cultures caused mitral cell axons to stall. These results suggest that LOT cells function as guidepost cells for mitral cell axons (Sato *et al.*, 1998).

B. Intermediate Targets and Glial Guidepost Cells

Often, the final target of an axon is a long distance away. In order to navigate toward their final target, axons use intermediate targets to guide them along the correct path of growth. Such intermediate targets can be pioneering axon populations such as those described above or glial cells or structures present along the pathway that secrete guidance factors (Fig. 3).

Once a growth cone encounters such an intermediate target, it slows and transforms its morphology, apparently looking for further molecular "directions." Guidepost cells have been identified in *Drosophila* (Auld, 1999; Hidalgo, 2003) in which glia secrete guidance cues and express cellular cues on their surface that guide axonal outgrowth. Glia act as intermediate targets in growth cone guidance. Because guidepost cells were found in invertebrates, similar guiding mechanisms have been postulated in the mammalian CNS. In vertebrates, primitive glial cells are involved in guiding pioneering growth cones in the developing spinal cord (Kuwada, 1986; Singer *et al.*, 1979), the ventral roots (Nordlander *et al.*, 1981), the optic nerve (Silver and Sapiro, 1981; Silver and Sidman, 1980), the auditory system (Carney and Silver, 1983) and the developing corpus callosum (Silver *et al.*, 1982, 1993). During embryonic development, glia cells are required for the formation of the CNS (Fitch and Silver, 1997). They also define boundaries between different brain areas or between functional subdomains within the same area (Cooper and Steindler, 1986; Garcia-Abreu *et al.*, 1995; Mastick and Easter, 1996; Silver, 1994; Silver *et al.*, 1993; Yoshida and Colman, 2000). These glial boundaries serve to prevent axons from straying from their correct path of growth (Fitch and Silver, 1997). At the *Drosophila* midline, glia function as guidepost cells for commissural and ipsilaterally projecting axons to determine which axons cross the midline and which do not (Hidalgo and Booth, 2000; Jacobs and Goodman, 1989; Kidd *et al.*, 1999). To better understand the origin of commissures, a mutant screen was carried out for flies with defective commissure formation. Commissureless (Comm) is one of the mutations isolated. In Comm mutants, commissural growth cones initially orient toward the midline, but none actually cross it. Rather, any short medially oriented processes are retracted, and the axons remain exclusively on their own side, producing the commissureless axon guidance phenotype (Seeger *et al.*, 1993). Robo was isolated in the same screen, and leads to the opposite misrouting; some growth cones that normally extend only on their own side project across the midline in Robo mutants. The phenotype of these two genes suggests that they encode components of attractive and repulsive signaling systems at the midline. Comm is able to downregulate levels of the Robo protein on the cell surface, which is necessary for axons to cross the midline. Comm is expressed and required in both commissural neurons and midline cells for correct midline crossing (Georgiou and Tear, 2002). It is suggested that the presence of Comm in the commissural neurons may encourage midline crossing. Comm protein accumulates at the axon surface within the commissural region, using a mechanism that is likely to involve Comm expressed by midline glia (Couch and Condron, 2002; Georgiou and Tear, 2002; Keleman *et al.*, 2002; Keleman *et al.*, 2005; Myat *et al.*, 2002).

Three pairs of midline glia, as well as the medial precursor 1 (MP1) and ventral unpaired median (VUM) neurons, are present at the midline of the *Drosophila* ventral nerve cord (Klambt *et al.*, 1991). Analysis of mutants defective in midline cell development reveal essential roles for these cells in the formation of the *Drosophila* ventral nerve cord. In the single-minded (*sim*) mutant, midline cells fail to differentiate and ultimately die. Consequently, commissures do not form and the longitudinal connectives collapse into a single fused tract at the midline (Klambt *et al.*, 1991; Thomas *et al.*, 1988).

Glial cells may also function to sort ipsilaterally from contralaterally projecting axons. At the optic chiasm, retinal ganglion cell (RGC) axons contact glial cells known as the glial palisade. These glia may contribute to retinal axonal divergence at the chiasm (Erskine *et al.*, 2000; Marcus *et al.*, 1995). Glial populations are also associated with the formation of commissures such as the anterior commissure (Cummings *et al.*, 1997; Pires-Neto *et al.*, 1998), the corpus callosum (Shu and Richards, 2001; Silver *et al.*, 1993) (see Section V.B), and decussating axons in the hindbrain, brain stem, and the corticospinal tract (Joosten *et al.*, 1989; Mori *et al.*, 1990; Van Hartesveldt *et al.*, 1986). Glial cells in the floor plate of the spinal cord guide commissural axons of the dorsal spinal cord (Altman and Bayer, 1984; Tessier-Lavigne *et al.*, 1988a).

IV. Molecules Involved in Axonal Guidance

Neuronal growth cones are guided to their targets by both short- and long-range cues. Both attractive and repulsive cues are equally important for the guidance of growth cones to their appropriate targets. Most axon guidance molecules identified were discovered first in *Drosophila* or *C. elegans*. In some cases, guidance factors were simultaneously directly purified from vertebrate systems through massive purification efforts. There are four major families of classic axon guidance molecules: Netrins, Slits, semaphorins, and ephrins and their receptors. However, a number of other molecules have been shown to guide axons, including morphogens, steroids, and extracellular matrix and adhesion molecules. Here we briefly review these molecules and some of the intracellular signaling mechanisms they use.

A. Netrins and DCC

Netrin-1, whose name stems from the Sanskrit term "one who guides," was originally purified in vertebrates as a floorplate-derived chemoattractant by an exhaustive purification from over 10,000 chick brains that can promote commissural axon outgrowth (Kennedy *et al.*, 1994; Serafini *et al.*, 1994). Netrins are secreted proteins that act on neural cells through transmembrane

receptors of the Neogenin A2b receptor, DCC (deleted in colorectal carci-
nomas), and UNC5H families (Mehlen and Mazelin, 2003) (Fig. 4). Netrins
make up a small family of secreted, laminin-related molecules with multifunctional roles in axon guidance, acting as context-dependent chemoattractants or chemorepellents. Several members of the Netrin family have been
identified in a variety of species: the UNC-6 gene product in nematodes;

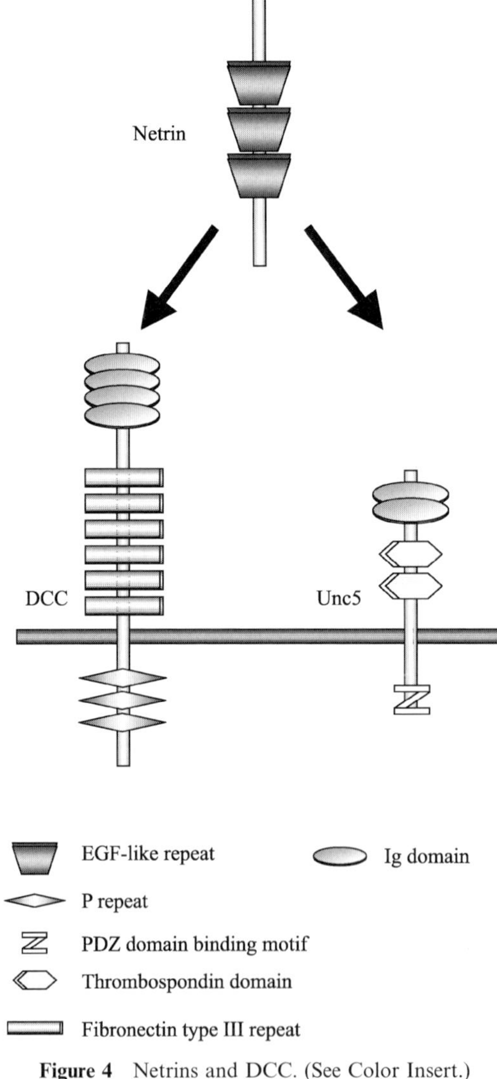

Figure 4   Netrins and DCC. (See Color Insert.)

netrin-A and netrin-B in *Drosophila;* netrin-1 and netrin-2 in chicks; netrin-1, netrin-3, netrin-G1, netrin-G2, and netrin-4/β-netrin in mice; and the NTN2L (netrin-2-like) in humans (Koch *et al.*, 2000; Nakashiba *et al.*, 2002; Serafini *et al.*, 1994, 1996; Wang *et al.*, 1999a). The Netrins encode ~60–80 kDa secreted proteins that share homologous domains with laminin (Banyai and Patthy, 1999; Serafini *et al.*, 1994). The *C. elegans* netrin (UNC6) was the first member of this family identified by examining mutants with uncoordinated (unc) phenotypes (Hedgecock *et al.*, 1990; Ishii *et al.*, 1992). Netrins regulate the development of commissural axons in both the spinal cord and brain, including the corpus callosum, hippocampal commissure, and the optic chiasm (Serafini *et al.*, 1996). Netrins are not only involved in axon guidance but also play central roles in the migration of neurons, glial oligodendrocyte precursors, and mesodermal cells during embryogenesis (Alcantara *et al.*, 2000; Bloch-Gallego *et al.*, 1999; Hamasaki *et al.*, 2001; Lim and Wadsworth, 2002; Spassky *et al.*, 2002; Su *et al.*, 2000; Sugimoto *et al.*, 2001; Tsai *et al.*, 2003; Yee *et al.*, 1999). More recently, it has been shown that the Netrin-1 receptor DCC is phosphorylated by Fyn and that phosphorylation is required for DCC function. This suggests that Fyn is essential to initiate the responses of axons to Netrin-1 (Meriane *et al.*, 2004).

B. Semaphorins, Neuropilins, and Plexins

Semaphorins (Semas) were originally identified in invertebrates (Kolodkin *et al.*, 1992). Sema3A, the prototype vertebrate member of the semaphorin family, was initially purified from chick brain extracts on the basis of its collapse-inducing activity on cultured dorsal root ganglion (DRG) growth cones (Luo *et al.*, 1993). The semaphorin family contains both secreted and membrane-bound members, divided into eight classes (Semaphorin Nomenclature Committee, 1999). The first two classes represent invertebrate semaphorins, classes 3–7 represent vertebrate semaphorins, and the eighth class comprises viral semaphorins. Of the vertebrate classes (Fig. 5), class 3 contains secreted semaphorins and classes 4–7 contain transmembrane or membrane-anchored semaphorins. All semaphorins share a conserved, 500-amino-acid motif, termed the sema domain (Kolodkin *et al.*, 1993; Luo *et al.*, 1993). The first semaphorin receptor identified was neuropilin (Chen *et al.*, 1997; Giger *et al.*, 1998; He and Tessier-Lavigne, 1997; Kolodkin *et al.*, 1997). The two main protein families now known to be involved in mediating semaphorin responses are the neuropilins (two members, Npn1 and Npn2; Fig. 5) and plexins (nine members), Plexin A1–4, Plexin B1–3, Plexin C1, and Plexin D1 (de Wit and Verhaagen, 2003; Fujisawa and Kitsukawa, 1998; Pasterkamp and Kolodkin, 2003; Raper, 2000; Tamagnone *et al.*, 1999; Winberg *et al.*, 1998). Invertebrate semaphorins, membrane-associated

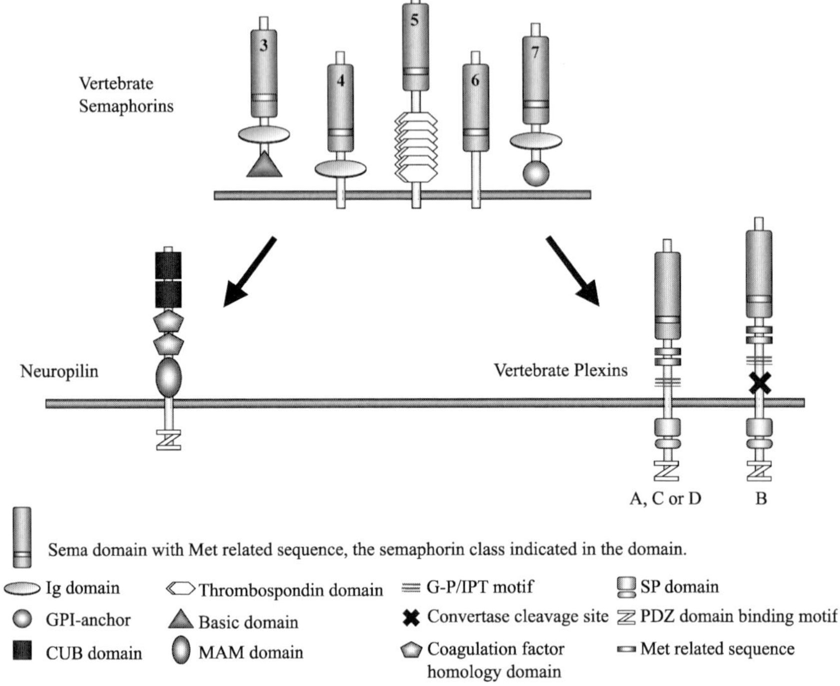

Figure 5 Semaphorins, Neuropilins, and Plexins. (See Color Insert.)

semaphorins in vertebrates, and viral semaphorins have been shown to interact directly with plexins (Comeau *et al.*, 1998; Tamagnone *et al.*, 1999; Winberg *et al.*, 1998). Vertebrate class 3 secreted semaphorins, however, utilize neuropilin proteins as ligand-binding obligate co-receptors, which assemble a semaphorin/neuropilin/plexin signaling complex (Chen *et al.*, 1997; He and Tessier-Lavigne, 1997; Takahashi *et al.*, 1999; Tamagnone *et al.*, 1999).

All four members of the plexin-A subfamily can function as a signal-transducing component in class 3 semaphorin receptor complexes. Plexin A1, Plexin A2, and Plexin A4, when expressed in a complex with Npn-1 or Npn-2, induce a contraction of the cytoskeleton of COS cells in response to class 3 semaphorins (Suto *et al.*, 2003; Takahashi and Strittmatter, 2001). Cultured sympathetic axons derived from Plexin A3 knockout mice, however, are no longer repelled by Sema3F and partially lose their responsiveness to Sema3A (Cheng *et al.*, 2001). The response to class 3 semaphorins is determined by the complement of neuropilins on the cell surface, and additional specificity is achieved by the combination of plexin-As expressed in the neuron.

Npn1 can bind another ligand, vascular endothelial growth factor (VEGF) (Soker *et al.*, 1998), and several proteins other than neuropilins and plexins can participate in semaphorin receptor complexes. These include the cell adhesion molecule L1, which transduces a chemorepulsive response to Sema3A in cortical neurons, together with Npn1 (Castellani *et al.*, 2000, 2002). To study the role of Npn1-Sema signaling independent of VEGF/Npn1 signaling, Gu and colleagues specifically mutated the Sema-binding domain of Npn1, while leaving the VEGF-binding domain intact (Gu *et al.*, 2003). These mice have axonal guidance defects in the formation of the entorhino-hippocampal pathway, cranial and spinal nerves, sensory projections to the inner ear, and the corpus callosum (Gu *et al.*, 2003).

Recently, integrins have been shown to function as receptors for Sema7A, which induces olfactory axon growth without the need for plexins (Pasterkamp *et al.*, 2003). Finally, recent data suggest that the attractive/repulsive signaling of Sema5A can be modulated by interactions with the extracellular matrix (Kantor *et al.*, 2004; see further below).

C. Slits and Robos

Slit proteins make up a family of multifunctional guidance cues with putative roles in regulating neuronal migration (Wu *et al.*, 1999), axonal and dendritic branching (Ozdinler and Erzurumlu, 2002; Wang *et al.*, 1999b; Whitford *et al.*, 2002), and axon guidance (Brose *et al.*, 1999; Kidd *et al.*, 1999). These large glycoproteins are conserved across species with three family members (Slit1, Slit2, and Slit3) identified in the developing and adult mammalian nervous systems (Brose *et al.*, 1999; Marillat *et al.*, 2002). They contain several protein motifs: leucine-rich repeats, EFG repeats, and a laminin G domain (Rothberg *et al.*, 1990) (Fig. 6). The three vertebrate Slits have overlapping, but distinct, patterns of expression throughout development and in adulthood (Holmes *et al.*, 1998; Itoh *et al.*, 1998; Marillat *et al.*, 2002; Piper *et al.*, 2000; Yuan *et al.*, 1999). Knockout studies demonstrate that Slits 1 and 2 play critical roles in the formation of several mammalian fiber tracts, including corticofugal, thalamocortical, callosal, optic, and the lateral olfactory tract (Bagri *et al.*, 2002; Keleman *et al.*, 2002; Nguyen-Ba-Charvet *et al.*, 2002; Plump *et al.*, 2002). The Robo family of transmembrane proteins are the receptors for Slits (Brose *et al.*, 1999; Fricke *et al.*, 2001; Kidd *et al.*, 1998; Zallen *et al.*, 1998). Three mammalian homologues of *Drosophila* Robo have been identified (Robo1, Robo2, and Rig-1). Robos are members of the immunoglobulin (Ig) superfamily; their ectodomain contains five Ig-like repeats followed by three fibronectin type III

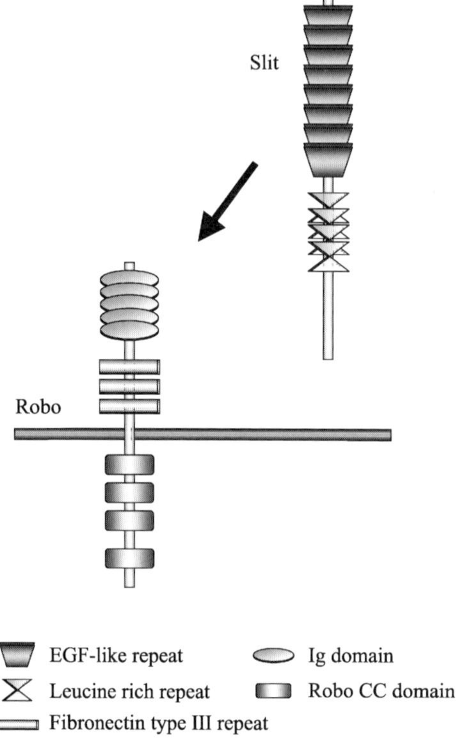

Figure 6   Slits and Robos. (See Color Insert.)

repeats (Kidd *et al.*, 1998). In *Drosophila*, ipsilaterally projecting axons and decussated commissural axons expressing Robo are prevented from inappropriate crossing/re-crossing of the midline via interaction with Slit (Kidd *et al.*, 1999). Genetic mutations in either Robo or Slit lead to aberrant crossing and re-crossing, or failure of these axons to leave the midline (Kidd *et al.*, 1999). In *Drosophila*, commissural axons acquire a postcrossing sensitivity to Slit resulting from increased surface expression of Robo. This occurs via a mechanism that includes inactivation of Comm, an intracellular sorting protein that normally targets Robo for endosomal degradation (Keleman *et al.*, 2002). In Comm mutants, commissural growth cones initially orient toward the midline, but none actually cross. Rather, any short medially oriented processes are retracted, and the axons remain exclusively on their own side, producing the commissureless phenotype (Seeger *et al.*, 1993). Downregulation of Robo is necessary for axons to cross the midline. Comm is expressed and required in both commissural neurons and midline cells for correct midline crossing (Georgiou and Tear, 2002). It is suggested that the presence of Comm in commissural neurons may encourage midline crossing. Comm protein accumulates at the axon surface within the commissural region, using a mechanism that is likely to involve Comm in the midline glia. However, Comm activity does not extend beyond the midline, allowing Robo levels to increase at the growth cone surface and initiate sensitivity to the midline inhibitor Slit that encourages axon growth away from the midline and prevents re-crossing (Couch and Condron, 2002; Georgiou and Tear, 2002). Thus, Comm controls axon guidance at the midline by regulating surface levels of Robo. Two different models have been proposed to explain how Comm regulates Robo. The first model proposes that Comm controls the sorting of Robo at the trans-Golgi network (Keleman *et al.*, 2002). The second model proposes that Comm controls Robo by acting at the plasma membrane. In this model, Comm does not block the delivery of Robo to the growth cone but instead rapidly removes it by endocytosis (Myat *et al.*, 2002). In a genetic rescue assay for Comm, Dickson and colleagues showed that midline crossing does not require the presence of Comm in midline cells (as proposed in the second model). They also showed by monitoring the trafficking of Robo that Comm prevents the delivery of Robo to the growth cone (as predicted in the first model) (Keleman *et al.*, 2005).

Slit proteins are also alternatively spliced in both mouse and human, implying that multiple Slit protein isoforms may exist (Little *et al.*, 2002). Recent studies of the Slit1 protein show an alternatively spliced mRNA product for *slit1* found specifically in the vertebrate nervous system (Tanno *et al.*, 2004). This variant was designated *slit1α*. *Slit1α* is specifically expressed in rat brain, but not heart or kidney, suggesting that *slit1α* plays a role in nervous system development. *Slit1α* expression was found in the cerebral cortex and the hippocampus. *In vivo*, *slit1α* acts as a chemorepellent of olfactory bulbs axons (Tanno *et al.*, 2004).

D. Ephrins and Eph

Ephrins and their tyrosine kinase receptors, the Eph molecules, are divided into two classes: ephrin-As, which are anchored to the membrane via glycosyl-phosphatidytinositol (GPI) linkage, and ephrin-Bs, which are transmembrane proteins (Fig. 7). The ephrin-A subclass contains ephrins A1 to A5 and the ephrin-B subclass has three members, ephrins B1 to B3 (Kullander and Klein, 2002). Eph receptors are divided into an A subclass that contains eight members (EphA1–EphA8), and a B subclass that contains five members (EphB1–EphB4, EphB6) (Cutforth and Harrison, 2002; Huot, 2004; Wilkinson, 2001). A-type receptors bind to most or all A-type ligands, and

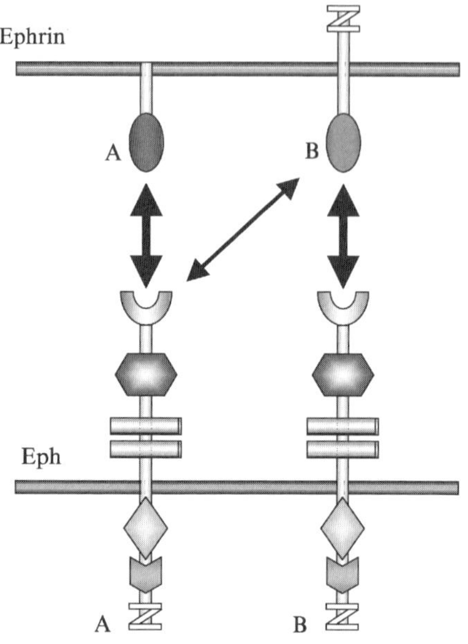

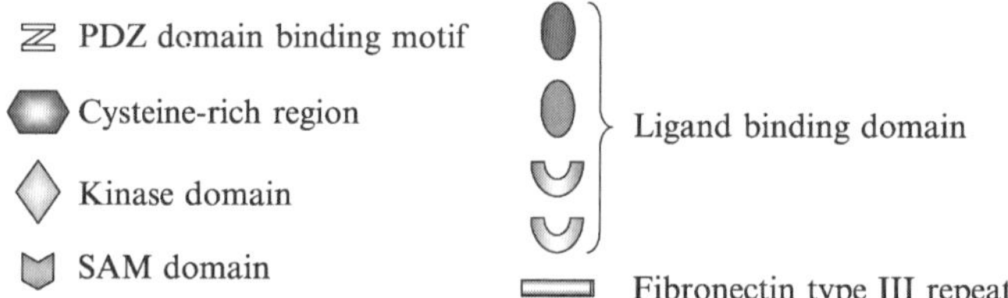

Figure 7   Ephrins and Eph. (See Color Insert.)

B-type receptors bind to most or all B-type ligands. The primary exception is the EphA4 receptor, which has been shown to interact with members of both class A and class B ephrins. The extracellular domain of the Eph receptors contains the ligand-binding domain, a cysteine-rich region, and two fibronectin type III repeats. The cytoplasmic domain of the Eph receptors can be divided into four functional units: the juxtamembrane domain that contains two conserved tyrosine residues, a classical protein tyrosine kinase domain, a sterile-α-motif, and a PDZ domain binding motif (Huot, 2004; Kullander and Klein, 2002). Since both the ephrin ligands and the Eph receptors are membrane bound, the interactions between ephrins and the Eph receptors require intracellular contact. An Eph receptor can also act as a ligand in the same way that an ephrin ligand can act as a receptor (Mellitzer *et al.*, 1999).

Ligand binding induces forward signaling, but ephrins can also signal via the receptor, which is called reverse signaling (Bruckner *et al.*, 1997; Holland *et al.*, 1996).

The ephrin–Eph receptor system regulates many cellular functions that depend upon cytoskeletal remodeling, such as axon guidance and synaptic plasticity. The influence of Eph–ephrin activation on cell behavior differs depending on the cell type. They are generally associated with repulsion of neighboring cells or of cellular processes, such as the neuronal growth cone. However, in some cases, Eph–ephrin activation leads to increased adhesion or attraction. Both classes have been implicated as regulators of axon guidance. Eph/ephrin molecules have also been implicated in guiding commissural axons in the spinal cord and brain (Henkemeyer *et al.*, 1996; Hu *et al.*, 2003; Imondi and Kaprielian, 2001; Imondi *et al.*, 2000; Orioli *et al.*, 1996; Palmer and Klein, 2003; Yokoyama *et al.*, 2001). Ephrin–Eph complexes also regulate axon guidance in the visual system. In particular, ephrin-A ligands and Eph A receptors mediate repulsion that is typically involved in regulating the mapping of retinal axons along the anterior-posterior axis (reviewed in Huot, 2004; see also Section VIII.B for the stripe assay). Topographic mapping of the anterior-posterior tectal/superior collicular axis is dependent upon expression patterns of EphA/ephrin-A (Brown *et al.*, 2000; Feldheim *et al.*, 2000), while EphB/ephrin-B signaling is important for mapping along the dorsoventral axis (Hindges *et al.*, 2002; Mann *et al.*, 2002b).

E. Morphogens

Morphogens are known for their effects on cell fate determination during development. Local concentration gradients convey positional information used during organization of the major body axes, limb development, and patterning of the nervous system. Recent studies suggest that in addition to these roles, morphogens, including members of the Wnt family, bone morphogenetic protein (BMP) family, and Sonic Hedgehog (Shh), might also function in axon guidance. For instance, a knockout of the Wnt receptor, Frizzled-3, results in severe defects in several major fiber tracts in the vertebrate forebrain (Wang *et al.*, 2002c), and Wnt-3 slows axon outgrowth and may mediate terminal branching of vertebrate muscle afferents (Krylova *et al.*, 2002). Wnt proteins are a large family of diffusible factors (7 members in *Drosophila* and 19 in humans) that play several important roles in both embryonic development and adult function. Wnt proteins have well-established roles in early cell fate decisions and embryonic patterning (Wodarz and Nusse, 1998), but they have also been implicated in synaptic remodeling and terminal arborization within the developing CNS (Hall *et al.*, 2000;

Krylova *et al.*, 2002; Packard *et al.*, 2002) and in planar cell polarity (Bhanot *et al.*, 1996; Vinson and Adler, 1987). In the *Drosophila* ventral nerve cord, Wnt-5 binding to the receptor tyrosine kinase Derailed is required for targeting axons to the appropriate midline commissure (Yoshikawa *et al.*, 2003). Genetic and biochemical data indicate that Wnt5 binds Derailed and prevents Derailed-expressing axons from entering the posterior commissure (Garrity, 2003). The dorsal spinal cord commissural neurons form several ascending somatosensory pathways. During embryonic development, they project axons to the ventral midline. At the floor plate, commissural axons cross the midline, enter the contralateral side of the spinal cord, and make a sharp anterior turn toward the brain (Bovolenta and Dodd, 1990). *In situ* hybridizations of Wnts in developing mouse embryos revealed that Wnt4, Wnt7b, and Wnt5a are expressed in areas where postcrossing axons turn anteriorly. Wnt4 was found specifically enriched in the floor plate and the ventricular zone, exhibiting a decreasing anterior-to-posterior gradient along the entire length of the floor plate (Lyuksyutova *et al.*, 2003). A similar Wnt4b gradient in the floor plate was found in zebrafish embryos at similar stages (Liu *et al.*, 2000). A directed source of Wnt4 protein attracted postcrossing commissural axons (Lyuksyutova *et al.*, 2003). Commissural axons in mice lacking the Wnt receptor Frizzled3 displayed anterior-posterior guidance defects after midline crossing. Thus, Wnt-Frizzled signaling guides commissural axons along the anterior-posterior axis of the spinal cord (Lyuksyutova *et al.*, 2003).

A role for BMPs in axonal guidance in vertebrates has emerged from studies of commissural axon trajectories in the developing spinal cord. Many commissural neurons differentiate adjacent to the roof plate at the dorsal midline of the spinal cord and extend axons ventrally (Dodd *et al.*, 1988; Holley, 1982). Signals derived from the floor plate contribute to the ventral trajectory of commissural axons (Colamarino and Tessier-Lavigne, 1995). The initiation of ventral growth of commissural axons may be mediated by a chemorepellent signal emanating from the roof plate. *In vitro* studies have shown that the roof plate is the source of a diffusible repellent activity that orients commissural axons in explants and that this repellent activity can be blocked by antagonists of BMPs (Augsburger *et al.*, 1999). At the time that commissural axon extension is initiated, *Bmp6, Bmp7*, and *Gdf7* are expressed in the rodent roof plate (Augsburger *et al.*, 1999; Lee *et al.*, 1998). BMP7 can mimic the repellent activity of the roof plate on commissural axons in explants *in vitro*, whereas BMP6 has only a low level of repellent activity, and GDF7 is inactive. Moreover, BMP7 elicits commissural growth cone collapse, illustrating the direct nature of its action. Roof plate tissue isolated from *Bmp7* mutant mice exhibits a marked reduction in roof plate-repellent activity *in vitro* (Augsburger *et al.*, 1999). Together, these findings suggest that BMPs can act as axon guidance signals that contribute to the chemorepellent activity of the roof plate. A more recent study analyzing roof plate-repellent activity in mice lacking Bmp7, Bmp6, and *Gdf7* alone and in pair-wise combinations show that both Gdf7 and Bmp7 but not Bmp6 are required for the ability of the roof plate to orient commissural axons. GDF7 and BMP7 heterodimerize, and the heterodimer is a more potent repellent than the BMP7 homodimer for commissural axons These results suggest that the GDF7:BMP7 heterodimer functions as a roof plate-derived repellent that establishes the initial ventral trajectory of commissural axons (Butler and Dodd, 2003).

Other evidence for morphogen involvement in axon guidance includes defects in retinal ganglion cell projections in mice deficient in the BMP receptor, BMPR-IB (Liu *et al.*, 2003). Shh can inhibit the outgrowth of neurons from retinal explants *in vitro* (Trousse *et al.*, 2001). Shh is a secreted protein that interacts with two transmembrane proteins, Patched (ptc) and Smoothened (smo). Ptc binds to shh, whereas smo is involved in signal transduction. In the absence of shh, ptc inhibits smo. A recent study by Charron and colleagues (2003) shows that Shh from midline structures collaborates with netrin-1 to guide commissural axons. This new role for the morphogen Shh raises the possibility that principles similar to those used to establish positional information in embryonic patterning are also employed during axon navigation (reviewed in Salinas, 2003).

F. Steroids

Steroid hormones may also induce directed neurite outgrowth. Estrogen elicits a significant enhancement of neurite outgrowth and differentiation within organotypic explant cultures of hypothalamus, preoptic area, and cerebral cortex (Toran-Allerand, 1976, 1980, 1984). Forebrain neurons coexpress nerve growth factor (NGF) receptors and estrogen receptor mRNA (Miranda *et al.*, 1993; Toran-Allerand *et al.*, 1992), and NGF significantly increases nuclear estrogen binding in cortical but not basal forebrain explants (Miranda *et al.*, 1996). Steroid/neurotrophin interactions may stimulate the synthesis of proteins required for neuronal differentiation, survival, and maintenance of function (Toran-Allerand, 1996). Estrogen and neurotrophin may regulate the same broad array of cytoskeletal and growth-associated genes involved in neurite growth and differentiation (Singh *et al.*, 1999).

The estrogen receptor, a nuclear transcription factor, is widely expressed in the developing forebrain (Toran-Allerand, 2004). Two mammalian estrogen receptors have been described, ER-α (White *et al.*, 1987), mediating most of the transcriptional action of estrogen in the brain, and ER-β (Kuiper *et al.*, 1996; Tremblay *et al.*, 1997). Mice lacking ER-β exhibit a reduction in the expression of genes involved in neuronal migration and axonal guidance, such as semaphorin G, syndecan 3, and reelin. Therefore, it appears that ER-β influences migration of neurons during development and neuronal survival throughout life (Wang *et al.*, 2002a, 2003).

The principal nucleus of the bed nuclei of the stria terminalis (BSTp) pathway to the anteroventral periventricular nucleus of the preoptic region (AVPV) develops in a sexually dimorphic pattern, suggesting a directed mechanism of axonal guidance (Hutton *et al.*, 1998). *In vitro*, addition of testosterone to BSTp to AVPV co-cultures induces neurite extension (Ibanez *et al.*, 2001). Testosterone induces a target-derived, diffusible chemotropic activity that results in a sexually dimorphic pattern of connectivity (reviewed in Simerly, 2002). The AVPV projects to the gonadotropin-releasing hormone (GnRH) neurons and the tuberoinfundibular dopaminergic (TIDA) neurons. Expression of ER-α by GnRH neurons (Skynner *et al.*, 1999) could suggest that estrogen may also direct the development of projections from the anteroventral periventricular nucleus to the GnRH neurons (Simerly, 2002).

The role of estrogen in branching has also been reported in invertebrates. Ecdysteroids, the insect steroids that trigger metamorphosis, control both regression and outgrowth *in vivo* and stimulate neuritic growth in cultured pupal leg motor neurons. Ecdysteroid enhances neuritic branching by altering growth cone structure and function, suggesting that hormonal modulation of cytoskeletal interactions contributes significantly to neuritic remodeling during metamorphosis (Matheson and Levine, 1999).

G. Intracellular Signaling Mechanisms

1. Ena/VASP

Ena/VASP proteins are a conserved family of actin regulatory proteins made up of Ena/VASP homology domain 1 (EVH1) and EVH2 domains and a proline-rich central region (Fig. 8). Mammalian Ena/VASP members are Mena, n-Mena, Ena/VASP like (EVL), and VASP. The members are 60–70% identical to each other. The Ena/VASP family is involved in Abl and/or cyclic nucleotide-dependent protein kinase signaling pathways. They have been implicated in actin-based processes such as fibroblast migration, axon guidance, and T-cell polarization and are important for the actin-based motility of the intracellular pathogen *Listeria monocytogenes* (Chakraborty *et al.*, 1995; Gerstel *et al.*, 1996; Gertler *et al.*, 1996; Pistor *et al.*, 1995; Smith *et al.*, 1996). Vertebrate Ena/VASP proteins are substrates for PKA/PKG serine/threonine kinases. Phosphorylation by these kinases appears to modulate Ena/VASP function within cells, although the mechanism underlying this regulation remains to be determined. Ena/VASP are also crucial factors in regulating actin dynamics and associated processes such as cell–cell adhesion.

Ena/VASP Family

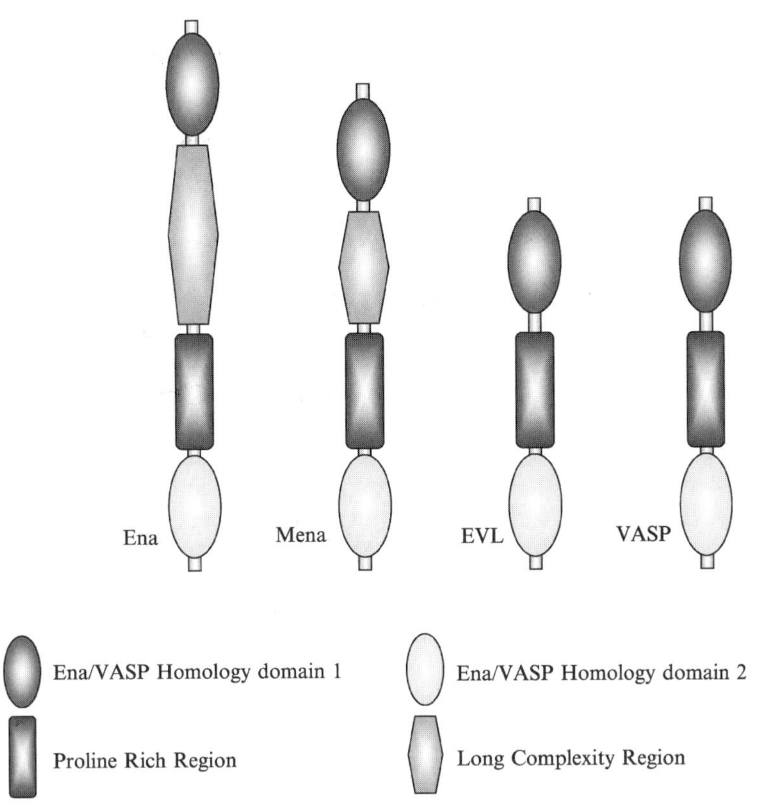

Figure 8   Ena/VASP. (See Color Insert.)

Evidence also suggests that these have inhibitory functions in integrin regulation, cell motility, and axon guidance (Krause *et al.*, 2003).

Mena and Mena/VASP mouse mutants have defects in several major axonal tracts (Lanier *et al.*, 1999), including the corpus callosum, the hippocampal commissure, the anterior commissure, and the pontocerebellar pathway. Ena/VASP proteins are highly expressed in the developing cortical plate in cells bordering reelin-expressing Cajal-Retzius cells and in the intermediate zone. Inhibition of Ena/VASP function through retroviral injections *in utero* leads to the aberrant migration of early-born pyramidal neurons in the superficial layers of both the embryonic and the postnatal cortex in a cell-autonomous fashion. The results demonstrate that Ena/VASP proteins play a key role in regulating neuronal migration and layering within the developing mouse neocortex (Goh *et al.*, 2002).

2. Rho GTPases

Rac, Rho, and Cdc42 are small GTPases of the Rho family. They have been shown to regulate actin organization in non-neuronal cells (Hall, 1998), as well as cytoskeletal dynamics in neuronal growth cones (Luo *et al.*, 1997). A model has been proposed (Hall, 1998; Mueller, 1999) in which attractive guidance cues activate Rac and Cdc42 to promote growth cone advance, whereas repulsive guidance cues activate Rho to inhibit growth and induce retraction (Dickson, 2001). For example, ephrin-A5 activates Rho and inhibits Rac in cultures of retinal ganglion cells. Both Rho and its downstream effector Rho kinase are required for growth cone collapse (Wahl *et al.*, 2000). Rac proteins mediate axon guidance, outgrowth, and branching as well as suppress the formation of ectopic axon growth (Lundquist, 2003). Calcium-dependent regulation of Rho GTPases triggers turning of nerve growth cones (Jin *et al.*, 2005). The regulators of the Rho GTPases, GTPase-activating proteins and guanine exchange factors, play important roles in axon guidance. Cross GTPase-activating proteins (CrGAPs) are involved in Robo-mediated repulsive axon guidance. Too much or too little CrGAP activity leads to defects in Robo-mediated repulsion at the midline. CrGAP directly interacts with Robo both biochemically and genetically and acts as a GTPase-activating protein specifically for Rac to regulate midline crossing (Hu *et al.*, 2005).

H. Extracellular Matrix Molecules

The extracellular matrix (ECM) is an important source of extrinsic cues that influence the response of growth cones to guidance cues (Condic *et al.*, 1999; Diefenbach *et al.*, 2000; Hopker *et al.*, 1999; Nguyen-Ba-Charvet *et al.*, 2001). They can act to promote or inhibit neurite outgrowth and modulate the response of axons to particular guidance cues. Laminin, tenascin, collagen, fibronectin, and a number of proteoglycans have been implicated in modulating axonal outgrowth. For example, laminin can promote, while tenascin can inhibit, neurite extension. Receptors for ECM molecules include integrins as well as Ig family members. For example, Ig CAM F3 can function as a receptor for a type of tenascin. The laminin family and its receptors are one of the best-studied examples of ECM molecules with regard to neuronal development. The laminins are heterotrimers, in which different subunits combine to form at least 10 different isoforms with growth-promoting or -inhibiting effects depending on the cell type. The axonal receptors for the laminins are the integrins. Integrins are heterodimers whose subunit composition determines their laminin binding specificity. The integrins link the ECM signals to the cytoskeleton and various signal transduction pathways. The exact role of laminins and other ECM molecules in neuronal development is not clear, although most evidence suggests a role in axonal guidance. Integrin signaling regulates cytoskeletal dynamics, adhesion, and migration events, through associated proteins such as talin, vinculin, integrin-linked kinase (ILK), focal adhesion kinase (FAK), paxillin, p130Cas, Abl kinase, and many other signaling or cytoskeletal proteins (Hynes, 2002).

In the developing nervous system, proteoglycans predominantly carry either chondroitin sulfate or heparan sulfate glycosaminoglycans (GAGs) (Bovolenta and Fernaud-Espinosa, 2000). Heparan sulfate proteoglycans (HSPGs) are a group of extracellular and cell surface proteins essential for proper axonal pathfinding during nervous system development (Bulow and Hobert, 2004; Walz *et al.*, 1997; Wang and Denburg, 1992), and it is increasingly evident that the major mechanism by which HSPGs influence axon pathfinding is by regulating the function of axon guidance cues. HSPGs affect several axon guidance cues, including fibroblast growth factor (FGF), heparin-binding growth associated molecule (HB-GAM), Slits, and Anosmin/Kallman syndrome gene (KAL-1) (Bulow and Hobert, 2004; Hu, 2001; Inatani *et al.*, 2003; Irie *et al.*, 2002; Johnson *et al.*, 2004; Kinnunen *et al.*, 1998; Steigemann *et al.*, 2004; Walz *et al.*, 1997). Chondroitin sulfate proteoglycans (CSPGs) are a heterogeneous set of proteins bearing GAGs of the chondroitin sulfate class (Lander, 1998). The CSPGs are also ECM molecules involved in the regulation of axon growth as demonstrated by *in vitro* studies on CSPGs such as NG2 (Dou and Levine, 1994), neurocan, and phosphacan (Margolis *et al.*, 1996). They influence the behavior of neuronal growth cones during development and, importantly, following CNS injury (Bovolenta and Fernaud-Espinosa, 2000; Morgenstern *et al.*, 2002). CSPGs are known to modulate the response of growth cones to other matrix components such as laminin (Condic *et al.*, 1999). This raises the possibility that CSPGs are components of the developmental environment capable of regulating how growth cones respond to surrounding guidance cues. The biological activity of CSPGs may also be determined by distinct proteins that bind to glycosaminoglycans and interact with receptors on the surface of neuronal growth cones (Anderson *et al.*, 1998; Brittis and Silver, 1994; Emerling and Lander, 1996; Golding *et al.*, 1999). Although CSPGs are known to interact with growth factors, adhesion molecules, and other matrix components, the specific binding proteins capable of mediating the effects of CSPGs on neuronal growth cones remain to be identified (Bovolenta and Fernaud-Espinosa, 2000; Morgenstern *et al.*, 2002).

A recent study showed that the thrombospondin repeats of Sema5A physically interact with the glycosaminoglycan portion of both CSPGs and HSPGs. CSPGs function as bound localized extrinsic cues that convert Sema5A from an attractive to an inhibitory guidance cue. Therefore, glycosaminoglycans provide a molecular mechanism for CSPG-mediated inhibition of axonal extension. Further, axonal HSPGs are required for Sema5A-mediated attraction, suggesting that HSPGs are components of functional Sema5A receptors. Therefore, the nature of a growth cone's response to Sema5A depends on the types of sulfated proteoglycans present in the developmental environment (Kantor *et al.*, 2004).

I. Adhesion Molecules

Several cell adhesion molecules (CAMs) of the Ig superfamily have also been implicated in regulating axon guidance at the midline, including mammalian L1CAM, NrCAM, and TAG-1 (mammalian ortholog of chick Axonin-1). The superfamily includes several subfamilies that are found in a number of tissues during development and in the adult (reviewed in Edelman and Crossin, 1991). Homologues of most of these have also been found in invertebrate animals. Each CAM has an extracellular region containing six Ig domains, as well as two (NrCAM), four (L1-CAM), or five (TAG-1) fibronectin type III extracellular domains (Walsh and Doherty, 1997) (Fig. 9). The cytoplasmic domain is highly conserved among individual members of the L1 subfamily and between invertebrate and vertebrate species (Hortsch, 1996). L1 is linked to the cytoskeleton through two regions in the cytoplasmic domain. An ankyrin binding site is located in the C terminus of the cytoplasmic domain (Davis and Bennett, 1994). CAMs,

Adhesion Molecules

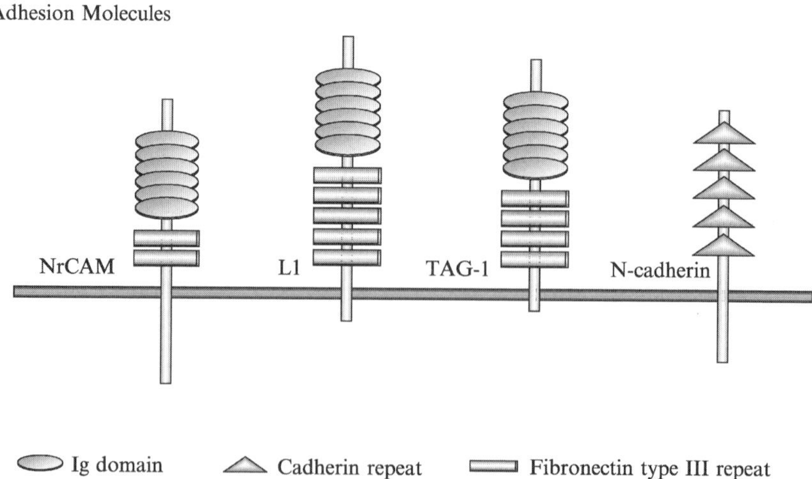

Figure 9   Adhesion molecules. (See Color Insert.)

including N-CAM, L1, and N-cadherin, promote neurite outgrowth (Doherty *et al.*, 1989; Doherty *et al.*, 1990).

L1 is one of the most intensely studied adhesion molecules expressed in the developing central and peripheral nervous system (Kamiguchi *et al.*, 1998). L1 is important in neuronal migration, axon growth, guidance, fasciculation, and synaptic plasticity. L1 is also expressed in non-neuronal cells such as the immune system, kidney, pigment cells, and a variety of cancers. Commissural axons extending toward the ventral midline of the rodent spinal cord express TAG-1, but not L1, while axonal segments on the contralateral side of the floor plate express L1, but not TAG-1 (Dodd *et al.*, 1988; Imondi *et al.*, 2000; Tran and Phelps, 2000). Consistent with an altered-responsiveness mechanism, it has been postulated that the switch in expression from TAG-1 to L1, presumably triggered by contact with the floor plate, delays the rostral turn exhibited by commissural axons until after they cross the floor plate (Dodd *et al.*, 1988). NrCAM, another commissural axon-associated IgCAM, is expressed at low levels on commissural axons both as they extend toward the floor plate on the ipsilateral side of the midline and as they project in the longitudinal direction on the contralateral side of the floor plate in the chick and mouse spinal cord (Matise *et al.*, 1999; Stoeckli and Landmesser, 1995).

Knockouts of L1CAM (Cohen *et al.*, 1998; Dahme *et al.*, 1997; Fransen *et al.*, 1998) display a reduced corticospinal tract, an abnormal pyramidal decussation, a decreased axonal association with non-myelinating Schwann cells, ventricular dilatation, and hypoplasia of the cerebellar vermis. Demyanenko and colleagues (1999) reported abnormal morphogenesis of cortical dendrites, showing that pyramidal neurons in layer V exhibited undulating apical dendrites that did not reach layer I. They also found that L1 mutants had a smaller hippocampus with fewer pyramidal and granule cells (Demyanenko *et al.*, 1999) and an altered distribution of dopaminergic neurons in the brain of L1 null mice (Demyanenko *et al.*, 2001). There is also a reduced size of the corpus callosum because of the failure of many callosal axons to cross the midline, as well as the formation of other commissural tracts in the brain. L1 has been shown to interact with neuropilin via the first Ig domain (Castellani *et al.*, 2002) and to participate in Sema3a signaling (Castellani *et al.*, 2000) to mediate axonal repulsion.

V. Axon Guidance Mechanisms at the Midline of the Nervous System

A. Commissural Axons in Spinal Cord

The spinal cord has proven to be a useful system for identifying molecules that guide axons to their appropriate targets. During spinal cord development, commissural neurons, which differentiate in the dorsal neural tube, send axons that project toward and subsequently across the floor plate, forming axon commissures (Colamarino and Tessier-Lavigne, 1995). These commissural axons project toward the midline in part because they are attracted by Netrin-1 (Kennedy *et al.*, 1994; Placzek *et al.*, 1990; Serafini *et al.*, 1996, 1994; Tessier-Lavigne *et al.*, 1988b). Once on the contralateral side, axons are no longer attracted but are repelled by Slit expressed by ventral midline cells (Brose and Tessier-Lavigne, 2000). This change in the response of axons is due to the silencing of Netrin-1 attraction by signaling through DCC interacting with the Slit receptor, Robo (Stein and Tessier-Lavigne, 2001). In *Netrin-1* or *DCC* mutant mice, many commissural axon trajectories fail to invade the ventral spinal cord and are misguided (Fazeli *et al.*, 1997; Serafini *et al.*, 1996). However, some of them do reach the midline, indicating that other guidance cues cooperate with Netrin-1 to guide these axons (Serafini *et al.*, 1996). Sonic hedgehog (Shh), a morphogen secreted by the floor plate, functions as a gradient signal for the generation of distinct classes of ventral neurons along the dorsoventral axis (Ingham and McMahon, 2001; Jessell, 2000; Marti and Bovolenta, 2002). Shh is an axonal chemoattractant that provides the Netrin-1-independent chemoattractant activity of the floor plate. Shh collaborates with Netrin-1 in commissural axon attraction *in vitro* and is required for normal guidance of these axons *in vivo* (Charron *et al.*, 2003). Several studies show that other members of the morphogen family, Wnt and BMP, are also involved in the spinal cord development. A directed source of Wnt4 protein attracts postcrossing commissural axons and Frizzled3 mutant mice display anterior-posterior guidance defects after midline crossing. This indicates that Wnt-Frizzled signaling guides commissural axons along the anterior-posterior axis of the spinal cord (Lyuksyutova *et al.*, 2003). Bmp7, Bmp6, and growth differentiation factor 7 (Gdf7) are expressed by the roof plate and are potential dorsal repellent cues for commissural axons (Augsburger *et al.*, 1999).

The *Slit* genes are also expressed in the floor plate at the ventral midline of the spinal cord, and *Robo1* and *2* are expressed in regions that include commissural neuron cell bodies (Brose *et al.*, 1999; Itoh *et al.*, 1998; Kidd *et al.*, 1998; Li *et al.*, 1999). *In vitro* commissural axons are repelled by Slit2 only after they have crossed the floor plate (Zou *et al.*, 2000). In *Slit1/Slit2* double mutant mice, although the formation of several major forebrain tracts (corticofugal, callosal, and the thalamocortical tracts) and the optic chiasm are defective (Bagri *et al.*, 2002; Plump *et al.*, 2002), no obvious commissural axon guidance defects were observed in the spinal cord (Plump *et al.*, 2002). Another member of the Slit family, Slit3, is also expressed by floor plate cells (Brose *et al.*, 1999), and analysis of triple Slit mutant revealed that many commissural axons stalled at the floor plate and failed to cross (Long *et al.*, 2004). This indicates that Slits contribute to the repulsion of axons away from the midline. Rig-1 is a divergent member of the Robo family (Yuan *et al.*, 1999) that is highly expressed before midline crossing and is downregulated after crossing (Sabatier *et al.*, 2004). Rig-1 prevents commissural axons from sensing Slit in the floor plate through their cognate receptor Robo1 as they grow toward the floor plate, allowing them to enter and cross to the contralateral side. At the midline, the downregulation of Rig-1 protein expression helps the axons to sense the floor plate as a repulsive environment, thus preventing them from re-crossing the midline (Sabatier *et al.*, 2004). Commissural axons also express another receptor, Neuropilin-2, mediating the repulsive effects of Sema3B, found in the floor plate, and Sema3F, expressed widely in the spinal cord, except in the floor plate (Zou *et al.*, 2000; Fig. 10). Analysis of homozygous *neuropilin-2* mutant mice shows disorganized axons at the midline while crossing (Zou *et al.*, 2000), suggesting a role for neuropilin-2 in commissural axon pathfinding.

Once the axons have crossed the midline, they execute a rostral turn at the contralateral floor plate margin and extend for a short distance within the ventral funiculus, a longitudinal fiber tract that forms in close apposition to the floor plate (Bovolenta and Dodd, 1990). EphA2 and EphB expression is upregulated on contralateral commissural axons (Brittis *et al.*, 2002; Imondi *et al.*, 2000), and class B ephrins are expressed in the floor plate as well as in the dorsal part of the spinal cord. A subset of decussated commissural axons takes a more dorsal trajectory before turning at the border of class B ephrin expression, indicating a role in excluding these axons from the dorsal spinal cord (Imondi and Kaprielian, 2001; Fig. 10).

B. Guidance of Cortical Axons at the Midline

Contralateral cerebral cortical projections through the corpus callosum integrate sensory and motor information between the two brain hemispheres. In split-brain animals and in people whose corpus callosum has been severed, interhemispheric transfer of sensory and motor information is deficient (Gazzaniga, 1995). In these individuals, visual and tactile information presented to one hemisphere is not available for analysis by the other hemisphere. In addition, perceptual interactions between the two hemispheres are absent in these individuals. These observations, pioneered by Roger Sperry in the 1960s (Sperry, 1968, 1982), defined the critical roles of contralateral cortical projections in human consciousness and behavior. Data in humans and in mice suggest the possibility that different mechanisms may regulate the development of the corpus callosum across its rostro-caudal extent (Richards *et al.*, 2004). The complex developmental processes required for formation of the corpus callosum may provide some insight into why such a large number of human congenital syndromes are associated with agenesis of this structure (Richards *et al.*, 2004). Anatomical studies

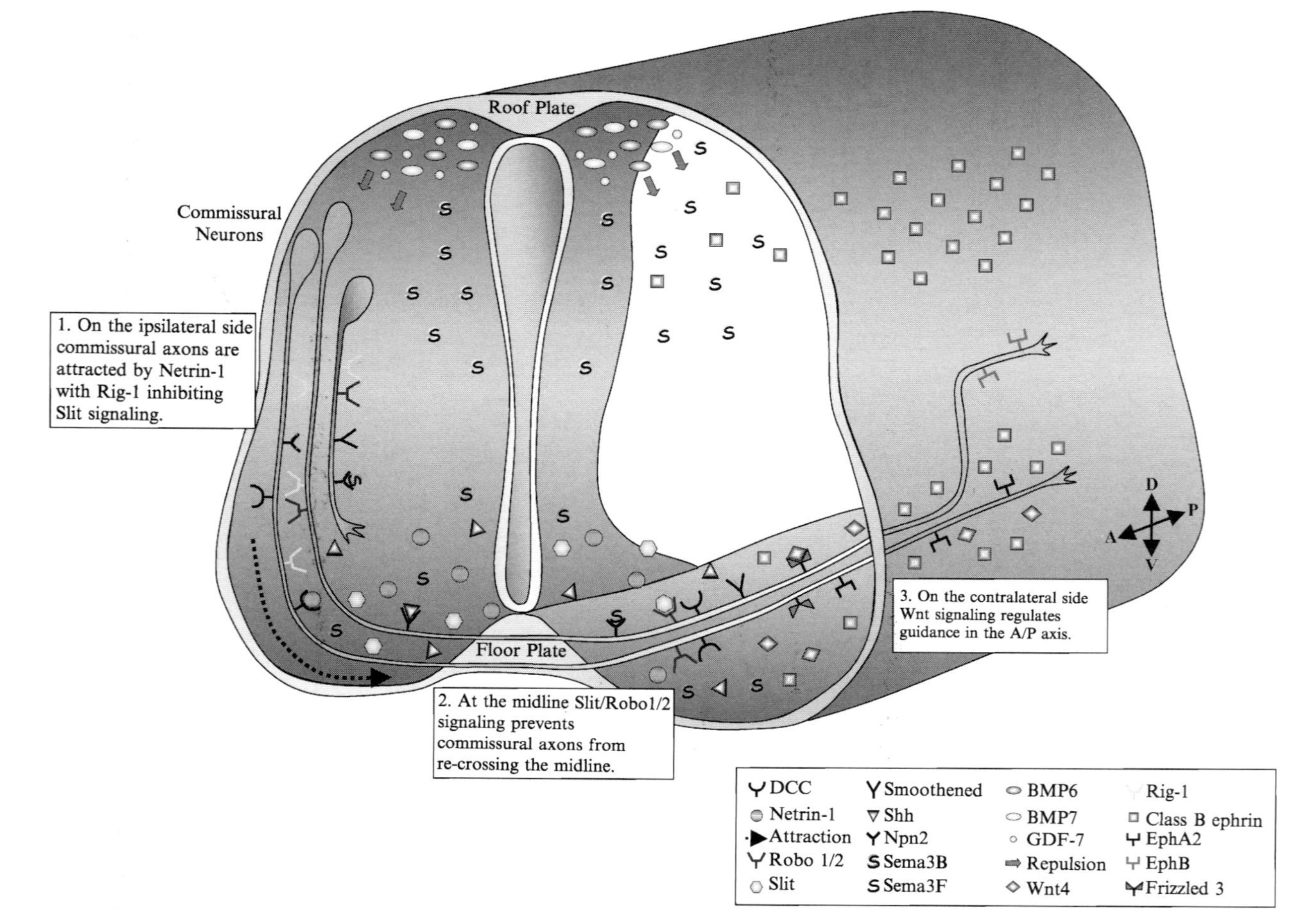

Roof Plate
Commissural Neurons
1. On the ipsilateral side commissural axons are attracted by Netrin-1 with Rig-1 inhibiting Slit signaling.
Floor Plate
2. At the midline Slit/Robo1/2 signaling prevents commissural axons from re-crossing the midline.
3. On the contralateral side Wnt signaling regulates guidance in the A/P axis.
D
P
A
V
DCC
Netrin-1
Attraction
Robo 1/2
Slit
Smoothened
Shh
Npn2
Sema3B
Sema3F
BMP6
BMP7
GDF-7
Repulsion
Wnt4
Rig-1
Class B ephrin
EphA2
EphB
Frizzled 3

have demonstrated that the majority of contralaterally projecting (callosal) neurons are located in layers 2/3 and 5 (Innocenti, 1986; Wise and Jones, 1976). In rodents callosal axons project to corresponding, homotopic areas in the contralateral cortex. During development, callosal axons grow ventrally to the intermediate zone (the future cortical white matter) and then turn medially, cross the midline, and re-enter the appropriate contralateral cortical area to form synapses with their targets (Fig. 11). The development of the corpus callosum depends on guidance by midline glial populations, and their expression of specific molecules, and fasciculation pioneering axons derived from neurons in the cingulate cortex (as described in Section III.A.2). Four midline populations at the corticoseptal boundary have been described: the glial wedge, the indusium griseum glia (Shu and Richards, 2001; Shu et al., 2003c), the midline zipper glia (Silver et al., 1993), and the subcallosal (glial) sling (Silver et al., 1982) (see Section III.B). The sling is a glial fibrillary acidic protein (GFAP)-negative population of cells (in rodents) with neuronal properties (Shu et al., 2003b) that migrates from the lateral sub-ventricular zone to underlie the developing corpus callosum (Silver et al., 1982). Both ablation and rescue experiments (Silver and Ogawa, 1983; Silver et al., 1982) have shown that the glial sling is required for the development of the corpus callosum. The indusium griseum glia and the midline zipper glia have many phenotypic and molecular characteristics in common, indicating that they may represent a common population of glia that becomes spatially distinct by the formation of the corpus callosum (Shu et al., 2003c). The glial wedge is part of the radial glial scaffold (Shu et al., 2003c) and, together with the indusium griseum glia, expresses Slit2. In the brain, unlike the spinal cord, Slit2 mediates both precrossing and postcrossing axonal guidance (Shu et al., 2003d). Robo1 and Robo2 mRNAs are expressed in the neocortex during callosal axon targeting (Shu and Richards, 2001), and Robo proteins are expressed on callosal axons (Sundaresan et al., 2004). In the Slit2 mutant, the corpus callosum fails to form. Instead, axons grow into large ectopic bundles of fibers on either side of the midline that resemble Probst bundles (Bagri et al., 2002). Taken together, these data

---

Figure 10 Molecules involved in the guidance of commissural axons in the spinal cord. Commissural neurons send their axons toward and across the midline. Bmp6 and 7 and GDF-7 expressed by the roof plate act as dorsal repellents for commissural axons. Axons express DCC, Robo1/2, Smoothened, and Npn2. The axons are attracted by Netrin-1 and Shh but initially are not responsive to Slit because Rig-1 inhibits Slit/Robo1/2 signaling. Sema/Npn2 signaling is required to avoid inappropriate targeting. Once the axons cross the midline, they are then repelled by Slit, expressed by the floor plate, and lose their attraction to Netrin-1, allowing them to leave the floor plate and preventing them from re-crossing the midline. Wtn4/Frizzled3 signaling regulates guidance in the anterior-posterior axis. A subset of commissural axons expressing EphB takes a more dorsal trajectory but grows between regions of Class 3 ephrin-B expression both dorsally and ventrally. (See Color Insert.)

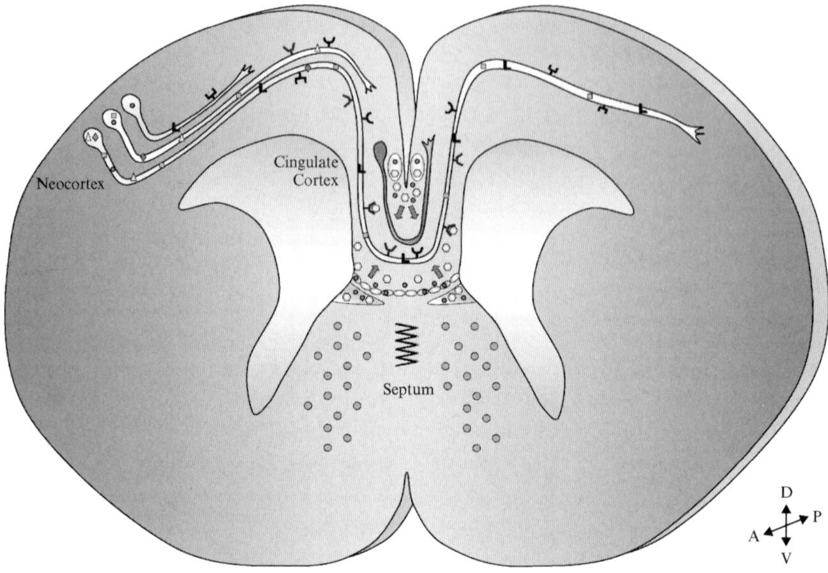

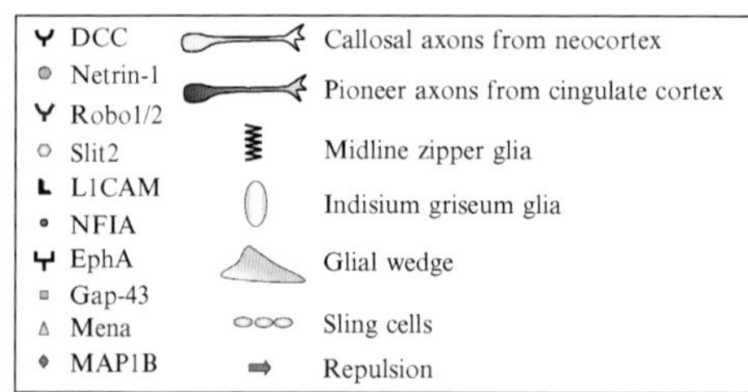

Figure 11 Molecules involved in the formation of the cortical commissural axons. Axons from the cingulate cortex pioneer a path across the midline. They are followed by neocortical callosal axons that probably fasciculate with them to cross the midline. Slit2, expressed by the glial wedge and the indusium griseum glia, provides a surround repulsion mechanism that keeps callosal axons within the tract, causing them to turn and cross the midline and preventing them from entering the septum. After axons have crossed the midline, Slit2 then repels the postcrossing axons away from the midline area. Callosal axons express several different molecules, including DCC, Robo1/2, EphA, NFIA, Gap-43, L1, Mena, and MAP1B. Netrin-1 is expressed within the septum under the corpus callosum; thus, unlike commissural axons in the spinal cord, callosal axons do not grow through the region of Slit and Netrin expression. NFIA is present in the sling cells, glial wedge, and the indusium griseum. Mouse mutants for DCC, Netrin-1, EphA, NFIA, Gap-43, L1, Mena, and MAP1B have defects in callosal formation (see Section V.B for more details) but how these genes regulate callosal axon pathfinding has not yet been elucidated. (See Color Insert.)

suggest an important role for Slit/Robo signaling in corpus callosum formation.

In the mammalian brain, Netrin-1 mutants and DCC knockout mice do not develop a corpus callosum or a hippocampal commissure and have a greatly reduced or absent anterior commissure (Fazeli *et al.*, 1997; Serafini *et al.*, 1996). In the developing forebrain, Netrin-1 has been shown to attract laterally directed cortical axons *in vitro* (Metin *et al.*, 1997; Richards *et al.*, 1997), indicating that these molecules are important for additional axonal guidance systems in the forebrain other than commissural axon guidance. Netrin-1 acts *in vitro* as an attractant and growth promoter for dorsal thalamic axons and is required for the proper development of the thalamo-cortical axon (TCA) projection *in vivo* (Braisted *et al.*, 2000). DCC protein is expressed predominantly in large fiber tracts such as the lateral olfactory tract, the internal capsule, the corpus callosum, the anterior commissure, the fimbria/fornix, the fasciculus retroflexus, and the stria medularis (Shu *et al.*, 2000), as well as in dorsal thalamus (Braisted *et al.*, 2000). DCC knockout mice have defects in multiple commissures, including the corpus callosum, the hippocampal commissure, and the anterior commissure (Fazeli *et al.* 1997). The basilar pons is absent in both Netrin-1 and DCC knockout mice (Fazeli *et al.*, 1997; Yee *et al.*, 1999).

The nuclear factor I (Nfi) family of transcription factors regulates both adenoviral DNA replication and viral and cellular gene expression, including the control of olfactory-specific genes (Baumeister *et al.*, 1999; Behrens *et al.*, 2000; Gronostajski *et al.*, 1985; Hennighausen *et al.*, 1985; Leegwater *et al.*, 1985; Nagata *et al.*, 1982, 1983; Nowock *et al.*, 1985). The Nfi family is made up of four members, Nfia, Nfib, Nfic, and Nfix. The Nfia mutant exhibits both agenesis of the corpus callosum and a reduction in GFAP expression (das Neves *et al.*, 1999). Nfia is expressed in midline glial structures, and the development of these structures is severely impaired in Nfia mutant mice. These data indicate that Nfia regulates both commissural development and the development of midline glia (Shu *et al.*, 2003a). Nfib mutant mice also display agenesis of the corpus callosum and abnormalities in midline glial development, as well as enlargement of the lateral ventricles (Steele-Perkins *et al.*, 2005).

In knockouts of the L1 gene (Cohen *et al.*, 1998; Dahme *et al.*, 1997; Fransen *et al.*, 1998) the corpus callosum failed to form properly due to the failure of many callosal axons to cross the midline (Demyanenko *et al.*, 1999, 2001). These findings suggest a variety of biological roles for L1 that are critical in brain development in different brain regions. Nr-CAM, a member of the L1 subfamily of cell adhesion molecules, is not expressed on callosal axons until postnatal day zero (P0), suggesting that Nr-CAM may be involved in the later stages of axonal growth or tract maintenance (Lustig *et al.*, 2001).

Mice lacking p35, an activator of cdk5 in the CNS, exhibit defects in a variety of CNS structures, most prominently characterized by a disruption in the laminar structure of the neocortex (Chae *et al.*, 1997). In these mutant mice, the corpus callosum appears bundled at the midline, but dispersed lateral to the midline. After crossing the midline, cortical axons defasciculate prematurely from the corpus callosum and take similarly oblique paths through the cortex. These results suggest that defective axonal fasciculation and guidance may be primary responses to the loss of p35 in the cortex (Kwon *et al.*, 1999).

In EphA5 mutant mice, callosal axons failed to grow into the corpus callosum, indicating that the EphA receptors and their ligands, the A-ephrins, play critical roles in the development of callosal axon projection to their contralateral targets (Hu *et al.*, 2003). EphA4 is expressed in the developing corpus callosum, with an interesting differential expression of EphA4 within different parts of the corpus callosum. In rostral regions, the entire corpus callosum was EphA4 positive whereas more caudally (around the hippocampal commissure) EphA4 was restricted to the most dorsal part of the corpus callosum (Greferath *et al.*, 2002). EphB2 (Nuk) and EphB3 (Sek4) mutant mice have been described (Henkemeyer *et al.*, 1996; Orioli *et al.*, 1996). EphB2 and EphB3 are members of the Eph-related family of receptor protein-tyrosine kinases. These receptors interact with a set of cell surface ligands that have recently been implicated in axon guidance and fasciculation. Whereas mice deficient in EphB2 exhibit defects in pathfinding of anterior commissure axons, EphB3 mutants have defects in corpus callosum formation. The phenotype in both axon tracts is markedly more severe in EphB2/EphB3 double mutants, indicating that the two receptors act in a partially redundant fashion (Orioli *et al.*, 1996).

MAP1B, a microtubule-associated protein, is expressed in axons, dendrites, and growth cones throughout the CNS during development. MAP1B is implicated in the crosstalk between microtubules and actin filaments. Homozygous MAP1B mutant mice display agenesis of the corpus callosum (Meixner *et al.*, 2000). A recent study showed that MAP1B phosphorylation is controlled by Netrin-1 (Del Rio *et al.*, 2004). Map1B mutant mice have severe abnormalities, similar to those described in netrin-1-deficient mice, in axonal tracts and in the pontine nuclei. These data indicate MAP1B may be a downstream effector in the Netrin-1-signaling pathway (Del Rio *et al.*, 2004).

A number of other genes are associated with agenesis of the corpus callosum in mice (reviewed in Richards *et al.*, 2004). The most common phenotype observed when cortical axons fail to reach the midline was that the axons do not stop growing but instead form swirled ipsilateral bundles of axons, called Probst bundles. Probst bundles form in mutants such as *Vax-1* (Bertuzzi *et al.*, 1999), *Gap-43* (Shen *et al.*, 2002), or heparan sulfate (Inatani *et al.*, 2003), indicating that these genes may regulate callosal axon guidance at the midline. However, in *GAP-43* mutant mice, callosal axons respond normally to Slit-2, although glial abnormalities may contribute to the phenotype (Shen *et al.*, 2002).

C. Guidance of Retinal Ganglion Cell Axons at the Optic Chiasm

In animals with binocular vision (such as mammals), retinal ganglion cell (RGC) axons originating from the nasal retina cross the midline to project into the contralateral optic tract, while a population of RGC axons from the temporal retina do not cross, but project away from the midline into the ipsilateral optic tract (Fig. 12). In mouse, ipsilaterally projecting RGCs are found in the ventro-temporal crescent of the retina, whereas contralaterally projecting RGCs are found throughout the retina (Guillery *et al.*, 1995; Mason and Sretavan, 1997; Sretavan, 1993).

After retinal ganglion cell axons exit each eye at the optic nerve head, forming the optic nerve, they traverse the ventral diencephalon toward the midline. Axons from the two eyes cross over each other to form the chiasm (X shape). The proportion of uncrossed to crossed retinal fibers varies across species. The ipsilateral projection represents about 40% of all RGCs in humans (Kandel *et al.*, 2000), less than 15% in ferrets (Cucchiaro, 1991; Thompson and Morgan, 1993), and about 3–5% in mice (Rice *et al.*, 1995). Adult birds and fish do not have an ipsilateral projection (O'Leary *et al.*, 1983) and thus lack binocular vision.

Neuronal cells within the chiasm are postulated to provide guidance cues for the earliest axons (Marcus and Mason, 1995; Sretavan *et al.*, 1994). When these neurons are destroyed by complement-mediated cytolysis, the growth of all axons entering the chiasm is halted (Sretavan *et al.*, 1995). Later axons, however, derive guidance cues from a midline palisade of radial glia. The cells in this region provide a generalized negative signal, as the growth cones of both crossed and uncrossed axons pause when entering this region *in vivo* (Godement, 1994) and *in vitro* (Mason and Wang, 1997; Wang *et al.*, 1996). After the optic chiasm, axons then continue through the optic tract to their targets: the superior colliculus in mammals or the optic tectum in fish, frogs, or birds. The main visual nuclei that receive retinal input are the lateral geniculate nucleus (LGN), the superior colliculus (SC), and the pretectal nuclei (Wassle, 1982). They also project to the superchiasmatic nucleus and the accessory optic system (Zhang and Hoffmann, 1993). Retinal axons express DCC (as well as the repulsive receptor UNC-5), but Netrin is absent from the chiasm region (Anderson and Holt, 2002; Deiner and Sretavan, 1999; Shewan *et al.*, 2002). Instead, Netrin is expressed at the optic nerve head, and in Netrin or DCC mutants RGC axons fail to exit this

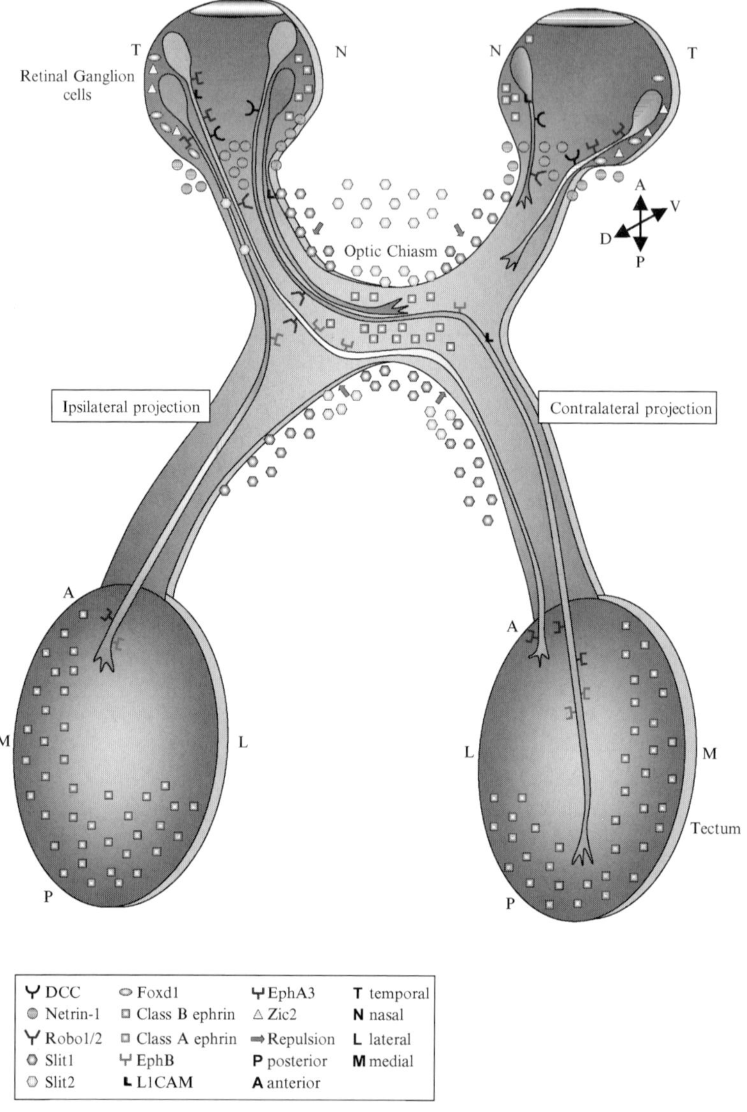

Figure 12 Molecules involved in the guidance of RGC axons to the tectum. The majority of RGC axons projects into the contralateral optic tract, while a small population of temporal RGC axons does not cross and projects ipsilaterally in mice. A diverse group of molecules acts to guide the RGC axons toward their final target in the tectum. Netrin-1 is expressed in the optic nerve head and Slits are expressed in the optic chiasm region, channeling the axons through the tectum and determining the position of the chiasm. RGC axons express DCC, Robo1/2, L1, and EphB and, as they enter the tectum, they express EphA3. Class B ephrins are expressed in the medial part of the tectum and mediate dorsoventral targeting, whereas Class A ephrins are expressed in the posterior tectum and mediate anterior-posterior topographic targeting. (See Color Insert.)

region (Deiner *et al.*, 1997). In the retina, *robo2* is expressed by RGCs before any axons have reached the ventral midline of the diencephalon and continues to be strongly expressed during later stages of development. In contrast, *robo1* is not detected until after a number of axons have started to cross the midline and then only in a subset of cells. This suggests that Robo2 is likely to be the principle receptor in RGC axon guidance (Erskine *et al.*, 2000). Two slit genes are expressed around the chiasm: Slit1 and Slit2 in mouse, Slit2 and Slit3 in zebrafish. Furthermore, Slit1 and Slit2 both repel retinal axons in culture (Erskine *et al.*, 2000; Niclou *et al.*, 2000; Ringstedt *et al.*, 2000). As shown by experiments in zebrafish and mouse, Robo/Slit signaling controls the formation of the optic chiasm (Fricke *et al.*, 2001; Hutson and Chien, 2002; Plump *et al.*, 2002). Zebrafish mutant for the astray/robo2 receptor show multiple guidance errors near the chiasm, including ipsilateral, anterior, and retinoretinal projections. Slit1 or Slit2 single knockout mice have little or no retinal projection phenotype, presumably because of the partial overlap in the expression of the two genes. However, Slit1 and Slit2 double knockouts show a strong phenotype, most notably the expansion of the chiasm more anteriorly. Developmental and time-lapse analyses in zebrafish embryos show that, even in wild-type animals, RGC growth cones occasionally leave their pathway and misproject, but these errors are quickly corrected (Hutson and Chien, 2002). In astray mutants, many more errors occur and persist. Thus, Slit/Robo signaling has two functions: preventing errors in the first place and correcting them if they do occur (Rasband *et al.*, 2003).

Optic nerve fibers grow out from the eye in fasciculated bundles, with contact between them mediated in part by membrane glycoproteins. The first of these glycoproteins to be identified was neural cell adhesion molecule (NCAM) (Jessell, 1988). NCAM plays a crucial role in axonal fasciculation and substrate adhesion. The localization of the cell adhesion molecules L1, NCAM, and myelin-associated glycoprotein (MAG) was studied in the developing and adult mouse optic nerve and retina. At all stages of development, NCAM is expressed by fasciculating axons, growth cones, and their contact sites with glial cells, and contacts between glial cells. MAG is first associated with the endoplasmic reticulum and Golgi apparatus in oligodendrocytes and then moves at cell surface. L1 is expressed by optic axons that are grouped into fascicular bundles and by their growth cones, but not by the growth cones that contact glial cells (Bartsch *et al.*, 1989). L1 was differentially expressed on unmyelinated axons, but absent on myelinated axons. This finding supports the notion that L1 may be involved in the stabilization of axonal fascicles but not of axon-myelin contacts (Bartsch *et al.*, 1989). Recent reports have shown that the absence of L1 in the optic pathway in L1 knockout mice has no obvious effect on the development pattern of axon divergence in the chiasm (Cohen *et al.*, 1998; Demyanenko and Maness, 2003). L1 and the polysialic acid-associated form of NCAM

(PSA-NCAM) are dynamically expressed in a regionally specific pattern in the retinotectal pathway (Chung *et al.*, 2004). At the chiasm, the level of L1 expression is high, whereas that of PSA-NCAM is low. However, within the tract, intense expression of both molecules is found predominantly on axons from the dorsal but not ventral retina. These changes are observed when axons arrive at the junction of the chiasm and the optic tract, indicating a site-specific switch in expression of cell adhesion molecules on the optic axons. Moreover a population of PSA-NCAM-rich cells also projects axons to the TPOC. These results suggest that CAM expression on the optic axons may control formation of the partial retinotopic axon order within the optic tract (Chung *et al.*, 2004).

CSPGs provide an unfavorable environment for axon growth and have been implicated in the changing patterns of fiber order in the developing retinotectal pathway (Brittis and Silver, 1995). In mouse, CSPGs are expressed by an early population of neurons in the ventral diencephalon, and enzymatic removal of the chondroitin moieties affects retinal ganglion cell guidance (Chung *et al.*, 2000a,b). CSPGs are expressed in the retina as well (Chung *et al.*, 2000a) and are also important for maintaining fiber order as axons approach the chiasm (Leung *et al.*, 2003).

Heparan sulfates are also involved in the formation of the optic chiasm (Inatani *et al.*, 2003). Analysis of mutant mice for EXT1, an enzyme indispensable for HS synthesis, revealed that retinal axons projected ectopically into the contralateral optic nerve, similar to Slit1/Slit2 double-knockout mice (Plump *et al.*, 2002).

B-class ephrins are required for the sorting of axons at the optic chiasm. In *Xenopus laevis*, ephrin-B2 is present at the chiasm coincident with the formation of the uncrossed component at metamorphosis, and premature misexpression of ephrin-B2 in the ventral diencephalon induced an ectopic ipsilateral projection (Nakamura *et al.*, 2000). EphB receptors are expressed in the retina (Birgbauer *et al.*, 2000; Braisted *et al.*, 1997; Hindges *et al.*, 2002; Mann *et al.*, 2002b; Williams *et al.*, 2003). A receptor for ephrin-B2, EphB1, is found exclusively in regions of retina that give rise to the ipsilateral projection. EphB1 null mice exhibit a dramatic reduction in the ipsilateral projection, suggesting that this receptor contributes to the formation of the ipsilateral retinal projection, most likely through its repulsive interaction with ephrin-B2 (Williams *et al.*, 2003).

Several regulatory genes expressed in the developing retina have been reported to play a role in retinal axon guidance at the optic chiasm. Mice lacking Vax1, Vax2, Pax2, or Brn3b exhibit different defects in retinal axon pathfinding at the chiasm (Barbieri *et al.*, 2002; Bertuzzi *et al.*, 1999; Mui *et al.*, 2002; Torres *et al.*, 1996; Wang *et al.*, 2002b). It has also been shown that mutual regulation of Pax-2 and Shh are important for the formation of the chiasm region (Alvarez-Bolado *et al.*, 1997). Overexpression of Pax-2 affects axon navigation through the chiasm and in the ascending postchiasmatic pathway, including the optic tract up to the tectum (Thanos *et al.*, 2004).

The zinc finger transcription factor Zic2 is expressed in retinal ganglion cells (Fig. 12). Loss- and gain-of-function analyses indicate that Zic2 is necessary and sufficient to regulate RGC axon repulsion by cues at the optic chiasm and to determine the ipsilateral projection (Herrera *et al.*, 2003).

The winged helix transcription factor Foxd1 (previously known as BF-2, brain factor 2) is expressed in the ventrotemporal retina, as well as in the ventral diencephalon during the formation of the optic chiasm. Both retinal development and chiasm morphogenesis are disrupted in embryos lacking Foxd1. In the Foxd1-deficient retina, proteins designating the ipsilateral projection, such as Zic2 and EphB1, are missing. In addition, in the Foxd1-deficient ventral diencephalon, Foxg1 expression invades the Foxd1 domain, Zic2 and Islet1 expression are minimized, and Slit2 expression prematurely expands, changes that could contribute to axon projection errors. Foxd1 plays a dual role in the establishment of the binocular visual pathways: first, in specification of the ventrotemporal retina, acting upstream of proteins directing the ipsilateral pathway, and second, in the patterning of the developing ventral diencephalon where the optic chiasm forms (Herrera *et al.*, 2004).

After forming the optic chiasm, RGC axons extend laterally and dorsally to establish the optic tract along the lateral wall of the diencephalon to reach visual targets in the thalamus and midbrain.

VI. Axon Guidance in the Retinotectal System

The retinotectal projection is one of the best-studied model systems for the examining of the mechanisms regulating precise topographic connectivity in the embryonic CNS. In 1963, Roger Sperry showed that after optic nerve section, retinal ganglion cell axons regenerate to their normal topographic positions in the tectum regardless of whether the entire retina is present (Attardi and Sperry, 1963). He proposed the chemoaffinity hypothesis: that each point in the target area has a unique molecular address determined by a specific distribution of cell surface molecules (Sperry, 1963). The retinotectal projection from the temporal retina is connected to the anterior tectum and the retinotectal projection from the nasal retina is connected to the posterior tectum (Fig. 12). Dorsal retina is connected to the dorsal tectum and ventral retina is connected to the ventral tectum (Holt and Harris, 1993; Mey and Thanos, 1992; Thanos and Mey, 2001; van Horck *et al.*, 2004). Time-lapse analyses of the retinotectal projection have demonstrated that retinal ganglion cell axons grow rapidly within the optic tract but move slowly after reaching the tectum, suggesting the presence of target-derived cues that suppress axonal growth (Harris *et al.*, 1987).

A combination of *in vitro* and genetic studies has revealed that the mapping of RGC axons along the anterio-posterior axis of the optic tectum is controlled in large part by the matched gradients of EphA receptor tyrosine kinases in the retina and the GPI-linked ephrin ligands in the tectum (McLaughlin *et al.*, 2003a). Activation of EphA by ligand binding leads to axon repulsion (Drescher *et al.*, 1997; Flanagan and Vanderhaeghen, 1998) and inhibition of axon branching (Yates *et al.*, 2001). Ephrin-A2 and ephrin-A5 are expressed in overlapping gradients in the tectum, while the corresponding receptors are expressed in complementary gradient across the retina (Cheng *et al.*, 1995). Axons from temporal retinal ganglion cells, where EphA expression is high, are inhibited from innervating the caudal tectum, where ephrin-A levels peak (Ruthazer and Cline, 2004). Ephrin-As control the temporal-nasal mapping of the retina in the optic tectum/superior colliculus by regulating the topographically specific interstitial branching of retinal axons along the anterior-posterior tectal axis. This branching is mediated by relative levels of EphA receptor repellent signaling (McLaughlin *et al.*, 2003a). In ephrin-A2/A5 double knockout mice the retinotectal map is severely disrupted (Feldheim *et al.*, 2000), indicating complementary functions of these two molecules in retinotectal mapping.

Members of the EphB family of receptor tyrosine kinases and their transmembrane ligands ephrin-Bs have been implicated in dorsoventral patterning of the vertebrate retinotectal projection (Flanagan and Vanderhaeghen, 1998). EphB/ephrin-B interactions mediate axon attraction (Hindges *et al.*, 2002; Mann *et al.*, 2002a). It has been shown in mice that EphB in ventral retina contributes to axon targeting to the medial part of the superior colliculus, where ephrin-B levels are high (Hindges *et al.*, 2002). In *Xenopus*, high levels of ephrin-B in dorsal RGCs guide axons to the EphB-rich ventral tectum (Mann *et al.*, 2002a). EphB/ephrin-B interactions are known to result in bidirectional signaling, characterized by signaling into cells expressing EphB receptors (i.e., forward signaling) and into cells expressing ephrin-B (i.e., reverse signaling) (Holland *et al.*, 1996). Both reverse and forward ephrin-B/EphB signaling are involved in regulating dorso-ventral topography (Hindges *et al.*, 2002; Mann *et al.*, 2002a; Pittman and Chien, 2002). EphB receptor forward signaling and ephrin-B reverse signaling mediate axon attraction to control dorsal-ventral retinal mapping along the lateral-medial tectal axis (McLaughlin *et al.*, 2003b).

VII. Axon Guidance in the Olfactory System

The sense of smell is a primal sense for humans as well as animals. For both humans and animals, it is an important means with which we sense our environment. Compared to other sensory systems, the olfactory system is unique in that new olfactory receptor neurons are continuously produced throughout life (Graziadei and Monti Graziadei, 1978). This process is essential for the replacement of mature olfactory receptor neurons that have a limited life span. Axons derived from newly differentiating olfactory receptor neurons in the olfactory epithelium project to glomeruli in the main olfactory bulb, where they synapse on the dendrites of mitral and tufted cells, whose neurons then send axons to the primary olfactory cortex (Shipley and Ennis, 1996). Therefore, all new olfactory axons exiting the olfactory epithelium, in embryos and adults, must be guided to their specific glomerular targets in the olfactory bulb. The precise olfactory axon targeting depends on the combination of guidance cues they encounter along their pathway.

A. Guidance Cues in the Main Olfactory System

The main olfactory system is formed by the olfactory epithelium, the olfactory nerve, and the olfactory bulbs. Olfactory sensory neurons located in the epithelium of the principal nasal cavity project their axons to the main olfactory bulb (MOB), where they synapse with second-order neurons (mitral and tufted cells) within specialized compartments of neuropil called glomeruli (reviewed in Dulac, 2000; Halpern, 1987). In rodents, axons extending from the olfactory epithelium grow initially through mesenchyme rich in NCAM (Croucher and Tickle, 1989; Key and Akeson, 1990; Miragall *et al.*, 1989), retinoic acid, FGF-8 (LaMantia *et al.*, 2000, 1993), L1, laminin, and HSPG and CSPG (Gong and Shipley, 1996; Raabe *et al.*, 1997; Treloar *et al.*, 1996). Several additional candidate guidance factors have been identified in the developing and mature olfactory system, but functional evidence for their involvement in the precise connectivity of olfactory sensory neurons is limited. Directional outgrowth toward distinct regions in the developing bulb involves a variety of signaling molecules, including semaphorins and neuropilins (Kobayashi *et al.*, 1997; Pasterkamp *et al.*, 1998, 1999; Renzi *et al.*, 2000; Schwarting *et al.*, 2000; Walz *et al.*, 2002; Williams-Hogarth *et al.*, 2000), adhesion molecules (Kafitz and Greer, 1998; Puche and Key, 1996; Yoshihara *et al.*, 1997), and Eph/ephrins (Maisonpierre *et al.*, 1993; St John and Key, 2001; St John *et al.*, 2000, 2002; Zhang *et al.*, 1996). Olfactory sensory neurons expressing a given odorant receptor project with precision to specific glomeruli in the olfactory bulb, generating a topographic map. This implies a tight linkage between the choice of a specific odorant receptor and the choice of a glomerular target. Mombaerts *et al.* (1996) developed a genetic approach to visualize these projections and examine the relationship between receptor expression and axon targeting. They observed that neurons expressing P2, an olfactory receptor gene, project with precision to two of the 1800 glomeruli within the olfactory bulb. These data provide direct support for a model in which a topographic map of the receptor topographically fixed glomeruli in the mouse olfactory bulb. Substitution of the olfactory receptor gene leads to the conclusion that the olfactory receptor may be one determinant in the guidance process (Mombaerts *et al.*, 1996; reviewed in Mombaerts, 1996). Neurons expressing different odorant receptors express different levels of ephrin-A protein on their axons. Moreover, alterations in the level of ephrin-A alter the glomerular map. Deletion of the ephrin-A5 and ephrin-A3 genes posteriorizes the glomerular locations for neurons expressing either the P2 or SR1 receptor, whereas overexpression of ephrin-A5 in P2 neurons results in an anterior shift in their glomeruli. Thus the ephrin-As are differentially expressed in distinct subpopulations of neurons and are likely to participate, along with the odorant receptors, in governing the targeting of like axons to precise locations in the olfactory bulb (Cutforth *et al.*, 2003).

Expression of dominant-negative neuropilin-1 protein causes chick olfactory neurons to enter the telencephalon prematurely (Renzi *et al.*, 2000). Mutations in galectin-1, Sema3A, neuropilin-2, p75NTR, CHL1, or N-CAM-180 (Montag-Sallaz *et al.*, 2002; Puche *et al.*, 1996; Schwarting *et al.*, 2000; Tisay *et al.*, 2000; Treloar *et al.*, 1997; Walz *et al.*, 2002) result in subtle perturbations in the overall pattern of olfactory axon projections; however, the role of these molecules in the precision of glomerular targeting has not been examined.

Netrin-1 expression has also been observed within the olfactory epithelium, and its receptor DCC was detected in cells under the neuroepithelium (Livesey and Hunt, 1997). DCC has also been found in the embryonic rodent olfactory bulb, on the granule cells, and on pioneer axons from output neurons (Gad *et al.*, 1997; Shu *et al.*, 2000). A detailed analysis of the spatio-temporal expression patterns of both Netrin-1 and DCC proteins in the developing rat olfactory system showed that the association of Netrin-1 expression near DCC-expressing olfactory axons is restricted to the initial period of olfactory nerve pathfinding, which suggests that Netrin-1 may play a role in the directed outgrowth of the nascent olfactory axons toward the telencephalon (Astic *et al.*, 2002). Another member of the netrin family, Netrin-4, is localized within the lateral olfactory tract and may be involved in promoting neurite elongation (Koch *et al.*, 2000).

Diffusible factors secreted from the bulb that attract axons of olfactory sensory neurons have been proposed in the past (Ressler *et al.*, 1994; Vassar *et al.*, 1994), and putative guidance molecules in olfactory tissue have been identified (Kafitz and Greer, 1997, 1998; Key and Akeson, 1990; Puche and Key, 1996; Schwarting *et al.*, 2000; St John *et al.*, 2000; Tisay and Key, 1999; Yoshihara *et al.*, 1997), but their direct involvement in axon outgrowth and pathfinding is still unknown. In addition, *in vitro* experiments suggest that a diffusible signal of unknown identity from the olfactory bulb may also direct receptor axons toward the bulb (Goetze *et al.*, 2002).

As the axons grow toward and into the rostral telencephalon, they encounter a variety of ECM molecules, some of which are associated with both axons and glial cells. These include NCAM, L1, laminin, OCAM (olfactory cell adhesion molecule), chondroitin sulfate proteoglycan (Gong and Shipley, 1996; Miragall and Dermietzel, 1992; Treloar *et al.*, 1996; Yoshihara *et al.*, 1997), and galectin-1 (Crandall *et al.*, 2000).

B. Guidance Cues in the Accessory Olfactory System

In many mammalian species, a second olfactory system, the vomeronasal system, is thought to be specialized in the perception of stimuli related to social and reproductive behaviors (Halpern, 1987; Keverne, 1999). The sensory neurons reside within an oblong-shaped structure, the vomeronasal organ (VNO), situated at the rostral end of the nasal cavity. The vomeronasal receptor neurons are bipolar neurons with a dendrite reaching the surface of the epithelium and an axon projecting through the vomeronasal nerves to the accessory olfactory bulb (AOB). The AOB is an oval structure situated on the dorso-posterior surface of the main olfactory bulb. The incoming vomeronasal nerve axons make synaptic contact with dendrites of AOB neurons in specialized structures called glomeruli (Farbman, 1992). Sensory neurons located in the apical portion of the vomeronasal organ innervate glomeruli restricted to the anterior region of the AOB. In contrast, neurons of the basal region of the VNO project their axons to glomeruli located in the posterior half of the AOB (Jia and Halpern, 1996). The molecular mechanisms that orchestrate the segregation of vomeronasal projections to the anterior and posterior halves of the AOB are just beginning to be understood. Two independent families of vomeronasal receptor genes have been characterized, the V1Rs and V2Rs, which encode seven-transmembrane domain proteins thought to represent the mammalian pheromone receptors (Dulac and Axel, 1995; Herrada and Dulac, 1997; Matsunami and Buck, 1997; Ryba and Tirindelli, 1997). The V1R and the V2R receptor families are expressed by two spatially segregated populations of VNO sensory neurons, such that neurons lining the apical half of the VNO neuroepithelium co-express V1Rs and the G-protein α-subunit, Gαi2, whereas neurons of the basal half of the VNO are both V2R and Gαo positive (Dulac, 2000).

One member of the ephrin family, ephrin-A5, is required for the targeting of V1R-expressing vomeronasal axons to the anterior AOB, presumably through interactions with the Eph receptor EphA6 (Knoll *et al.*, 2001). The targeting of apical vomeronasal axons to the anterior AOB is also dependent on repulsive forces from the posterior AOB that signal, at least in part, through neuropilin-2 (Npn-2) (Cloutier *et al.*, 2002). Npn-2 is required for the targeting of apical vomeronasal axons to the anterior AOB as well as fasciculation of the vomeronasal nerve and also for separation of the main and accessory olfactory projections during development (Cloutier *et al.*, 2002; Walz *et al.*, 2002).

Secreted semaphorins and slits also participate in establishing the glomerular map in the accessory olfactory system. It is likely that a combination of attractive and repulsive cues in the AOB, and possibly in the vomeronasal sensory neurons themselves, allows for the precise targeting of these sensory afferent projections to either posterior or anterior regions of the AOB. The Slit and Robo families of axon guidance molecules function as cues for basal vomeronasal axons (Knoll *et al.*, 2003). Vomeronasal neurons are repelled by Slit proteins *in vitro*, and Robo2 is expressed in vomeronasal neurons located in the basal region of the VNO (Knoll *et al.*, 2003; Marillat *et al.*, 2002).

Sema3F-Npn-2 signaling is essential for fasciculation of the vomeronasal nerve as it courses past the MOB but is largely dispensable for the targeting of apical vomeronasal axons to the anterior AOB. Moreover, Sema3F is required for the segregation of sensory neuron projections within the main and accessory olfactory systems and for accurate laminar targeting of main olfactory sensory neuron axons (Cloutier *et al.*, 2004). Slit-1 is not required for fasciculation of the vomeronasal nerve but is critical for the targeting of basal vomeronasal sensory neuron axons to the posterior AOB (Cloutier *et al.*, 2004). These results show that two families of secreted repellents play complementary roles in the development of primary sensory neuron projections in the accessory olfactory system.

VIII. Analysis of Axon Growth and Guidance

Several techniques have been used to study axonal guidance. The collagen gel assay, the stripe assay, the pipette assay, and the pump assay are very powerful techniques that have been used to study the effects (attraction, repulsion, or collapse) of different molecules on given axonal populations. In order to study the formation of axon tracts *in vivo,* researchers have used both tract tracing techniques using fluorescent dyes such as DiI or *in vivo* imaging by diffusion tensor magnetic resonance imaging (DTI-MRI). This nondestructive technique can be used to survey the development of multiple axonal tracts in three dimensions in the embryonic, postnatal, and adult brains.

A. The Collagen Gel Assay

The three-dimensional collagen gel assay is one of the most widely used techniques for analyzing axon growth from explants. Collagen is a physiologically relevant biological matrix that allows neurites to grow and soluble factors to diffuse. Generally, two explants or one explant and a source of a putative guidance factor are placed side-by-side in the gel. One explant contains the cells being studied (for example, the dorsal spinal cord containing commissural neurons). As these axons grow into the collagen, they encounter guidance factors released from a second explant (such as the floor plate). The number of axons emanating from the side toward the second, or target, explant can be compared with the number of axons projecting from the side away from the target. The number of axon turning within the gel can also be analyzed. The collagen gel assay was first used by Lumsden and Davies (1983), using co-cultured embryonic mouse sensory neurons, but has since been used to identify and characterize a number of different axonal guidance molecules.

B. The Stripe Assay

In order to elucidate the mechanisms involved in axonal guidance by membrane-bound cues, especially with regard to the topographic targeting of retinal ganglion cell axons within the tectum, Friedrich Bonhoeffer and colleagues developed the "stripe assay" (Walter *et al.*, 1987a,b). In this assay, retinal explants are grown on membrane stripes made from alternating rostral and caudal parts of the tectum. Membrane fragments of two different sources are arranged as a carpet of very narrow alternating strips. Axons growing on such striped carpets are simultaneously confronted with the two substrates at the stripe borders. If there is a preference of axons for one or the other substrate, axons become re-oriented and grow within the lanes of the preferred substrate. Such preferential growth can be due to an increased affinity for attractive factors on the preferred stripes or avoidance of repulsive factors on the alternate stripes, and thus adequate controls must be used to differentiate between these two possibilities. Tectal cell membranes are an excellent substrate for the growth of retinal axons. Results using tissues from developing chicks or rodents show that temporal RGC axons avoid stripes made up of caudal tectal tissue, while nasal RGC axons grow equally well on either rostral or caudal tectal tissue stripes (Godement and Bonhoeffer, 1989; Roskies and O'Leary, 1994; Walter *et al.*, 1987a,b), or show a preference for nasal stripes providing specific pre-treatments (von Boxberg *et al.*, 1993).

These experiments suggest the existence of biochemical labels for the specification of axon–target interactions and provide the conceptual basis for the work of Friedrich Bonhoeffer and others that ultimately led to the identification of specific molecular guidance factors for retinotectal topography.

C. The Pipette Assay

Originally pioneered by Mu-Ming Poo and colleagues, the "*Xenopus* growth cone turning assay" has recently become a popular assay for assessing growth cone turning responses to gradients of guidance factors. The turning assay uses cultured *Xenopus* spinal neurons to examine the cytoplasmic events associated with neurite growth and the response of the growth cone toward extracellular guidance cues. This method uses repetitive pulse application to create reproducible chemical gradients. Microscopic gradients are generated by repetitive pulsatile ejection of picoliters of chemical solution near the growth cone. The gradient generated is a stable gradient over a distance of tens of microns within several minutes. Growth cones of isolated *Xenopus* spinal neurons in culture exhibit chemotactic turning responses when exposed to a gradient of Netrin-1 (Ming *et al.*, 2002). The gradient is produced by repetitive pulsatile ejection of picoliters of solution containing Netrin-1 from a micropipette, positioned at a 45° angle with respect to the direction of initial neurite extension and a distance of 100 μm away from the center of the growth cone. By applying defined extracellular gradients of guidance molecules, early responses of the growth cone to specific guidance cues can be examined as well as the involvement of various cytoplasmic signaling pathways in regulating the turning decision of the growth cone.

Using this assay, Poo and colleagues had previously demonstrated that growth cones from young *Xenopus* retinal explants extend toward a gradient of Netrin-1 protein and away from a gradient of Sema3A (de la Torre *et al.*, 1997; Song *et al.*, 1998). In addition to inducing transient collapse and branching, Sema3A can act as a directional guidance cue in the growth cone turning assay for retinal growth cones (Campbell *et al.*, 2001).

D. The Pump Assay

Geoffrey Goodhill and colleagues designed a technique for generating precise and reproducible gradients of diffusible molecules in collagen gels (Narasimhan, 2004; Rosoff *et al.*, 2004). The assay establishes molecular gradients by printing drops of solution onto the surface of a thin collagen gel. Based on the mechanics of diffusion, the shape and steepness of the gradient can be controlled, and the actual concentration of molecules produced by this method can be measured quantitatively with fluorescence imaging. Using this technology, they showed that growth cones are capable of detecting a concentration difference as small as about one molecule across their spatial extent. Furthermore, this sensitivity exists across only a relatively small range of ligand concentrations, indicating that adaptation in these growth cones is limited. Their gradient generation method allows for the production of large numbers of identical gradients that require only limited quantities of chemotropic molecules. Moreover, the gradients are stable for at least a day or two after they have been generated. This technique could also be used to generate gradients of multiple factors with different shapes and arbitrary spatial relationships. This powerful new technology can be applied to quantitative studies of other biological processes controlled by molecular gradients, such as cell migration, as well axonal regeneration following injury.

E. The Organotypic Slice Assay

In mammals, surgical manipulation is difficult before birth. Thus, many researchers have turned to using slice preparations to study axonal growth and guidance. Slices maintained in stationary culture with the interface method are ideally suited for manipulations that demonstrate the guidance properties of a prospective target tissue (Gahwiler *et al.*, 1997). For example, to better understand the projection of the commissural axons in the spinal cord, an "open book" organotypic slice assay was used, in which sections of spinal cord are split at the midline, flattened, and cultured in collagen to analyze the trajectories of the commissural axons. Advances in live optical imaging, including confocal and two-photon microscopy using GFP-based markers, allow the direct observation of axonal growth within the slice. Thus, using this assay, live growing axons can be studied *in situ* with maximum resolution and precision.

F. MRI/DTI

Magnetic resonance imaging (MRI) uses a strong electromagnetic field to align hydrogen atoms in the body in parallel with the magnetic field, in either the same or the opposite direction of the field. Hydrogen atoms become excited and resonate with the exciting wave. As the hydrogen atoms return to their original energy state, energy is released in the form of radio waves, which are detected by the MRI machine. Diffusion tensor imaging (DTI) measures the alignment of water molecules in the brain. Water molecules preferentially align within ordered structures in the brain such as axonal tract. The vector orientation of their alignment can be observed and color-coded to give a visual image of the position and direction of axon tracts within the brain. DTI can be combined with conventional MRI to capture T1- and T2-weighted images such that a three-dimensional map of the brain can be generated (Zhang *et al.*, 2003).

Recent advances in magnetic resonance research have opened up new opportunities for gathering functional information about the brain. The development of DTI has offered the possibility to go beyond anatomical imaging to study tissue structure at a microscopic level *in vivo*. Currently, DTI provides an invaluable tool for the study and diagnosis of white-matter diseases. An important application of DTI is called "fiber tracking," which follows the vector trajectories from given coordinates with the brain to "trace" axonal pathways. Combined with fMRI, information about white-matter tracts reveals important information about neurocognitive networks and may improve our understanding of brain function. Future advances could involve the use of magnetically labeled molecules to analyze axons expressing specific molecules *in vivo*, possibly in living brains.

IX. Conclusions

The establishment of correct neuronal connections is crucial for proper functioning of the vertebrate nervous system. Such precise wiring of neuronal connections is also required in the adult brain for functional recovery after brain injury and disease. To reach their proper targets, axons rely upon the expression of highly conserved families of attractive and repulsive guidance molecules, including the Netrins, Slits, semaphorins, and ephrins. These guidance systems are used to generate an astonishingly varied set of neuronal circuits. It is difficult to understand how such complexity in brain wiring is derived from such a relatively small set of molecules. The regulation of guidance receptors and ligands allows a single guidance system to generate a variety of different responses, but even these cannot account for the huge number of different connections made in the nervous system. What is clear is that the same molecules are used over and over in establishing different projections within the brain and spinal cord. Furthermore, although they may function in slightly different ways, the molecules themselves are conserved from flies and worms to humans. Recent evidence suggests that additional layers of complexity may exist where molecules and receptors from different molecular families interact in signaling to create more complicated responses. We do not know exactly how growth cones are able to respond to molecular gradients within the environment, nor exactly how these signals are relayed intracellularly. Furthermore, little is known about how axons are able to respond to multiple gradients simultaneously. Because neuronal regeneration is, to a certain extent, a recapitulation of development, understanding the mechanisms regulating axon guidance will not only help delineate the pathoetiology of various neurological disorders, due to erroneous axon pathfinding, but also shed light on potential ways of clinically repairing the injured nervous system.

References

Ahmad, F. J., Pienkowski, T. P., and Baas, P. W. (1993). Regional differences in microtubule dynamics in the axon. *J. Neurosci.* **13**, 856–866.

Aizawa, H., Wakatsuki, S., Ishii, A., Moriyama, K., Sasaki, Y., Ohashi, K., Sekine-Aizawa, Y., Sehara-Fujisawa, A., Mizuno, K., Goshima, Y., and Yahara, I. (2001). Phosphorylation of cofilin by LIM-kinase is necessary for semaphorin 3A-induced growth cone collapse. *Nat. Neurosci.* **4**, 367–373.

Alcantara, S., Ruiz, M., De Castro, F., Soriano, E., and Sotelo, C. (2000). Netrin 1 acts as an attractive or as a repulsive cue for distinct migrating neurons during the development of the cerebellar system. *Development* **127**, 1359–1372.

Altman, J., and Bayer, S. A. (1984). The development of the rat spinal cord. *Adv Anat. Embryol. Cell Biol.* **85**, 1–164.

Alvarez-Bolado, G., Schwarz, M., and Gruss, P. (1997). Pax-2 in the chiasm. *Cell Tissue Res.* **290**, 197–200.

Anderson, R. B., and Holt, C. E. (2002). Expression of UNC-5 in the developing Xenopus visual system. *Mech. Dev.* **118**, 157–160.

Anderson, R. B., Walz, A., Holt, C. E., and Key, B. (1998). Chondroitin sulfates modulate axon guidance in embryonic Xenopus brain. *Dev. Biol.* **202**, 235–243.

Astic, L., Pellier-Monnin, V., Saucier, D., Charrier, C., and Mehlen, P. (2002). Expression of netrin-1 and netrin-1 receptor, DCC, in the rat olfactory nerve pathway during development and axonal regeneration. *Neuroscience* **109**, 643–656.

Attardi, D. G., and Sperry, R. W. (1963). Preferential selection of central pathways by regenerating optic fibers. *Exp. Neurol.* **7**, 46–64.

Augsburger, A., Schuchardt, A., Hoskins, S., Dodd, J., and Butler, S. (1999). BMPs as mediators of roof plate repulsion of commissural neurons. *Neuron* **24**, 127–141.

Auld, V. (1999). Glia as mediators of growth cone guidance: Studies from insect nervous systems. *Cell Mol. Life Sci.* **55**, 1377–1385.

Bagri, A., Marin, O., Plump, A. S., Mak, J., Pleasure, S. J., Rubenstein, J. L., and Tessier-Lavigne, M. (2002). Slit proteins prevent midline crossing and determine the dorsoventral position of major axonal pathways in the mammalian forebrain. *Neuron* **33**, 233–248.

Bak, M., and Fraser, S. E. (2003). Axon fasciculation and differences in midline kinetics between pioneer and follower axons within commissural fascicles. *Development* **130**, 4999–5008.

Bamburg, J. R. (2003). Introduction to cytoskeletal dynamics and pathfinding of neuronal growth cones. *J. Histochem. Cytochem.* **51**, 407–409.

Banyai, L., and Patthy, L. (1999). The NTR module: Domains of netrins, secreted frizzled related proteins, and type I procollagen C-proteinase enhancer protein are homologous with tissue inhibitors of metalloproteases. *Protein Sci.* **8**, 1636–1642.

Barbieri, A. M., Broccoli, V., Bovolenta, P., Alfano, G., Marchitiello, A., Mocchetti, C., Crippa, L., Bulfone, A., Marigo, V., Ballabio, A., and Banfi, S. (2002). Vax2 inactivation in mouse determines alteration of the eye dorsal-ventral axis, misrouting of the optic fibres and eye coloboma. *Development* **129**, 805–813.

Bartsch, U., Kirchhoff, F., and Schachner, M. (1989). Immunohistological localization of the adhesion molecules L1, N-CAM, and MAG in the developing and adult optic nerve of mice. *J. Comp. Neurol.* **284**, 451–462.

Bastiani, M. J., Raper, J. A., and Goodman, C. S. (1984). Pathfinding by neuronal growth cones in grasshopper embryos. III. Selective affinity of the G growth cone for the P cells within the A/P fascicle. *J. Neurosci.* **4**, 2311–2328.

Bastiani, M. J., du Lac, S., and Goodman, C. S. (1986). Guidance of neuronal growth cones in the grasshopper embryo. I. Recognition of a specific axonal pathway by the pCC neuron. *J. Neurosci.* **6**, 3518–3531.

Bate, C. M. (1976). Pioneer neurones in an insect embryo. *Nature* **260**, 54–56.

Baumeister, H., Gronostajski, R. M., Lyons, G. E., and Margolis, F. L. (1999). Identification of NFI-binding sites and cloning of NFI-cDNAs suggest a regulatory role for NFI transcription factors in olfactory neuron gene expression. *Brain Res. Mol. Brain Res.* **72**, 65–79.

Bear, J. E., Svitkina, T. M., Krause, M., Schafer, D. A., Loureiro, J. J., Strasser, G. A., Maly, I. V., Chaga, O. Y., Cooper, J. A., Borisy, G. G., and Gertler, F. B. (2002). Antagonism between Ena/VASP proteins and actin filament capping regulates fibroblast motility. *Cell* **109**, 509–521.

Behrens, M., Venkatraman, G., Gronostajski, R. M., Reed, R. R., and Margolis, F. L. (2000). NFI in the development of the olfactory neuroepithelium and the regulation of olfactory marker protein gene expression. *Eur. J. Neurosci.* **12**, 1372–1384.

Bentley, D., and Caudy, M. (1983). Pioneer axons lose directed growth after selective killing of guidepost cells. *Nature* **304**, 62–65.

Bentley, D., and O'Connor, T. P. (1994). Cytoskeletal events in growth cone steering. *Curr. Opin. Neurobiol.* **4**, 43–48.

Bertuzzi, S., Hindges, R., Mui, S. H., O'Leary, D. D., and Lemke, G. (1999). The homeodomain protein vax1 is required for axon guidance and major tract formation in the developing forebrain. *Genes Dev.* **13**, 3092–3105.

Bhanot, P., Brink, M., Samos, C. H., Hsieh, J. C., Wang, Y., Macke, J. P., Andrew, D., Nathans, J., and Nusse, R. (1996). A new member of the frizzled family from *Drosophila* functions as a Wingless receptor. *Nature* **382**, 225–230.

Birgbauer, E., Cowan, C. A., Sretavan, D. W., and Henkemeyer, M. (2000). Kinase independent function of EphB receptors in retinal axon pathfinding to the optic disc from dorsal but not ventral retina. *Development* **127**, 1231–1241.

Bloch-Gallego, E., Ezan, F., Tessier-Lavigne, M., and Sotelo, C. (1999). Floor plate and netrin-1 are involved in the migration and survival of inferior olivary neurons. *J. Neurosci.* **19**, 4407–4420.

Bovolenta, P., and Dodd, J. (1990). Guidance of commissural growth cones at the floor plate in embryonic rat spinal cord. *Development* **109**, 435–447.

Bovolenta, P., and Fernaud-Espinosa, I. (2000). Nervous system proteoglycans as modulators of neurite outgrowth. *Prog. Neurobiol.* **61**, 113–132.

Braisted, J. E., McLaughlin, T., Wang, H. U., Friedman, G. C., Anderson, D. J., and O'Leary, D, D. (1997). Graded and lamina-specific distributions of ligands of EphB receptor tyrosine kinases in the developing retinotectal system. *Dev. Biol.* **191**, 14–28.

Braisted, J. E., Catalano, S. M., Stimac, R., Kennedy, T. E., Tessier-Lavigne, M., Shatz, C. J., and O'Leary, D. D. (2000). Netrin-1 promotes thalamic axon growth and is required for proper development of the thalamocortical projection. *J. Neurosci.* **20**, 5792–5801.

Bray, D., and Chapman, K. (1985). Analysis of microspike movements on the neuronal growth cone. *J. Neurosci.* **5**, 3204–3213.

Bridgman, P. C., and Dailey, M. E. (1989). The organization of myosin and actin in rapid frozen nerve growth cones. *J. Cell Biol.* **108,** 95–109.

Brittis, P. A., and Silver, J. (1994). Exogenous glycosaminoglycans induce complete inversion of retinal ganglion cell bodies and their axons within the retinal neuroepithelium. *Proc. Natl. Acad. Sci. USA* **91,** 7539–7542.

Brittis, P. A., and Silver, J. (1995). Multiple factors govern intraretinal axon guidance: A time-lapse study. *Mol. Cell Neurosci.* **6,** 413–432.

Brittis, P. A., Lu, Q., and Flanagan, J. G. (2002). Axonal protein synthesis provides a mechanism for localized regulation at an intermediate target. *Cell* **110,** 223–235.

Brose, K., and Tessier-Lavigne, M. (2000). Slit proteins: Key regulators of axon guidance, axonal branching, and cell migration. *Curr. Opin. Neurobiol.* **10,** 95–102.

Brose, K., Bland, K. S., Wang, K. H., Arnott, D., Henzel, W., Goodman, C. S., Tessier-Lavigne, M., and Kidd, T. (1999). Slit proteins bind Robo receptors and have an evolutionarily conserved role in repulsive axon guidance. *Cell* **96,** 795–806.

Brown, A., Yates, P. A., Burrola, P., Ortuno, D., Vaidya, A., Jessell, T. M., Pfaff, S. L., O'Leary, D. D., and Lemke, G. (2000). Topographic mapping from the retina to the midbrain is controlled by relative but not absolute levels of EphA receptor signaling. *Cell* **102,** 77–88.

Bruckner, K., Pasquale, E. B., and Klein, R. (1997). Tyrosine phosphorylation of transmembrane ligands for Eph receptors. *Science* **275,** 1640–1643.

Brunjes, P. C., and Frazier, L. L. (1986). Maturation and plasticity in the olfactory system of vertebrates. *Brain Res.* **396,** 1–45.

Bulow, H. E., and Hobert, O. (2004). Differential sulfations and epimerization define heparan sulfate specificity in nervous system development. *Neuron* **41,** 723–736.

Butler, S. J., and Dodd, J. (2003). A role for BMP heterodimers in roof plate-mediated repulsion of commissural axons. *Neuron* **38,** 389–401.

Campbell, D. S., Regan, A. G., Lopez, J. S., Tannahill, D., Harris, W. A., and Holt, C. E. (2001). Semaphorin 3A elicits stage-dependent collapse, turning, and branching in Xenopus retinal growth cones. *J. Neurosci.* **21,** 8538–8547.

Carlier, M. F., Laurent, V., Santolini, J., Melki, R., Didry, D., Xia, G. X., Hong, Y., Chua, N. H., and Pantaloni, D. (1997). Actin depolymerizing factor (ADF/cofilin) enhances the rate of filament turnover: Implication in actin-based motility. *J. Cell Biol.* **136,** 1307–1322.

Carney, P. R., and Silver, J. (1983). Studies on cell migration and axon guidance in the developing distal auditory system of the mouse. *J. Comp. Neurol.* **215,** 359–369.

Castellani, V., Chedotal, A., Schachner, M., Faivre-Sarrailh, C., and Rougon, G. (2000). Analysis of the L1-deficient mouse phenotype reveals cross-talk between Sema3A and L1 signaling pathways in axonal guidance. *Neuron* **27,** 237–249.

Castellani, V., De Angelis, E., Kenwrick, S., and Rougon, G. (2002). Cis and trans interactions of L1 with neuropilin-1 control axonal responses to semaphorin 3A. *EMBO J.* **21,** 6348–6357.

Chae, T., Kwon, Y. T., Bronson, R., Dikkes, P., Li, E., and Tsai, L. H. (1997). Mice lacking p35, a neuronal specific activator of Cdk5, display cortical lamination defects, seizures, and adult lethality. *Neuron* **18,** 29–42.

Chakraborty, T., Ebel, F., Domann, E., Niebuhr, K., Gerstel, B., Pistor, S., Temm-Grove, C. J., Jockusch, B. M., Reinhard, M., Walter, U., *et al.* (1995). A focal adhesion factor directly linking intracellularly motile Listeria monocytogenes and Listeria ivanovii to the actin-based cytoskeleton of mammalian cells. *EMBO J.* **14,** 1314–1321.

Challacombe, J. F., Snow, D. M., and Letourneau, P. C. (1996). Actin filament bundles are required for microtubule reorientation during growth cone turning to avoid an inhibitory guidance cue. *J. Cell Sci.* **109**(Pt. 8), 2031–2040.

Challacombe, J. F., Snow, D. M., and Letourneau, P. C. (1997). Dynamic microtubule ends are required for growth cone turning to avoid an inhibitory guidance cue. *J. Neurosci.* **17,** 3085–3095.

Charron, F., Stein, E., Jeong, J., McMahon, A. P., and Tessier-Lavigne, M. (2003). The morphogen sonic hedgehog is an axonal chemoattractant that collaborates with netrin-1 in midline axon guidance. *Cell* **113,** 11–23.

Chen, H., Chedotal, A., He, Z., Goodman, C. S., and Tessier-Lavigne, M. (1997). Neuropilin-2, a novel member of the neuropilin family, is a high affinity receptor for the semaphorins Sema E and Sema IV but not Sema III. *Neuron* **19,** 547–559.

Cheng, H. J., Nakamoto, M., Bergemann, A. D., and Flanagan, J. G. (1995). Complementary gradients in expression and binding of ELF-1 and Mek4 in development of the topographic retinotectal projection map. *Cell* **82,** 371–381.

Cheng, H. J., Bagri, A., Yaron, A., Stein, E., Pleasure, S. J., and Tessier-Lavigne, M. (2001). Plexin-A3 mediates semaphorin signaling and regulates the development of hippocampal axonal projections. *Neuron* **32,** 249–263.

Chitnis, A. B., and Kuwada, J. Y. (1990). Axonogenesis in the brain of zebrafish embryos. *J. Neurosci.* **10,** 1892–1905.

Chung, K. Y., Shum, D. K., and Chan, S. O. (2000a). Expression of chondroitin sulfate proteoglycans in the chiasm of mouse embryos. *J. Comp. Neurol.* **417,** 153–63.

Chung, K. Y., Taylor, J. S., Shum, D. K., and Chan, S. O. (2000b). Axon routing at the optic chiasm after enzymatic removal of chondroitin sulfate in mouse embryos. *Development* **127,** 2673–2683.

Chung, K. Y., Leung, K. M., Lin, C. C., Tam, K. C., Hao, Y. L., Taylor, J. S., and Chan, S. O. (2004). Regionally specific expression of L1 and sialylated NCAM in the retinofugal pathway of mouse embryos. *J. Comp. Neurol.* **471,** 482–498.

Cloutier, J. F., Giger, R. J., Koentges, G., Dulac, C., Kolodkin, A. L., and Ginty, D. D. (2002). Neuropilin-2 mediates axonal fasciculation, zonal segregation, but not axonal convergence, of primary accessory olfactory neurons. *Neuron* **33,** 877–892.

Cloutier, J. F., Sahay, A., Chang, E. C., Tessier-Lavigne, M., Dulac, C., Kolodkin, A. L., and Ginty, D. D. (2004). Differential requirements for semaphorin 3F and Slit-1 in axonal targeting, fasciculation, and segregation of olfactory sensory neuron projections. *J. Neurosci.* **24,** 9087–9096.

Cohen, N. R., Taylor, J. S., Scott, L. B., Guillery, R. W., Soriano, P., and Furley, A. J. (1998). Errors in corticospinal axon guidance in mice lacking the neural cell adhesion molecule L1. *Curr. Biol.* **8,** 26–33.

Colamarino, S. A., and Tessier-Lavigne, M. (1995). The role of the floor plate in axon guidance. *Annu. Rev. Neurosci.* **18,** 497–529.

Colavita, A., Krishna, S., Zheng, H., Padgett, R. W., and Culotti, J. G. (1998). Pioneer axon guidance by UNC-129, a C. elegans TGF-beta. *Science* **281,** 706–709.

Comeau, M. R., Johnson, R., Du Bose, R. F., Petersen, M., Gearing, P., Vanden Bos, T., Park, L., Farrah, T., Buller, R. M., Cohen, J. I., Strockbine, L. D., Rauch, C., and Spriggs, M. K. (1998). A poxvirus-encoded semaphorin induces cytokine production from monocytes and binds to a novel cellular semaphorin receptor, VESPR. *Immunity* **8,** 473–482.

Condic, M. L., Snow, D. M., and Letourneau, P. C. (1999). Embryonic neurons adapt to the inhibitory proteoglycan aggrecan by increasing integrin expression. *J. Neurosci.* **19,** 10036–10043.

Cooper, N. G., and Steindler, D. A. (1986). Monoclonal antibody to glial fibrillary acidic protein reveals a parcellation of individual barrels in the early postnatal mouse somatosensory cortex. *Brain Res.* **380,** 341–348.

Couch, J., and Condron, B. (2002). Axon guidance: Comm hither, Robo. *Curr. Biol.* **12,** R741–R742.

Crandall, J. E., Dibble, C., Butler, D., Pays, L., Ahmad, N., Kostek, C., Puschel, A. W., and Schwarting, G. A. (2000). Patterning of olfactory sensory connections is mediated by extracellular matrix proteins in the nerve layer of the olfactory bulb. *J. Neurobiol.* **45,** 195–206.

Croucher, S. J., and Tickle, C. (1989). Characterization of epithelial domains in the nasal passages of chick embryos: Spatial and temporal mapping of a range of extracellular matrix and cell surface molecules during development of the nasal placode. *Development* **106,** 493–509.

Cucchiaro, J. B. (1991). Early development of the retinal line of decussation in normal and albino ferrets. *J. Comp. Neurol.* **312,** 193–206.

Cummings, D. M., Malun, D., and Brunjes, P. C. (1997). Development of the anterior commissure in the opossum: Midline extracellular space and glia coincide with early axon decussation. *J. Neurobiol.* **32,** 403–414.

Cutforth, T., and Harrison, C. J. (2002). Ephs and ephrins close ranks. *Trends Neurosci.* **25,** 332–334.

Cutforth, T., Moring, L., Mendelsohn, M., Nemes, A., Shah, N. M., Kim, M. M., Frisen, J., and Axel, R. (2003). Axonal ephrin-As and odorant receptors: Coordinate determination of the olfactory sensory map. *Cell* **114,** 311–322.

Dahme, M., Bartsch, U., Martini, R., Anliker, B., Schachner, M., and Mantei, N. (1997). Disruption of the mouse L1 gene leads to malformations of the nervous system. *Nat. Genet.* **17,** 346–349.

das Neves, L., Duchala, C. S., Tolentino-Silva, F., Haxhiu, M. A., Colmenares, C., Macklin, W. B., Campbell, C. E., Butz, K. G., and Gronostajski, R. M. (1999). Disruption of the murine nuclear factor I-A gene (Nfia) results in perinatal lethality, hydrocephalus, and agenesis of the corpus callosum. *Proc. Natl. Acad. Sci. USA* **96,** 11946–11951.

Davis, J. Q., and Bennett, V. (1994). Ankyrin binding activity shared by the neurofascin/L1/ NrCAM family of nervous system cell adhesion molecules. *J. Biol. Chem* **269,** 27163–27166.

De Carlos, J. A., and O'Leary, D. D. (1992). Growth and targeting of subplate axons and establishment of major cortical pathways. *J. Neurosci.* **12,** 1194–1211.

de la Torre, J. R., Hopker, V. H., Ming, G. L., Poo, M. M., Tessier-Lavigne, M., Hemmati-Brivanlou, A., and Holt, C. E. (1997). Turning of retinal growth cones in a netrin-1 gradient mediated by the netrin receptor DCC. *Neuron* **19,** 1211–1224.

de Wit, J., and Verhaagen, J. (2003). Role of semaphorins in the adult nervous system. *Prog. Neurobiol.* **71,** 249–267.

Deiner, M. S., and Sretavan, D. W. (1999). Altered midline axon pathways and ectopic neurons in the developing hypothalamus of netrin-1- and DCC-deficient mice. *J. Neurosci.* **19,** 9900–9912.

Deiner, M. S., Kennedy, T. E., Fazeli, A., Serafini, T., Tessier-Lavigne, M., and Sretavan, D. W. (1997). Netrin-1 and DCC mediate axon guidance locally at the optic disc: Loss of function leads to optic nerve hypoplasia. *Neuron* **19,** 575–589.

Del Rio, J. A., Heimrich, B., Borrell, V., Forster, E., Drakew, A., Alcantara, S., Nakajima, K., Miyata, T., Ogawa, M., Mikoshiba, K., Derer, P., Frotscher, M., and Soriano, E. (1997). A role for Cajal-Retzius cells and reelin in the development of hippocampal connections. *Nature* **385,** 70–74.

Del Rio, J. A., Gonzalez-Billault, C., Urena, J. M., Jimenez, E. M., Barallobre, M. J., Pascual, M., Pujadas, L., Simo, S., La Torre, A., Wandosell, F., Avila, J., and Soriano, E. (2004). MAP1B is required for Netrin 1 signaling in neuronal migration and axonal guidance. *Curr. Biol* **14,** 840–850.

Demyanenko, G. P., and Maness, P. F. (2003). The L1 cell adhesion molecule is essential for topographic mapping of retinal axons. *J. Neurosci.* **23,** 530–538.

Demyanenko, G. P., Tsai, A. Y., and Maness, P. F. (1999). Abnormalities in neuronal process extension, hippocampal development, and the ventricular system of L1 knockout mice. *J. Neurosci.* **19,** 4907–4920.

Demyanenko, G. P., Shibata, Y., and Maness, P. F. (2001). Altered distribution of dopaminergic neurons in the brain of L1 null mice. *Brain Res. Dev. Brain Res.* **126,** 21–30.

Dent, E. W., and Gertler, F. B. (2003). Cytoskeletal dynamics and transport in growth cone motility and axon guidance. *Neuron* **40,** 209–227.

Dent, E. W., and Kalil, K. (2001). Axon branching requires interactions between dynamic microtubules and actin filaments. *J. Neurosci.* **21,** 9757–69.

Dent, E. W., Callaway, J. L., Szebenyi, G., Baas, P. W., and Kalil, K. (1999). Reorganization and movement of microtubules in axonal growth cones and developing interstitial branches. *J. Neurosci.* **19,** 8894–8908.

Dent, E. W., Barnes, A. M., Tang, F., and Kalil, K. (2004). Netrin-1 and semaphorin 3A promote or inhibit cortical axon branching, respectively, by reorganization of the cytoskeleton. *J. Neurosci.* **24,** 3002–3012.

Dickson, B. J. (2001). Rho GTPases in growth cone guidance. *Curr. Opin. Neurobiol.* **11,** 103–110.

Diefenbach, T. J., Guthrie, P. B., and Kater, S. B. (2000). Stimulus history alters behavioral responses of neuronal growth cones. *J. Neurosci.* **20,** 1484–1494.

Dodd, J., Morton, S. B., Karagogeos, D., Yamamoto, M., and Jessell, T. M. (1988). Spatial regulation of axonal glycoprotein expression on subsets of embryonic spinal neurons. *Neuron* **1,** 105–116.

Doherty, P., Barton, C. H., Dickson, G., Seaton, P., Rowett, L. H., Moore, S. E., Gower, H. J., and Walsh, F. S. (1989). Neuronal process outgrowth of human sensory neurons on monolayers of cells transfected with cDNAs for five human N-CAM isoforms. *J. Cell Biol.* **109,** 789–798.

Doherty, P., Cohen, J., and Walsh, F. S. (1990). Neurite outgrowth in response to transfected N-CAM changes during development and is modulated by polysialic acid. *Neuron* **5,** 209–219.

Dou, C. L., and Levine, J. M. (1994). Inhibition of neurite growth by the NG2 chondroitin sulfate proteoglycan. *J. Neurosci.* **14,** 7616–7628.

Drescher, U., Bonhoeffer, F., and Muller, B. K. (1997). The Eph family in retinal axon guidance. *Curr. Opin. Neurobiol.* **7,** 75–80.

du Lac, S., Bastiani, M. J., and Goodman, C. S. (1986). Guidance of neuronal growth cones in the grasshopper embryo. II. Recognition of a specific axonal pathway by the aCC neuron. *J. Neurosci.* **6,** 3532–3541.

Dulac, C. (2000). Sensory coding of pheromone signals in mammals. *Curr. Opin. Neurobiol.* **10,** 511–518.

Dulac, C., and Axel, R. (1995). A novel family of genes encoding putative pheromone receptors in mammals. *Cell* **83,** 195–206.

Easter, S. S., Jr., Ross, L. S., and Frankfurter, A. (1993). Initial tract formation in the mouse brain. *J. Neurosci.* **13,** 285–299.

Edelman, G. M., and Crossin, K. L. (1991). Cell adhesion molecules: Implications for a molecular histology. *Annu. Rev. Biochem.* **60,** 155–190.

Eickholt, B. J., Walsh, F. S., and Doherty, P. (2002). An inactive pool of GSK-3 at the leading edge of growth cones is implicated in Semaphorin 3A signaling. *J. Cell Biol.* **157,** 211–217.

Emerling, D. E., and Lander, A. D. (1996). Inhibitors and promoters of thalamic neuron adhesion and outgrowth in embryonic neocortex: Functional association with chondroitin sulfate. *Neuron* **17,** 1089–1100.

Erskine, L., Williams, S. E., Brose, K., Kidd, T., Rachel, R. A., Goodman, C. S., Tessier-Lavigne, M., and Mason, C. A. (2000). Retinal ganglion cell axon guidance in the mouse optic chiasm: Expression and function of robos and slits. *J. Neurosci.* **20,** 4975–4982.

Fan, J., Mansfield, S. G., Redmond, T., Gordon-Weeks, P. R., and Raper, J. A. (1993). The organization of F-actin and microtubules in growth cones exposed to a brain-derived collapsing factor. *J. Cell Biol.* **121,** 867–878.

Farbman, A. I. (1992). "Cell Biology of Olfaction." Cambridge University Press, Cambridge, UK.

Fazeli, A., Dickinson, S. L., Hermiston, M. L., Tighe, R. V., Steen, R. G., Small, C. G., Stoeckli, E. T., Keino-Masu, K., Masu, M., Rayburn, H., Simons, J., Bronson, R. T., Gordon, J. I., Tessier-Lavigne, M., and Weinberg, R. A. (1997). Phenotype of mice lacking functional Deleted in colorectal cancer (Dcc) gene. *Nature* **386,** 796–804.

Feldheim, D. A., Kim, Y. I., Bergemann, A. D., Frisen, J., Barbacid, M., and Flanagan, J. G. (2000). Genetic analysis of ephrin-A2 and ephrin-A5 shows their requirement in multiple aspects of retinocollicular mapping. *Neuron* **25,** 563–574.

Fitch, M. T., and Silver, J. (1997). Glial cell extracellular matrix: Boundaries for axon growth in development and regeneration. *Cell Tissue Res.* **290,** 379–384.

Flanagan, J. G., and Vanderhaeghen, P. (1998). The ephrins and Eph receptors in neural development. *Annu. Rev. Neurosci.* **21,** 309–345.

Forscher, P., and Smith, S. J. (1988). Actions of cytochalasins on the organization of actin filaments and microtubules in a neuronal growth cone. *J. Cell Biol.* **107,** 1505–1516.

Fournier, A. E., Nakamura, F., Kawamoto, S., Goshima, Y., Kalb, R. G., and Strittmatter, S. M. (2000). Semaphorin3A enhances endocytosis at sites of receptor-F-actin colocalization during growth cone collapse. *J. Cell Biol.* **149,** 411–422.

Fransen, E., D'Hooge, R., Van Camp, G., Verhoye, M., Sijbers, J., Reyniers, E., Soriano, P., Kamiguchi, H., Willemsen, R., Koekkoek, S. K., De Zeeuw, C. I., De Deyn, P. P., Van der Linden, A., Lemmon, V., Kooy, R. F., and Willems, P. J. (1998). L1 knockout mice show dilated ventricles, vermis hypoplasia and impaired exploration patterns. *Hum. Mol. Genet.* **7,** 999–1009.

Fricke, C., Lee, J. S., Geiger-Rudolph, S., Bonhoeffer, F., and Chien, C. B. (2001). astray, a zebrafish roundabout homolog required for retinal axon guidance. *Science* **292,** 507–510.

Fujisawa, H., and Kitsukawa, T. (1998). Receptors for collapsin/semaphorins. *Curr. Opin. Neurobiol.* **8,** 587–592.

Gad, J. M., Keeling, S. L., Wilks, A. F., Tan, S. S., and Cooper, H. M. (1997). The expression patterns of guidance receptors, DCC and Neogenin, are spatially and temporally distinct throughout mouse embryogenesis. *Dev. Biol.* **192,** 258–273.

Gahwiler, B. H., Capogna, M., Debanne, D., McKinney, R. A., and Thompson, S. M. (1997). Organotypic slice cultures: A technique has come of age. *Trends Neurosci.* **20,** 471–477.

Garcia-Abreu, J., Moura Neto, V., Carvalho, S. L., and Cavalcante, L. A. (1995). Regionally specific properties of midbrain glia: I. Interactions with midbrain neurons. *J. Neurosci. Res.* **40,** 471–477.

Garrity, P. A. (2003). Developmental biology: How neurons avoid derailment. *Nature* **422,** 570–571.

Gazzaniga, M. S. (1995). Principles of human brain organization derived from split-brain studies. *Neuron* **14,** 217–228.

Georgiou, M., and Tear, G. (2002). Commissureless is required both in commissural neurones and midline cells for axon guidance across the midline. *Development* **129,** 2947–2956.

Gerstel, B., Grobe, L., Pistor, S., Chakraborty, T., and Wehland, J. (1996). The ActA polypeptides of Listeria ivanovii and Listeria monocytogenes harbor related binding sites for host microfilament proteins. *Infect. Immun.* **64,** 1929–1936.

Gertler, F. B., Niebuhr, K., Reinhard, M., Wehland, J., and Soriano, P. (1996). Mena, a relative of VASP and Drosophila Enabled, is implicated in the control of microfilament dynamics. *Cell* **87,** 227–239.

Ghosh, A., and Shatz, C. J. (1992). Pathfinding and target selection by developing geniculocortical axons. *J. Neurosci.* **12,** 39–55.

Ghosh, A., and Shatz, C. J. (1993). A role for subplate neurons in the patterning of connections from thalamus to neocortex. *Development* **117,** 1031–1047.

Ghosh, A., Antonini, A., McConnell, S. K., and Shatz, C. J. (1990). Requirement for subplate neurons in the formation of thalamocortical connections. *Nature* **347,** 179–181.

Giger, R. J., Urquhart, E. R., Gillespie, S. K., Levengood, D. V., Ginty, D. D., and Kolodkin, A. L. (1998). Neuropilin-2 is a receptor for semaphorin IV: Insight into the structural basis of receptor function and specificity. *Neuron* **21,** 1079–1092.

Godement, P. (1994). Specific guidance and modulation of growth cone motility during *in vivo* development. *J. Physiol. Paris* **88,** 259–264.

Godement, P., and Bonhoeffer, F. (1989). Cross-species recognition of tectal cues by retinal fibers *in vitro*. *Development* **106,** 313–320.

Goetze, B., Breer, H., and Strotmann, J. (2002). A long-term culture system for olfactory explants with intrinsically fluorescent cell populations. *Chem. Senses* **27,** 817–824.

Goh, K. L., Cai, L., Cepko, C. L., and Gertler, F. B. (2002). Ena/VASP proteins regulate cortical neuronal positioning. *Curr. Biol.* **12,** 565–569.

Goldberg, D. J., and Burmeister, D. W. (1986). Stages in axon formation: Observations of growth of Aplysia axons in culture using video-enhanced contrast-differential interference contrast microscopy. *J. Cell Biol.* **103,** 1921–1931.

Goldberg, D. J., and Burmeister, D. W. (1989). Looking into growth cones. *Trends Neurosci.* **12,** 503–506.

Golding, J. P., Tidcombe, H., Tsoni, S., and Gassmann, M. (1999). Chondroitin sulphate-binding molecules may pattern central projections of sensory axons within the cranial mesenchyme of the developing mouse. *Dev. Biol.* **216,** 85–97.

Gong, Q., and Shipley, M. T. (1996). Expression of extracellular matrix molecules and cell surface molecules in the olfactory nerve pathway during early development. *J. Comp. Neurol.* **366,** 1–14.

Goodman, C. S., Bastiani, M. J., Doe, C. Q., du Lac, S., Helfand, S. L., Kuwada, J. Y., and Thomas, J. B. (1984). Cell recognition during neuronal development. *Science* **225,** 1271–1279.

Gordon-Weeks, P. R. (1991). Microtubule organization in growth cones. *Biochem. Soc. Trans.* **19,** 1080–1085.

Goshima, Y., Nakamura, F., Strittmatter, P., and Strittmatter, S. M. (1995). Collapsin-induced growth cone collapse mediated by an intracellular protein related to UNC-33. *Nature* **376,** 509–514.

Graziadei, P. P. C., and Monti Graziadei, G. A. (1978). The olfactory system: A model for the study of neurogenesis and axon regeneration in mammals. *In* "Neuronal Plasticity" (C. W. Cotman, Ed.), pp. 131–153. Raven Press, New York.

Greferath, U., Canty, A. J., Messenger, J., and Murphy, M. (2002). Developmental expression of EphA4-tyrosine kinase receptor in the mouse brain and spinal cord. *Gene Expr. Patterns* **2,** 267–274.

Grenningloh, G., Soehrman, S., Bondallaz, P., Ruchti, E., and Cadas, H. (2004). Role of the microtubule destabilizing proteins SCG10 and stathmin in neuronal growth. *J. Neurobiol.* **58,** 60–69.

Gronostajski, R. M., Adhya, S., Nagata, K., Guggenheimer, R. A., and Hurwitz, J. (1985). Site-specific DNA binding of nuclear factor I: Analyses of cellular binding sites. *Mol. Cell. Biol.* **5,** 964–971.

Gu, C., Rodriguez, E. R., Reimert, D. V., Shu, T., Fritzsch, B., Richards, L. J., Kolodkin, A. L., and Ginty, D. D. (2003). Neuropilin-1 conveys semaphorin and VEGF signaling during neural and cardiovascular development. *Dev. Cell* **5**, 45–57.

Guillery, R. W., Mason, C. A., and Taylor, J. S. (1995). Developmental determinants at the mammalian optic chiasm. *J. Neurosci.* **15**, 4727–4737.

Gundersen, R. W., and Barrett, J. N. (1979). Neuronal chemotaxis: Chick dorsal-root axons turn toward high concentrations of nerve growth factor. *Science* **206**, 1079–1080.

Hall, A. (1998). Rho GTPases and the actin cytoskeleton. *Science* **279**, 509–514.

Hall, A. C., Lucas, F. R., and Salinas, P. C. (2000). Axonal remodeling and synaptic differentiation in the cerebellum is regulated by WNT-7a signaling. *Cell* **100**, 525–535.

Halpern, M. (1987). The organization and function of the vomeronasal system. *Annu. Rev. Neurosci.* **10**, 325–362.

Hamasaki, T., Goto, S., Nishikawa, S., and Ushio, Y. (2001). A role of netrin-1 in the formation of the subcortical structure striatum: Repulsive action on the migration of late-born striatal neurons. *J. Neurosci.* **21**, 4272–4280.

Harris, W. A., Holt, C. E., and Bonhoeffer, F. (1987). Retinal axons with and without their somata, growing to and arborizing in the tectum of Xenopus embryos: A time-lapse video study of single fibres *in vivo*. *Development* **101**, 123–133.

He, Z., and Tessier-Lavigne, M. (1997). Neuropilin is a receptor for the axonal chemorepellent Semaphorin III. *Cell* **90**, 739–751.

Hedgecock, E. M., Culotti, J. G., and Hall, D. H. (1990). The unc-5, unc-6, and unc-40 genes guide circumferential migrations of pioneer axons and mesodermal cells on the epidermis in C. elegans. *Neuron* **4**, 61–85.

Henkemeyer, M., Orioli, D., Henderson, J. T., Saxton, T. M., Roder, J., Pawson, T., and Klein, R. (1996). Nuk controls pathfinding of commissural axons in the mammalian central nervous system. *Cell* **86**, 35–46.

Hennighausen, L., Siebenlist, U., Danner, D., Leder, P., Rawlins, D., Rosenfeld, P., and Kelly, T., Jr. (1985). High-affinity binding site for a specific nuclear protein in the human IgM gene. *Nature* **314**, 289–292.

Herrada, G., and Dulac, C. (1997). A novel family of putative pheromone receptors in mammals with a topographically organized and sexually dimorphic distribution. *Cell* **90**, 763–773.

Herrera, E., Brown, L., Aruga, J., Rachel, R. A., Dolen, G., Mikoshiba, K., Brown, S., and Mason, C. A. (2003). Zic2 patterns binocular vision by specifying the uncrossed retinal projection. *Cell* **114**, 545–557.

Herrera, E., Marcus, R., Li, S., Williams, S. E., Erskine, L., Lai, E., and Mason, C. (2004). Foxd1 is required for proper formation of the optic chiasm. *Development* **131**, 5727–5739.

Hidalgo, A. (2003). Neuron-glia interactions during axon guidance in Drosophila. *Biochem. Soc. Trans.* **31**, 50–55.

Hidalgo, A., and Booth, G. E. (2000). Glia dictate pioneer axon trajectories in the Drosophila embryonic CNS. *Development* **127**, 393–402.

Hidalgo, A., and Brand, A. H. (1997). Targeted neuronal ablation: The role of pioneer neurons in guidance and fasciculation in the CNS of Drosophila. *Development* **124**, 3253–3262.

Hindges, R., McLaughlin, T., Genoud, N., Henkemeyer, M., and O'Leary, D. D. (2002). EphB forward signaling controls directional branch extension and arborization required for dorsal-ventral retinotopic mapping. *Neuron* **35**, 475–487.

Hjorth, J., and Key, B. (2002). Development of axon pathways in the zebrafish central nervous system. *Int. J. Dev. Biol.* **46**, 609–619.

Holland, S. J., Gale, N. W., Mbamalu, G., Yancopoulos, G. D., Henkemeyer, M., and Pawson, T. (1996). Bidirectional signalling through the EPH-family receptor Nuk and its transmembrane ligands. *Nature* **383**, 722–725.

Holley, J. A. (1982). Early development of the circumferential axonal pathway in mouse and chick spinal cord. *J. Comp. Neurol.* **205**, 371–382.

Holmes, G. P., Negus, K., Burridge, L., Raman, S., Algar, E., Yamada, T., and Little, M. H. (1998). Distinct but overlapping expression patterns of two vertebrate slit homologs implies functional roles in CNS development and organogenesis. *Mech. Dev.* **79**, 57–72.

Holt, C. E., and Harris, W. A. (1993). Position, guidance, and mapping in the developing visual system. *J. Neurobiol.* **24**, 1400–1422.

Hopker, V. H., Shewan, D., Tessier-Lavigne, M., Poo, M., and Holt, C. (1999). Growth-cone attraction to netrin-1 is converted to repulsion by laminin-1. *Nature* **401**, 69–73.

Hortsch, M. (1996). The L1 family of neural cell adhesion molecules: Old proteins performing new tricks. *Neuron* **17**, 587–593.

Hu, H. (2001). Cell-surface heparan sulfate is involved in the repulsive guidance activities of Slit2 protein. *Nat. Neurosci.* **4**, 695–701.

Hu, H., Li, M., Labrador, J. P., McEwen, J., Lai, E. C., Goodman, C. S., and Bashaw, G. J. (2005). Cross GTPase-activating protein (CrossGAP)/Vilse links the Roundabout receptor to Rac to regulate midline repulsion. *Proc. Natl. Acad. Sci. USA.*.

Hu, Z., Yue, X., Shi, G., Yue, Y., Crockett, D. P., Blair-Flynn, J., Reuhl, K., Tessarollo, L., and Zhou, R. (2003). Corpus callosum deficiency in transgenic mice expressing a truncated ephrin-A receptor. *J. Neurosci.* **23**, 10963–10970.

Huot, J. (2004). Ephrin signaling in axon guidance. *Prog. Neuropsychopharmacol. Biol. Psychiat.* **28**, 813–818.

Hutson, L. D., and Chien, C. B. (2002). Pathfinding and error correction by retinal axons: The role of astray/robo2. *Neuron* **33**, 205–217.

Hutton, L. A., Gu, G., and Simerly, R. B. (1998). Development of a sexually dimorphic projection from the bed nuclei of the stria terminalis to the anteroventral periventricular nucleus in the rat. *J. Neurosci.* **18**, 3003–3013.

Hynes, R. O. (2002). Integrins: Bidirectional, allosteric signaling machines. *Cell* **110**, 673–687.

Ibanez, M. A., Gu, G., and Simerly, R. B. (2001). Target-dependent sexual differentiation of a limbic-hypothalamic neural pathway. *J. Neurosci.* **21**, 5652–5659.

Imondi, R., and Kaprielian, Z. (2001). Commissural axon pathfinding on the contralateral side of the floor plate: A role for B-class ephrins in specifying the dorsoventral position of longitudinally projecting commissural axons. *Development* **128**, 4859–4871.

Imondi, R., Wideman, C., and Kaprielian, Z. (2000). Complementary expression of transmembrane ephrins and their receptors in the mouse spinal cord: A possible role in constraining the orientation of longitudinally projecting axons. *Development* **127**, 1397–1410.

Inatani, M., Irie, F., Plump, A. S., Tessier-Lavigne, M., and Yamaguchi, Y. (2003). Mammalian brain morphogenesis and midline axon guidance require heparan sulfate. *Science* **302**, 1044–1046.

Ingham, P. W., and McMahon, A. P. (2001). Hedgehog signaling in animal development: Paradigms and principles. *Genes Dev.* **15**, 3059–3087.

Innocenti, G. M. (1986). General organization of callosal connections in the cerebral cortex. *In* "Cerebral Cortex" (E. G. Jones and A. Peters, Eds.), Vol. 5, pp. 291–353. Plenum Press, New York.

Irie, A., Yates, E. A., Turnbull, J. E., and Holt, C. E. (2002). Specific heparan sulfate structures involved in retinal axon targeting. *Development* **129**, 61–70.

Ishii, N., Wadsworth, W. G., Stern, B. D., Culotti, J. G., and Hedgecock, E. M. (1992). UNC-6, a laminin-related protein, guides cell and pioneer axon migrations in C. elegans. *Neuron* **9**, 873–881.

Itoh, A., Miyabayashi, T., Ohno, M., and Sakano, S. (1998). Cloning and expressions of three mammalian homologues of Drosophila slit suggest possible roles for Slit in the formation and maintenance of the nervous system. *Brain Res. Mol. Brain Res.* **62**, 175–186.

Jacobs, J. R., and Goodman, C. S. (1989). Embryonic development of axon pathways in the Drosophila CNS. I. A glial scaffold appears before the first growth cones. *J. Neurosci.* **9,** 2402–2411.

Jenzora, A., Behrendt, B., Small, J. V., Wehland, J., and Stradal, T. E. (2005). PREL1 provides a link from Ras signalling to the actin cytoskeleton via Ena/VASP proteins. *FEBS Lett.* **579,** 455–463.

Jessell, T. M. (1988). Adhesion molecules and the hierarchy of neural development. *Neuron* **1,** 3–13.

Jessell, T. M. (2000). Neuronal specification in the spinal cord: Inductive signals and transcriptional codes. *Nat. Rev. Genet.* **1,** 20–29.

Jia, C., and Halpern, M. (1996). Subclasses of vomeronasal receptor neurons: Differential expression of G proteins (Gi alpha 2 and G(o alpha)) and segregated projections to the accessory olfactory bulb. *Brain Res.* **719,** 117–128.

Jin, M., Guan, C. B., Jiang, Y. A., Chen, G., Zhao, C. T., Cui, K., Song, Y. Q., Wu, C. P., Poo, M. M., and Yuan, X. B. (2005). Ca2+-dependent regulation of rho GTPases triggers turning of nerve growth cones. *J. Neurosci.* **25,** 2338–2347.

Jin, Z., and Strittmatter, S. M. (1997). Rac1 mediates collapsin-1-induced growth cone collapse. *J. Neurosci.* **17,** 6256–6263.

Johnson, K. G., Ghose, A., Epstein, E., Lincecum, J., O'Connor, M. B., and Van Vactor, D. (2004). Axonal heparan sulfate proteoglycans regulate the distribution and efficiency of the repellent slit during midline axon guidance. *Curr. Biol* **14,** 499–504.

Joosten, E. A., Gribnau, A. A., and Dederen, P. J. (1989). Postnatal development of the corticospinal tract in the rat. An ultrastructural anterograde HRP study. *Anat. Embryol. (Berl.)* **179,** 449–456.

Kafitz, K. W., and Greer, C. A. (1997). Role of laminin in axonal extension from olfactory receptor cells. *J. Neurobiol.* **32,** 298–310.

Kafitz, K. W., and Greer, C. A. (1998). Differential expression of extracellular matrix and cell adhesion molecules in the olfactory nerve and glomerular layers of adult rats. *J. Neurobiol.* **34,** 271–282.

Kamiguchi, H., Long, K. E., Pendergast, M., Schaefer, A. W., Rapoport, I., Kirchhausen, T., and Lemmon, V. (1998). The neural cell adhesion molecule L1 interacts with the AP-2 adaptor and is endocytosed via the clathrin-mediated pathway. *J. Neurosci.* **18,** 5311–5321.

Kandel, E. R., Shwartz, J. H., and Jessell, T. M. (2000). "Principles of Neural Science." McGraw-Hill, New York.

Kantor, D. B., Chivatakarn, O., Peer, K. L., Oster, S. F., Inatani, M., Hansen, M. J., Flanagan, J. G., Yamaguchi, Y., Sretavan, D. W., Giger, R. J., and Kolodkin, A. L. (2004). Semaphorin 5A is a bifunctional axon guidance cue regulated by heparan and chondroitin sulfate proteoglycans. *Neuron* **44,** 961–975.

Keleman, K., Rajagopalan, S., Cleppien, D., Teis, D., Paiha, K., Huber, L. A., Technau, G. M., and Dickson, B. J. (2002). Comm sorts robo to control axon guidance at the Drosophila midline. *Cell* **110,** 415–427.

Keleman, K., Ribeiro, C., and Dickson, B. J. (2005). Comm function in commissural axon guidance: Cell-autonomous sorting of Robo *in vivo*. *Nat. Neurosci.* **8,** 156–163.

Kennedy, T. E., Serafini, T., de la Torre, J. R., and Tessier-Lavigne, M. (1994). Netrins are diffusible chemotropic factors for commissural axons in the embryonic spinal cord. *Cell* **78,** 425–435.

Keshishian, H., and Bentley, D. (1983). Embryogenesis of peripheral nerve pathways in grasshopper legs. I. The initial nerve pathway to the CNS. *Dev. Biol.* **96,** 89–102.

Keverne, E. B. (1999). The vomeronasal organ. *Science* **286,** 716–720.

Key, B., and Akeson, R. A. (1990). Olfactory neurons express a unique glycosylated form of the neural cell adhesion molecule (N-CAM). *J. Cell Biol.* **110,** 1729–1743.

Kidd, T., Brose, K., Mitchell, K. J., Fetter, R. D., Tessier-Lavigne, M., Goodman, C. S., and Tear, G. (1998). Roundabout controls axon crossing of the CNS midline and defines a novel subfamily of evolutionarily conserved guidance receptors. *Cell* **92**, 205–215.

Kidd, T., Bland, K. S., and Goodman, C. S. (1999). Slit is the midline repellent for the robo receptor in Drosophila. *Cell* **96**, 785–794.

Kinnunen, A., Kinnunen, T., Kaksonen, M., Nolo, R., Panula, P., and Rauvala, H. (1998). N-syndecan and HB-GAM (heparin-binding growth-associated molecule) associate with early axonal tracts in the rat brain. *Eur. J. Neurosci.* **10**, 635–648.

Klambt, C., Jacobs, J. R., and Goodman, C. S. (1991). The midline of the Drosophila central nervous system: A model for the genetic analysis of cell fate, cell migration, and growth cone guidance. *Cell* **64**, 801–815.

Knoll, B., Zarbalis, K., Wurst, W., and Drescher, U. (2001). A role for the EphA family in the topographic targeting of vomeronasal axons. *Development* **128**, 895–906.

Knoll, B., Schmidt, H., Andrews, W., Guthrie, S., Pini, A., Sundaresan, V., and Drescher, U. (2003). On the topographic targeting of basal vomeronasal axons through Slit-mediated chemorepulsion. *Development* **130**, 5073–5082.

Kobayashi, H., Koppel, A. M., Luo, Y., and Raper, J. A. (1997). A role for collapsin-1 in olfactory and cranial sensory axon guidance. *J. Neurosci.* **17**, 8339–8352.

Koch, M., Murrell, J. R., Hunter, D. D., Olson, P. F., Jin, W., Keene, D. R., Brunken, W. J., and Burgeson, R. E. (2000). A novel member of the netrin family, beta-netrin, shares homology with the beta chain of laminin: Identification, expression, and functional characterization. *J. Cell Biol.* **151**, 221–234.

Koester, S. E., and O'Leary, D. D. (1994). Development of projection neurons of the mammalian cerebral cortex. *Prog. Brain Res.* **102**, 207–215.

Kolodkin, A. L., Matthes, D. J., O'Connor, T. P., Patel, N. H., Admon, A., Bentley, D., and Goodman, C. S. (1992). Fasciclin IV: Sequence, expression, and function during growth cone guidance in the grasshopper embryo. *Neuron* **9**, 831–845.

Kolodkin, A. L., Matthes, D. J., and Goodman, C. S. (1993). The semaphorin genes encode a family of transmembrane and secreted growth cone guidance molecules. *Cell* **75**, 1389–1399.

Kolodkin, A. L., Levengood, D. V., Rowe, E. G., Tai, Y. T., Giger, R. J., and Ginty, D. D. (1997). Neuropilin is a semaphorin III receptor. *Cell* **90**, 753–762.

Kornack, D. R., and Giger, R. J. (2005). Probing microtubule +TIPs: Regulation of axon branching. *Curr. Opin. Neurobiol.* **15**, 58–66.

Krause, M., Dent, E. W., Bear, J. E., Loureiro, J. J., and Gertler, F. B. (2003). Ena/VASP proteins: Regulators of the actin cytoskeleton and cell migration. *Annu. Rev. Cell Dev. Biol.* **19**, 541–564.

Krause, M., Leslie, J. D., Stewart, M., Lafuente, E. M., Valderrama, F., Jagannathan, R., Strasser, G. A., Rubinson, D. A., Liu, H., Way, M., Yaffe, M. B., Boussiotis, V. A., and Gertler, F. B. (2004). Lamellipodin, an Ena/VASP ligand, is implicated in the regulation of lamellipodial dynamics. *Dev. Cell* **7**, 571–583.

Krylova, O., Herreros, J., Cleverley, K. E., Ehler, E., Henriquez, J. P., Hughes, S. M., and Salinas, P. C. (2002). WNT-3, expressed by motoneurons, regulates terminal arborization of neurotrophin-3-responsive spinal sensory neurons. *Neuron* **35**, 1043–1056.

Kuhn, T. B., Brown, M. D., Wilcox, C. L., Raper, J. A., and Bamburg, J. R. (1999). Myelin and collapsin-1 induce motor neuron growth cone collapse through different pathways: Inhibition of collapse by opposing mutants of rac1. *J. Neurosci.* **19**, 1965–1975.

Kuiper, G. G., Enmark, E., Pelto-Huikko, M., Nilsson, S., and Gustafsson, J. A. (1996). Cloning of a novel receptor expressed in rat prostate and ovary. *Proc. Natl. Acad. Sci. USA* **93**, 5925–5930.

Kullander, K., and Klein, R. (2002). Mechanisms and functions of Eph and ephrin signalling. *Nat. Rev. Mol. Cell. Biol.* **3**, 475–486.

Kuwada, J. Y. (1986). Cell recognition by neuronal growth cones in a simple vertebrate embryo. *Science* **233**, 740–746.

Kwon, Y. T., Tsai, L. H., and Crandall, J. E. (1999). Callosal axon guidance defects in p35(-/-) mice. *J. Comp. Neurol.* **415**, 218–229.

Lafuente, E. M., van Puijenbroek, A. A., Krause, M., Carman, C. V., Freeman, G. J., Berezovskaya, A., Constantine, E., Springer, T. A., Gertler, F. B., and Boussiotis, V. A. (2004). RIAM, an Ena/VASP and Profilin ligand, interacts with Rap1-GTP and mediates Rap1-induced adhesion. *Dev. Cell* **7**, 585–595.

LaMantia, A. S., Colbert, M. C., and Linney, E. (1993). Retinoic acid induction and regional differentiation prefigure olfactory pathway formation in the mammalian forebrain. *Neuron* **10**, 1035–1048.

LaMantia, A. S., Bhasin, N., Rhodes, K., and Heemskerk, J. (2000). Mesenchymal/epithelial induction mediates olfactory pathway formation. *Neuron* **28**, 411–425.

Lander, A. D. (1998). Proteoglycans: Master regulators of molecular encounter? *Matrix Biol.* **17**, 465–472.

Lanier, L. M., Gates, M. A., Witke, W., Menzies, A. S., Wehman, A. M., Macklis, J. D., Kwiatkowski, D., Soriano, P., and Gertler, F. B. (1999). Mena is required for neurulation and commissure formation. *Neuron* **22**, 313–325.

Lebrand, C., Dent, E. W., Strasser, G. A., Lanier, L. M., Krause, M., Svitkina, T. M., Borisy, G. G., and Gertler, F. B. (2004). Critical role of Ena/VASP proteins for filopodia formation in neurons and in function downstream of netrin-1. *Neuron* **42**, 37–49.

Lee, K. J., Mendelsohn, M., and Jessell, T. M. (1998). Neuronal patterning by BMPs: A requirement for GDF7 in the generation of a discrete class of commissural interneurons in the mouse spinal cord. *Genes Dev.* **12**, 3394–3407.

Leegwater, P. A., van Driel, W., and van der Vliet, P. C. (1985). Recognition site of nuclear factor I, a sequence-specific DNA-binding protein from HeLa cells that stimulates adenovirus DNA replication. *EMBO J.* **4**, 1515–1521.

Letourneau, P. C., and Ressler, A. H. (1984). Inhibition of neurite initiation and growth by taxol. *J. Cell Biol.* **98**, 1355–1362.

Letourneau, P. C., Shattuck, T. A., and Ressler, A. H. (1986). Branching of sensory and sympathetic neurites *in vitro* is inhibited by treatment with taxol. *J. Neurosci.* **6**, 1912–1917.

Letourneau, P. C., Shattuck, T. A., and Ressler, A. H. (1987). "Pull" and "push" in neurite elongation: Observations on the effects of different concentrations of cytochalasin B and taxol *Cell Motil. Cytoskeleton* **8**, 193–209.

Leung, K. M., Taylor, J. S., and Chan, S. O. (2003). Enzymatic removal of chondroitin sulphates abolishes the age-related axon order in the optic tract of mouse embryos. *Eur. J. Neurosci.* **17**, 1755–1767.

Lewis, A. K., and Bridgman, P. C. (1992). Nerve growth cone lamellipodia contain two populations of actin filaments that differ in organization and polarity. *J. Cell Biol.* **119**, 1219–1243.

Li, H. S., Chen, J. H., Wu, W., Fagaly, T., Zhou, L., Yuan, W., Dupuis, S., Jiang, Z. H., Nash, W., Gick, C., Ornitz, D. M., Wu, J. Y., and Rao, Y. (1999). Vertebrate slit, a secreted ligand for the transmembrane protein roundabout, is a repellent for olfactory bulb axons. *Cell* **96**, 807–818.

Lim, Y. S., and Wadsworth, W. G. (2002). Identification of domains of netrin UNC-6 that mediate attractive and repulsive guidance and responses from cells and growth cones. *J. Neurosci.* **22**, 7080–7087.

Lin, C. H., and Forscher, P. (1993). Cytoskeletal remodeling during growth cone-target interactions. *J. Cell Biol.* **121**, 1369–1383.

Lin, C. H., and Forscher, P. (1995). Growth cone advance is inversely proportional to retrograde F-actin flow. *Neuron* **14**, 763–771.

Lin, C. H., Thompson, C. A., and Forscher, P. (1994). Cytoskeletal reorganization underlying growth cone motility. *Curr. Opin. Neurobiol.* **4,** 640–647.

Little, M., Rumballe, B., Georgas, K., Yamada, T., and Teasdale, R. D. (2002). Conserved modularity and potential for alternate splicing in mouse and human Slit genes. *Int J. Dev. Biol.* **46,** 385–391.

Liu, A., Majumdar, A., Schauerte, H. E., Haffter, P., and Drummond, I. A. (2000). Zebrafish wnt4b expression in the floor plate is altered in sonic hedgehog and gli-2 mutants. *Mech. Dev.* **91,** 409–413.

Liu, B. P., and Strittmatter, S. M. (2001). Semaphorin-mediated axonal guidance via Rho-related G proteins. *Curr. Opin. Cell Biol.* **13,** 619–626.

Liu, S. F., Jiang, Y. L., and Du, L. X. (2003). [Studies of BMPR-IB and BMP15 as candidate genes for fecundity in little tailed han sheep]. *Yi Chuan Xue Bao* **30,** 755–760.

Livesey, F. J., and Hunt, S. P. (1997). Netrin and netrin receptor expression in the embryonic mammalian nervous system suggests roles in retinal, striatal, nigral, and cerebellar development. *Mol. Cell Neurosci.* **8,** 417–429.

Long, H., Sabatier, C., Ma, L., Plump, A., Yuan, W., Ornitz, D. M., Tamada, A., Murakami, F., Goodman, C. S., and Tessier-Lavigne, M. (2004). Conserved roles for Slit and Robo proteins in midline commissural axon guidance. *Neuron* **42,** 213–223.

Luduena, M. A., and Wessells, N. K. (1973). Cell locomotion, nerve elongation, and microfilaments. *Dev. Biol.* **30,** 427–440.

Lumsden, A. G., and Davies, A. M. (1983). Earliest sensory nerve fibres are guided to peripheral targets by attractants other than nerve growth factor. *Nature* **306,** 786–788.

Lundquist, E. A. (2003). Rac proteins and the control of axon development. *Curr. Opin. Neurobiol.* **13,** 384–390.

Luo, L., Jan, L. Y., and Jan, Y. N. (1997). Rho family GTP-binding proteins in growth cone signalling. *Curr. Opin. Neurobiol.* **7,** 81–86.

Luo, Y., Raible, D., and Raper, J. A. (1993). Collapsin: A protein in brain that induces the collapse and paralysis of neuronal growth cones. *Cell* **75,** 217–227.

Lustig, M., Erskine, L., Mason, C. A., Grumet, M., and Sakurai, T. (2001). Nr-CAM expression in the developing mouse nervous system: Ventral midline structures, specific fiber tracts, and neuropilar regions. *J. Comp. Neurol.* **434,** 13–28.

Lyuksyutova, A. I., Lu, C. C., Milanesio, N., King, L. A., Guo, N., Wang, Y., Nathans, J., Tessier-Lavigne, M., and Zou, Y. (2003). Anterior-posterior guidance of commissural axons by Wnt-frizzled signaling. *Science* **302,** 1984–1988.

Maisonpierre, P. C., Barrezueta, N. X., and Yancopoulos, G. D. (1993). Ehk-1 and Ehk-2: Two novel members of the Eph receptor-like tyrosine kinase family with distinctive structures and neuronal expression. *Oncogene* **8,** 3277–3288.

Mallavarapu, A., and Mitchison, T. (1999). Regulated actin cytoskeleton assembly at filopodium tips controls their extension and retraction. *J. Cell Biol.* **146,** 1097–1106.

Mann, F., Peuckert, C., Dehner, F., Zhou, R., and Bolz, J. (2002a). Ephrins regulate the formation of terminal axonal arbors during the development of thalamocortical projections. *Development* **129,** 3945–3955.

Mann, F., Ray, S., Harris, W., and Holt, C. (2002b). Topographic mapping in dorsoventral axis of the Xenopus retinotectal system depends on signaling through ephrin-B ligands. *Neuron* **35,** 461–473.

Marcus, R. C., and Mason, C. A. (1995). The first retinal axon growth in the mouse optic chiasm: Axon patterning and the cellular environment. *J. Neurosci.* **15,** 6389–6402.

Marcus, R. C., Blazeski, R., Godement, P., and Mason, C. A. (1995). Retinal axon divergence in the optic chiasm: Uncrossed axons diverge from crossed axons within a midline glial specialization. *J. Neurosci.* **15,** 3716–3729.

Margolis, R. K., Rauch, U., Maurel, P., and Margolis, R. U. (1996). Neurocan and phosphacan: Two major nervous tissue-specific chondroitin sulfate proteoglycans. *Perspect. Dev. Neurobiol.* **3,** 273–290.

Marillat, V., Cases, O., Nguyen-Ba-Charvet, K. T., Tessier-Lavigne, M., Sotelo, C., and Chedotal, A. (2002). Spatiotemporal expression patterns of slit and robo genes in the rat brain. *J. Comp. Neurol.* **442,** 130–155.

Marsh, L., and Letourneau, P. C. (1984). Growth of neurites without filopodial or lamellipodial activity in the presence of cytochalasin B. *J. Cell Biol.* **99,** 2041–2047.

Marti, E., and Bovolenta, P. (2002). Sonic hedgehog in CNS development: One signal, multiple outputs. *Trends Neurosci.* **25,** 89–96.

Mason, C. A., and Sretavan, D. W. (1997). Glia, neurons, and axon pathfinding during optic chiasm development. *Curr. Opin. Neurobiol.* **7,** 647–653.

Mason, C. A., and Wang, L. C. (1997). Growth cone form is behavior-specific and, consequently, position-specific along the retinal axon pathway. *J. Neurosci.* **17,** 1086–1100.

Mastick, G. S., and Easter, S. S., Jr. (1996). Initial organization of neurons and tracts in the embryonic mouse fore- and midbrain. *Dev. Biol.* **173,** 79–94.

Matheson, S. F., and Levine, R. B. (1999). Steroid hormone enhancement of neurite outgrowth in identified insect motor neurons involves specific effects on growth cone form and function. *J. Neurobiol.* **38,** 27–45.

Matise, M. P., Lustig, M., Sakurai, T., Grumet, M., and Joyner, A. L. (1999). Ventral midline cells are required for the local control of commissural axon guidance in the mouse spinal cord. *Development* **126,** 3649–3659.

Matsunami, H., and Buck, L. B. (1997). A multigene family encoding a diverse array of putative pheromone receptors in mammals. *Cell* **90,** 775–784.

McConnell, S. K., Ghosh, A., and Shatz, C. J. (1989). Subplate neurons pioneer the first axon pathway from the cerebral cortex. *Science* **245,** 978–982.

McConnell, S. K., Ghosh, A., and Shatz, C. J. (1994). Subplate pioneers and the formation of descending connections from cerebral cortex. *J. Neurosci.* **14,** 1892–1907.

McLaughlin, T., Hindges, R., and O'Leary, D. D. (2003a). Regulation of axial patterning of the retina and its topographic mapping in the brain. *Curr. Opin. Neurobiol.* **13,** 57–69.

McLaughlin, T., Hindges, R., Yates, P. A., and O'Leary, D. D. (2003b). Bifunctional action of ephrin-B1 as a repellent and attractant to control bidirectional branch extension in dorsal-ventral retinotopic mapping. *Development* **130,** 2407–2418.

Mehlen, P., and Mazelin, L. (2003). The dependence receptors DCC and UNC5H as a link between neuronal guidance and survival. *Biol. Cell* **95,** 425–436.

Meixner, A., Haverkamp, S., Wassle, H., Fuhrer, S., Thalhammer, J., Kropf, N., Bittner, R. E., Lassmann, H., Wiche, G., and Propst, F. (2000). MAP1B is required for axon guidance and Is involved in the development of the central and peripheral nervous system. *J. Cell Biol.* **151,** 1169–1178.

Mellitzer, G., Xu, Q., and Wilkinson, D. G. (1999). Eph receptors and ephrins restrict cell intermingling and communication. *Nature* **400,** 77–81.

Meriane, M., Tcherkezian, J., Webber, C. A., Danek, E. I., Triki, I., McFarlane, S., Bloch-Gallego, E., and Lamarche-Vane, N. (2004). Phosphorylation of DCC by Fyn mediates Netrin-1 signaling in growth cone guidance. *J. Cell Biol.* **167,** 687–698.

Metin, C., Deleglise, D., Serafini, T., Kennedy, T. E., and Tessier-Lavigne, M. (1997). A role for netrin-1 in the guidance of cortical efferents. *Development* **124,** 5063–5074.

Mey, J., and Thanos, S. (1992). Development of the visual system of the chick—a review. *J. Hirnforsch.* **33,** 673–702.

Ming, G. L., Wong, S. T., Henley, J., Yuan, X. B., Song, H. J., Spitzer, N. C., and Poo, M. M. (2002). Adaptation in the chemotactic guidance of nerve growth cones. *Nature* **417,** 411–418.

Miragall, F., and Dermietzel, R. (1992). Immunocytochemical localization of cell adhesion molecules in the developing and mature olfactory system. *Microsc. Res. Tech.* **23,** 157–172.

Miragall, F., Kadmon, G., and Schachner, M. (1989). Expression of L1 and N-CAM cell adhesion molecules during development of the mouse olfactory system. *Dev. Biol.* **135,** 272–286.

Miranda, R., Sohrabji, F., Singh, M., and Toran-Allerand, D. (1996). Nerve growth factor (NGF) regulation of estrogen receptors in explant cultures of the developing forebrain. *J. Neurobiol.* **31,** 77–87.

Miranda, R. C., Sohrabji, F., and Toran-Allerand, C. D. (1993). Neuronal colocalization of mRNAs for neurotrophins and their receptors in the developing central nervous system suggests a potential for autocrine interactions. *Proc. Natl. Acad. Sci. USA* **90,** 6439–6443.

Mitchison, T., and Kirschner, M. (1988). Cytoskeletal dynamics and nerve growth. *Neuron* **1,** 761–772.

Mombaerts, P. (1996). Targeting olfaction. *Curr. Opin. Neurobiol.* **6,** 481–486.

Mombaerts, P., Wang, F., Dulac, C., Chao, S. K., Nemes, A., Mendelsohn, M., Edmondson, J., and Axel, R. (1996). Visualizing an olfactory sensory map. *Cell* **87,** 675–686.

Montag-Sallaz, M., Schachner, M., and Montag, D. (2002). Misguided axonal projections, neural cell adhesion molecule 180 mRNA upregulation, and altered behavior in mice deficient for the close homolog of L1. *Mol. Cell. Biol* **22,** 7967–7981.

Morgenstern, D. A., Asher, R. A., and Fawcett, J. W. (2002). Chondroitin sulphate proteoglycans in the CNS injury response. *Prog. Brain Res.* **137,** 313–332.

Mori, K., Ikeda, J., and Hayaishi, O. (1990). Monoclonal antibody R2D5 reveals midsagittal radial glial system in postnatally developing and adult brainstem. *Proc. Natl. Acad. Sci. USA* **87,** 5489–5493.

Mueller, B. K. (1999). Growth cone guidance: First steps towards a deeper understanding. *Annu. Rev. Neurosci.* **22,** 351–388.

Mui, S. H., Hindges, R., O'Leary, D. D., Lemke, G., and Bertuzzi, S. (2002). The homeodomain protein Vax2 patterns the dorsoventral and nasotemporal axes of the eye. *Development* **129,** 797–804.

Mullins, R. D., Heuser, J. A., and Pollard, T. D. (1998). The interaction of Arp2/3 complex with actin: Nucleation, high affinity pointed end capping, and formation of branching networks of filaments. *Proc. Natl. Acad. Sci. USA* **95,** 6181–6186.

Myat, A., Henry, P., McCabe, V., Flintoft, L., Rotin, D., and Tear, G. (2002). Drosophila Nedd4, a ubiquitin ligase, is recruited by Commissureless to control cell surface levels of the roundabout receptor. *Neuron* **35,** 447–459.

Nagata, K., Guggenheimer, R. A., Enomoto, T., Lichy, J. H., and Hurwitz, J. (1982). Adenovirus DNA replication *in vitro*: Identification of a host factor that stimulates synthesis of the preterminal protein-dCMP complex. *Proc. Natl. Acad. Sci. USA* **79,** 6438–6442.

Nagata, K., Guggenheimer, R. A., and Hurwitz, J. (1983). Specific binding of a cellular DNA replication protein to the origin of replication of adenovirus DNA. *Proc. Natl. Acad. Sci. USA* **80,** 6177–6181.

Nakagawa, H., Miki, H., Ito, M., Ohashi, K., Takenawa, T., and Miyamoto, S. (2001). N-WASP, WAVE and Mena play different roles in the organization of actin cytoskeleton in lamellipodia. *J. Cell Sci.* **114,** 1555–1565.

Nakamura, F., Kalb, R. G., and Strittmatter, S. M. (2000). Molecular basis of semaphorin-mediated axon guidance. *J. Neurobiol.* **44,** 219–229.

Nakashiba, T., Nishimura, S., Ikeda, T., and Itohara, S. (2002). Complementary expression and neurite outgrowth activity of netrin-G subfamily members. *Mech. Dev.* **111,** 47–60.

Narasimhan, K. (2004). Assaying axon sensitivity. *Nat. Neurosci.* **7,** 574.

Neely, M. D., and Gesemann, M. (1994). Disruption of microfilaments in growth cones following depolarization and calcium influx. *J. Neurosci.* **14,** 7511–7520.

Nguyen-Ba-Charvet, K. T., Brose, K., Marillat, V., Sotelo, C., Tessier-Lavigne, M., and Chedotal, A. (2001). Sensory axon response to substrate-bound Slit2 is modulated by laminin and cyclic GMP. *Mol. Cell Neurosci.* **17,** 1048–1058.

Nguyen-Ba-Charvet, K. T., Plump, A. S., Tessier-Lavigne, M., and Chedotal, A. (2002). Slit1 and slit2 proteins control the development of the lateral olfactory tract. *J. Neurosci.* **22,** 5473–5480.

Niclou, S. P., Jia, L., and Raper, J. A. (2000). Slit2 is a repellent for retinal ganglion cell axons. *J. Neurosci.* **20,** 4962–4974.

Nordlander, R. H., Singer, J. F., Beck, R., and Singer, M. (1981). An ultrastructural examination of early ventral root formation in amphibia. *J. Comp. Neurol.* **199,** 535–551.

Nowock, J., Borgmeyer, U., Puschel, A. W., Rupp, R. A., and Sippel, A. E. (1985). The TGGCA protein binds to the MMTV-LTR, the adenovirus origin of replication, and the BK virus enhancer. *Nucleic Acids Res.* **13,** 2045–2061.

Nural, H. F., and Mastick, G. S. (2004). Pax6 guides a relay of pioneer longitudinal axons in the embryonic mouse forebrain. *J. Comp. Neurol.* **479,** 399–409.

Oakley, R. A., and Tosney, K. W. (1991). Peanut agglutinin and chondroitin-6-sulfate are molecular markers for tissues that act as barriers to axon advance in the avian embryo. *Dev. Biol.* **147,** 187–206.

O'Connor, T. P., and Bentley, D. (1993). Accumulation of actin in subsets of pioneer growth cone filopodia in response to neural and epithelial guidance cues in situ. *J. Cell Biol.* **123,** 935–948.

O'Leary, D. D., Bicknese, A. R., De Carlos, J. A., Heffner, C. D., Koester, S. E., Kutka, L. J., and Terashima, T. (1990). Target selection by cortical axons: Alternative mechanisms to establish axonal connections in the developing brain. *Cold Spring Harb. Symp. Quant. Biol.* **55,** 453–468.

O'Leary, D. M., Gerfen, C. R., and Cowan, W. M. (1983). The development and restriction of the ipsilateral retinofugal projection in the chick. *Brain Res.* **312,** 93–109.

Orioli, D., Henkemeyer, M., Lemke, G., Klein, R., and Pawson, T. (1996). Sek4 and Nuk receptors cooperate in guidance of commissural axons and in palate formation. *EMBO J.* **15,** 6035–6049.

Ozaki, H. S., and Wahlsten, D. (1998). Timing and origin of the first cortical axons to project through the corpus callosum and the subsequent emergence of callosal projection cells in mouse. *J. Comp. Neurol.* **400,** 197–206.

Ozdinler, P. H., and Erzurumlu, R. S. (2002). Slit2, a branching-arborization factor for sensory axons in the Mammalian CNS. *J. Neurosci.* **22,** 4540–4549.

Packard, M., Koo, E. S., Gorczyca, M., Sharpe, J., Cumberledge, S., and Budnik, V. (2002). The Drosophila Wnt, wingless, provides an essential signal for pre- and postsynaptic differentiation. *Cell* **111,** 319–330.

Palmer, A., and Klein, R. (2003). Multiple roles of ephrins in morphogenesis, neuronal networking, and brain function. *Genes Dev.* **17,** 1429–5140.

Pasterkamp, R. J., and Kolodkin, A. L. (2003). Semaphorin junction: Making tracks toward neural connectivity. *Curr. Opin. Neurobiol.* **13,** 79–89.

Pasterkamp, R. J., De Winter, F., Holtmaat, A. J., and Verhaagen, J. (1998). Evidence for a role of the chemorepellent semaphorin III and its receptor neuropilin-1 in the regeneration of primary olfactory axons. *J. Neurosci.* **18,** 9962–9976.

Pasterkamp, R. J., Ruitenberg, M. J., and Verhaagen, J. (1999). Semaphorins and their receptors in olfactory axon guidance. *Cell Mol. Biol. (Noisy-le-grand)* **45,** 763–779.

Pasterkamp, R. J., Peschon, J. J., Spriggs, M. K., and Kolodkin, A. L. (2003). Semaphorin 7A promotes axon outgrowth through integrins and MAPKs. *Nature* **424,** 398–405.

Paves, H., and Saarma, M. (1997). Neurotrophins as *in vitro* growth cone guidance molecules for embryonic sensory neurons. *Cell Tissue Res.* **290,** 285–297.

Piper, M., Georgas, K., Yamada, T., and Little, M. (2000). Expression of the vertebrate Slit gene family and their putative receptors, the Robo genes, in the developing murine kidney. *Mech. Dev.* **94**, 213–217.

Pires-Neto, M. A., Braga-De-Souza, S., and Lent, R. (1998). Molecular tunnels and boundaries for growing axons in the anterior commissure of hamster embryos. *J. Comp. Neurol.* **399**, 176–188.

Pistor, S., Chakraborty, T., Walter, U., and Wehland, J. (1995). The bacterial actin nucleator protein ActA of Listeria monocytogenes contains multiple binding sites for host microfilament proteins. *Curr. Biol* **5**, 517–525.

Pittman, A., and Chien, C. B. (2002). Understanding dorsoventral topography: Backwards and forwards. *Neuron* **35**, 409–411.

Placzek, M., Tessier-Lavigne, M., Jessell, T., and Dodd, J. (1990). Orientation of commissural axons *in vitro* in response to a floor plate-derived chemoattractant. *Development* **110**, 19–30.

Plump, A. S., Erskine, L., Sabatier, C., Brose, K., Epstein, C. J., Goodman, C. S., Mason, C. A., and Tessier-Lavigne, M. (2002). Slit1 and Slit2 cooperate to prevent premature midline crossing of retinal axons in the mouse visual system. *Neuron* **33**, 219–232.

Puche, A. C., and Key, B. (1996). N-acetyl-lactosamine in the rat olfactory system: Expression and potential role in neurite growth. *J. Comp. Neurol.* **364**, 267–278.

Puche, A. C., Poirier, F., Hair, M., Bartlett, P. F., and Key, B. (1996). Role of galectin-1 in the developing mouse olfactory system. *Dev. Biol.* **179**, 274–287.

Raabe, E. H., Yoshida, K., and Schwarting, G. A. (1997). Differential laminin isoform expression in the developing rat olfactory system. *Brain Res. Dev. Brain Res.* **101**, 187–196.

Raper, J. A. (2000). Semaphorins and their receptors in vertebrates and invertebrates. *Curr. Opin. Neurobiol.* **10**, 88–94.

Raper, J. A., Bastiani, M., and Goodman, C. S. (1983a). Pathfinding by neuronal growth cones in grasshopper embryos. I. Divergent choices made by the growth cones of sibling neurons. *J. Neurosci.* **3**, 20–30.

Raper, J. A., Bastiani, M., and Goodman, C. S. (1983b). Pathfinding by neuronal growth cones in grasshopper embryos. II. Selective fasciculation onto specific axonal pathways. *J. Neurosci.* **3**, 31–41.

Raper, J. A., Bastiani, M. J., and Goodman, C. S. (1984). Pathfinding by neuronal growth cones in grasshopper embryos. IV. The effects of ablating the A and P axons upon the behavior of the G growth cone. *J. Neurosci.* **4**, 2329–2345.

Rasband, K., Hardy, M., and Chien, C. B. (2003). Generating X: Formation of the optic chiasm. *Neuron* **39**, 885–888.

Rash, B. G., and Richards, L. J. (2001). A role for cingulate pioneering axons in the development of the corpus callosum. *J. Comp. Neurol.* **434**, 147–157.

Renzi, M. J., Wexler, T. L., and Raper, J. A. (2000). Olfactory sensory axons expressing a dominant-negative semaphorin receptor enter the CNS early and overshoot their target. *Neuron* **28**, 437–447.

Ressler, K. J., Sullivan, S. L., and Buck, L. B. (1994). A molecular dissection of spatial patterning in the olfactory system. *Curr. Opin. Neurobiol.* **4**, 588–596.

Rice, D. S., Williams, R. W., and Goldowitz, D. (1995). Genetic control of retinal projections in inbred strains of albino mice. *J. Comp. Neurol.* **354**, 459–469.

Richards, L. J., Koester, S. E., Tuttle, R., and O'Leary, D. D. (1997). Directed growth of early cortical axons is influenced by a chemoattractant released from an intermediate target. *J. Neurosci.* **17**, 2445–2458.

Richards, L. J., Plachez, C., and Ren, T. (2004). Mechanisms regulating the development of the corpus callosum and its agenesis in mouse and human. *Clin. Genet.* **66**, 276–289.

Ringstedt, T., Braisted, J. E., Brose, K., Kidd, T., Goodman, C., Tessier-Lavigne, M., and O'Leary, D. D. (2000). Slit inhibition of retinal axon growth and its role in retinal axon pathfinding and innervation patterns in the diencephalon. *J. Neurosci.* **20,** 4983–4991.

Rivas, R. J., and Hatten, M. E. (1995). Motility and cytoskeletal organization of migrating cerebellar granule neurons. *J. Neurosci.* **15,** 981–989.

Rochlin, M. W., Wickline, K. M., and Bridgman, P. C. (1996). Microtubule stability decreases axon elongation but not axoplasm production. *J. Neurosci.* **16,** 3236–3246.

Rochlin, M. W., Dailey, M. E., and Bridgman, P. C. (1999). Polymerizing microtubules activate site-directed F-actin assembly in nerve growth cones. *Mol. Biol. Cell* **10,** 2309–2327.

Rodriguez, O. C., Schaefer, A. W., Mandato, C. A., Forscher, P., Bement, W. M., and Waterman-Storer, C. M. (2003). Conserved microtubule-actin interactions in cell movement and morphogenesis. *Nat. Cell Biol.* **5,** 599–609.

Roskies, A. L., and O'Leary, D. D. (1994). Control of topographic retinal axon branching by inhibitory membrane-bound molccules. *Science* **265,** 799–803.

Rosoff, W. J., Urbach, J. S., Esrick, M. A., McAllister, R. G., Richards, L. J., and Goodhill, G. J. (2004). A new chemotaxis assay shows the extreme sensitivity of axons to molecular gradients. *Nat. Neurosci.* **7,** 678–682.

Ross, L. S., Parrett, T., and Easter, S. S., Jr. (1992). Axonogenesis and morphogenesis in the embryonic zebrafish brain. *J. Neurosci.* **12,** 467–482.

Rothberg, J. M., Jacobs, J. R., Goodman, C. S., and Artavanis-Tsakonas, S. (1990). slit: An extracellular protein necessary for development of midline glia and commissural axon pathways contains both EGF and LRR domains. *Genes Dev.* **4,** 2169–2187.

Ruthazer, E. S., and Cline, H. T. (2004). Insights into activity-dependent map formation from the retinotectal system: A middle-of-the-brain perspective. *J. Neurobiol.* **59,** 134–146.

Ryba, N. J., and Tirindelli, R. (1997). A new multigene family of putative pheromone receptors. *Neuron* **19,** 371–379.

Sabatier, C., Plump, A. S., Le, M., Brose, K., Tamada, A., Murakami, F., Lee, E. Y., and Tessier-Lavigne, M. (2004). The divergent Robo family protein rig-1/Robo3 is a negative regulator of slit responsiveness required for midline crossing by commissural axons. *Cell* **117,** 157–169.

Sabry, J. H., O'Connor, T. P., Evans, L., Toroian-Raymond, A., Kirschner, M., and Bentley, D. (1991). Microtubule behavior during guidance of pioneer neuron growth cones in situ. *J. Cell Biol.* **115,** 381–395.

Salinas, P. C. (2003). The morphogen sonic hedgehog collaborates with netrin-1 to guide axons in the spinal cord. *Trends Neurosci.* **26,** 641–643.

Sanchez, D., Ganfornina, M. D., and Bastiani, M. J. (1995). Developmental expression of the lipocalin Lazarillo and its role in axonal pathfinding in the grasshopper embryo. *Development* **121,** 135–147.

Sanchez, D., Ganfornina, M. D., and Bastiani, M. J. (2000). Lazarillo, a neuronal lipocalin in grasshoppers with a role in axon guidance. *Biochim. Biophys. Acta* **1482,** 102–109.

Sasaki, Y., Cheng, C., Uchida, Y., Nakajima, O., Ohshima, T., Yagi, T., Taniguchi, M., Nakayama, T., Kishida, R., Kudo, Y., Ohno, S., Nakamura, F., and Goshima, Y. (2002). Fyn and Cdk5 mediate semaphorin-3A signaling, which is involved in regulation of dendrite orientation in cerebral cortex. *Neuron* **35,** 907–920.

Sato, Y., Hirata, T., Ogawa, M., and Fujisawa, H. (1998). Requirement for early-generated neurons recognized by monoclonal antibody lot1 in the formation of lateral olfactory tract. *J. Neurosci.* **18,** 7800–7810.

Schnapp, B. J., Vale, R. D., Sheetz, M. P., and Reese, T. S. (1986). Microtubules and the mechanism of directed organelle movement. *Ann. N Y Acad. Sci.* **466,** 909–918.

Schwarting, G. A., Kostek, C., Ahmad, N., Dibble, C., Pays, L., and Puschel, A. W. (2000). Semaphorin 3A is required for guidance of olfactory axons in mice. *J. Neurosci.* **20,** 7691–7697.

Schwob, J. E., and Price, J. L. (1984). The development of lamination of afferent fibers to the olfactory cortex in rats, with additional observations in the adult. *J. Comp. Neurol.* **223,** 203–222.

Seeger, M., Tear, G., Ferres-Marco, D., and Goodman, C. S. (1993). Mutations affecting growth cone guidance in Drosophila: Genes necessary for guidance toward or away from the midline. *Neuron* **10,** 409–426.

Semaphorin Nomenclature Committee (1999). Unified nomenclature for the semaphorins/ collapsins. *Cell* **97,** 551–552.

Serafini, T., Kennedy, T. E., Galko, M. J., Mirzayan, C., Jessell, T. M., and Tessier-Lavigne, M. (1994). The netrins define a family of axon outgrowth-promoting proteins homologous to C. elegans UNC-6. *Cell* **78,** 409–424.

Serafini, T., Colamarino, S. A., Leonardo, E. D., Wang, H., Beddington, R., Skarnes, W. C., and Tessier-Lavigne, M. (1996). Netrin-1 is required for commissural axon guidance in the developing vertebrate nervous system. *Cell* **87,** 1001–1014.

Shen, K., and Bargmann, C. I. (2003). The immunoglobulin superfamily protein SYG-1 determines the location of specific synapses in C. elegans. *Cell* **112,** 619–630.

Shen, K., Fetter, R. D., and Bargmann, C. I. (2004). Synaptic specificity is generated by the synaptic guidepost protein SYG-2 and its receptor, SYG-1. *Cell* **116,** 869–881.

Shen, Y., Mani, S., Donovan, S. L., Schwob, J. E., and Meiri, K. F. (2002). Growth-associated protein-43 is required for commissural axon guidance in the developing vertebrate nervous system. *J. Neurosci.* **22,** 239–247.

Shewan, D., Dwivedy, A., Anderson, R., and Holt, C. E. (2002). Age-related changes underlie switch in netrin-1 responsiveness as growth cones advance along visual pathway. *Nat. Neurosci.* **5,** 955–962.

Shipley, M. T., and Ennis, M. (1996). Functional organization of olfactory system. *J. Neurobiol.* **30,** 123–176.

Shipley, M. Y., McLean, J. H., and Ennis, M. (1995). Olfactory system. *In* "The Rat Nervous System" (G. Paxinos, Ed.), 2nd edn. Academic Press, San Diego.

Shu, T., and Richards, L. J. (2001). Cortical axon guidance by the glial wedge during the development of the corpus callosum. *J. Neurosci.* **21,** 2749–2758.

Shu, T., Valentino, K. M., Seaman, C., Cooper, H. M., and Richards, L. J. (2000). Expression of the netrin-1 receptor, deleted in colorectal cancer (DCC), is largely confined to projecting neurons in the developing forebrain. *J. Comp. Neurol.* **416,** 201–212.

Shu, T., Butz, K. G., Plachez, C., Gronostajski, R. M., and Richards, L. J. (2003a). Abnormal development of forebrain midline glia and commissural projections in Nfia knock-out mice. *J. Neurosci.* **23,** 203–212.

Shu, T., Li, Y., Keller, A., and Richards, L. J. (2003b). The glial sling is a migratory population of developing neurons. *Development* **130,** 2929–2937.

Shu, T., Puche, A. C., and Richards, L. J. (2003c). Development of midline glial populations at the corticoseptal boundary. *J. Neurobiol.* **57,** 81–94.

Shu, T., Sundaresan, V., McCarthy, M. M., and Richards, L. J. (2003d). Slit2 guides both precrossing and postcrossing callosal axons at the midline *in vivo. J. Neurosci.* **23,** 8176–8184.

Silver, J. (1994). Inhibitory molecules in development and regeneration. *J. Neurol.* **242,** S22–24.

Silver, J., and Ogawa, M. Y. (1983). Postnatally induced formation of the corpus callosum in acallosal mice on glia-coated cellulose bridges. *Science* **220,** 1067–1069.

Silver, J., and Sapiro, J. (1981). Axonal guidance during development of the optic nerve: The role of pigmented epithelia and other extrinsic factors. *J. Comp. Neurol.* **202,** 521–538.

Silver, J., and Sidman, R. L. (1980). A mechanism for the guidance and topographic patterning of retinal ganglion cell axons. *J. Comp. Neurol.* **189,** 101–111.

Silver, J., Lorenz, S. E., Wahlsten, D., and Coughlin, J. (1982). Axonal guidance during development of the great cerebral commissures: Descriptive and experimental studies, *in vivo,* on the role of preformed glial pathways. *J. Comp. Neurol.* **210,** 10–29.

Silver, J., Edwards, M. A., and Levitt, P. (1993). Immunocytochemical demonstration of early appearing astroglial structures that form boundaries and pathways along axon tracts in the fetal brain. *J. Comp. Neurol.* **328,** 415–436.

Simerly, R. B. (2002). Wired for reproduction: Organization and development of sexually dimorphic circuits in the mammalian forebrain. *Annu. Rev. Neurosci.* **25,** 507–536.

Singer, M., Nordlander, R. H., and Egar, M. (1979). Axonal guidance during embryogenesis and regeneration in the spinal cord of the newt: The blueprint hypothesis of neuronal pathway patterning. *J. Comp. Neurol.* **185,** 1–21.

Singh, M., Setalo, G., Jr., Guan, X., Warren, M., and Toran-Allerand, C. D. (1999). Estrogen-induced activation of mitogen-activated protein kinase in cerebral cortical explants: Convergence of estrogen and neurotrophin signaling pathways. *J. Neurosci.* **19,** 1179–1188.

Skynner, M. J., Sim, J. A., and Herbison, A. E. (1999). Detection of estrogen receptor alpha and beta messenger ribonucleic acids in adult gonadotropin-releasing hormone neurons. *Endocrinology* **140,** 5195–5201.

Smith, G. A., Theriot, J. A., and Portnoy, D. A. (1996). The tandem repeat domain in the Listeria monocytogenes ActA protein controls the rate of actin-based motility, the percentage of moving bacteria, and the localization of vasodilator-stimulated phosphoprotein and profilin. *J. Cell Biol.* **135,** 647–660.

Soker, S., Takashima, S., Miao, H. Q., Neufeld, G., and Klagsbrun, M. (1998). Neuropilin-1 is expressed by endothelial and tumor cells as an isoform-specific receptor for vascular endothelial growth factor. *Cell* **92,** 735–745.

Song, H., Ming, G., He, Z., Lehmann, M., McKerracher, L., Tessier-Lavigne, M., and Poo, M. (1998). Conversion of neuronal growth cone responses from repulsion to attraction by cyclic nucleotides. *Science* **281,** 1515–1518.

Song, H. J., Ming, G. L., and Poo, M. M. (1997). cAMP-induced switching in turning direction of nerve growth cones. *Nature* **388,** 275–279.

Soriano, E., Del Rio, J. A., Martinez, A., and Super, H. (1994). Organization of the embryonic and early postnatal murine hippocampus. I. Immunocytochemical characterization of neuronal populations in the subplate and marginal zone. *J. Comp. Neurol.* **342,** 571–595.

Spassky, N., de Castro, F., Le Bras, B., Heydon, K., Queraud-Le Saux, F., Bloch-Gallego, E., Chedotal, A., Zalc, B., and Thomas, J. L. (2002). Directional guidance of oligodendroglial migration by class 3 semaphorins and netrin-1. *J. Neurosci.* **22,** 5992–6004.

Sperry, R. (1982). Some effects of disconnecting the cerebral hemispheres. *Science* **217,** 1223–1226.

Sperry, R. W. (1963). Chemoaffinity In The Orderly Growth Of Nerve Fiber Patterns And Connections. *Proc. Natl. Acad. Sci. USA* **50,** 703–710.

Sperry, R. W. (1968). Hemisphere deconnection and unity in conscious awareness. *Am. Psychol.* **23,** 723–733.

Sretavan, D. W. (1993). Pathfinding at the mammalian optic chiasm. *Curr. Opin. Neurobiol.* **3,** 45–52.

Sretavan, D. W., and Reichardt, L. F. (1993). Time-lapse video analysis of retinal ganglion cell axon pathfinding at the mammalian optic chiasm: Growth cone guidance using intrinsic chiasm cues. *Neuron* **10,** 761–777.

Sretavan, D. W., Feng, L., Pure, E., and Reichardt, L. F. (1994). Embryonic neurons of the developing optic chiasm express L1 and CD44, cell surface molecules with opposing effects on retinal axon growth. *Neuron* **12,** 957–975.

Sretavan, D. W., Pure, E., Siegel, M. W., and Reichardt, L. F. (1995). Disruption of retinal axon ingrowth by ablation of embryonic mouse optic chiasm neurons. *Science* **269**, 98–101.

St John, J. A., and Key, B. (2001). EphB2 and two of its ligands have dynamic protein expression patterns in the developing olfactory system. *Brain Res. Dev. Brain Res.* **126**, 43–56.

St John, J. A., Tisay, K. T., Caras, I. W., and Key, B. (2000). Expression of EphA5 during development of the olfactory nerve pathway in rat. *J. Comp. Neurol.* **416**, 540–550.

St John, J. A., Pasquale, E. B., and Key, B. (2002). EphA receptors and ephrin-A ligands exhibit highly regulated spatial and temporal expression patterns in the developing olfactory system. *Brain Res. Dev. Brain Res.* **138**, 1–14.

Steele-Perkins, G., Plachez, C., Butz, K. G., Yang, G., Bachurski, C. J., Kinsman, S. L., Litwack, E. D., Richards, L. J., and Gronostajski, R. M. (2005). The transcription factor gene Nfib is essential for both lung maturation and brain development. *Mol. Cell. Biol.* **25**, 685–698.

Steigemann, P., Molitor, A., Fellert, S., Jackle, H., and Vorbruggen, G. (2004). Heparan sulfate proteoglycan syndecan promotes axonal and myotube guidance by slit/robo signaling. *Curr. Biol.* **14**, 225–230.

Stein, E., and Tessier-Lavigne, M. (2001). Hierarchical organization of guidance receptors: Silencing of netrin attraction by slit through a Robo/DCC receptor complex. *Science* **291**, 1928–1938.

Stoeckli, E. T., and Landmesser, L. T. (1995). Axonin-1, Nr-CAM, and Ng-CAM play different roles in the *in vivo* guidance of chick commissural neurons. *Neuron* **14**, 1165–1179.

Stossel, T. P. (1993). On the crawling of animal cells. *Science* **260**, 1086–94.

Su, M., Merz, D. C., Killeen, M. T., Zhou, Y., Zheng, H., Kramer, J. M., Hedgecock, E. M., and Culotti, J. G. (2000). Regulation of the UNC-5 netrin receptor initiates the first reorientation of migrating distal tip cells in Caenorhabditis elegans. *Development* **127**, 585–594.

Sugimoto, Y., Taniguchi, M., Yagi, T., Akagi, Y., Nojyo, Y., and Tamamaki, N. (2001). Guidance of glial precursor cell migration by secreted cues in the developing optic nerve. *Development* **128**, 3321–3330.

Sugisaki, N., Hirata, T., Naruse, I., Kawakami, A., Kitsukawa, T., and Fujisawa, H. (1996). Positional cues that are strictly localized in the telencephalon induce preferential growth of mitral cell axons. *J. Neurobiol.* **29**, 127–137.

Sundaresan, V., Mambetisaeva, E., Andrews, W., Annan, A., Knoll, B., Tear, G., and Bannister, L. (2004). Dynamic expression patterns of Robo (Robo1 and Robo2) in the developing murine central nervous system. *J. Comp. Neurol.* **468**, 467–481.

Super, H., Martinez, A., Del Rio, J. A., and Soriano, E. (1998). Involvement of distinct pioneer neurons in the formation of layer-specific connections in the hippocampus. *J. Neurosci.* **18**, 4616–4626.

Suto, F., Murakami, Y., Nakamura, F., Goshima, Y., and Fujisawa, H. (2003). Identification and characterization of a novel mouse plexin, plexin-A4. *Mech. Dev.* **120**, 385–396.

Szebenyi, G., Callaway, J. L., Dent, E. W., and Kalil, K. (1998). Interstitial branches develop from active regions of the axon demarcated by the primary growth cone during pausing behaviors. *J. Neurosci.* **18**, 7930–7940.

Takahashi, T., and Strittmatter, S. M. (2001). Plexina1 autoinhibition by the plexin sema domain. *Neuron* **29**, 429–439.

Takahashi, T., Fournier, A., Nakamura, F., Wang, L. H., Murakami, Y., Kalb, R. G., Fujisawa, H., and Strittmatter, S. M. (1999). Plexin-neuropilin-1 complexes form functional semaphorin-3A receptors. *Cell* **99**, 59–69.

Tamagnone, L., Artigiani, S., Chen, H., He, Z., Ming, G. I., Song, H., Chedotal, A., Winberg, M. L., Goodman, C. S., Poo, M., Tessier-Lavigne, M., and Comoglio, P. M. (1999). Plexins are a large family of receptors for transmembrane, secreted, and GPI-anchored semaphorins in vertebrates. *Cell* **99,** 71–80.

Tanaka, E., and Sabry, J. (1995). Making the connection: Cytoskeletal rearrangements during growth cone guidance. *Cell* **83,** 171–176.

Tanaka, E., Ho, T., and Kirschner, M. W. (1995). The role of microtubule dynamics in growth cone motility and axonal growth. *J. Cell Biol.* **128,** 139–155.

Tanaka, E. M., and Kirschner, M. W. (1991). Microtubule behavior in the growth cones of living neurons during axon elongation. *J. Cell Biol.* **115,** 345–363.

Tanno, T., Takenaka, S., and Tsuyama, S. (2004). Expression and Function of Slit1{alpha}, a Novel Alternative Splicing Product for Slit1. *J. Biochem. (Tokyo)* **136,** 575–581.

Tessier-Lavigne, M., and Goodman, C. S. (1996). The molecular biology of axon guidance. *Science* **274,** 1123–1133.

Tessier-Lavigne, M., Attwell, D., Mobbs, P., and Wilson, M. (1988a). Membrane currents in retinal bipolar cells of the axolotl. *J. Gen. Physiol.* **91,** 49–72.

Tessier-Lavigne, M., Placzek, M., Lumsden, A. G., Dodd, J., and Jessell, T. M. (1988b). Chemotropic guidance of developing axons in the mammalian central nervous system. *Nature* **336,** 775–778.

Thanos, S., and Mey, J. (2001). Development of the visual system of the chick. II. Mechanisms of axonal guidance. *Brain Res. Brain Res. Rev.* **35,** 205–245.

Thanos, S., Puttmann, S., Naskar, R., Rose, K., Langkamp-Flock, M., and Paulus, W. (2004). Potential role of Pax-2 in retinal axon navigation through the chick optic nerve stalk and optic chiasm. *J. Neurobiol.* **59,** 8–23.

Thomas, J. B., Crews, S. T., and Goodman, C. S. (1988). Molecular genetics of the single-minded locus: A gene involved in the development of the Drosophila nervous system. *Cell* **52,** 133–141.

Thompson, I. D., and Morgan, J. E. (1993). The development of retinal ganglion cell decussation patterns in postnatal pigmented and albino ferrets. *Eur. J. Neurosci.* **5,** 341–356.

Tisay, K. T., and Key, B. (1999). The extracellular matrix modulates olfactory neurite outgrowth on ensheathing cells. *J. Neurosci.* **19,** 9890–9899.

Tisay, K. T., Bartlett, P. F., and Key, B. (2000). Primary olfactory axons form ectopic glomeruli in mice lacking p75NTR. *J. Comp. Neurol.* **428,** 656–670.

Toran-Allerand, C. D. (1976). Sex steroids and the development of the newborn mouse hypothalamus and preoptic area *in vitro*: Implications for sexual differentiation. *Brain Res.* **106,** 407–412.

Toran-Allerand, C. D. (1980). Sex steroids and the development of the newborn mouse hypothalamus and preoptic area *in vitro*. II. Morphological correlates and hormonal specificity. *Brain Res.* **189,** 413–427.

Toran-Allerand, C. D. (1984). Gonadal hormones and brain development: Implications for the genesis of sexual differentiation. *Ann. N Y Acad. Sci.* **435,** 101–111.

Toran-Allerand, C. D. (1996). The estrogen/neurotrophin connection during neural development: Is co-localization of estrogen receptors with the neurotrophins and their receptors biologically relevant? *Dev. Neurosci.* **18,** 36–48.

Toran-Allerand, C. D. (2004). Estrogen and the brain: Beyond ER-alpha and ER-beta. *Exp. Gerontol.* **39,** 1579–1586.

Toran-Allerand, C. D., Miranda, R. C., Bentham, W. D., Sohrabji, F., Brown, T. J., Hochberg, R. B., and Mac Lusky, N. J. (1992). Estrogen receptors colocalize with low-affinity nerve growth factor receptors in cholinergic neurons of the basal forebrain. *Proc. Natl. Acad. Sci. USA* **89,** 4668–4672.

Torres, M., Gomez-Pardo, E., and Gruss, P. (1996). Pax2 contributes to inner ear patterning and optic nerve trajectory. *Development* **122,** 3381–3391.

Tran, T. S., and Phelps, P. E. (2000). Axons crossing in the ventral commissure express L1 and GAD65 in the developing rat spinal cord. *Dev. Neurosci.* **22,** 228–236.

Treloar, H., Tomasiewicz, H., Magnuson, T., and Key, B. (1997). The central pathway of primary olfactory axons is abnormal in mice lacking the N-CAM-180 isoform. *J. Neurobiol.* **32,** 643–658.

Treloar, H. B., Nurcombe, V., and Key, B. (1996). Expression of extracellular matrix molecules in the embryonic rat olfactory pathway. *J. Neurobiol.* **31,** 41–55.

Tremblay, G. B., Tremblay, A., Copeland, N. G., Gilbert, D. J., Jenkins, N. A., Labrie, F., and Giguere, V. (1997). Cloning, chromosomal localization, and functional analysis of the murine estrogen receptor beta. *Mol. Endocrinol.* **11,** 353–365.

Trousse, F., Marti, E., Gruss, P., Torres, M., and Bovolenta, P. (2001). Control of retinal ganglion cell axon growth: A new role for Sonic hedgehog. *Development* **128,** 3927–3936.

Tsai, H. H., Tessier-Lavigne, M., and Miller, R. H. (2003). Netrin 1 mediates spinal cord oligodendrocyte precursor dispersal. *Development* **130,** 2095–2105.

Van Hartesveldt, C., Moore, B., and Hartman, B. K. (1986). Transient midline raphe glial structure in the developing rat. *J. Comp. Neurol.* **253,** 174–184.

van Horck, F. P., Weinl, C., and Holt, C. E. (2004). Retinal axon guidance: Novel mechanisms for steering. *Curr. Opin. Neurobiol.* **14,** 61–66.

Vassar, R., Chao, S. K., Sitcheran, R., Nunez, J. M., Vosshall, L. B., and Axel, R. (1994). Topographic organization of sensory projections to the olfactory bulb. *Cell* **79,** 981–991.

Vinson, C. R., and Adler, P. N. (1987). Directional non-cell autonomy and the transmission of polarity information by the frizzled gene of Drosophila. *Nature* **329,** 549–551.

von Boxberg, Y., Deiss, S., and Schwarz, U. (1993). Guidance and topographic stabilization of nasal chick retinal axons on target-derived components *in vitro. Neuron* **10,** 345–357.

Wadsworth, W. G. (2002). Moving around in a worm: Netrin UNC-6 and circumferential axon guidance in C. elegans. *Trends Neurosci.* **25,** 423–429.

Wadsworth, W. G., Bhatt, H., and Hedgecock, E. M. (1996). Neuroglia and pioneer neurons express UNC-6 to provide global and local netrin cues for guiding migrations in C. elegans. *Neuron* **16,** 35–46.

Wahl, S., Barth, H., Ciossek, T., Aktories, K., and Mueller, B. K. (2000). Ephrin-A5 induces collapse of growth cones by activating Rho and Rho kinase. *J. Cell Biol.* **149,** 263–270.

Walsh, F. S., and Doherty, P. (1997). Neural cell adhesion molecules of the immunoglobulin superfamily: Role in axon growth and guidance. *Annu. Rev. Cell Dev. Biol.* **13,** 425–456.

Walter, J., Henke-Fahle, S., and Bonhoeffer, F. (1987a). Avoidance of posterior tectal membranes by temporal retinal axons. *Development* **101,** 909–913.

Walter, J., Kern-Veits, B., Huf, J., Stolze, B., and Bonhoeffer, F. (1987b). Recognition of position-specific properties of tectal cell membranes by retinal axons *in vitro. Development* **101,** 685–696.

Walz, A., McFarlane, S., Brickman, Y. G., Nurcombe, V., Bartlett, P. F., and Holt, C. E. (1997). Essential role of heparan sulfates in axon navigation and targeting in the developing visual system. *Development* **124,** 2421–2430.

Walz, A., Rodriguez, I., and Mombaerts, P. (2002). Aberrant sensory innervation of the olfactory bulb in neuropilin-2 mutant mice. *J. Neurosci.* **22,** 4025–4035.

Wang, H., Copeland, N. G., Gilbert, D. J., Jenkins, N. A., and Tessier-Lavigne, M. (1999a). Netrin-3, a mouse homolog of human NTN2L, is highly expressed in sensory ganglia and shows differential binding to netrin receptors. *J. Neurosci.* **19,** 4938–4947.

Wang, K. H., Brose, K., Arnott, D., Kidd, T., Goodman, C. S., Henzel, W., and Tessier-Lavigne, M. (1999b). Biochemical purification of a mammalian slit protein as a positive regulator of sensory axon elongation and branching. *Cell* **96,** 771–784.

Wang, L., and Denburg, J. L. (1992). A role for proteoglycans in the guidance of a subset of pioneer axons in cultured embryos of the cockroach. *Neuron* **8,** 701–714.

Wang, L., Andersson, S., Warner, M., and Gustafsson, J. A. (2002a). Estrogen actions in the brain. *Sci. STKE* **2002a,** PE29.

Wang, L., Andersson, S., Warner, M., and Gustafsson, J. A. (2003). Estrogen receptor (ER) beta knockout mice reveal a role for ERbeta in migration of cortical neurons in the developing brain. *Proc. Natl. Acad. Sci. USA* **100,** 703–708.

Wang, L. C., Rachel, R. A., Marcus, R. C., and Mason, C. A. (1996). Chemosuppression of retinal axon growth by the mouse optic chiasm. *Neuron* **17,** 849–862.

Wang, S. W., Mu, X., Bowers, W. J., Kim, D. S., Plas, D. J., Crair, M. C., Federoff, H. J., Gan, L., and Klein, W. H. (2002b). Brn3b/Brn3c double knockout mice reveal an unsuspected role for Brn3c in retinal ganglion cell axon outgrowth. *Development* **129,** 467–477.

Wang, Y., Thekdi, N., Smallwood, P. M., Macke, J. P., and Nathans, J. (2002c). Frizzled-3 is required for the development of major fiber tracts in the rostral CNS. *J. Neurosci.* **22,** 8563–73.

Wassle, H. (1982). Morphological types and central projections of ganglion cells in the cat retina. *Prog. Ret. Res.* **1,** 125–152.

White, R., Lees, J. A., Needham, M., Ham, J., and Parker, M. (1987). Structural organization and expression of the mouse estrogen receptor. *Mol. Endocrinol.* **1,** 735–744.

Whitford, K. L., Marillat, V., Stein, E., Goodman, C. S., Tessier-Lavigne, M., Chedotal, A., and Ghosh, A. (2002). Regulation of cortical dendrite development by Slit-Robo interactions. *Neuron* **33,** 47–61.

Whitington, P. M. (1993). Axon guidance factors in invertebrate development. *Pharmacol. Ther.* **58,** 263–299.

Whitington, P. M., Quilkey, C., and Sink, H. (2004). Necessity and redundancy of guidepost cells in the embryonic Drosophila CNS. *Int J. Dev. Neurosci.* **22,** 157–163.

Wilkinson, D. G. (2001). Multiple roles of EPH receptors and ephrins in neural development. *Nat. Rev. Neurosci.* **2,** 155–164.

Williams, S. E., Mann, F., Erskine, L., Sakurai, T., Wei, S., Rossi, D. J., Gale, N. W., Holt, C. E., Mason, C. A., and Henkemeyer, M. (2003). Ephrin-B2 and EphB1 mediate retinal axon divergence at the optic chiasm. *Neuron* **39,** 919–935.

Williams-Hogarth, L. C., Puche, A. C., Torrey, C., Cai, X., Song, I., Kolodkin, A. L., Shipley, M. T., and Ronnett, G. V. (2000). Expression of semaphorins in developing and regenerating olfactory epithelium. *J. Comp. Neurol.* **423,** 565–578.

Williamson, T., Gordon-Weeks, P. R., Schachner, M., and Taylor, J. (1996). Microtubule reorganization is obligatory for growth cone turning. *Proc. Natl. Acad. Sci. USA* **93,** 15221–15226.

Wilson, S. W., Ross, L. S., Parrett, T., and Easter, S. S., Jr. (1990). The development of a simple scaffold of axon tracts in the brain of the embryonic zebrafish, Brachydanio rerio. *Development* **108,** 121–145.

Winberg, M. L., Noordermeer, J. N., Tamagnone, L., Comoglio, P. M., Spriggs, M. K., Tessier-Lavigne, M., and Goodman, C. S. (1998). Plexin A is a neuronal semaphorin receptor that controls axon guidance. *Cell* **95,** 903–916.

Wise, S. P., and Jones, E. G. (1976). The organization and postnatal development of the commissural projection of the rat somatic sensory cortex. *J. Comp. Neurol.* **168,** 313–343.

Wodarz, A., and Nusse, R. (1998). Mechanisms of Wnt signaling in development. *Annu. Rev. Cell Dev. Biol.* **14,** 59–88.

Wu, W., Wong, K., Chen, J., Jiang, Z., Dupuis, S., Wu, J. Y., and Rao, Y. (1999). Directional guidance of neuronal migration in the olfactory system by the protein Slit. *Nature* **400,** 331–336.

Yamada, K. M., Spooner, B. S., and Wessells, N. K. (1970). Axon growth: Roles of microfilaments and microtubules. *Proc. Natl. Acad. Sci. USA* **66,** 1206–1212.

Yamada, K. M., Spooner, B. S., and Wessells, N. K. (1971). Ultrastructure and function of growth cones and axons of cultured nerve cells. *J. Cell Biol.* **49,** 614–635.

Yates, P. A., Roskies, A. L., McLaughlin, T., and O'Leary, D. D. (2001). Topographic-specific axon branching controlled by ephrin-As is the critical event in retinotectal map development. *J. Neurosci.* **21,** 8548–8563.

Yee, K. T., Simon, H. H., Tessier-Lavigne, M., and O'Leary, D. M. (1999). Extension of long leading processes and neuronal migration in the mammalian brain directed by the chemoattractant netrin-1. *Neuron* **24,** 607–622.

Yokoyama, N., Romero, M. I., Cowan, C. A., Galvan, P., Helmbacher, F., Charnay, P., Parada, L. F., and Henkemeyer, M. (2001). Forward signaling mediated by ephrin-B3 prevents contralateral corticospinal axons from recrossing the spinal cord midline. *Neuron* **29,** 85–97.

Yoshida, M., and Colman, D. R. (2000). Glial-defined rhombomere boundaries in developing Xenopus hindbrain. *J. Comp. Neurol.* **424,** 47–57.

Yoshihara, Y., Kawasaki, M., Tamada, A., Fujita, H., Hayashi, H., Kagamiyama, H., and Mori, K. (1997). OCAM: A new member of the neural cell adhesion molecule family related to zone-to-zone projection of olfactory and vomeronasal axons. *J. Neurosci.* **17,** 5830–5842.

Yoshikawa, S., McKinnon, R. D., Kokel, M., and Thomas, J. B. (2003). Wnt-mediated axon guidance via the Drosophila Derailed receptor. *Nature* **422,** 583–588.

Yuan, W., Zhou, L., Chen, J. H., Wu, J. Y., Rao, Y., and Ornitz, D. M. (1999). The mouse SLIT family: Secreted ligands for ROBO expressed in patterns that suggest a role in morphogenesis and axon guidance. *Dev. Biol.* **212,** 290–306.

Zallen, J. A., Yi, B. A., and Bargmann, C. I. (1998). The conserved immunoglobulin superfamily member SAX-3/Robo directs multiple aspects of axon guidance in C. elegans. *Cell* **92,** 217–227.

Zhang, H. Y., and Hoffmann, K. P. (1993). Retinal projections to the pretectum, accessory optic system and superior colliculus in pigmented and albino ferrets. *Eur. J. Neurosci.* **5,** 486–500.

Zhang, J., Richards, L. J., Yarowsky, P., Huang, H., van Zijl, P. C., and Mori, S. (2003). Three-dimensional anatomical characterization of the developing mouse brain by diffusion tensor microimaging. *Neuroimage* **20,** 1639–1648.

Zhang, J. H., Cerretti, D. P., Yu, T., Flanagan, J. G., and Zhou, R. (1996). Detection of ligands in regions anatomically connected to neurons expressing the Eph receptor Bsk: Potential roles in neuron-target interaction. *J. Neurosci.* **16,** 7182–7192.

Zhou, F. Q., and Cohan, C. S. (2001). Growth cone collapse through coincident loss of actin bundles and leading edge actin without actin depolymerization. *J. Cell Biol.* **153,** 1071–1084.

Zhou, F. Q., Waterman-Storer, C. M., and Cohan, C. S. (2002). Focal loss of actin bundles causes microtubule redistribution and growth cone turning. *J. Cell Biol.* **157,** 839–849.

Zou, Y., Stoeckli, E., Chen, H., and Tessier-Lavigne, M. (2000). Squeezing axons out of the gray matter: A role for slit and semaphorin proteins from midline and ventral spinal cord. *Cell* **102,** 363–375.

Index

Contents of Previous Volumes

Contents of Previous Volumes

Volume 51

Volume 52

Volume 53

Volume 54

Volume 55

6 Growth Factors and Early Development of Otic Neurons: Interactions between Intrinsic and Extrinsic Signals

Berta Alsina, Fernando Giraldez, and Isabel Varela-Nieto

7 Neurotrophic Factors during Inner Ear Development

Ulla Pirvola and Jukka Ylikoski

8 FGF Signaling in Ear Development and Innervation

Tracy J. Wright and Suzanne L. Mansour

9 The Roles of Retinoic Acid during Inner Ear Development

Raymond Romand

10 Hair Cell Development in Higher Vertebrates

Wei-Qiang Gao

11 Cell Adhesion Molecules during Inner Ear and Hair Cell Development, Including Notch and Its Ligands

Matthew W. Kelley

12 Genes Controlling the Development of the Zebrafish Inner Ear and Hair Cells

Bruce B. Riley

13 Functional Development of Hair Cells

Ruth Anne Eatock and Karen M. Hurley

14 The Cell Cycle and the Development and Regeneration of Hair Cells

Allen F. Ryan

Volume 58

1 A Role for Endogenous Electric Fields in Wound Healing

Richard Nuccitelli

2 The Role of Mitotic Checkpoint in Maintaining Genomic Stability

Song-Tao Liu, Jan M. van Deursen, and Tim J. Yen

3 The Regulation of Oocyte Maturation

Ekaterina Voronina and Gary M. Wessel

4 Stem Cells: A Promising Source of Pancreatic Islets for Transplantation in Type 1 Diabetes

Cale N. Street, Ray V. Rajotte, and Gregory S. Korbutt

Volume 62

Volume 63

Volume 65

Volume 66

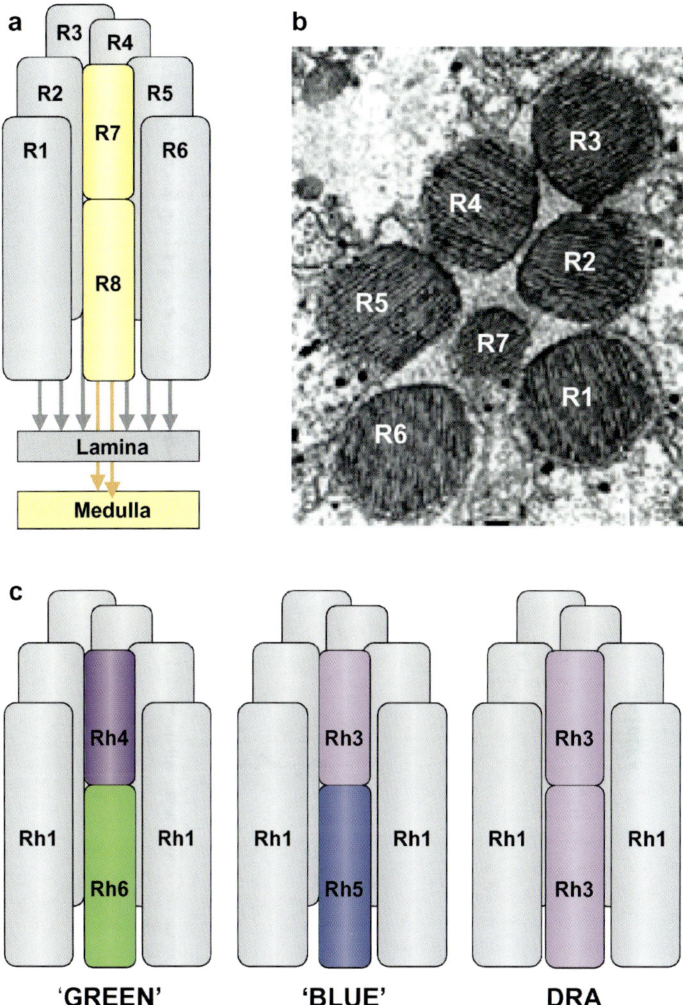

Chapter 1, Figure 1 The three subtypes of ommatidia present in the fly retina. (a): schematic representation of the position, morphology and axonal projection of the outer photoreceptors (R1 to R6) and of the inner photoreceptors (R7 and R8). (b): electron micrograph of a cross-section through an ommatidium. (c): schematic representation of the three ommatidial subtypes present in the retina. In the 'green' subtype (left) R7 expresses UV-*rh4* and R8 Green-*rh6*; in the 'blue' subtype (center) R7 expresses UV-*rh3* and R8 Blue-*rh5*, in the Dorsal Rim Area ommatidia (DRA, right) both R7 and R8 express *rh3*.

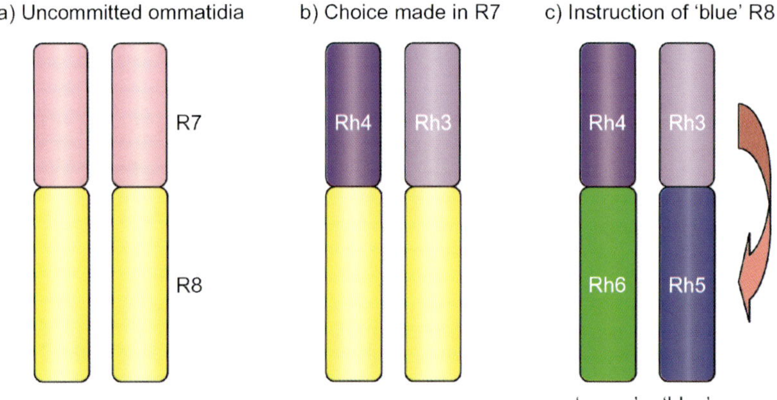

Chapter 1, Figure 2   A two-step model for the cell fate decision 'green' *versus* 'blue' in inner photoreceptors. (a): in ommatidia that have not yet committed to the 'green' or 'blue' fate, R7 and R8 do not express any rhodopsin. (b): a stochastic event in R7 induces expression of either *rh4* or *rh3*. (c): an *rh3* expressing R7 cell sends a signal to the underlying R8 cell inducing *rh5* expression. In the absence of signal (*i.e.*, when R7 expresses *rh4*), the R8 cell expresses *rh6*.

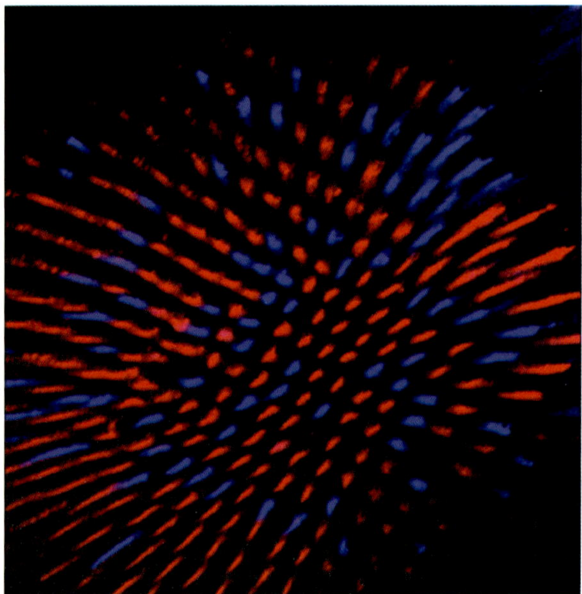

Chapter 1, Figure 3   Stochastic distribution of the 'green' and 'blue' subtypes in the retina. Confocal image of a wild type whole mount retina stained with anti-Rh5 in blue ('blue' subtype) and anti-Rh6 in red ('green' subtype). No pattern or rule can be found in the distribution of the two subtypes.

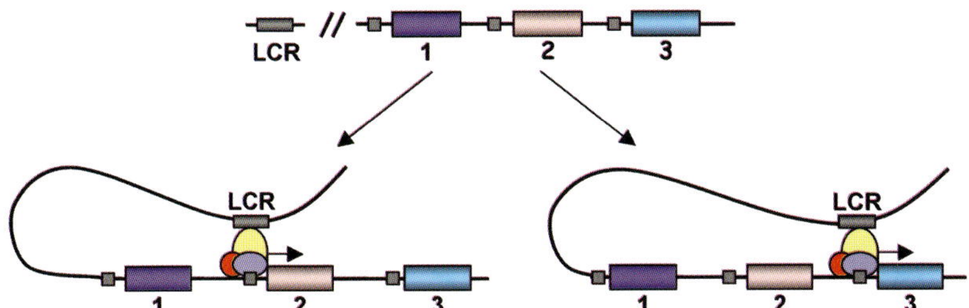

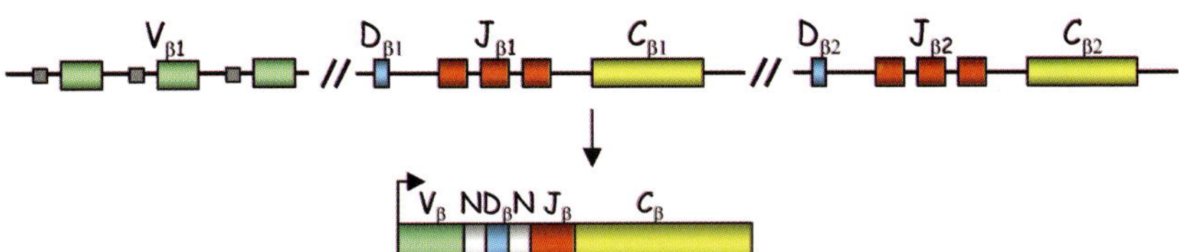

Chapter 1, Figure 4   How to choose one gene from many and how to make many variants from one gene. (a): schematic drawing of a choice involving a Locus Control Region model. Though a looping mechanism, the LCR can contact only one promoter at a time, inducing the expression of only one gene in a cluster. (b): Schematic drawing of the recombination model in the immune system. The V (Variable), D (Diversity), J (Junction) and C (constant) segments that compose the β chain of the T cell antigen receptor are brought together by DNA rearrangements. Nucleotide addiction and deletion (N) in the joining region further increases variability.

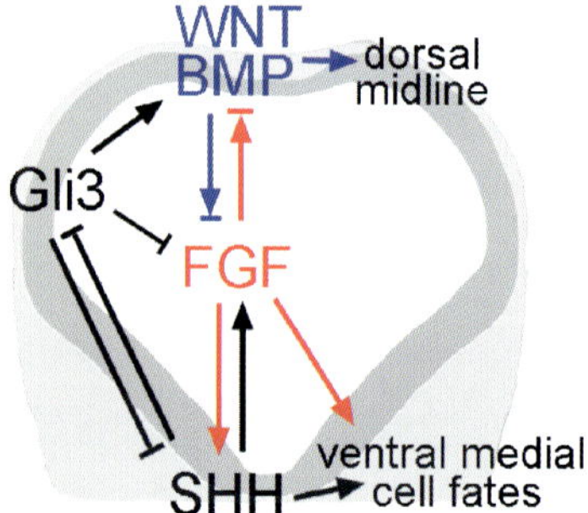

Chapter 2, Figure 2 Model of the interactions between SHH, FGF, and BMP/WNT in forming the telencephalic midline. Schematic of a coronal section through the early telencephalon with dorsal up. Blue and red arrows and bars represent hypothesized roles for BMP/Wnt and FGF signaling, respectively. BMPs and FGFs are hypothesized to interact antagonistically or protagonistically depending on the stages of development and their levels of expression (an early threshold level of BMP is required to induce FGF8 in the ANB [Fig. 1], whereas BMPs later appear to repress FGFs; conversely, FGFs regulate BMPs in a dose-dependent manner via *Foxg1*, not shown, with low and high levels inhibiting and promoting BMPs, respectively). SHH is required to maintain expression of at least one Fgf gene, *Fgf8*, and is required to specify ventral medial cell fates. *Gli3* has been implicated in regulating the function of all three signaling centers.

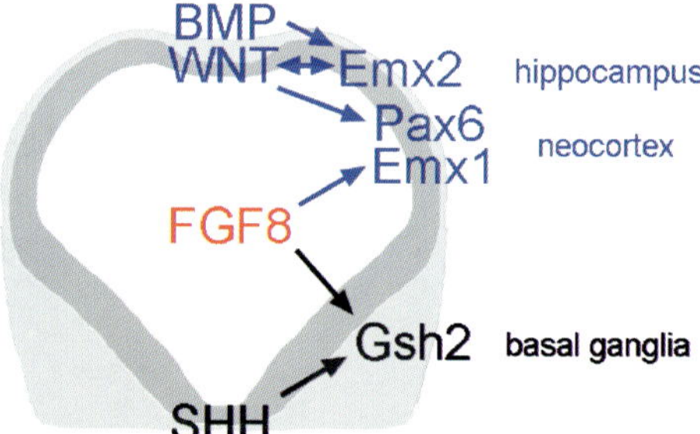

Chapter 2, Figure 3 Model of the likely interactions between certain dorsalizing (blue) and ventralizing (black) factors that generate the major subdivisions of the telencephalon. FGFs (red) may affect both dorsal and ventral processes. Interactions between the transcription factors themselves have been omitted, and only the interactions between midline-associated secreted factors and transcription factors are illustrated.

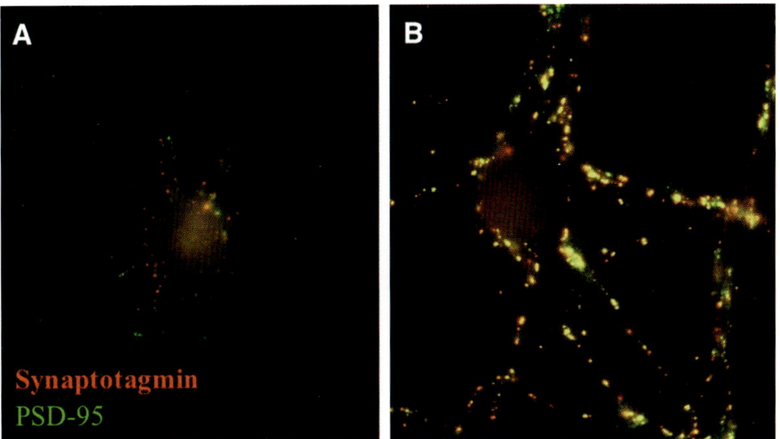

Chapter 3, Figure 1 Glia promote synaptogenesis. (A) Retinal ganglion cells (RGCs) cultured in the absence of glia, stained using anti-synaptotagmin (a pre-synaptic marker; red) and anti-PSD-95 (a postsynaptic marker; green). Note few yellow puncta. (B) RGCs cultured in the presence of glia. Note large increase in yellow puncta. (Image courtesy of Erik Ullian and Ben Barres.)

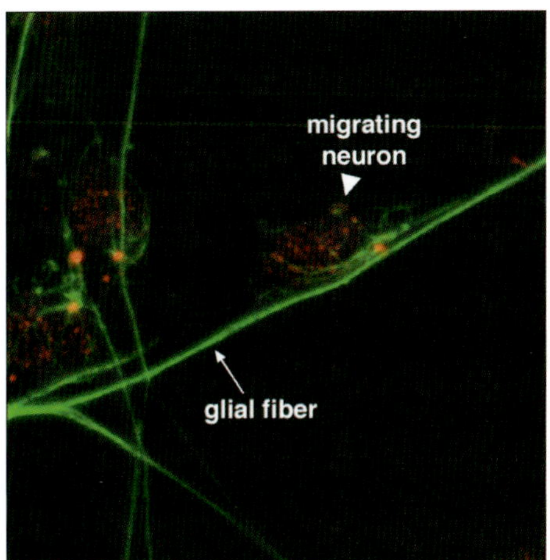

Chapter 3, Figure 2 Cerebellar granule cell migrating on a Bergmann glia fiber. Green, ß-tubulin in a neuron, marking the glial fiber track; red, dynein intermediate chain in the nucleus and centrosome of the migrating neuron. (Image courtesy of David Solecki and Mary Beth Hatten.)

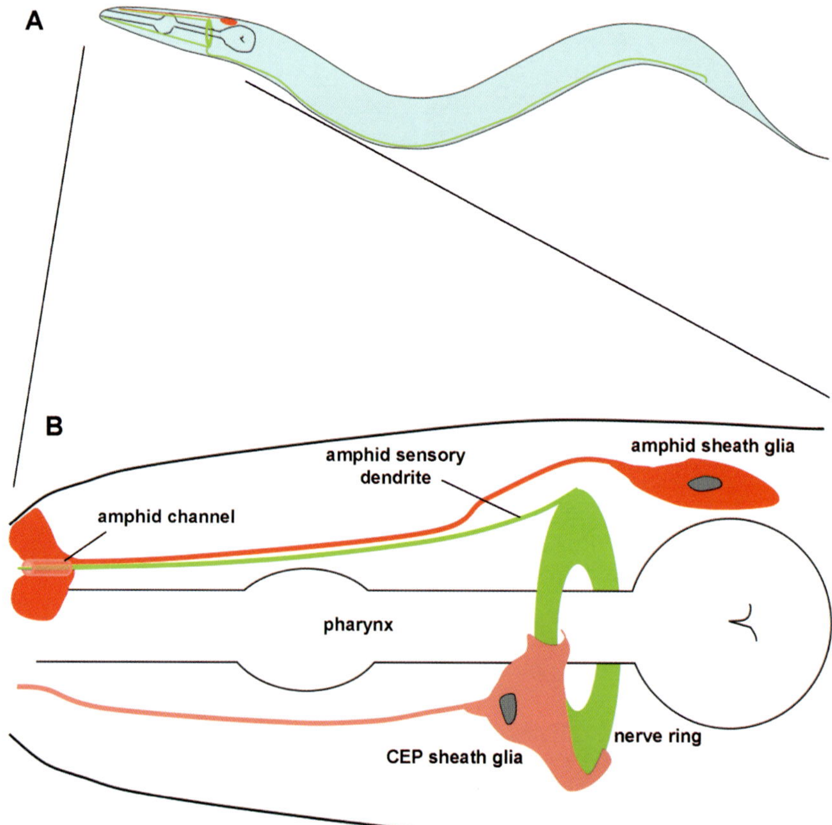

Chapter 3, Figure 3   Glia in *C. elegans*. (A) Schematic of a *C. elegans* adult hermaphrodite. Anterior, left; dorsal, up. Major neural tracts (green) and an amphid sheath glia (red) are depicted. The outline of the pharynx of the animal is also shown. (B) Enlarged view of the anterior region. The nerve ring and an amphid channel neuron dendrite (green), the amphid sheath glia and channel (red), and the CEP sheath glia (pink) are depicted.

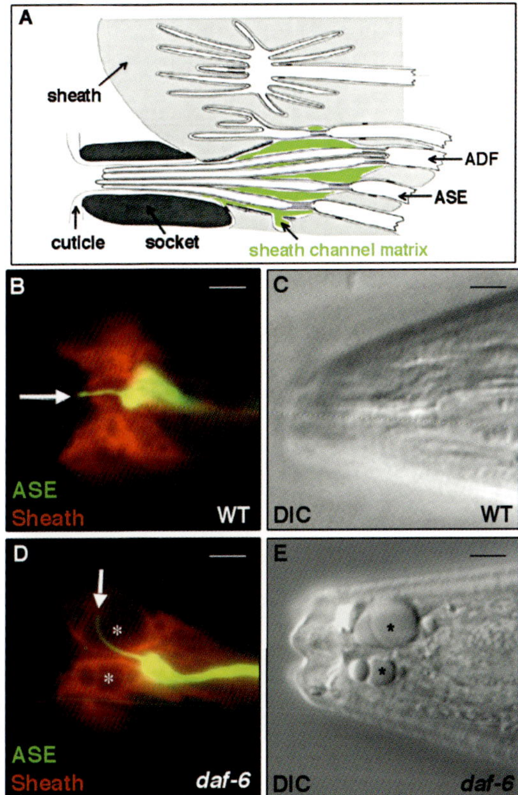

Chapter 3, Figure 4   The glial channel of the *C. elegans* amphid sensory organ is closed in *daf-6* mutants. (A) Schematic showing the dendritic tip region of the amphid (adapted from Ward *et al.*, 1975). A representative neuron embedded within the sheath glia is shown. Two channel neurons, ADF and ASE, are labeled, as are the sheath and socket glia. The sheath secretes a matrix into the channel (green). The socket glia secrete cuticle, which is contiguous with the cuticle on the animal's exterior. This image is an enlarged view of the anterior of Figure 3B. (B) Fluorescence image of a wild-type amphid. Sheath (red) and ASE channel neuron (green) are shown. Arrow points to ASE cilium in the amphid channel. (C) Differential interference contrast (DIC) image of animal in (B). (D) Fluorescence image of a *daf-6* mutant amphid. Note absence of exposed channel. (E) DIC image of animal in (D). Asterisks indicate vacuoles accumulating within the sheath glia.

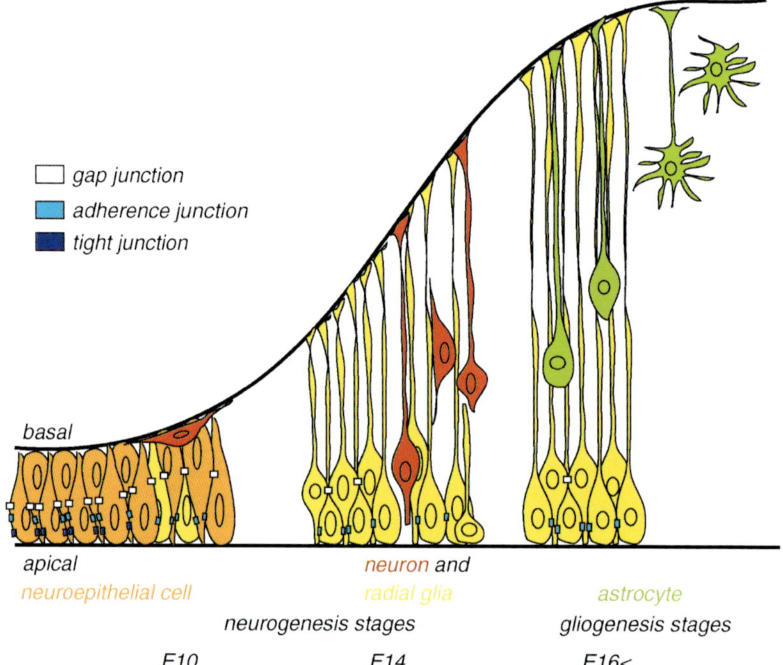

Chapter 4, Figure 1 Schematic drawing of the developing mouse cerebral cortex. At early stages, all precursors are neuroepithelial cells depicted in orange to the left. Radial glial cells (yellow) appear around the onset of neurogenesis and exhibit many cell biological or molecular similarities with astrocytes. Postmitotic neurons are depicted in red, and the schematic drawing reflects the observation that when the first neurons appear some neuroepithelial precursors start to acquire the first radial glial features. At this time tight junctions (dark blue) are converted to adherens junctions (light blue). Radial glial cells generate neurons that either migrate by somal translocation, with the basal process attached to the basal surface, or they migrate basally along radial glial cells as depicted in the drawing. At the end of neurogenesis, radial glia transform into astrocytes (green). Tight junctions, adherence junctions, and gap junctions are dark blue, light blue, and white, respectively.

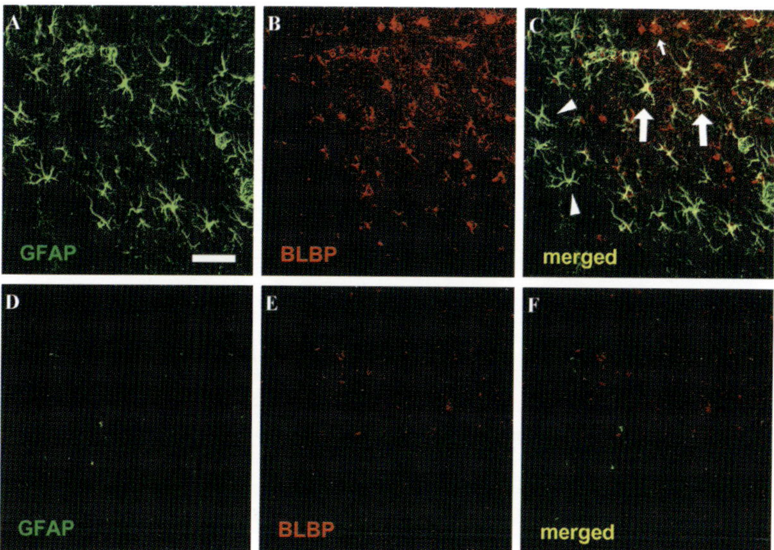

Chapter 4, Figure 2   Heterogeneity of reactive astroglia of the adult cerebral cortex. Three days after a stab-wound lesion, strong GFAP and BLBP immunoreactivity is detectable in reactive astrocytes of the hemisphere subjected to the lesion (A, B, C), but not in astrocytes of the intact cortical parenchyma in the other hemisphere (D, E, F). Interestingly, the astroglia response to lesion is not homogeneous, as cells that express GFAP (arrowheads) or BLBP (small arrow) only can be identified, whereas others co-express the examined markers (large arrows). Scale bar: 50μm.

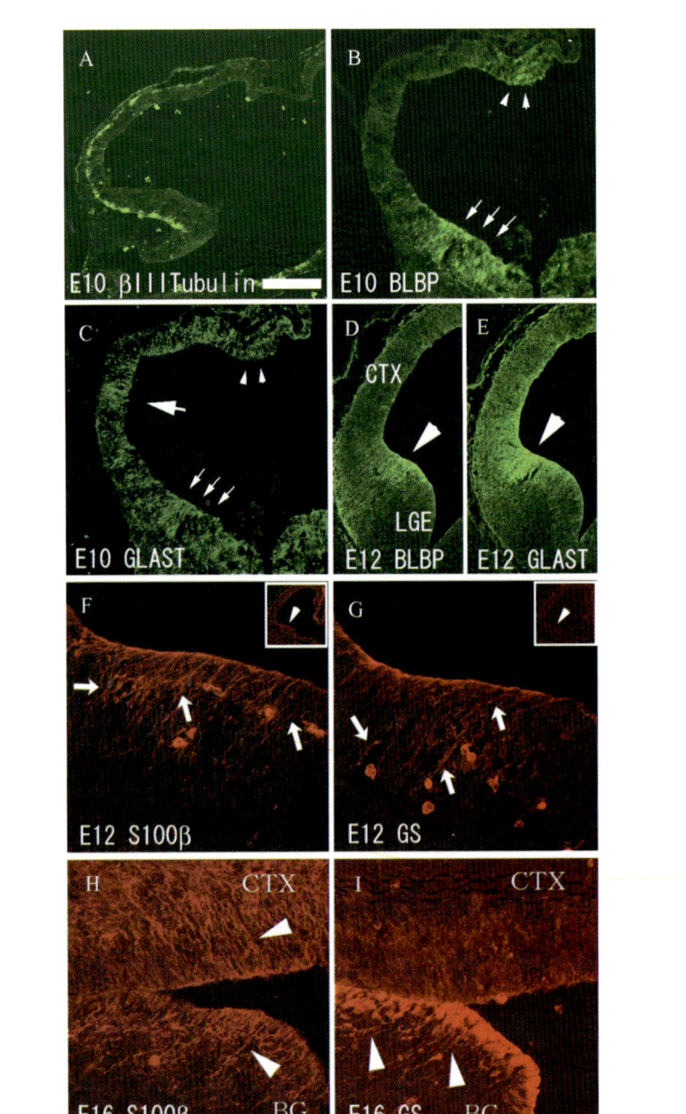

A
E10 βIIITubulin
B
E10 BLBP
C
E10 GLAST
D
CTX
LGE
E12 BLBP
E
E12 GLAST
F
E12 S100β
G
E12 GS
H
CTX
BG
E16 S100β
I
CTX
BG
E16 GS

Chapter 4, Figure 3 Astrocyte markers in radial glia. Fluorescent micrographs of coronal sections through the embryonic mouse cortex are depicted. At E10, postmitotic neurons immunolabeled for βIII-tubulin are already detectable in the basal part of the telencephalon (green cells in A). At this stage, BLBP (B) and GLAST (C) immunoreactivity become detectable in subsets of radial glia. Small arrows in (B) and (C) indicate the basal telencephalon; arrowheads indicate the cortical hem. Some radial glial cells express GLAST in the middle part of the telencephalon (large arrow in C). Panels (D) and (E) depict BLBP (D) and GLAST (E) immunoreactivity in the dorsal-most part of the lateral ganglionic eminence, the border between dorsal and ventral telencephalon, at E12 [arrowhead in (D) and (E)]. S100β (F) and GS (G) immunoreactivity starts to be weakly detectable at E12; arrows indicate radial glial processes. The insets in (F) and (G) show a low-power view of the telencephalon, and the arrowhead indicates where the high-power view is taken (at the border between the dorsal and the ventral telencephalon). Note that GS immunoreactivity is strongly upregulated at E16 with stronger signal in the ventral telencephalon generating the basal ganglia (BG) than the dorsal telencephalon, the future cortex (CTX). Arrowheads in (H) and (I) indicate radial glial processes. CTX; cortex, LGE; lateral ganglionic eminence, BG; basal ganglia. Scale bar: 200 μm (A, B, C, D, E); 50 μm (F, G, H, I).

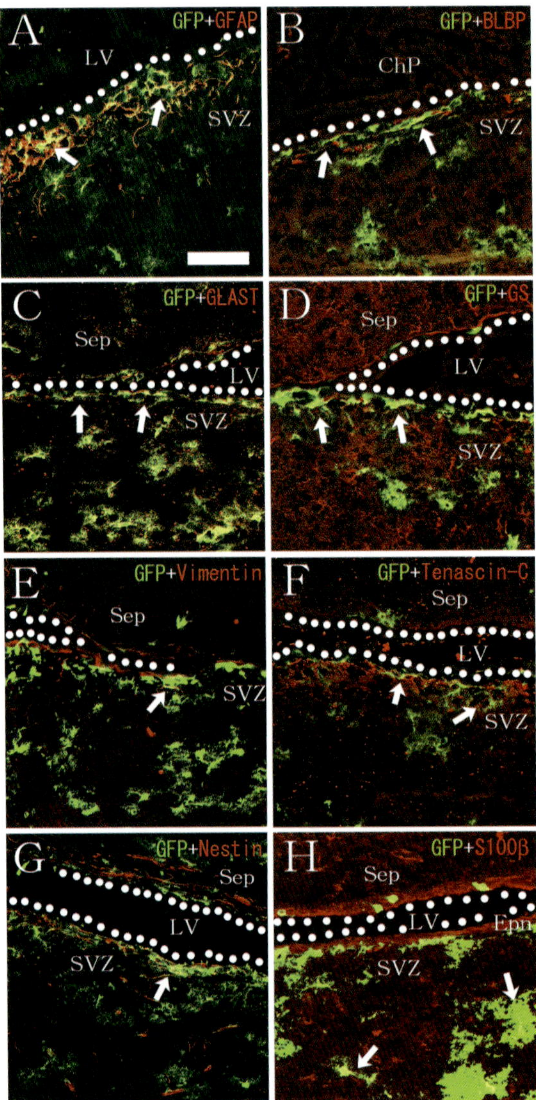

Chapter 4, Figure 5 Adult SVZ cells express astrocyte markers. The panels depict single optical sections of confocal microscope micrographs taken from coronal sections of adult SVZ zone lining the lateral ventricle in transgenic mice expressing GFP under control of the human GFAP promoter. As indicated in the panels, GFP is depicted in green and the respective astrocyte markers are in red. Colocalization gives rise to yellow as indicated by arrows. The ventricle is indicated by the dotted lines. Note that GFP-positive astrocytes in the adult SVZ are also GFAP- (A), BLBP- (B), GLAST- (C), and GS- (D) immunopositive. Subsets of GFP-positive astrocytes were double-stained with antiserum directed against vimentin (E), Tenascin-C (F), or nestin (G), while S100β immunoreactivity was detected in ependymal cells and parenchymal astrocytes, but hardly in SVZ astrocytes (H). ChP, choroid plexus; LV, lateral ventricle; Sep, septum; SVZ, subventricular zone. Scale bar: 50 μm.

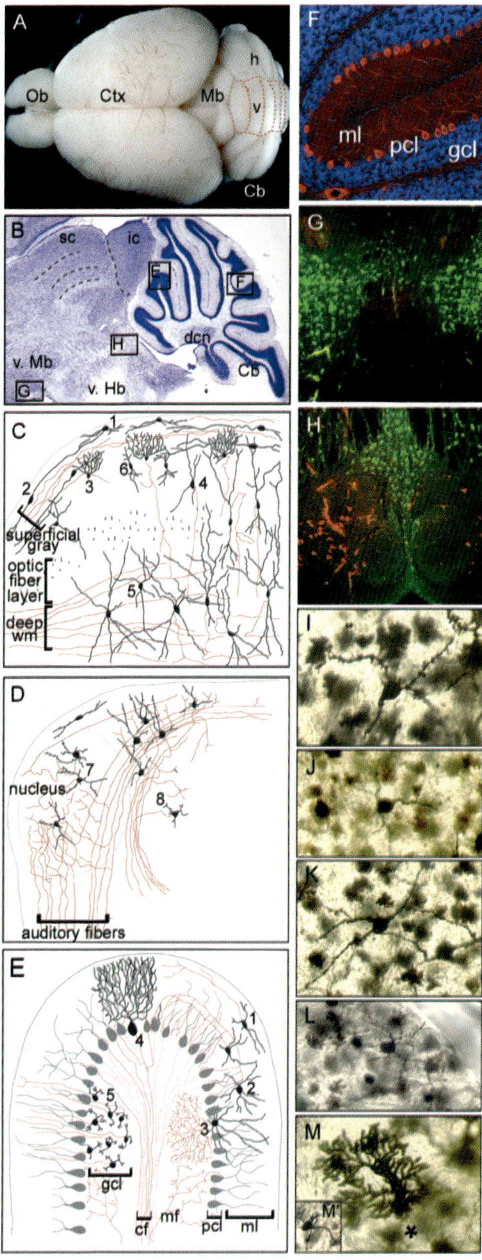

Chapter 5, Figure 1 *Continued*

Chapter 5, Figure 1 Midbrain and cerebellum morphology and cell types in the adult mouse. (A) The adult mouse brain showing the olfactory bulbs (Ob), cerebral cortex (Ctx), midbrain (Mb), as well as vermis (v) and hemispheres (h) of the cerebellum (Cb). (B) Nissl-stained sagittal section showing the superior (sc) and inferior (ic) colliculus of the dorsal Mb and Cb; also shown is the ventral Mb (v. Mb), hindbrain (v. Hb), and the deep cerebellar nuclei (dcn). The lettered marques indicate regions shown in illustrations or at higher magnification. (C–E) Semi-diagrammatic illustrations of the superior and inferior colliculi and the Cb as modified from Ramon y Cajal (Ramon y Cajal, 1995). (C, D) A diverse array of cell morphologies and the highly laminated versus globular cytoarchitecture of the sc (C) and ic (D) can be seen in these coronal views. Cell types: 1, marginal cell; 2, horizontal fusiform cell; 3, cell with complex dendritic bouquet; 4, large vertical fusiform cell; 5, large cells in transverse fiber layer; 6, radial fusiform cell; 7, spine laden triangular cells in ic nucleus; 8, multipolar cells in the central gray area. (E) The Cb folia displays a laminar arrangement and contain the following cell types: 1, stellate cells in molecular layer (ml); 2, basket cells in the molecular layer; 3, Golgi cells in Purkinje cell layer (pcl); 4, Purkinje cells in the pcl; 5, granule cells in granule cell layer (gcl); cf, climbing fibers; mf, mossy fibers. Note: solid and dashed red lines in cf region indicate cfs and Purkinje cell axons, respectively. (F) Calbindin-immunoreactive (IR, red) Purkinje cells in the Cb folium with dendrites in the ml; the gcl contains densely packed granule cells that can be observed with Hoechst staining (blue). (G, H) Horizontal section of v. Mb (G) or v. Hb (H). (G) Tyrosine hydroxylase-IR dopaminergic neurons of the v. Mb. (H) 5-hydroxy tryptophan-IR serotonergic (green) and choline acetyl transferase-IR cholinergic neurons (red) of the v. Hb. (I–M) Golgi impregnated neurons from adult mouse brain regions: (I) large triangular neuron with bulbous spines from the deep layer of sc; (J) small triangular neuron with few spines from superficial gray layer of ic; (K) pyramidal neuron from the substantia nigra of the v. Mb; (L) stellate cell from the Cb ml; (M, M′) Purkinje cell dendritic arbor (* indicates cell body that is out of focal plane) and granule cell, respectively. All cells in I-M were obtained at the same magnification for direct comparison.

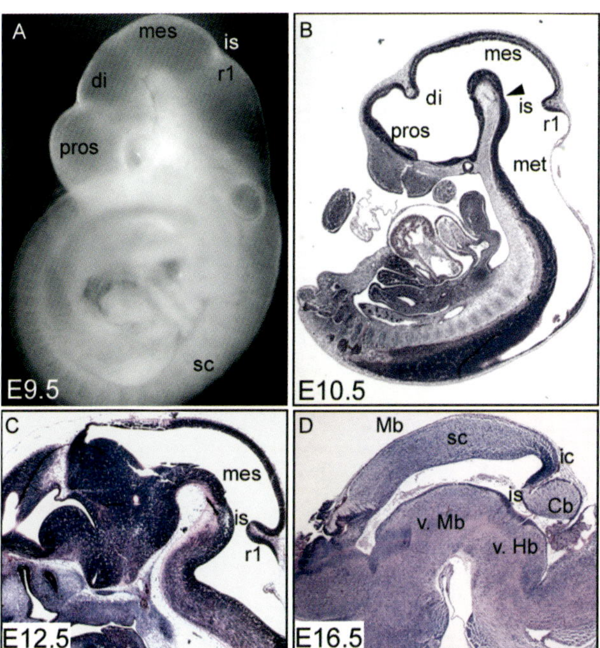

Chapter 5, Figure 2   Mes/r1 morphogenesis in the developing mouse embryo. (A) E9.5 embryo showing the prosencephalon (pros), diencephalon (di), mesencephalon (mes), isthmus (is), rhombomere 1 (r1), and spinal cord (sc). (B–D) Hematoxylin/eosin-stained sagittal sections of an E10.5 embryo (B), E12.5 head (C), and E16.5 brain (D). Additional abbreviations: ic, inferior colliculus; sc, superior colliculus; v. Mb, ventral midbrain; v. Hb, ventral hindbrain.

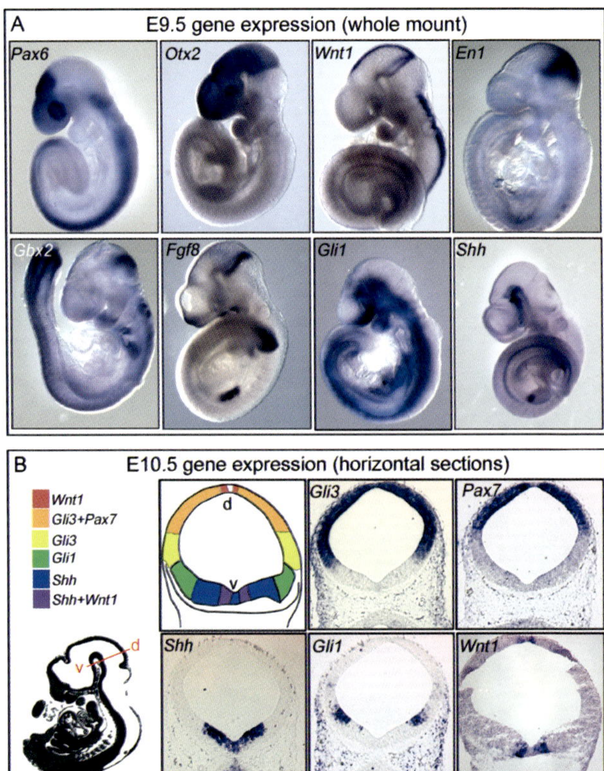

Chapter 5, Figure 3   Gene expression in mes/r1 at embryonic stages. (A) Whole mount *in situ* hybridization with the indicated probes showing gene expression patterns at E9.5. *Pax6* is expressed in the prosencephalon, rhombomeres 2–7, and spinal cord. *Otx2* is expressed throughout the pros and mes. *Wnt1* is expressed in a semi-circle at the posterior limit of the mes and in a row at the dorsal and ventral midline of the diencephalon and mes as well as in r2–7 and sc. *En1* traverses both the posterior mes and r1. *Gbx2* is localized to r1 and the tail bud. *Fgf8* is expressed in signaling centers including r1 in a semi-circular pattern and in the anterior neural ridge, limb apical ectodermal ridge, branchial arches, and tail bud. *Gli1* is expressed adjacent to the floor plate and extends along the entire A-P axis. *Shh* is expressed in the floor plate at the ventral midline and becomes broader in the mes/r1 region. Figure 2A shows anatomical subdivisions and Figure 4B shows mes/r1 spatial relationships. (B) Gene expression along dorsal-ventral axes at E10.5. *In situ* hybridization on horizontal sections as shown by the red line through the embryo on the left. Dorsal (d) is at the top, ventral (v) at the bottom. The spatial relationship of the indicated genes is shown in the schematic in the upper left panel.

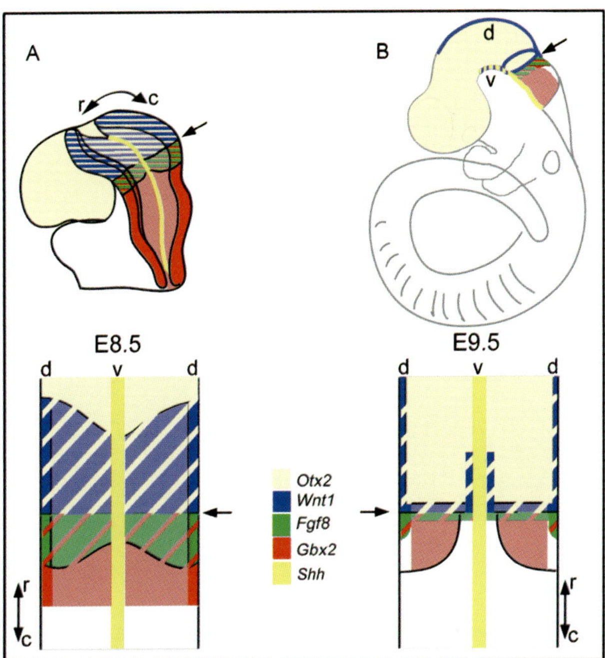

Chapter 5, Figure 4 Illustrations of mouse embryos at E8.5 and E9.5 highlighting changes in morphology and gene expression patterns. Drawings depict whole mount embryos (top) or flat mount schematic (bottom). (A) Overlapping gene expression is shown at E8.5 when the neural tube is open; only the anterior region of the embryo is drawn. The bottom schematic shows the neural tube spread flat. The *Shh* domain is located in the midline (labeled as ventral, v) while *Wnt1* is strongest in the lateral neuroepithelium (labeled as dorsal, d) to indicate the M-L transition to D-V once the neural tube closes. *Otx2* and *Wnt1* are expressed in the mes versus *Gbx2* and *Fgf8*, which are expressed in r1; *Shh* is expressed in the floor plate. The rostral-caudal axis (r-c) is indicated; the arrow indicates the mes/r1 interface. (B) Shortly after neural tube closure, gene expression undergoes a dynamic change. *Wnt1* becomes restricted to a semi-circle at the posterior mes as well as the dorsal and ventral midline. *Fgf8* also becomes restricted to a semi-circle in the isthmus that is posterior and juxtaposed to *Wnt1*. The flat mount illustration depicts the neural tube cut along the dorsal midline; ventral and dorsal are therefore illustrated medially and laterally, respectively. This facilitates a direct comparison of the gene expression domains at E9.5 to E8.5.

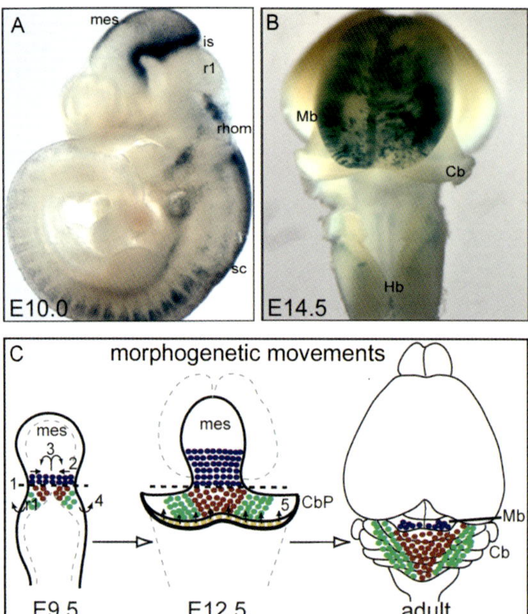

Chapter 5, Figure 5 Morphogenetic movements in Mb and Cb development. (A) The initial population of *Wnt1*-derived cells marked by tamoxifen administration to Wnt1-CreERT; R26R mouse embryos at E8.5 is shown at E10.0. In this sagittal view, marked cells can be seen in the mes, but not in r1. (B) By E14.5 the mes and r1 have undergone growth and a dramatic change in morphology (compare A to B) to give rise to the Mb and Cb, respectively. During this time *Wnt1*-derived cells marked by tamoxifen at E8.5 are retained in mes and cannot migrate into r1, resulting in the Mb being marked, but not the Cb. (C) Schematic of dorsal views illustrating complex morphogenetic movements underlying Mb and Cb development. E9.5: A lineage boundary restricts mes-derived cells (blue circles) from posterior movement into r1 (1); mes-derived cells migrate from lateral to medial (2); but primarily expand anteriorly (3); r1 begins to undergo an orthogonal rotation (4). E12.5: Mes-derived cells have expanded to fill the posterior mes and r1 has rotated such that anterior r1 (red circles) becomes medial while posterior r1 (green circles) becomes lateral; granule cell precursors from the upper rhombic lip (orange circles) migrate over the surface of the Cb primordium (CbP) (5). Adult: The final distribution of the indicated cells is shown in the adult; granule cells have settled deep within the Cb during early postnatal development and are not shown (see Section III in text for details).

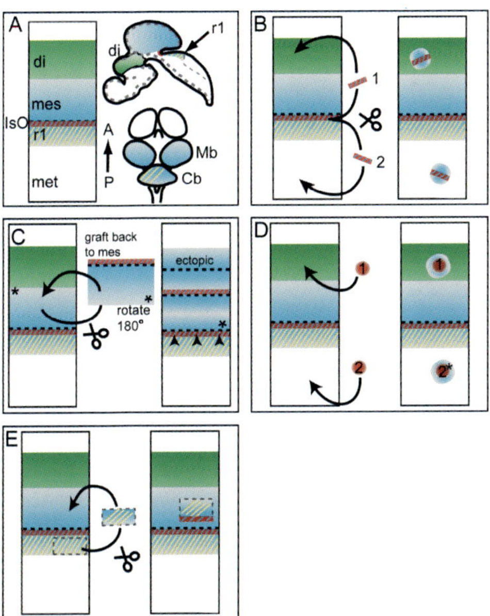

Chapter 5, Figure 6   Schematic of IsO function in A-P patterning. (A) Schematic illustrating *Pax6* in the diencephalon (di) (green), *En2* in the mes/isthmus/r1 (blue), and *Fgf8* in the isthmus (red); r1 is indicated by yellow hatched lines and the isthmus constriction is depicted by black dashed line. Note that *En2* is expressed in two opposing gradients with highest levels at the isthmus and fading into the mes anteriorly and into r1 posteriorly. *Fgf8* is nested in the *En2* domain in the isthmus organizer (IsO). The chick brain is depicted at HH24 (top line drawing) and in the adult (bottom line drawing). (B) Ectopically transplanting small pieces of isthmus tissue into either diencephalon (1) or caudal metencephalon (met, 2) induces an ectopic *En2* gradient in host. The grafted isthmus induced host diencephalon to become an ectopic Mb and induces host metencephalon to become an ectopic Cb. (C) Removing the mes, rotating it 180°, and transplanting it back into mes region initially reverses the *En2* gradient in the mes, which is quickly reverted to a normal *En2* gradient. (* is for reference of graft orientation); presumably an *Fgf8* domain is re-established at the mes/r1 interface (arrowheads). The isthmus (*Fgf8*) and *En2* gradient of the transplanted graft is presumptively maintained in its new location. An ectopic *En2* gradient (ectopic) is established in the diencephalon and isthmus (black dashed line, ectopic). The presence of two organizers in the mes (*Fgf8*) results in the formation of a bi-caudal Mb. (D) Fgf8-soaked beads placed into the posterior diencephalon (1) induced En2 expression and an ectopic Mb, but when placed into metencephalon (2) did not induce an ectopic Cb (2*) although *En2* was induced; this is unlike isthmus transplants into metencephalon (B, 2). (E) Transplanting r1 into the mes results in the formation of a new mes/r1 interface complete with the induction of *Fgf8* and *En2*.

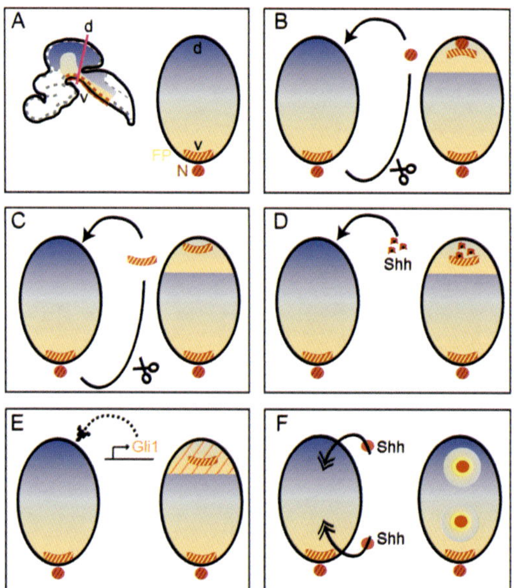

Chapter 5, Figure 7 Schematic of organizer function in D-V patterning. (A) Schematic drawing of an embryonic brain (left) and a simplified cross-section (right) through the mes as depicted by the pink line. Dorsal (d) is at the top, ventral (v) at the bottom. The dorsal or alar plate of the mes/r1 is defined by the purple gradient, the ventral or basal plate by the orange gradient. Shh expression domains (red): notochord (N, dark red) and the floor plate (FP, yellow). Transplants of N (B) and FP (C), grafts of Shh-expressing cells (D), or ectopic expression of Gli1 in the dorsal mes (E) induce an ectopic floor plate as well as ventral markers and cell types. (F) Precisely localized expression of Shh by electroporation in either dorsal or ventral-lateral mes induces the graded expression of transcription factors normally observed along the D-V axis around the ectopic source of Shh.

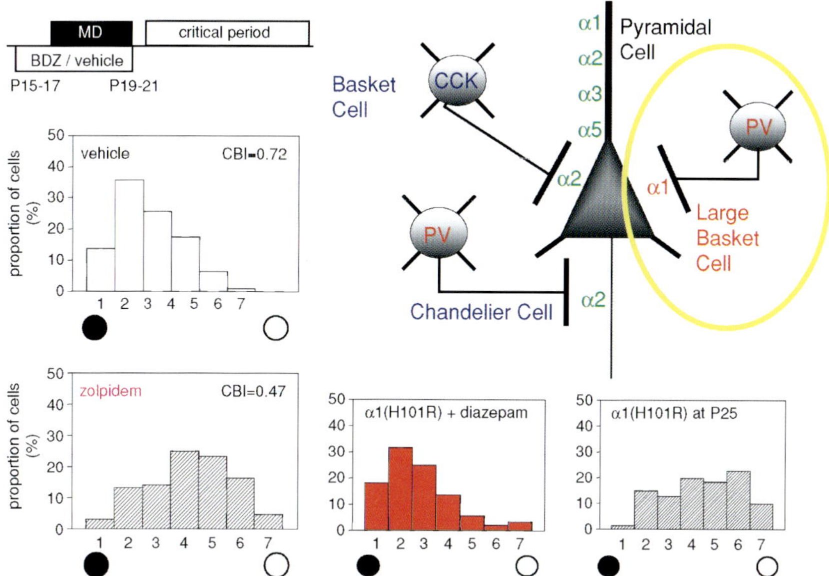

Chapter 8, Figure 3 An essential inhibitory subcircuit for critical period plasticity in the visual cortex. Premature ocular dominance plasticity is triggered by the $GABA_A$ receptor α1-subunit-selective benzodiazepine agonist zolpidem. Large parvalbumin (PV)-positive basket cells make somatic synapses that utilize $GABA_A$ receptors containing the α1-subunit. Knock-in of a point mutation rendering only the α1-receptors insensitive to diazepam prevents critical period acceleration by these drugs (red bars). Note that plasticity emerges naturally at the proper time (P25, black bars), since these are still functional GABA receptors. Point mutation of other α-subunits does not interfere with drug-induced premature plasticity. Basket cells extend a wide, horizontal axonal plexus across ocular dominance columns in cats ideally suited for comparing input from the two eyes (Buzas *et al.*, 2001).

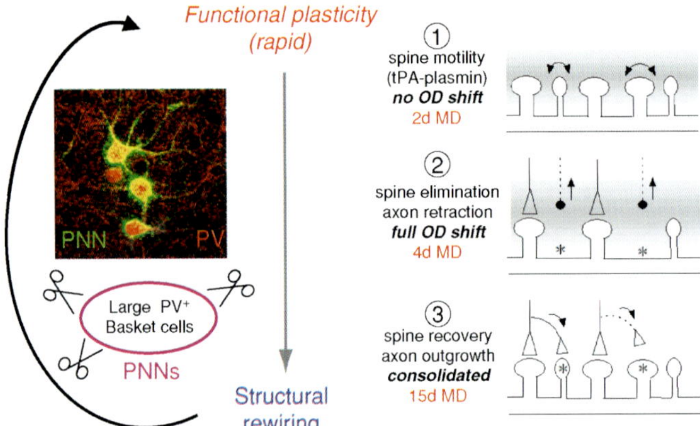

Chapter 8, Figure 4   Structural consolidation during the critical period. The structural events that link the functional detection of imbalanced sensory input by GABAergic circuits to anatomical rewiring. A three-step process of increased spine motility (Oray *et al.*, 2004), transient elimination, then regrowth (Mataga *et al.*, 2004) is mediated by a biochemical increase of proteolytic activity (tPA-plasmin; gray background) between 2 and 7 days of monocular deprivation (MD) (Mataga *et al.*, 2002). Spine pruning is the first anatomical correlate of the rapid physiological shifts in ocular dominance (OD) by brief MD. Taking this structural view, plasticity is successfully restored to adult visual cortex only by loosening up the extracellular matrix (ECM) by infusion of chondroitinases (left; Pizzorusso *et al.*, 2002). Interestingly, this treatment (unlike tPA) degrades peri-neuronal net (PNN) structures, which preferentially enwrap the large PV-positive basket cells believed to be the endogenous trigger for the critical period (see text).

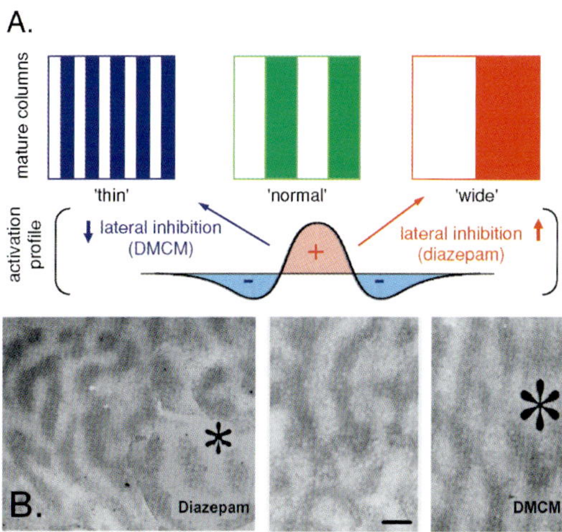

Chapter 8, Figure 6 Local circuit control of developing columnar architecture. Activity-dependent models of segregation predict a role for cortical GABAergic circuits in determining final column spacing from an initially overlapping mosaic of afferents (Miller *et al.*, 1989). (A) Neuronal activity from thalamic input serving the right or left eye is spread by local excitatory connections (red cell) within the neocortex but inhibited at farther distances (blue cell). When this "Mexican hat" activation profile is modulated during development by enhancing or reducing horizontal, long-range inhibition preferentially (Hensch and Stryker, 2004), columns emerge that are wider or thinner than normal, respectively. (B) This hypothesis was verified *in vivo* by modulating $GABA_A$ currents with benzodiazepine agonists (diazepam) or inverse agonists (DMCM) throughout the critical period (Hensch and Stryker, 2004). Asterisks, infusion sites; control, middle panel. Scale bar: 1 mm.

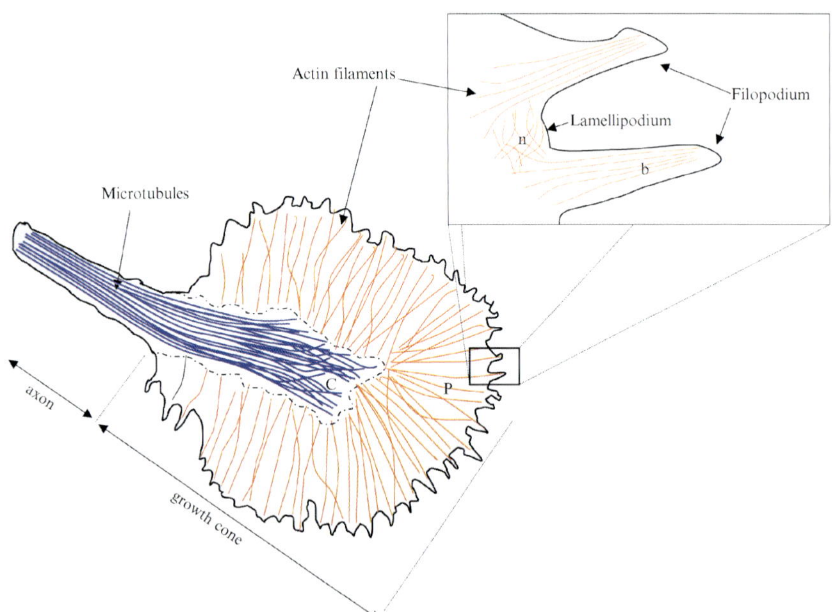

Chapter 10, Figure 1   Schematic of the growth cone. The growing tip of the axon consists of a central domain **C** and a peripheral domain **P** with filopodium and lamellopodium. The C domain contains a dense network of microtubules, but some microtubules can also be found in the base of filipodia. Actin filaments are predominant in the P domain. Filopodium move like fingers exploring their environment. These movements are based on actin polymerization and depolymerization. Filopodial actin is organized into bundles (**b**) whereas actin filaments in lamellopodia and in the C region form an intricate network (**n**) (see inset).

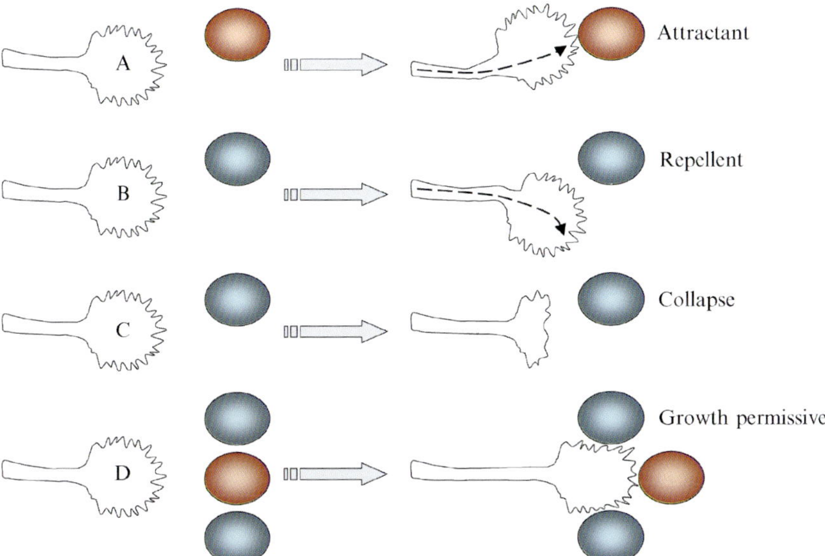

Chapter 10, Figure 2   Growth cone behavior. The growth cone is growing toward an attractive molecule (A) or is turning away from a repulsive molecule (B). Some molecules can produce a total collapse of the growth cone (C). Some substrates are growth permissive; the growth cone grows into it with repellent cues on either side (D).

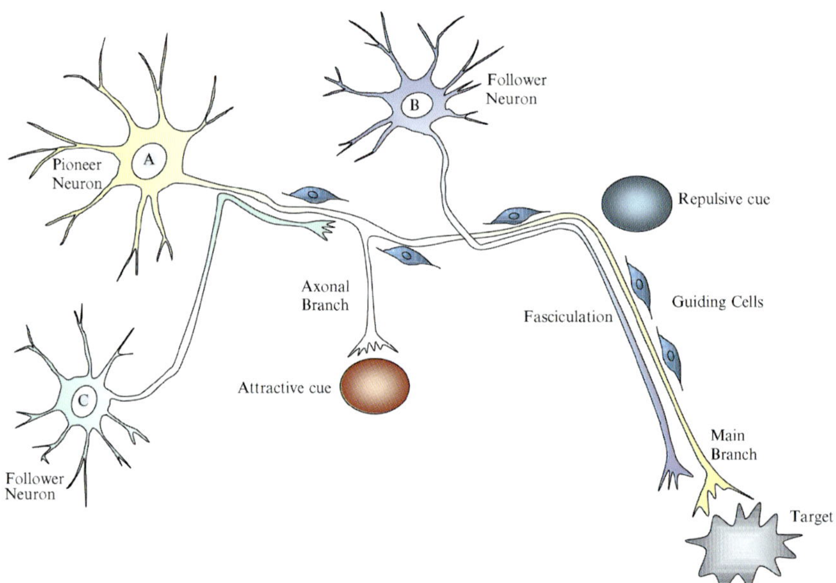

Chapter 10, Figure 3   The pioneers. Axon of the pioneer neuron (A) grows along the pathway formed by the presence of guidepost cells as well attractive and repulsive molecules. An axonal branch can form and grow toward an attractive molecule. This axonal branch may become a stable axon, resulting in the retraction of the main branch. As pioneer neurons establish their route, follower neurons (B and C) extend their neurites. Axons of the follower neurons fasciculate with the pioneer to project to the appropriate target.

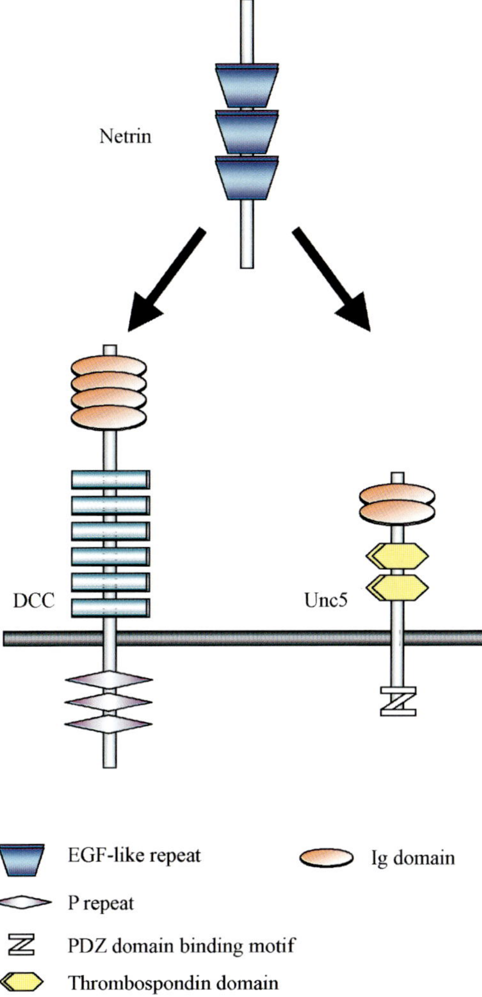

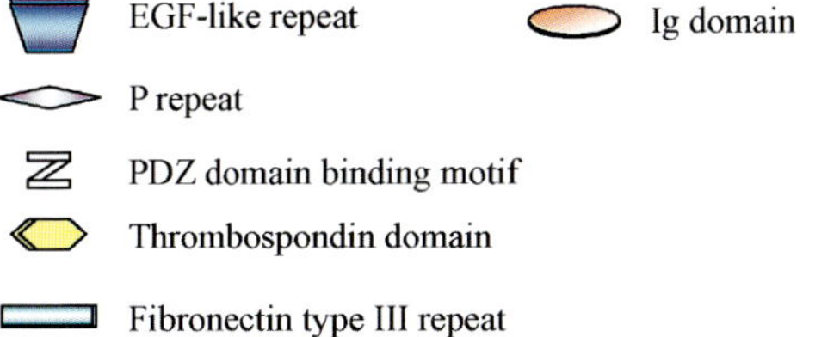

Chapter 10, Figure 4   Netrins and DCC.

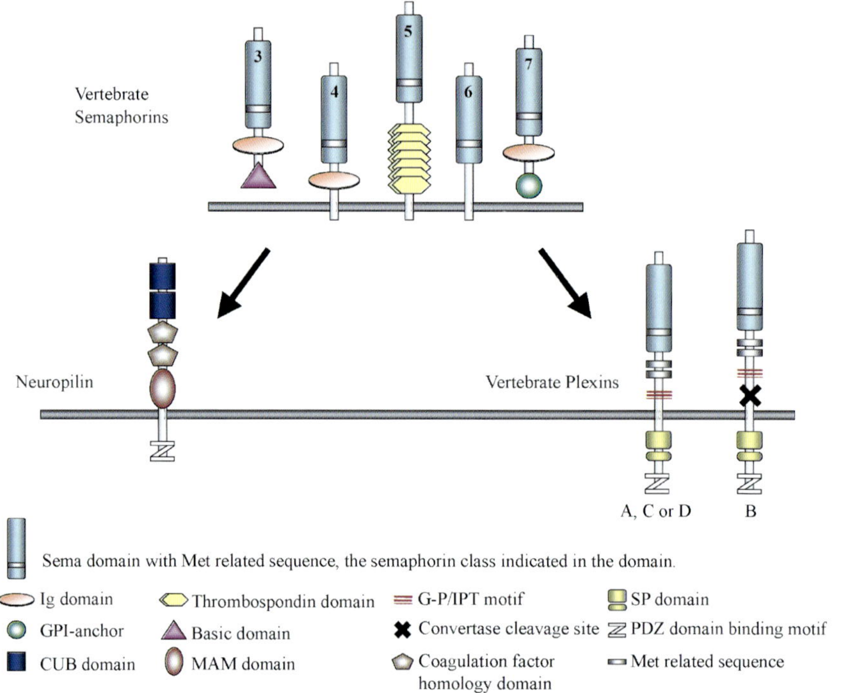

Chapter 10, Figure 5   Semaphorins, Neuropilins, and Plexins.

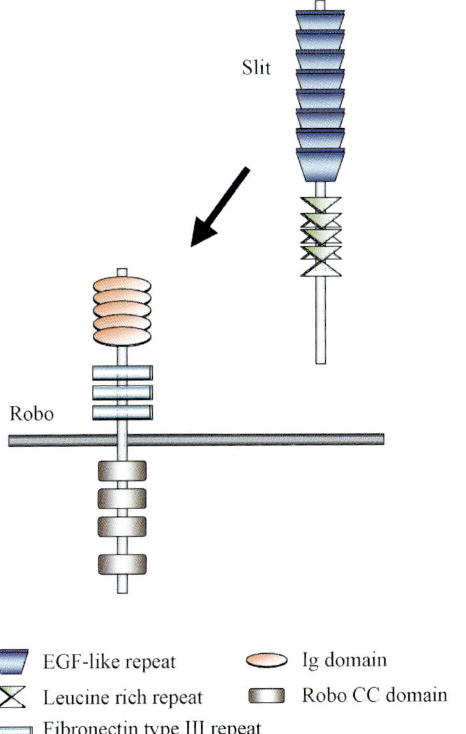

Chapter 10, Figure 6   Slits and Robos.

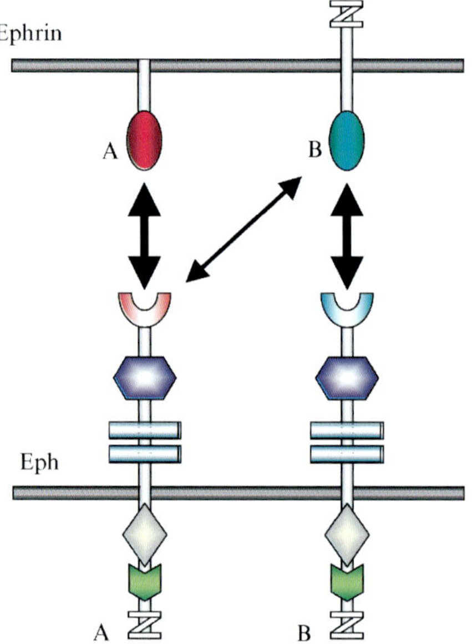

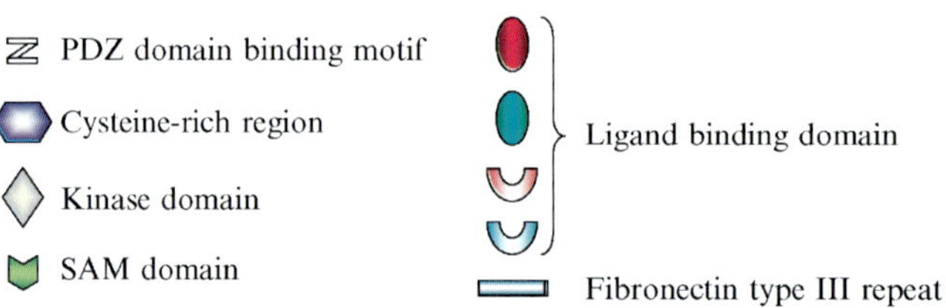

Chapter 10, Figure 7 Ephrins and Eph.

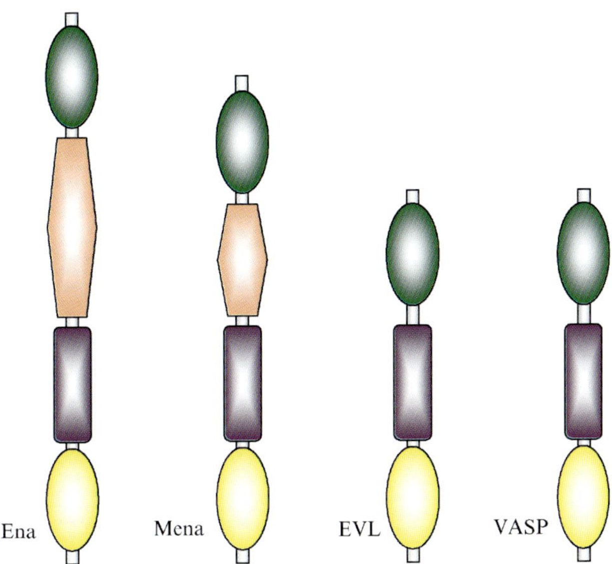

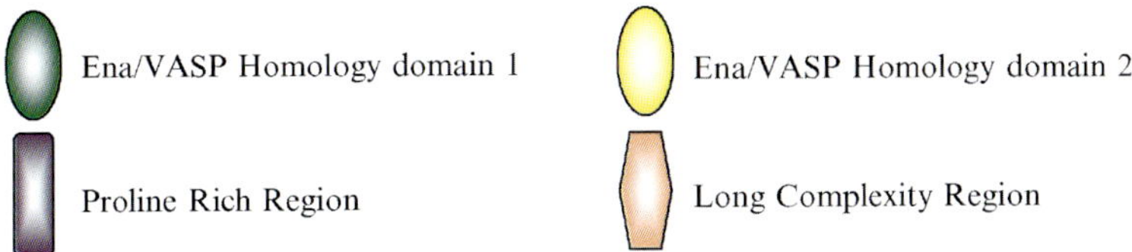

Chapter 10, Figure 8   Ena/VASP.

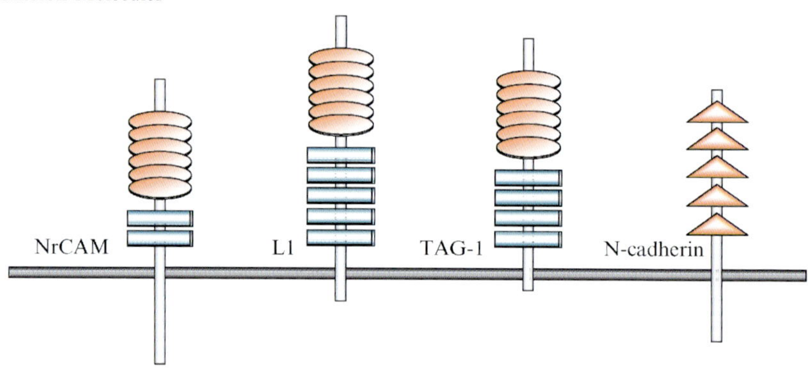

Chapter 10, Figure 9 Adhesion molecules.

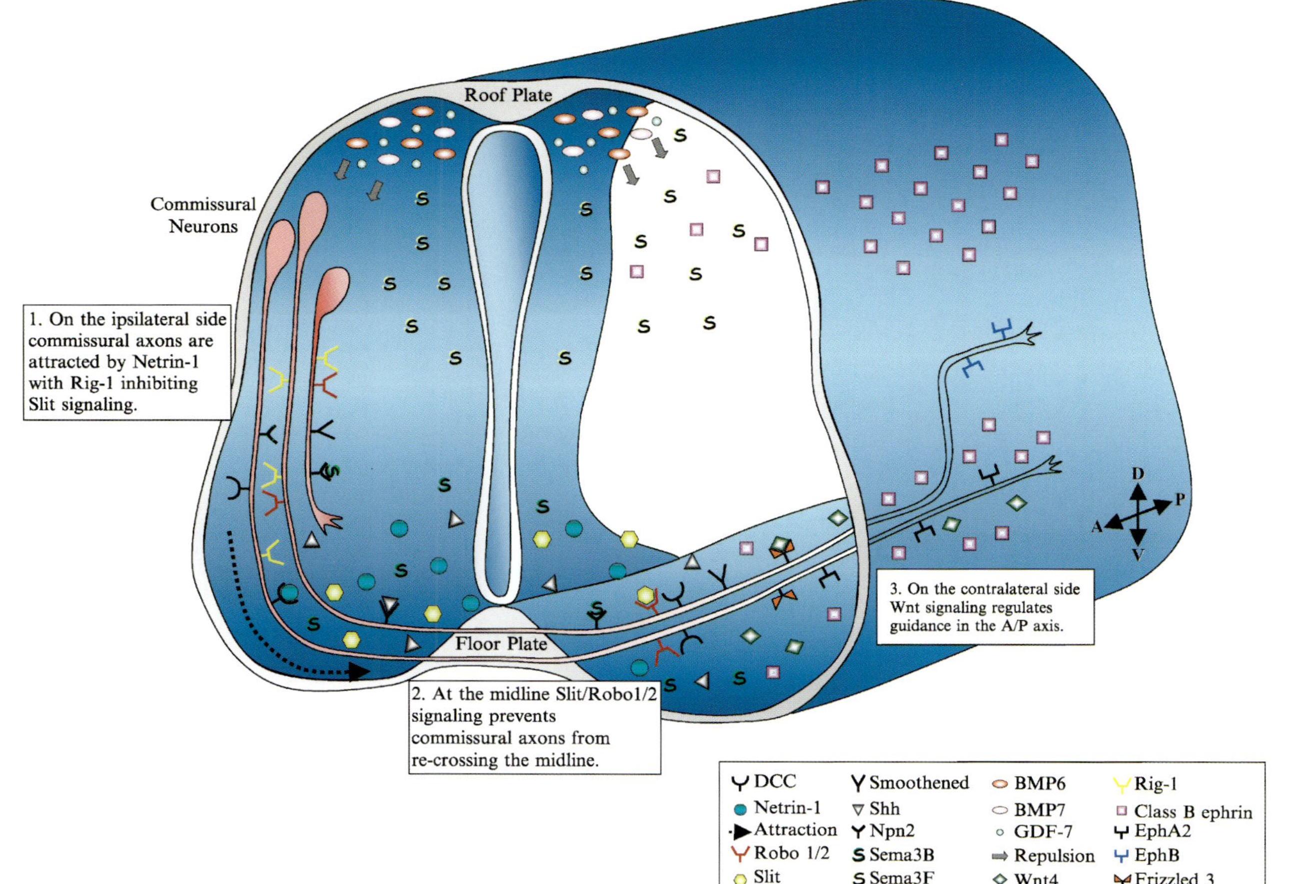

Chapter 10, Figure 10 *Continued*

Chapter 10, Figure 10   Molecules involved in the guidance of commissural axons in the spinal cord. Commissural neurons send their axons toward and across the midline. Bmp6 and 7 and GDF-7 expressed by the roof plate act as dorsal repellents for commissural axons. Axons express DCC, Robo1/2, Smoothened, and Npn2. The axons are attracted by Netrin-1 and Shh but initially are not responsive to Slit because Rig-1 inhibits Slit/Robo1/2 signaling. Sema/Npn2 signaling is required to avoid inappropriate targeting. Once the axons cross the midline, they are then repelled by Slit, expressed by the floor plate, and lose their attraction to Netrin-1, allowing them to leave the floor plate and preventing them from re-crossing the midline. Wtn4/Frizzled3 signaling regulates guidance in the anterior-posterior axis. A subset of commissural axons expressing EphB takes a more dorsal trajectory but grows between regions of Class 3 ephrin-B expression both dorsally and ventrally.

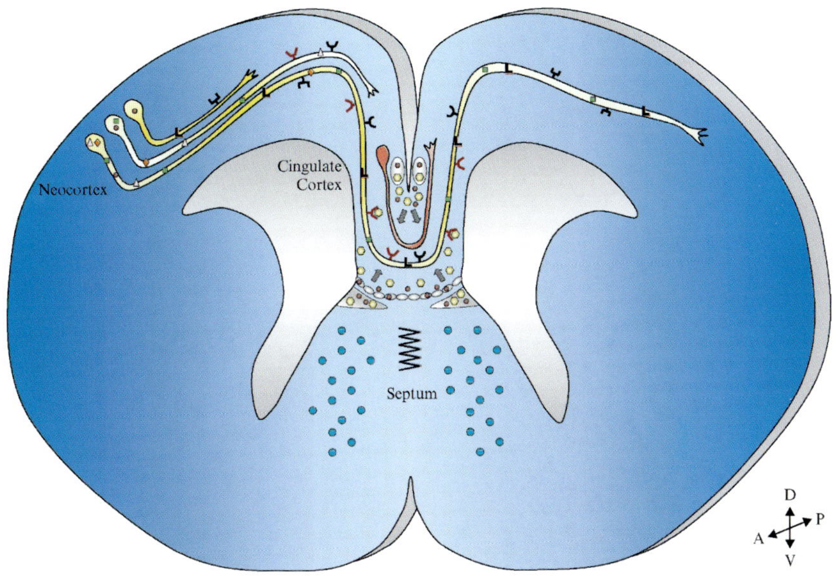

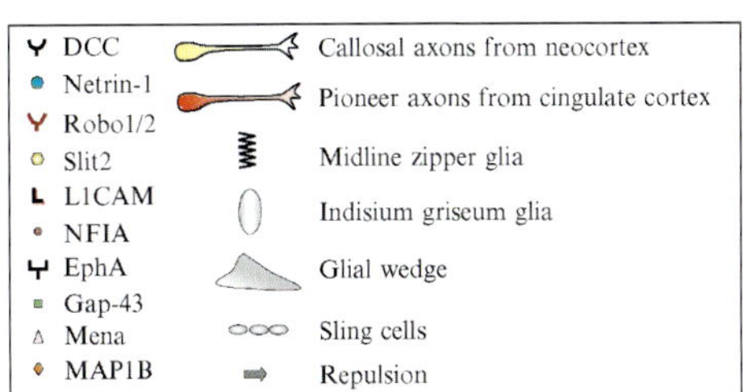

Chapter 10, Figure 11   Molecules involved in the formation of the cortical commissural axons. Axons from the cingulate cortex pioneer a path across the midline. They are followed by neocortical callosal axons that probably fasciculate with them to cross the midline. Slit2, expressed by the glial wedge and the indusium griseum glia, provides a surround repulsion mechanism that keeps callosal axons within the tract, causing them to turn and cross the midline and preventing them from entering the septum. After axons have crossed the midline, Slit2 then repels the postcrossing axons away from the midline area. Callosal axons express several different molecules, including DCC, Robo1/2, EphA, NFIA, Gap-43, L1, Mena, and MAP1B. Netrin-1 is expressed within the septum under the corpus callosum; thus, unlike commissural axons in the spinal cord, callosal axons do not grow through the region of Slit and Netrin expression. NFIA is present in the sling cells, glial wedge, and the indusium griseum. Mouse mutants for DCC, Netrin-1, EphA, NFIA, Gap-43, L1, Mena, and MAP1B have defects in callosal formation (see Section V.B for more details) but how these genes regulate callosal axon pathfinding has not yet been elucidated.

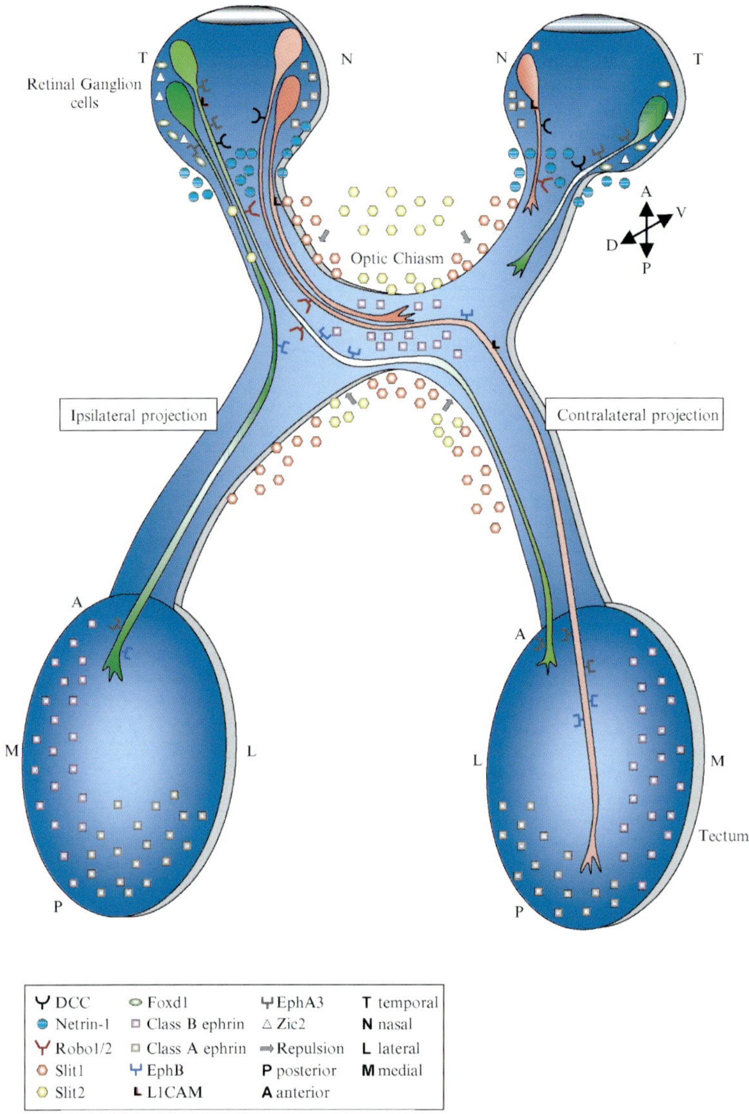

Chapter 10, Figure 12   Molecules involved in the guidance of RGC axons to the tectum. The majority of RGC axons projects into the contralateral optic tract, while a small population of temporal RGC axons does not cross and projects ipsilaterally in mice. A diverse group of molecules acts to guide the RGC axons toward their final target in the tectum. Netrin-1 is expressed in the optic nerve head and Slits are expressed in the optic chiasm region, channeling the axons through the tectum and determining the position of the chiasm. RGC axons express DCC, Robo1/2, L1, and EphB and, as they enter the tectum, they express EphA3. Class B ephrins are expressed in the medial part of the tectum and mediate dorsoventral targeting, whereas Class A ephrins are expressed in the posterior tectum and mediate anterior-posterior topographic targeting.